Mass Spectrometry for Lipidomics

Mass Spectrometry for Lipidomics

Methods and Applications

Edited by Michal Holčapek and Kim Ekroos

Volume 2

WILEY-VCH

Editors

Dr. Michal Holčapek
University of Pardubice
Faculty of Chemical Technology
Studentská 573
53210 Pardubice
Czech Republic

Dr. Kim Ekroos
Lipidomics Consulting Ltd.
Irisviksvägen 31D
02230 Espoo
Finland

Cover Design: Wiley
Cover Images: © Kateryna Kon/Shutterstock; Courtesy of Michaela Chocholoušková

Library of Congress Card No.: applied for

British Library Cataloguing-in-Publication Data
A catalogue record for this book is available from the British Library.

Bibliographic information published by the Deutsche Nationalbibliothek
The Deutsche Nationalbibliothek lists this publication in the Deutsche Nationalbibliografie; detailed bibliographic data are available on the Internet at <http://dnb.d-nb.de>.

Print ISBN: 978-3-527-35222-7
ePDF ISBN: 978-3-527-83649-9
ePub ISBN: 978-3-527-83650-5
oBook ISBN: 978-3-527-83651-2

Typesetting Straive, Chennai, India

Contents

Preface

The field of lipidomics has undergone an enormous growth in recent years, which can be illustrated by the number of published articles and other bibliometric parameters. This highlights the renewed interest in lipids, now driven by the enthusiasm to explore the world of lipidomes and how these, among others, impact health and disease. The excitement is enormous, prompting many newcomers to enter the field. However, training and education in lipidomics are still scarce or even lacking. A successful lipidomics study requires appropriate expertise in all aspects of the lipidomic workflow, covering experimental design, sample preparation, analytical measurement using mass spectrometry techniques, data processing, and finally correct reporting of lipidomic results. The large discrepancy in know-how and lipidomics assessments causes confusion in the field that is also mirrored in the literature. Recently, the International Lipidomics Society was established to fill this gap and to unite researchers around the world interested in all aspects of lipidomics research and collectively start creating urgently needed textbook chapters in lipidomics. This situation prompted us to start working on this book project, where we have assembled the content covering three sections: analytical methodologies in lipidomics, lipidomic analysis according to lipid categories and classes, and finally lipidomic applications. We invited leading experts for particular topics, and, after more than a year of tedious work, we are proud to present the result.

We believe that this book can serve as a valuable tool and resource for anyone interested in lipidomics, from beginners to field leaders, because everyone should be able to find something new in these 27 chapters. The methodological section describes the most common methods used in lipidomic analysis, such as the preanalytical phase, sample preparation, shotgun mass spectrometry, coupling with chromatography, mass spectrometry imaging, ion mobility, advanced tools for structural characterization, approaches for the right identification and quantitation, and finally bioinformatics, software, and databases. The second section is prepared from a different view, targeting selected lipid categories and classes and then sorting convenient methods for their analysis. We believe that this point of view is important for researchers looking for the best method for their lipids of interest. Finally, we present an application section to illustrate a wide range of lipidomics, which covers, for example, clinical diagnostics, biobanking, nutritional aspects, plant science, fluxomics, multiomics, cell biology, microbial lipidomics, and research on serious diseases, such as cancer, Alzheimer's disease, and aging. We hope that these chapters provide an interesting inspiration for new possible applications of lipidomics.

We greatly appreciate the great effort and the extensive time invested by all authors in the preparation of their chapters. Last but not least, we appreciate the support of the publisher in compiling this up-to-date book on lipidomic analysis. We hope that you enjoy reading and that the book will be an everyday companion rather than a dust-covered item on the bookshelf.

Michal Holčapek and Kim Ekroos
Pardubice and Esbo
31 July 2022

Part II

Lipidomic Analysis According to Lipid Categories and Classes

11

Fatty Acids: Structural and Quantitative Analysis

Dong Hao Wang[1] *and J. Thomas Brenna*[1,2]

[1] *School of Agriculture, Shenzhen Campus of Sun Yat-sen University, Shenzhen 518107, China*
[2] *Cornell University, Ithaca, NY, USA*

11.1 Fatty Acids/Acyl Groups as Analytical Targets

Fatty acids (FAs) composed of a carboxyl group with an exclusive hydrocarbon chain are a particular analytical challenge because any particular empirical formula can be present as any of dozens of isomers. Each isomer likely has specific biochemical function and/or nutritional effect. Structural heterogeneity is found in double-bond position and geometry, and chain branching, with less common structures such as triple bonds and cyclopropane rings. The tools of high-performance mass spectrometry, specifically high mass resolution and exact mass measurement, excel at resolving ions of unique empirical formula, but are unable to resolve isomers. As a result, chemical properties such as fragmentation due to collisional activation, reactivity, chromatographic retention times, and to some degree molecular shape, are required to characterize structure and identify unique signals for quantitative analysis. Importantly, the vast majority of FA structures are not available as fully characterized commercial standards. The ideal method for FA analysis is sufficiently specific to enable *de novo* structure elucidation and quantitative analysis *without the need for chemical standards.*

In most of the biomedical literature, the term "fatty acid" (FA) refers to hydrocarbon chains esterified to glycerolipids – phospholipids (PLs) and triacylglycerols (TAGs) – and other molecules bearing single hydroxyl groups, e.g. cholesteryl esters, wax esters, fatty acid esters of hydroxy fatty acids (FAHFAs). Free fatty acids (FFAs) or non-esterified fatty acids (NEFAs) are the terms generally reserved for the usually transient pool of these molecular forms that are usually no more than a few percent of the total acyl pool. The ester group functions as a covalent, rapid on–off connector of FFA to –OH. For many biological problems, FA structures are of interest independent of the lipid class to which they may be esterified. In these cases,

Mass Spectrometry for Lipidomics: Methods and Applications, First Edition.
Edited by Michal Holčapek and Kim Ekroos.

hydrolysis of FA from their native lipid class is not a compromise to limited analytical capability.

FA profiles are maintained by metabolism separate from the various processes that synthesize lipid classes required for assembly of lipid-rich supramolecular structures such as membranes. In mammals and most multicellular organisms, ingested fat is predominantly (>90%) TAG with a few percent PL and other esters, all of which is rapidly hydrolyzed (lipolyzed) with lipolytic enzymes [1] (lipases and phospholipases) secreted at the back of the tongue (lingual lipase), in the stomach (gastric lipase), and in the intestine (pancreatic lipase) to yield mixtures of FFA and 2-monoacylglycerols (2-MAGs) and lysoPL that are emulsified prior to absorption. Once absorbed, FFA are re-esterified into TAG and PL for transport via the lymph into the bloodstream, where they are again lipolyzed to FFA and taken up into tissue. In cells, lipolysis-re-esterification is a characteristic of the Lands cycle, Kennedy pathway, and other pathways that remodel membranes lipids, TAG, and all lipid structures, as well as providing substrates for synthesis of signaling molecules. FA cycle many times between FFA and acyl moieties, passing from lipid class to lipid class via a transient and low concentration FFA pool, prior to entering a catabolic pathway.

Independent of FA lipolysis-re-esterification, specific biochemical pathways synthesize FA via chain elongation (FAS, ELOVL1-7 in humans) and desaturation (SCD1, FADS1-2, humans) enzymes targeting at specific FA structures.

11.1.1 Fatty Acid Classification

FAs are the building blocks of glycerolipids. They are often classified based on their degree of unsaturation as saturated fatty acid (SFA), monounsaturated fatty acid (MUFA), and polyunsaturated fatty acid (PUFA). Trivial names are commonly used for specific FA and designate unique structures, e.g. oleic acid. IUPAC names are the recommended formal names octadec-9Z-enoic acid [2]. Shorthand names based on the IUPAC notation in the form *n*:*d* where *n* is the number of C atoms and *d* is the number of double bonds are also used. For unsaturates, the first double bond appearing from the methyl end is designated by $n - x$ (read "n minus x") where x is the C position numbered from the carboxyl group; for oleic acid the full shorthand designation is 18:1n–9. The omega (ω) notation, introduced by Holman, is functionally equivalent but numbers from the methyl end, e.g. 18:1ω9 [3]. Figure 11.1 presents the FA structure of oleic acid.

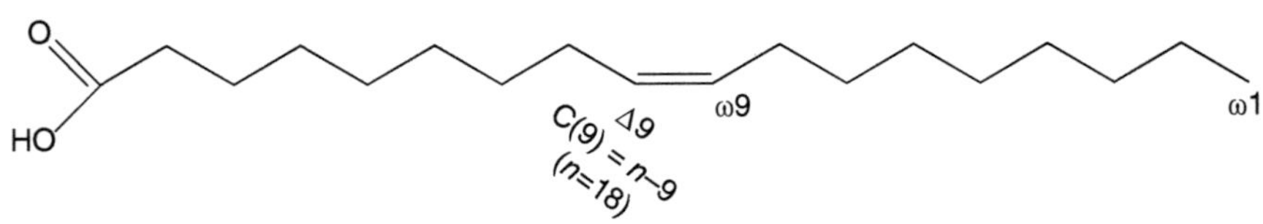

Figure 11.1 Structure and carbon numbering of 18:1n–9, also 18:1ω9, where n = 18 and d = 1 in the shorthand notation *n*:*d*. The n – 9 and ω9 are formally on opposite sides of the double bond closest to the methyl; however, they designate one unique structure.

From an analytical point of view, FAs can be analyzed directly as acids, e.g. by electrospray ionization (ESI) in negative mode, or derivatized to fatty acid methyl ester (FAME) or fatty acid ethyl ester (FAEE), and in specialized approaches, converted into other derivatives.

11.1.2 Conventional Gas Chromatography (GC) – Mass Spectrometry (MS)

11.1.2.1 High-Resolution GC

Compared to liquid chromatography (LC), gas chromatography (GC) is superior in separating FAs especially double-bond isomers. For example, a 25 m short GC column yields partial or full resolution for a variety of conjugated linolenic acids [4]. With the same GC conditions, myristic acid (14:0), for instance, has an area/height (A/H) ratio of ~1.3 at normal operating concentrations, while an A/H ratio of 11.6 was observed for a reverse-phase C18 column LC experiment [5]. In other words, high-resolution GC resolution has peak capacity typically more than 10-fold greater than LC, with higher selectivity toward FA double-bond isomers available with polar GC columns. A 200 m, ionic GC column can resolve nearly all 18:1 double bond and geometric isomers in complex food samples [6]. In addition, polar GC column has superior resolution for SFA of straight-chain and branched-chain configurations, and as importantly, have exceptionally low retention time variation from batch to batch [7].

FAME fragmentation by electron ionization (EI) can yield fragments specific to structure but is generally useful only when the range of possible FAME structures in the sample is well defined, e.g. an 18:3n−6 distinguished from an 18:3n−3. Double-bond positions are scrambled by EI due to charge-driven processes, and thus cannot be used for *de novo* sequencing. Re-esterification to a charge-localizing group is an effective approach that overcomes this limitation.

11.1.2.2 DMOX (4,4-Dimethyloxazoline) Derivatization

A successful system is derivatization to 4,4-dimethyloxazoline (DMOX) for the characterization of double-bond position of FAs by conventional EI GC/MS. GC conditions similar to those used for analyzing FAME can be employed for DMOX FA analysis. Upon EI activation, DMOX FA yields a number of fragments at each of the C=C single or double bond. A gap of 12 amu indicates the presence of a double bond (Figure 11.2). High abundance of m/z 113 and m/z 126 from McLafferty rearrangement is present in all electron ionization mass spectrometry (EIMS) spectra of DMOX FA, which are not characteristic of double-bond position but verify that the derivative

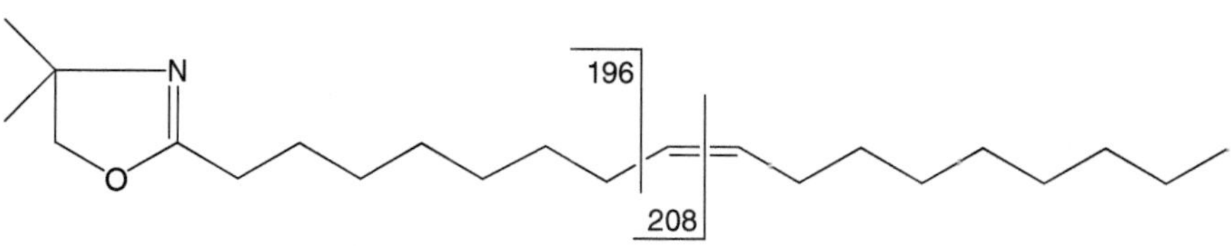

Figure 11.2 DMOX structure and diagnostic ions indicating double-bond position.

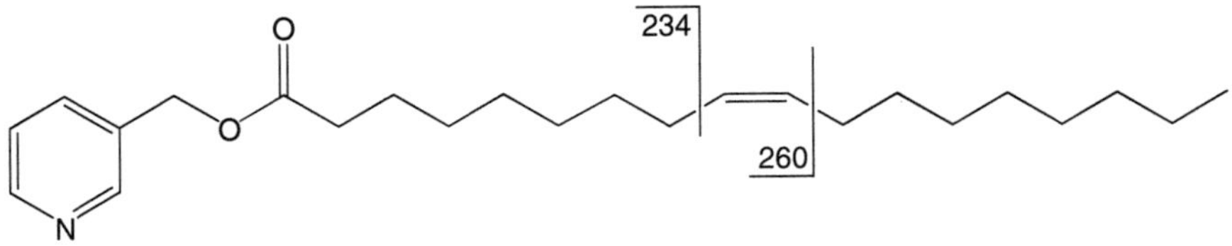

Figure 11.3 Picolinyl ester structure and diagnostic ions indicating double-bond position.

is as expected. Pairs of ions (12 amu difference) indicating a double bond are usually found at low abundance. As a result, the sensitivity of EIMS analysis of DMOX is generally poor and it is not suitable for the analysis of minor FA. In addition, DMOX derivatives of branched-chain fatty acid (BCFA) and straight-chain FA cannot be distinguished [8].

11.1.2.3 Picolinyl Ester (3-Pyridylcarbinol)

Derivatization of FA to picolinyl ester is another well-established strategy for FA structural identification. Upon EI activation, it follows a similar charge remote fragmentation pattern as DMOX, fragmenting between C=C single bonds. A gap of 26 amu indicates the presence of a double bond (Figure 11.3). One major advantage of picolinyl ester derivatization is that BCFA methyl branch position can be readily obtained by observing a gap of 28 amu. Similar to DMOX, spectra of picolinyl ester also contain many McLafferty ions of high abundance, compromising sensitivity to minor FAME. GC analysis of picolinyl ester requires column temperatures about 50 °C higher than FAME.

11.1.3 GC-Solvent-Mediated (SM) Covalent Adduct Chemical Ionization (CACI)-MS/MS

Methods for derivatization of double bonds to yield 4-membered heterocyclic rings have emerged for GC and for LC analysis of FAME and native (underivatized) lipids, respectively. The heteroatom carries the charge and collisional activation yields fragments at or near the erstwhile double bond. In GC, the process is entirely online with the reaction in the chemical ionization source. In LC, the Paternò–Büchi (PB) reaction is typically implemented prior to admitting the analyte to the mass spectrometer, reviewed later.

11.1.3.1 Assignment of Double-Bond Position

GC-solvent-mediated (SM) covalent adduct chemical ionization (CACI)-MS is a recent innovation implemented on a Shimadzu triple quadrupole mass spectrometer, representing a new generation of CACI-capable instrumentation. The triple quadrupole platform provides the CACI technique with higher sensitivity and quantitative capacity, compared with the original CACI-MS/MS implemented with internal ionization three-dimensional ion traps [9].

For the SM application, acetonitrile is externally pressurized into the ion source by a stream of argon gas, and in the ion source acetonitrile self-reacts to form the

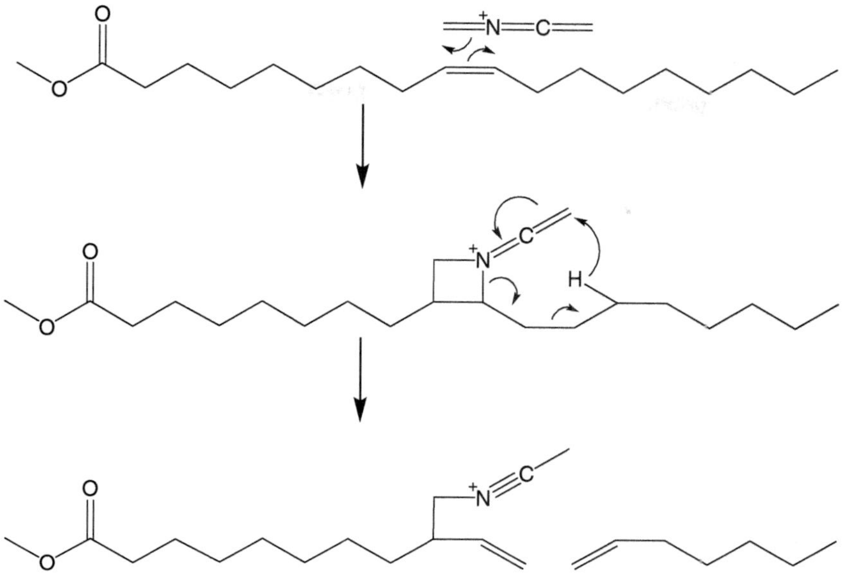

Figure 11.4 Formation of α diagnostic ion (*m/z* 252) by CID activation of oleic acid methyl ester and MIE adduct.

reagent ion, 1-methyleneimino-1-ethenylium (MIE). In Figure 11.4, (*Z*)-octadec-9-enoic acid (oleic acid) methyl ester is shown for CACI reaction. In the first step, MIE reagent ion reacts with the double bond at Δ9 position, forming covalent linkages. Then CID induces electron transferring and hydrogen migration leading to the bond breakage at the allylic position to the erstwhile double bond. As a result, an *m/z* 252 α ion and a neutral loss fragment (98 Da) are formed. When the MIE ion attaches to the Δ9 double bond in an antiparallel orientation and undergoes the similar process, an *m/z* 208 ω ion forms in MS/MS application (Figure 11.5a). For monounsaturated FAME, both the α and ω ions attract a hydrogen. Besides the prominent diagnostic ions, another major ion comes from the loss of methanol (−32 Da). SM-CACI-MS/MS is equally effective in localizing the double-bond position of polyunsaturated FAME. Figure 11.5b presents the spectra of (4*Z*,7*Z*,10*Z*,13*Z*,16*Z*,19*Z*)-docosa-4,7,10,13,16,19-hexaenoic acid (DHA) methyl ester. Both the α and ω diagnostic ions enclose five of the six double bonds at *m/z* 326 and *m/z* 268, respectively. An additional pair of α and ω ions enclosing four of the six double bonds is *m/z* 286 and *m/z* 228. Compared to monoenes and dienes, diagnostic ions of polyenes are more visible among other ions at lower CID energy (6 V under our conditions).

The SM-CACI technology can also identify double-bond position of unusual monounsaturated FA. Figure 11.6 presents the CACI-MS/MS spectra of various 14:1 and 15:1 monoene isomers identified in bovine milkfat [10]. Comparing the left to the right panels, we can easily notice that α diagnostic ions are the same because the double-bond positions are at the same Δ position. On the other hand, ω diagnostic ions differ in 14 Da reflecting an additional distal $-CH_2-$ for 15:1. Monoenes with a

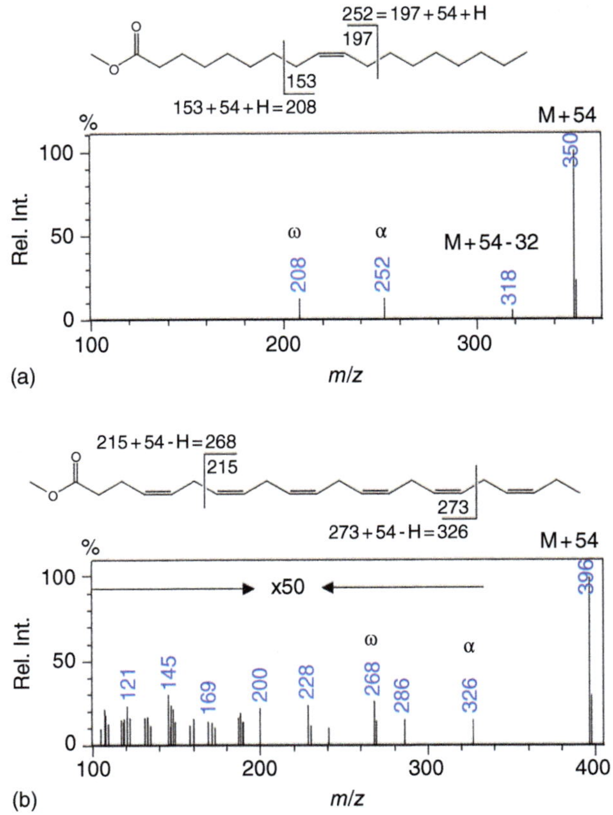

Figure 11.5 SM-CACI MS/MS spectra of (a) methyl oleate and (b) methyl DHA.

Δ5 double bond are rare and authentic standards are limited. In a previous study, we determined the unique fragmentation pattern for Δ5 monoenes analyzing 20:1 (*cis*-5) and 22:1 (*cis*-5) standards. While the α diagnostic ion follows the same allylic cleavage shown in Figure 11.4, the ω diagnostic ion is one more C away (i.e. C2–C3) than the normal allylic cleavage for monoenes. For example, 15:1(*cis*-5) would have a putative ω diagnostic ion at *m/z* 222 for an allylic cleavage and the actual cleavage at C2–C3 yields an ion at *m/z* 234. The difference of 12 Da between 234 and 222 instead of 14 Da is due to opposite H migration (diagnostic ion loses a H for C2–C3 cleavage).

Assignment of double-bond positions of conjugated FA is straightforward, and we recently reported low levels of conjugated linolenic acid (CLnA) in melons [4]. 18:3 with conjugated double bonds is readily distinguishable from 18:3 with methylene-interrupted double bonds by SM-CACI-MS1 by exploring the [M+54]/[M+54-32] ion ratio. The ratios are below 1 for all conjugated 18:3 while as high as 7–12 for α and γ linolenic acid (Figure 11.7). Similar rules apply to conjugated 18:2, as previously reported [11].

SM-CACI-MS/MS is capable of pinpointing the double-bond positions of all 18:3 regardless of the types of double-bond system. Figure 11.8a demonstrates that normal methylene interrupted C18 trienes yield their α and ω diagnostic ions from internal cleavage with a H loss (−1 Da). Figure 11.8b–d shows that CLnA follow an

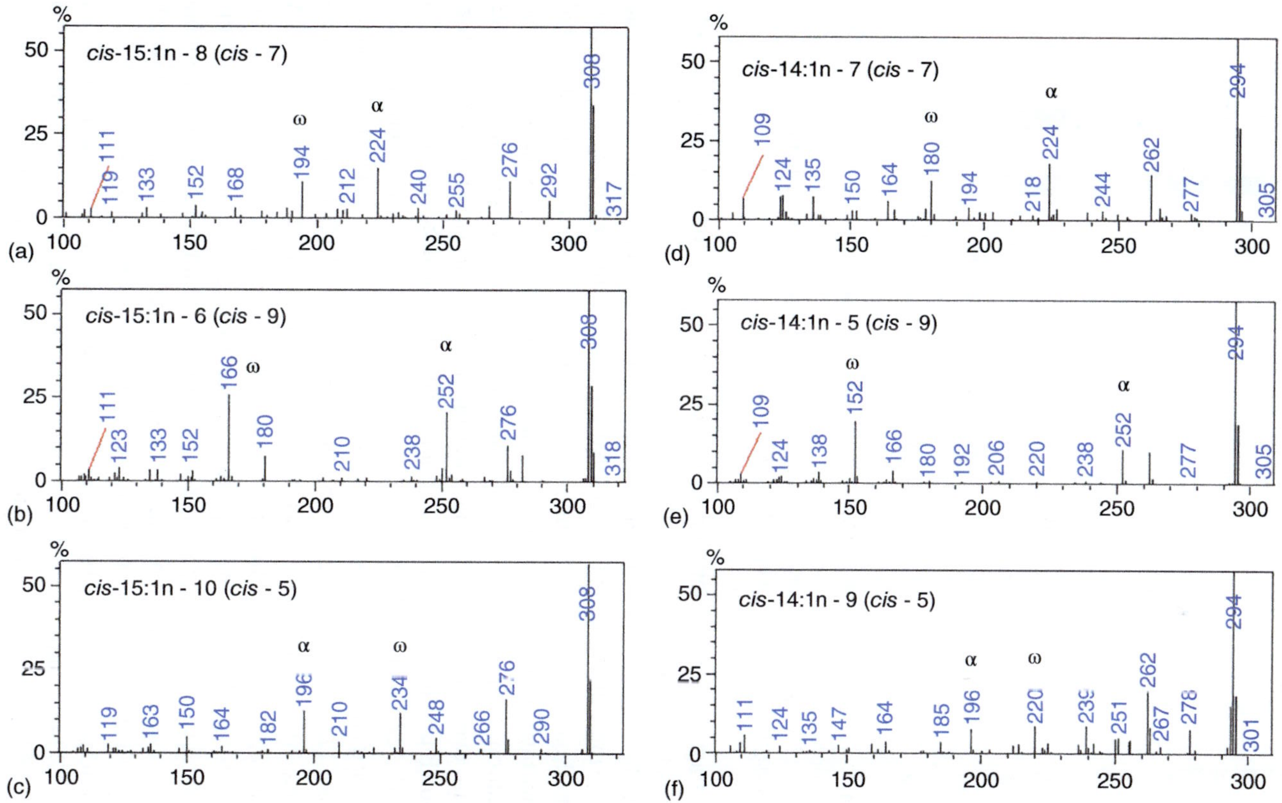

Figure 11.6 MS/MS spectra of unusual (a, b, c) 15 :1 and (d, e, f) 14 :1 FA including those with a carboxyl proximal **Δ**5 double bond; all [M + 54]→ products. Source: Reprinted with permission Wang et al. [10]. Copyright 2020 American Chemical Society.

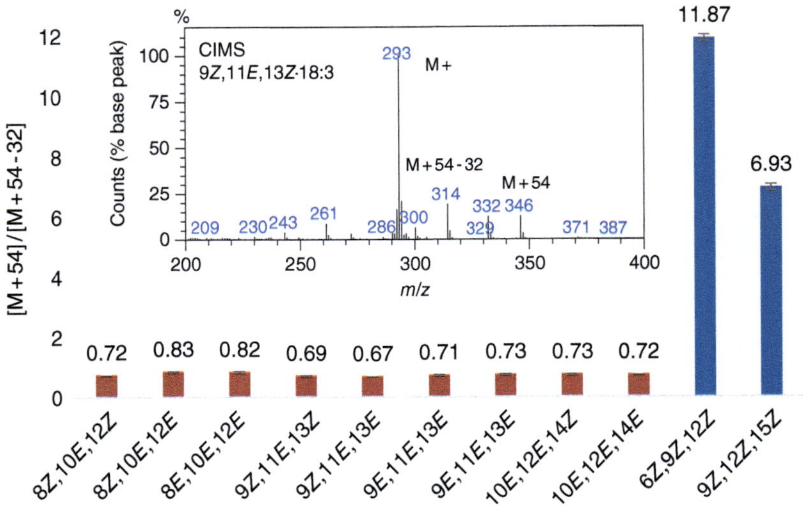

Figure 11.7 [M+54]/[M+54-32] ratios of CLnAs (ratio<1) and 18:3 with normal methylene-interrupted double bonds (ratio>6). Inset: CIMS1 spectrum of 9*Z*,11*E*,13*Z*-18:3 leading to a ratio of 0.69. Source: Reprinted with permission Wang et al. [4]. Copyright 2019 American Chemical Society.

external fragmentation pattern for both α and ω diagnostic ions, and in contrast to normal 18:3, the diagnostic ions gain a H (+1 Da).

Polymethylene interrupted (PMI)-PUFA with arbitrarily placed double bonds are among the more challenging FA to resolve and structurally identify by a single analytical method and without derivatization prior to instrumental analysis. In initial studies we investigated PMI-PUFA and established systematic rules for their *de novo* identification. Figure 11.9a–c presents the CACI-MS/MS spectra of one PMI 18:3 along with two common (methylene-interrupted) 18:3. An ω ion from C6–C7 fragmentation is unique to 18:3(5*Z*,9*Z*,12*Z*) and readily distinguishes this PMI 18:3 from 18:3(6*Z*,9*Z*,12*Z*) and 18:3(9*Z*,12*Z*,15*Z*), which have their characteristic high-abundance α and ω ions coming from internal cleavages at *m/z* 234 and *m/z* 190 for 18:3(6*Z*,9*Z*,12*Z*) and *m/z* 276 and *m/z* 148 for 18:3(9*Z*,12*Z*,15*Z*). Figure 11.9d–f presents three 20:3 isomers, 20:3(5*Z*,11*Z*,14*Z*) and 20:3(7*Z*,11*Z*,14*Z*) being PMI-PUFA and 20:3(8*Z*,11*Z*,14*Z*) a methylene-interrupted PUFA. 20:3(5*Z*,11*Z*,14*Z*) can be readily distinguished from the other two isomers by *m/z* 304 and *m/z* 246, coming from a distal vinylic cleavage and C6–C7 cleavage typical for a Δ5 isolated double bond, respectively. In addition, C10–C11 cleavage yields an ω ion at *m/z* 192 which gains a H (+1 Da), in contrast to 20:3(7*Z*,11*Z*,14*Z*) and 20:3(8*Z*,11*Z*,14*Z*) which have the H-deficient ion at *m/z* 190. High abundance of *m/z* 262 is characteristic of 20:3(7*Z*,11*Z*,14*Z*), although it shares the same chemical structure with 20:3(8*Z*,11*Z*,14*Z*) except the position of the first double bond differs in one carbon along the hydrocarbon chain. Another major spectral difference for the two isomers is at C7–C8, where 20:3(7*Z*,11*Z*,14*Z*) does not have any ions due to the presence of the Δ7 double bond and 20:3(8*Z*,11*Z*,14*Z*) yields an ω ion at *m/z* 232.

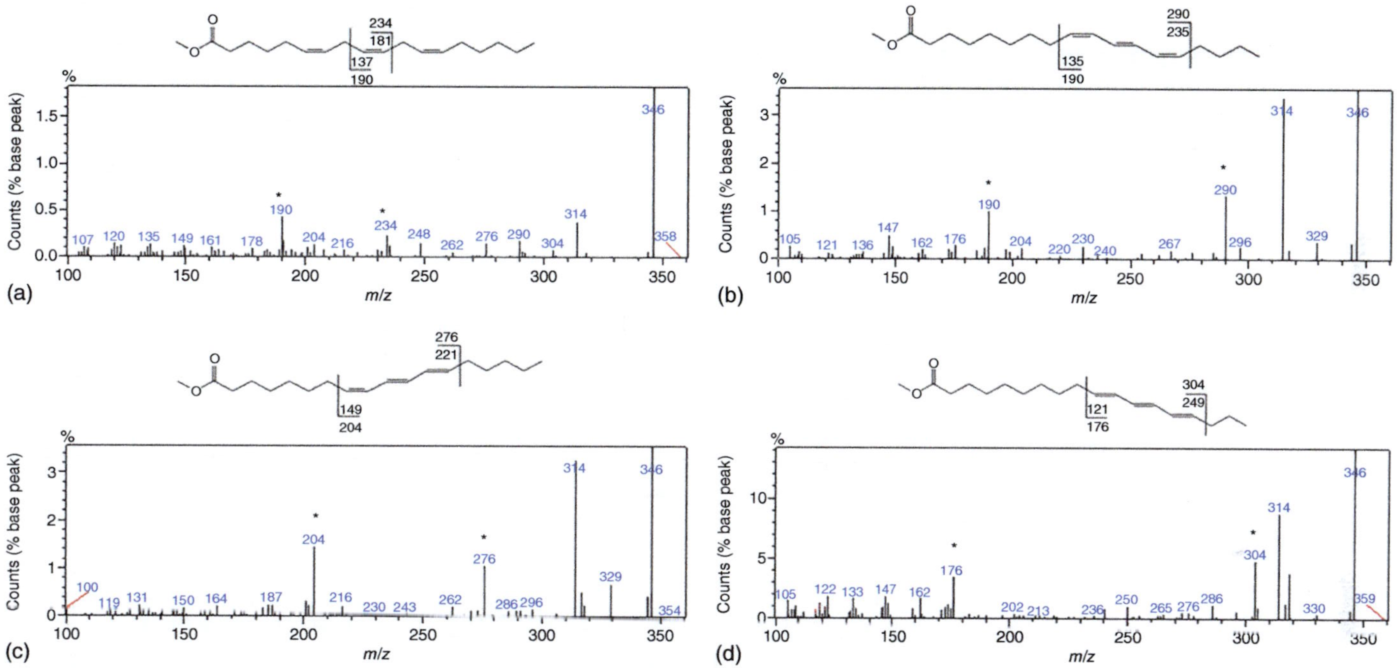

Figure 11.8 MS/MS spectra of (a) 6*Z*,9*Z*,12*Z*-18 : 3 and three CLnA, (b) 9*Z*,11*E*,13*Z*-18 : 3, (c) 8Z,10*E*,12*E*-18 : 3, (d) 10*E*,12*E*,14*E*-18 : 3. Source: Reproduced with permission Wang et al. [4]. Copyright 2019 American Chemical Society.

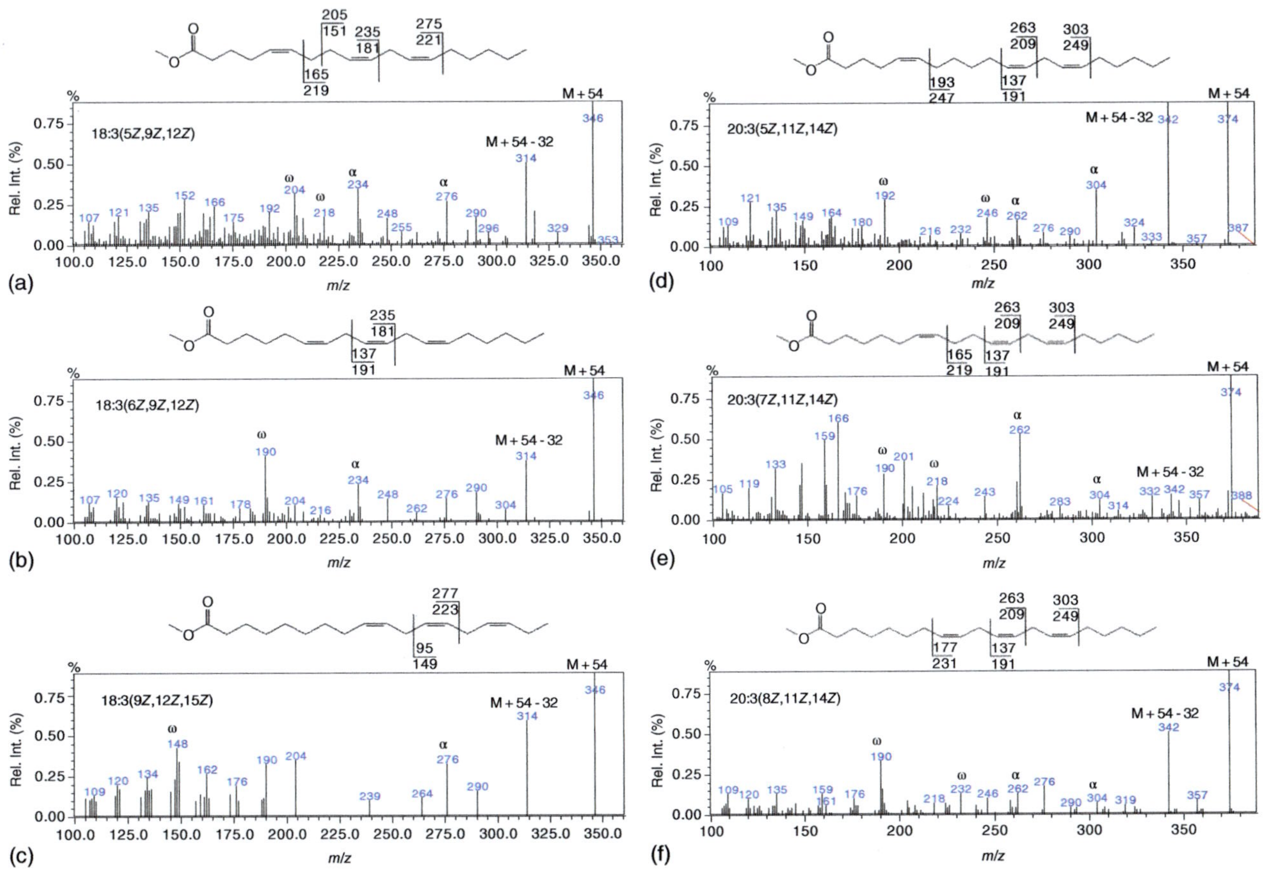

Figure 11.9 Comparisons of three 18:3 isomers and 20:3 isomers CACI-MS/MS spectra reveal unique diagnostic ions for PMI-PUFA. [M + 54]+→products. (a, d, e) are PMI-PUFA, (b, c, f) are methylene interrupted. Relative Intensity (Rel. Int.). Source: Reproduced with permission Wang et al. [12]. Copyright 2020 American Chemical Society.

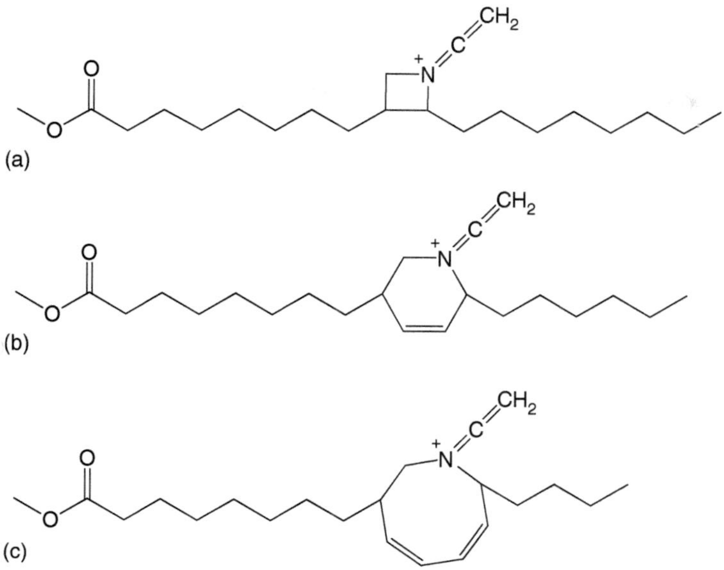

Figure 11.10 [M+54]+ CACI adducts of (a) monoenes, (b) conjugated dienes, and (c) conjugated trienes have four-, six-, and eight-membered rings, respectively.

Monoenes, conjugated dienes, and conjugated trienes fragment exclusively on both sides of the double bond(s), without any internal fragmentation. We established that the covalent adduct MIE ion is a four-membered ring [13] structure at the erstwhile double bonds and MIE ions for monoenes and isolated double bonds in PMI-PUFA, and a six-membered ring for conjugated dienes [11]. By analogy we propose an eight-membered ring structure for $[M+54]^+$ ions of conjugated trienes (Figure 11.10). Figure 11.11 summarizes the fragmentation and formation of α and ω diagnostic ions for monoenes ($\Delta \geq 6$), dienes, polyenes, conjugated and polymethylene-interrupted (PMI) FAME. Monoenes with a Δ5 double bond have an ω diagnostic ion coming from C2–C3 cleavage instead of an allylic cleavage typical for $\Delta \geq 6$ double-bond position. Polyenes with up to eight double bonds were found to follow similar fragmentation patterns with trienes shown here [14]. Prominent diagnostic ions are those enclosing $m-1$ double bonds (where m is the number of total double bonds) while other internal cleavage also occurs for highly unsaturated polyenes, usually leading to an array of additional diagnostic ions at slightly lower abundance. Conjugated FAME of two, three, and four double bonds all cleave at vinylic sites at both ends of the conjugated double-bond system. PMI dienes with an isolated Δ5 double bond, for instance, have a characteristic ω ion coming from C6–C7 bond breakage and an α ion following monoene fragmentation pattern. PMI FAME with extra normal methylene interrupted double bonds toward the ω methyl end (isolated double bond +2 or 3 methylene-interrupted double bonds) have some variation on the formation of the α ion(s). Details on the fragmentation pattern of these unusual FAME and reports of their respective spectra are elsewhere [12].

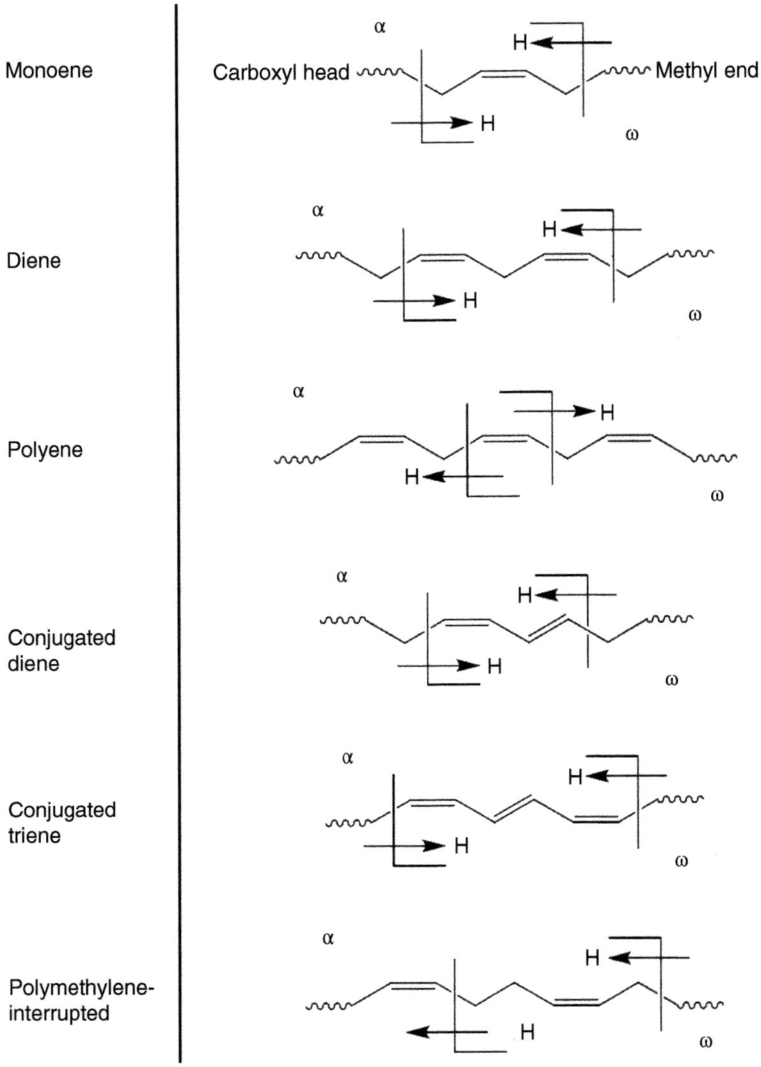

Figure 11.11 Fragmentation, formation of α and ω diagnostic ions. and H transfer rules for monoenes ($\Delta \geq 6$), dienes, polyenes, conjugated, and polymethylene-interrupted (PMI) FAME.

11.1.3.2 Geometry of Double Bonds in Conjugated Linoleic Acids

CACI product dissociation represents an early example of collisional dissociation that distinguishes between double-bond *Z* and *E* geometries. The original work was done with a three-dimensional ion trap for a series of conjugated diene FAME [11, 15]. Ratios of α and ω diagnostic ion abundance are indicative of double-bond geometry. Specifically, when α/ω ratio is >4.8, it is consistent with a *cis*/*trans* configuration; when α/ω ratio is <0.5, it is consistent with a *trans*/*cis* configuration. While *cis*/*cis* and *trans*/*trans* fall between 0.7–3.2 and cannot be distinguished by mass spectrometry alone.

11.1.3.3 Identification of Branched-Chain FA (BCFA)

Most FA in mammalian tissue are straight chain ("normal"); however, methyl branched FA are synthesized at substantial quantity in glands exposed to body surfaces, specifically the sebaceous glands and the meibomian glands of the eyelid. BCFA are also prominent in many bacterial species where they can constitute >90% of all lipids [16]. Methyl branching lowers melting point/glass transition temperature in a manner similar to *cis* double bonds while retaining resistance to attack by reactive oxygen species.

Electron ionization MS produces ions that are suggestive of branched chain position. Early work on branched-chain FAs used MS1 fragments to deduce structure, though in many cases interpretation is not straightforward.

Zirrolli and Murphy discovered that MS/MS spectra of FAME EI molecular ions (M^+) yield spectra that are very different from MS-1 spectra [17]; the classic *m/z* 74 McLafferty rearrangement ion is no longer present, and the spectrum is dominated by ions from breakage of C=C bonds and proposed that it should be used for BCFA analysis based on analysis of single BCFA. Much later we comprehensively investigated this approach, establishing that it works for 30+ BCFA from 12 to 31 carbons (Figure 11.12).

Chemical ionization, including SM ionization, generates protonated molecular ions. We recently investigated whether CID of the $[MH]^+$ ion behaves similar to the M^+. Consistent with EI-MS/MS, SM-CI-MS/MS of *iso*-BCFA generates higher $[M\text{-}43]^+$ and higher $[M\text{-}57]^+$ ions for *anteiso*-BCFA [19]. A three-ion monitoring method based on the $[MH]^+$, $[M\text{-}43]^+$, and $[M\text{-}57]^+$ unambiguously distinguishes straight-chain, *iso*-, and *anteiso*- BCFA below 0.01% of total FA [7]. It was also found that authentic BCFA were not necessary for response factor corrections since little variation was found between *iso*- and *anteiso*- BCFAME or various chain lengths [19]. This method therefore meets the criteria required for complete *de novo* analysis of BCFAME, enabling structural elucidation and quantitative profiling without verified chemical standards.

11.1.3.4 Quantitative Analysis by SM Chemical Ionization and SM-CACI-MS/MS

Quantification of lipids including FA remains a significant challenge for MS application when purified standards are unavailable. GC-flame ionization detector (FID) has uniform responses for various FA and thus is the most widely used method for FA quantification. However, GC-FID cannot resolve coeluted FA and trace FA concentrations can be overestimated by baseline fluctuation and other non-FAME contaminants. In this sense, MS1 and diagnostic ions of FA isomers from MS/MS are the desired technique to offset the limitations of GC-FID quantification.

MS quantification must overcome two major hurdles, (i) varied response factors (RFs) for different FA and (ii) linearity between concentration and peak area. The first hurdle is usually addressed by running authentic standards, usually an equal weight mixture along with the samples. Then response factors for each individual FA can be calculated from analyzing the standard mixture and applied to correct the sample FA concentrations. However, standards of unusual FA are often either

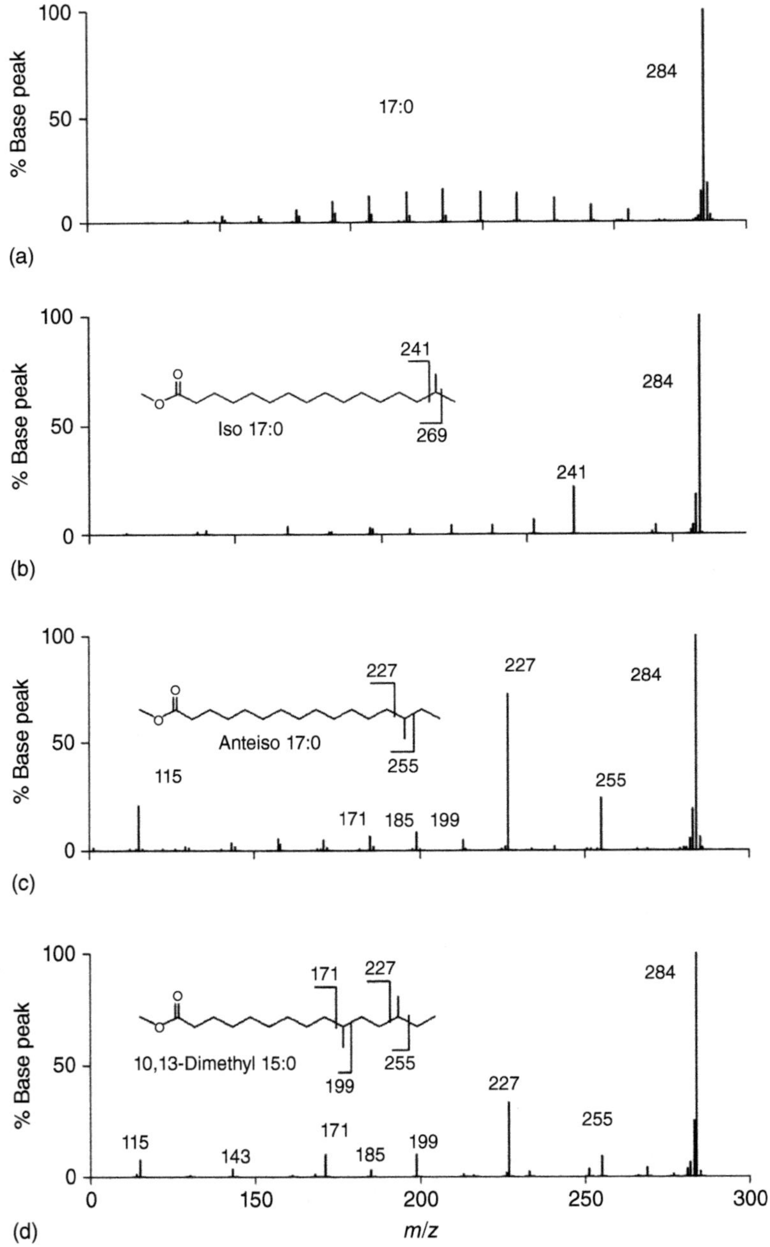

Figure 11.12 EI-MS/MS of the *m/z* 284 molecular ion of methyl heptadecanoate (*n*–17:0) isomers. (a) *normal* (straight chain) products reflect bond breakage between every C=C. (b) *iso*-17:0 (15-methyl hexadecanoate) yields a distinct loss of 43 corresponding to the terminal isopropyl group. (c) *anteiso*-17:0 (14 methyl hexadecanoate) yields loss of ethyl and sec-butyl. (d) 10,13-dimethyl pentadecanoate yields fragments based on losses from both sides of the methyl groups. Source: Ran-Ressler et al. [18]/Elsevier/CC BY 4.0.

unavailable or prohibitory, making comprehensive profiling by MS difficult. Fortunately, we discovered that SM-CACI technique allowed not only *de novo* structural identification but also quantitative analysis of unusual FAs without chemical standards. For example, α-linolenic acid, γ-linolenic acid, and various conjugated linolenic acid share the same RF and thus a single standard can calibrate all these 18:3 isomers, some of which are not commercially available [4]. As BCFA are garnering more and more attention, quantification of low levels of BCFA, e.g. in dairy foods, is highly desirable for researchers. However, not all labs can afford the expensive BCFA standards of various chain lengths and both *iso*- and *anteiso*- configurations. Quantification of BCFA on an EI-MS platform sounds reasonable; however, we found that some structurally similar BCFA such as *iso*-15:0 and *anteiso*-15:0 can have twofold RF difference and even greater discrepancy for different chain length. On the other hand, SM chemical ionization has uniform RF for C14–C19 straight-chain SFA and BCFA commonly seen in foods, and thus do not require BCFA standards for quantification purposes [19]. Very close RF were also found in SM-CACI-MS/MS quantification by diagnostic ions of MUFA of various chain lengths and double-bond positions. Accuracy is within a factor of two without authentic standards for calibration. It is noteworthy that the diagnostic ion approach is intended for coeluted isomers (same *m/z*) which in the majority of cases, reported as a coelution due to inability to resolve and quantify them.

There are several improvements of the recently developed SM-CACI system coupled with a triple quadrupole mass spectrometer compared with the previous home-built version on a three-dimensional ion trap. The SM-CACI system is better poised for quantitative analysis with wider dynamic range and provides superior sensitivity. This translates into a good range of linearity between concentration and peak area especially toward the low end. As a result, SM-CACI is capable of quantifying FA in more than two orders of concentration difference. This commercialized SM-CACI system has a promising future as a routine FAME identification and quantification tool.

11.1.4 Electrospray Ionization (ESI) Methods

11.1.4.1 Conventional ESI

ESI is one of the most widely used ionization methods for mass spectrometric applications. Chain length and degree of unsaturation of FAs can be readily obtained in negative mode; however, in-depth information such as double-bond position and geometry is usually unavailable. Hsu and Turk previously demonstrated that CID activation of $[M-H+2Li]^+$ yielded interpretable fragments for double-bond position localization [20]. However, a variety of other fragments including McLafferty ions exist and double-bond isomers sometimes contain mostly the same ions of varied abundance, making such an approach unsuitable for analyzing mixture of FA isomers.

11.1.4.2 Ozone-Induced Dissociation (OzID)

Ozone-induced dissociation (OzID) conducted offline on purified FAs, hydrocarbons, and other carbon chains with analysis of stable reaction products, typically by

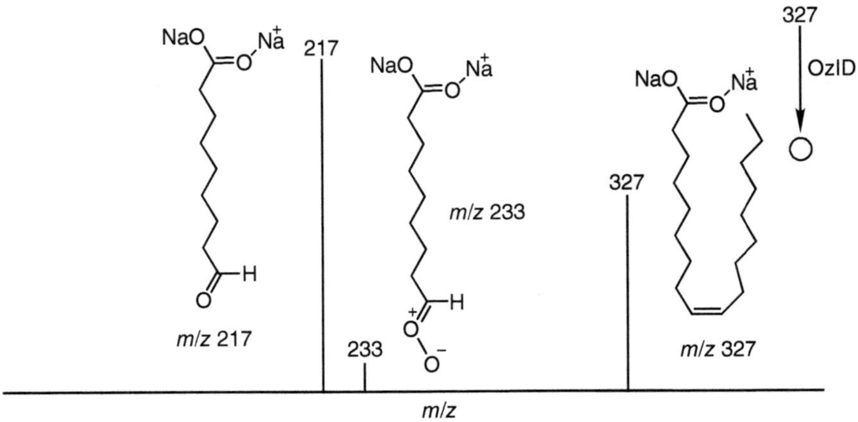

Figure 11.13 Ozone-induced dissociation spectra for oleic acid disodiated adduct.

GC, is a standard technique for double-bond assignment. Online OzID mass spectrometry was developed by Blanksby and coworkers for characterizing double-bond positions in glycerolipids such as PLs and triacylglycerols, and applies FAs [21]. A controlled stream of ozone gas is introduced directly into the ion trap where ozonolysis of FAs is induced. Figure 11.13 shows the diagnostic ion pair for localizing the double-bond position of oleic acid disodiated adduct. The two diagnostic ions differ in an oxygen atom (16 Da) and the aldehyde fragment (having only one O) is usually more abundant. FAME adducted with one sodium ion can also be analyzed in a similar manner including conjugated linoleic acid (CLA) methyl esters [22]. *N*-(4-aminomethylphenyl)pyridinium (AMPP) derivatives of FAs were also demonstrated as a sensitive approach for characterization of unsaturated FAs by OzID [23].

11.1.4.3 Paternò–Büchi (PB) Reaction

Derivatization of the double bond prior to introduction to the mass spectrometer, either offline or inline post-separation, using the PB reaction has proven a convenient and versatile technique for double-bond localization and characterization of the isotopic structure of FAME and other lipids [24–29]. FAs react with ketones containing the PB reaction moiety, e.g. acetone, 2-acetylpyridine, and trifluoroacetophenone inline after LC separation or offline as a sample pre-treatment. Ultraviolet light around 254 nm is used to activate the PB reaction resulting in covalent addition of the reagent to the double bonds. Upon CID dissociation, diagnostic ions indicative of double-bond position in high abundance appear with no rearrangement ions typically seen with EIMS. Figure 11.14 shows PB-MS/MS spectrum of 18:2(9*Z*,12*Z*) lithium adduct and illustrates its CID fragmentation and formation of diagnostic ions. Upon CID activation of lithiated adduct of 18:2(9*Z*,12*Z*) PB product, where the PB reagent is acetone, it yields two sets of diagnostic ions. Using a Δ9 double bond for illustration, PB addition yields a four-membered oxygen containing ring in place of the erstwhile double bond; CID induces ring fragmentation. Depending on the orientation of acetone molecules and FA, two diagnostic ions *m/z* 179 (aldehyde)

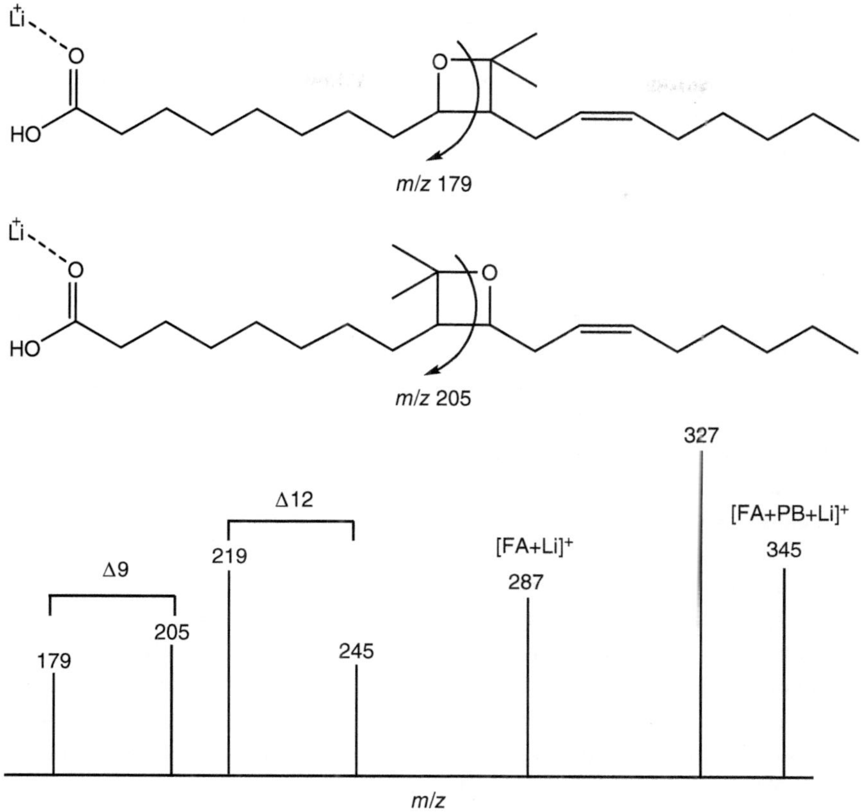

Figure 11.14 PB-MS/MS of 18:2(9*Z*,12*Z*) lithium adduct. One pair of diagnostic ions reveals the double-bond position of Δ9, while the other pair of Δ12. The mechanism of generating diagnostic ions indicative of the Δ9 double bond is shown above the spectrum.

and *m/z* 205 (olefin) are generated in high abundance. A similar schematic can be applied to Δ12 double bond, yielding unique *m/z* 219 and *m/z* 245 ions.

There are several advantages of PB-MS/MS workflow. It is easy to implement either inline or offline, requiring only a UV lamp and a PB reagent in addition to an ESI platform. It follows a straightforward fragmentation pattern and yields diagnostic ions in high abundance, allowing unambiguous assignment of double-bond positions.

11.1.4.4 Ion–Ion Chemistry

Ion–ion reaction involving tris-phenanthroline complexes is a very recent mass spectrometric innovation by McLuckey and coworkers as an online derivatization method for in-depth characterization of FA and lipid structures [30–34]. The negative charge of a FA is reversed to take advantage of fragmentation pathways favored for positive or negative ions, producing rich fragments along the hydrocarbon chain of a FA including saturated FA. Products are similar to the fragmentation pattern of DMOX on an EIMS platform; however, it does not require prior sample preparation as the ion–ion reactions take place online. Identification of double-bond position in monoene and diene FA is facilitated by a spectral curve, a distal allylic doublet and disruption of the 14 Da spacing (and presence of a 12 Da spacing) at the double-bond position. Identification of polyunsaturated FA, however, is not straightforward and relies on library matching. In reality, the main application of ion–ion chemistry MS is to characterize fatty acyl chain structure released from a glycerolipid by CID.

11.1.4.5 Epoxidation

Similar to OzID and PB reactions, epoxidation also utilizes the susceptibility of a double bond against oxidation and the weaker C=C bond around an oxygen heterocyclic center upon dissociation [35]. When the derivatized FA epoxide is subjected to CID activation, two diagnostic ions, i.e. an aldehyde and an olefin, are generated which differs in 16 Da. Although effective in localizing the double-bond position of monoenes and dienes, its application toward polyunsaturated FA is limited because their complicated spectra prevent straightforward double-bond position assignments. When dealing with polyunsaturated FA, SM-CACI/OzID/PB yield more interpretable spectra.

11.1.4.6 Silver Ion Liquid Chromatography-ESI

Generally speaking, LC has a lower resolving power than GC. Specialized LC columns such as silver-ion-impregnated column (Agilent HPLC ChromSpher Lipids) do yield comparable resolution toward double-bond and geometrical FA isomers. Application of silver ion LC for the separation of CLAs has been proved especially successful. Sun and Sehat separately reported that *trans, trans* CLA eluted first and *cis, cis*-CLA last, well resolved from *trans, cis* and *cis, trans* isomers, which eluted closely [36, 37]. When coupled with inline ozonolysis, both the geometrical and double-bond isomers can be fully resolved [37]. Although yet to be demonstrated, silver ion LC-PB-MS/MS should be able to fully resolve CLA geometrical and double-bond isomers.

11.1.5 Characterization of Deuteration in Fatty Acids

Isotopic FA naturally occur at low abundance and pure or synthetic FA isotopes are widely used in scientific research to investigate, e.g. lipid metabolism. Deuterated FA can have deuterium at various C sites depending on the synthetic methods, and can contain many isotopologues (isotope analogues) and isotopomers (isotope isomers). In this sense, the ability to accurately characterize these FA isotopes is the premise for any metabolic studies. Nuclear magnetic resonance (NMR) is conventionally used to quantify the level of deuteration in FA but is of very limited sensitivity, and it cannot resolve isotopologues and resolving individual C sites (e.g. the allylic sites in DHA) requires high fields and specialized techniques.

Recently, we adapted the PB-MS/MS workflow initially developed to locate double-bond positions to quantify level of deuteration in D-FA deuterated to allylic positions by Ru catalysis [27]. After PB derivatization, individual isotopologue (*m/z* value) was isolated and subjected to CID activation. With our careful developed

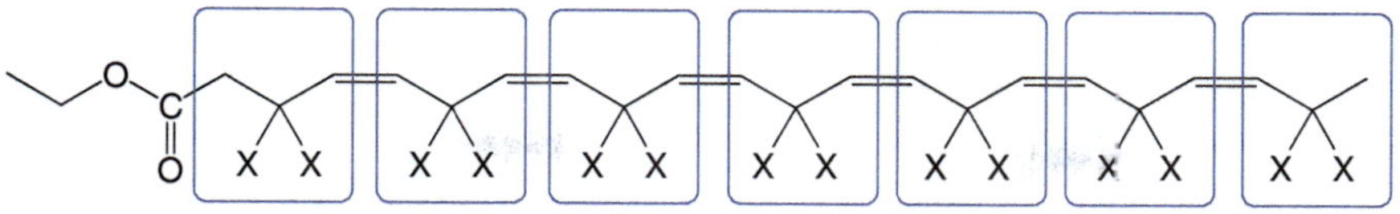

Figure 11.15 Six PB diagnostic ions by CID activation of D-DHA separate C sites into seven compartments for quantitative analysis. Each compartment contains one allylic (mono-allylic or bis-allylic) site (X = H or D).

algorithms, percent deuteration at each of the allylic (mono-allylic or bis-allylic) sites can be quantified for each isotopologue. Figure 11.15 illustrates that PB diagnostic ions can enclose each allylic site individually, and thus deuteration can be determined along the carbon chain. It is noteworthy that vinylic sites are not accessible for Ru catalyzed deuteration, and this knowledge allows a complete sequencing of deuteration.

Taking analyses of D9-DHA and D10-DHA isotopologues mass isolated from a Ru catalyzed deuterated D-DHA as an example, they were first derivatized by 2-acetylpyridine via PB reactions and subjected to CID activation. Two sets of diagnostic ions, one aldehyde (O containing) and the other with the pyridine ring were generated, e.g. *m/z* 362 (aldehyde) and *m/z* 451 (pyridine) both from the Δ19 double bond of ^{PB}D9-DHA. The aldehyde sets appeared to be more abundant and thus selected for D compositional calculations. In Figure 11.16, *m/z* 153 corresponds to zero D at the first compartment on the left of Figure 11.15. In Figure 11.16a inset, *m/z* 194 and *m/z* 195 corresponds to one and two D at the first/second compartments combined shown in Figure 11.15. ^{PB}D10-DHA has an additional *m/z* 196 corresponding to three D at the first/second compartments combined. Since D compositions at the first compartment and the first/second compartment combined are revealed by PB fragments, a straightforward calculation would result in the D composition at the second compartment. Detailed algorithms on determining the D levels at each compartment (or allylic site) can be found in elsewhere [27].

11.1.6 Conclusion

Characterization of FAs and their simple esters (methyl, ethyl) by mass spectrometry established the extraordinary diversity of structures synthesized by biology, from single cells to humans. Consideration of known biology further establishes that the structures of the FAs themselves, independent of the lipid class to which they may be esterified, is a key to health and disease. Now that intact lipid class analysis by mass spectrometry is reasonably straightforward, rapid progress is now being made on the much more challenging problem of total lipid structural and quantitative characterization. Chemical reactivity including unimolecular decomposition upon collisional activation, laser dissociation, and reagent chemistry is likely to be required for to resolve the scores of isomers possible for lipids which cannot be resolved by high mass resolution/exact mass measurements.

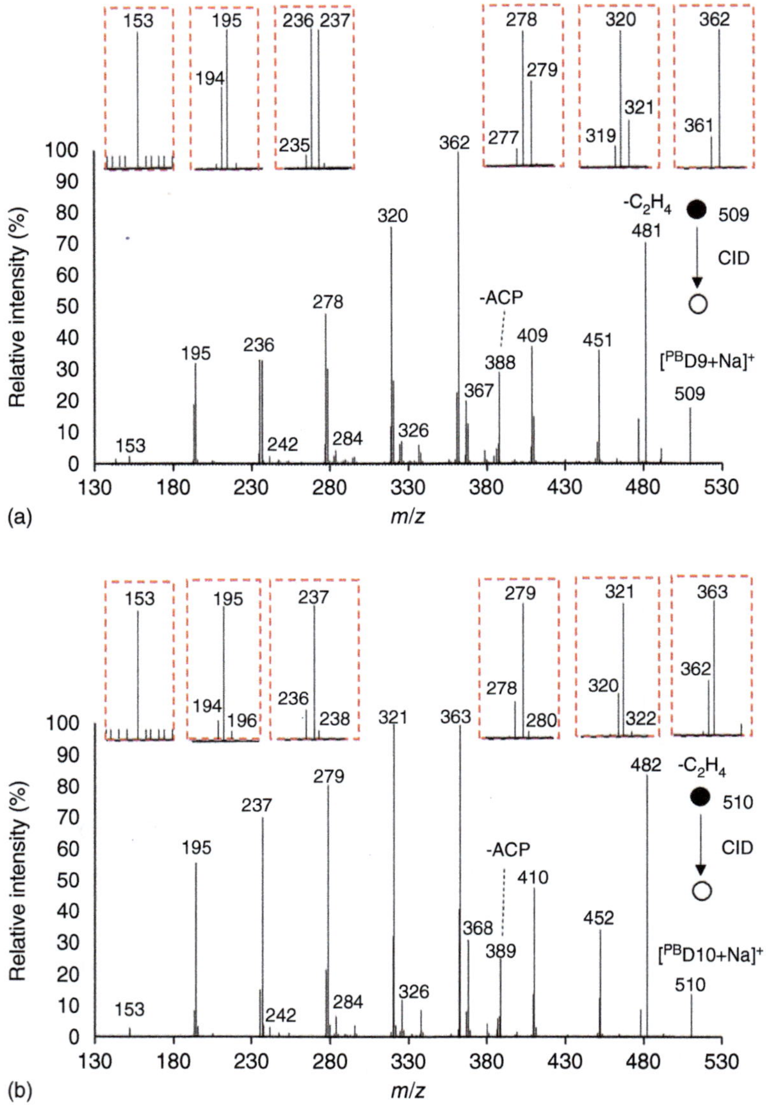

Figure 11.16 ^{PB}D9-DHA (a) and ^{PB}D10-DHA (b) CID spectra (M+Na$^+$→products) reveals H and D composition at allylic sites. Insets show fragmented isotopologues which correspond to various numbers of D in their compartment(s) at predictable levels. Source: Reproduced with permission [27]. Copyright 2021 American Chemical Society.

References

1 Brockerhoff, H. and Jensen, R.G. (1974). *Lipolytic Enzymes*. New York: Academic Press.

2 IUPAC (1997). *Compendium of Chemical Terminology*, 2end ed. (the "Gold Book"). Oxford: Blackwell Scientific Publications.

3 Holman, R.T. and Mohrhauer, H. (1963). A hypothesis involving competitive inhibitions in the metabolism of polyunsaturated fatty acids. *Acta Chem. Scand.* 17: 1.

4 Wang, D.H., Wang, Z., Le, K.P. et al. (2019). Potentially high value conjugated linolenic acids (CLnA) in melon seed waste. *J. Agric. Food Chem.* 67 (37): 10306–10312.

5 Hubert, F., Loiseau, C., Ergan, F. et al. (2017). Fast fatty acid analysis by core–shell reversed-phase liquid chromatography coupled to evaporative light-scattering detector. *Food Nutr. Sci.* 8 (12): 1051–1062.

6 Tyburczy, C., Delmonte, P., Fardin-Kia, A.R. et al. (2012). Profile of trans fatty acids (FAs) including trans polyunsaturated FAs in representative fast food samples. *J. Agric. Food Chem.* 60 (18): 4567–4577.

7 Wang, D.H., Wang, Z., Chen, R. et al. (2020). Very long chain branched chain fatty acids in chia seeds: implications for human use. *J. Agric. Food Chem.* 68 (47): 13871–13878.

8 Christie, W.W. (1998). Gas chromatography-mass spectrometry methods for structural analysis of fatty acids. *Lipids* 33 (4): 343–353.

9 Van Pelt, C.K. and Brenna, J.T. (1999). Acetonitrile chemical ionization tandem mass spectrometry to locate double bonds in polyunsaturated fatty acid methyl esters. *Anal. Chem.* 71 (10): 1981–1989.

10 Wang, D.H., Wang, Z., Chen, R., and Brenna, J.T. (2020). Characterization and semiquantitative analysis of novel Ultratrace C10-24 monounsaturated fatty acid in bovine milkfat by solvent-mediated covalent adduct chemical ionization (CACI) MS/MS. *J. Agric. Food Chem.* 68 (28): 7482–7489.

11 Michaud, A.L., Yurawecz, M.P., Delmonte, P. et al. (2003). Identification and characterization of conjugated fatty acid methyl esters of mixed double bond geometry by acetonitrile chemical ionization tandem mass spectrometry. *Anal. Chem.* 75 (18): 4925–4930.

12 Wang, D.H., Wang, Z., Cortright, J.R. et al. (2020). Identification of polymethylene-interrupted polyunsaturated fatty acids (PMI-PUFA) by solvent-mediated covalent adduct chemical ionization triple quadrupole tandem mass spectrometry. *Anal. Chem.* 92 (12): 8209–8217.

13 Van Pelt, C.K., Carpenter, B.K., and Brenna, J.T. (1999). Studies of structure and mechanism in acetonitrile chemical ionization tandem mass spectrometry of polyunsaturated fatty acid methyl esters. *J. Am. Soc. Mass Spectrom.* 10 (12): 1253–1262.

14 Van Pelt, C.K., Huang, M.C., Tschanz, C.L., and Brenna, J.T. (1999). An octaene fatty acid, 4,7,10,13,16,19,22,25-octacosaoctaenoic acid (28:8*n*−3), found in marine oils. *J. Lipid Res.* 40 (8): 1501–1505.

15 Michaud, A.L., Lawrence, P., Adlof, R., and Brenna, J.T. (2005). On the formation of conjugated linoleic acid diagnostic ions with acetonitrile chemical ionization tandem mass spectrometry. *Rapid Commun. Mass Spectrom.* 19 (3): 363–368.

16 Kaneda, T. (1991). *Iso*- and *anteiso*-fatty acids in bacteria: biosynthesis, function, and taxonomic significance. *Microbiol. Rev.* 55 (2): 288–302.

17 Zirrolli, J.A. and Murphy, R.C. (1993). Low-energy tandem mass spectrometry of the molecular ion derived from fatty acid methyl esters: a novel method for analysis of branched-chain fatty acids. *J. Am. Soc. Mass Spectrom.* 4 (3): 223–229.

18 Ran-Ressler, R., Lawrence, P., and Brenna, J.T. (2012). Structural characterization of saturated branched chain fatty acid methyl esters by collisional dissociation of molecular ions generated by electron ionization. *J. Lipid Res.* 53 (1): 195–203.

19 Wang, D.H., Wang, Z., and Brenna, J.T. (2020). Gas chromatography chemical ionization mass spectrometry and tandem mass spectrometry for identification and straightforward quantification of branched chain fatty acids in foods. *J. Agric. Food Chem.* 68 (17): 4973–4980.

20 Hsu, F.F. and Turk, J. (2008). Elucidation of the double-bond position of long-chain unsaturated fatty acids by multiple-stage linear ion-trap mass spectrometry with electrospray ionization. *J. Am. Soc. Mass Spectrom.* 19 (11): 1673–1680.

21 Poad, B.L., Pham, H.T., Thomas, M.C. et al. (2010). Ozone-induced dissociation on a modified tandem linear ion-trap: observations of different reactivity for isomeric lipids. *J. Am. Soc. Mass Spectrom.* 21 (12): 1989–1999.

22 Pham, H.T., Maccarone, A.T., Campbell, J.L. et al. (2013). Ozone-induced dissociation of conjugated lipids reveals significant reaction rate enhancements and characteristic odd-electron product ions. *J. Am. Soc. Mass Spectrom.* 24 (2): 286–296.

23 Poad, B.L.J., Marshall, D.L., Harazim, E. et al. (2019). Combining charge-switch derivatization with ozone-induced dissociation for fatty acid analysis. *J. Am. Soc. Mass Spectrom.* 30 (10): 2135–2143.

24 Esch, P. and Heiles, S. (2020). Investigating C=C positions and hydroxylation sites in lipids using Paternò–Büch functionalization mass spectrometry. *Analyst* 145 (6): 2256–2266.

25 Franklin, E.T. and Xia, Y. (2020). Structural elucidation of triacylglycerol using online acetone Paternò–Büch reaction coupled with reversed-phase liquid chromatography mass spectrometry. *Analyst* 145 (20): 6532–6540.

26 Ma, X., Chong, L., Tian, R. et al. (2016). Identification and quantitation of lipid C=C location isomers: a shotgun lipidomics approach enabled by photochemical reaction. *Proc. Natl. Acad. Sci. U.S.A.* 113 (10): 2573–2578.

27 Wang, D.H., Park, H.G., Wang, Z. et al. (2021). Toward quantitative sequencing of deuteration of unsaturated hydrocarbon chains in fatty acids. *Anal. Chem.* 93 (23): 8238–8247.

28 Xie, X. and Xia, Y. (2019). Analysis of conjugated fatty acid isomers by the Paternò–Büch reaction and trapped ion mobility mass spectrometry. *Anal. Chem.* 91 (11): 7173–7180.

29 Xie, X., Zhao, J., Lin, M. et al. (2020). Profiling of cholesteryl esters by coupling charge-tagging Paternò–Büch reaction and liquid chromatography-mass spectrometry. *Anal. Chem.* 92 (12): 8487–8496.

30 Randolph, C.E., Blanksby, S.J., and McLuckey, S.A. (2020). Toward complete structure elucidation of glycerophospholipids in the gas phase through charge inversion ion/ion chemistry. *Anal. Chem.* 92 (1): 1219–1227.

31 Randolph, C.E., Fabijanczuk, K.C., Blanksby, S.J., and McLuckey, S.A. (2020). Proton transfer reactions for the gas-phase separation, concentration, and identification of cardiolipins. *Anal. Chem.* 92 (15): 10847–10855.

32 Randolph, C.E., Foreman, D.J., Blanksby, S.J., and McLuckey, S.A. (2019). Generating fatty acid profiles in the gas phase: fatty acid identification and relative quantitation using ion/ion charge inversion chemistry. *Anal. Chem.* 91 (14): 9032–9040.

33 Randolph, C.E., Marshall, D.L., Blanksby, S.J., and McLuckey, S.A. (2020). Charge-switch derivatization of fatty acid esters of hydroxy fatty acids via gas-phase ion/ion reactions. *Anal. Chim. Acta* 1129: 31–39.

34 Randolph, C.E., Shenault, D.M., Blanksby, S.J., and McLuckey, S.A. (2020). Structural elucidation of ether glycerophospholipids using gas-phase ion/ion charge inversion chemistry. *J. Am. Soc. Mass Spectrom.* 31 (5): 1093–1103.

35 Feng, Y., Chen, B., Yu, Q., and Li, L. (2019). Identification of double bond position isomers in unsaturated lipids by *m*-CPBA epoxidation and mass spectrometry fragmentation. *Anal. Chem.* 91 (3): 1791–1795.

36 Sehat, N., Yurawecz, M.P., Roach, J.A. et al. (1998). Silver-ion high-performance liquid chromatographic separation and identification of conjugated linoleic acid isomers. *Lipids* 33 (2): 217–221.

37 Sun, C., Black, B.A., Zhao, Y.Y. et al. (2013). Identification of conjugated linoleic acid (CLA) isomers by silver ion-liquid chromatography/in-line ozonolysis/mass spectrometry (Ag^+-LC/O_3-MS). *Anal. Chem.* 85 (15): 7345–7352.

12

Quantitation of Oxylipins in Biological Samples, Focusing on Plasma, and Urine

Valerie B. O'Donnell[1], *Ginger L. Milne*[2], *Marina S. Nogueira*[2], *Martin Giera*[3], *and Nils Helge Schebb*[4]

[1] *Cardiff University, School of Medicine, Systems Immunity Research Institute, Heath Park, Cardiff, CF14 4XN, UK*
[2] *Vanderbilt University Medical Center, Division of Clinical Pharmacology, Department of Medicine, 2200 Pierce Ave, Nashville, TN, 37232-6602, USA*
[3] *Leiden University Medical Center, Center for Proteomics and Metabolomics, Albinusdreef 2, 2333ZA, Leiden, The Netherlands*
[4] *University of Wuppertal, Chair of Food Chemistry, Faculty of Mathematics and Natural Sciences, Gausstraße 20, 42119, Wuppertal*

12.1 Introduction

Oxylipins are structurally related oxygenated polyunsaturated fatty acids (PUFAs) generated by enzymes. Additionally, their stereo- or positional isomers can be generated non-enzymatically. In the LIPID MAPS classification/nomenclature, they are in the fatty acyl (FA) class represented by octadecanoids, eicosanoids, and docosanoids [1]. They arise by the oxygenation of PUFA of usually 18–22 carbons and between two and six double bonds and are formed by lipoxygenases (LOX), cyclooxygenases (COX), or cytochrome P450s (CYP450). Sequential metabolism by two or more enzymes can also occur. Here, formation of prostaglandin (PG) H_2 (PGH_2) by COX is followed by formation of thromboxane A2 (TXA_2) by the CYP450, thromboxane synthase [2]. Prostanoids/prostaglandins (PGs) represent oxylipins that contain a five-carbon prostanoid ring and include prostaglandins such as PGE_2, PGD_2, and $PGF_{2\alpha}$. Distinct products are formed from different PUFAs based on the substrate and enzyme. For example, while 15-LOX1 oxygenation of arachidonic acid (20:4) generates 15*S*-hydroperoxyeicosatetraenoic acid (15-HpETE), metabolism of docosahexaenoic acid (22:6) results in 17*S*-hydroperoxydocosahexaenoic acid (17-HpDOHE). The major oxylipin biosynthetic pathways are found on LIPID MAPS/Wikipathways (https://www.lipidmaps.org/resources/pathways/wikipathways), summarized in Figures 12.1 and 12.2.

Mass Spectrometry for Lipidomics: Methods and Applications, First Edition.
Edited by Michal Holčapek and Kim Ekroos.

Oxylipins are generated in virtually all organs and cells. Many maintain homeostasis, e.g. PGE_2 in the gastric system, or TXA_2 and prostacyclin (PGI_2) in the vasculature. PGE_2 is generated in inflammation, involved in immunity, cancer, infection, and pain. During homeostasis, PGE_2 is mainly from COX-1, while during inflammation, high amounts of PGs are generated via COX-2 induced by damage-associated molecular patterns (DAMPs) or pathogen-associated molecular patterns (PAMPs), such as lipopolysaccharide (LPS). Oxylipins signal by either (i) high-affinity binding and activation of G-protein-coupled receptors (GPCRs) or (ii) by low-affinity activation of the transcription factor, peroxisome proliferator-activated receptor PPARγ. In the case of GPCRs, specific oxylipins are recognized by their cognate receptors, activating highly controlled signaling in a tissue/cell-specific manner. In contrast, PPARγ is activated by many oxylipins, upregulating a transcriptional program to dampen inflammation and fibrosis.

Nowadays, oxylipin analysis generally utilizes high-sensitivity triple-quadrupole or Q-Trap MS instruments coupled via electrospray ionization (ESI) with LC, which have superseded GC/MS, LC/UV, and other older methods. MS is typically operated in the *multiple reaction monitoring (MRM)* mode, which may or may not be *scheduled*, allowing simultaneous detection of large numbers of lipids. The *scheduled* mode allows the instrument to analyze lipids only around the time when they are expected to elute from the LC column. This relies on the use of *precursor-to-product* ion transitions that are considered sufficiently specific for each lipid analyzed and is also termed selected (SRM). For example, during a 15–20 minute separation, each oxylipin will be measured only for up to 2 minutes, indicating that at any particular time, only a small number are simultaneously measured. This allows from 100 to 200 lipids (depending on the platform, chromatographic separation, etc.) to be quantified per sample as the instrument can acquire sufficient points across a peak (generally around 6–20 or more points per peak is captured) with sufficient dwell time. Here, isotopically labeled (typically deuterated) internal standards are added before extraction, with the most structurally similar being used for each primary (unlabeled) analyte. For all analytes quantified, a primary standard is required. Quantitation relies on generating standard curves for every analyte, plotted against a constant amount of internal standard in serial dilutions. Limit of detection (LOD) and limit of quantitation (LOQ) are defined based on signal:noise (S:N), calculated by height or area, and most vendors' software provide scripts for this. As a general rule, LOQ is used at a S:N of at least 5–10 in analytical laboratories, with LOD being around 3 : 1. A clearly discernible peak must be seen above the baseline noise, or the lipid is judged to be below the LOD/quantitation in the sample. This is in line with the guidelines from international bodies [3–12]. Approaches that define LOQ/LOD based on the minimum area of integration in the chromatogram or the presence of ions considered to be diagnostic in an MS/MS spectrum can incorrectly infer that lipids are present in baseline noise and should be avoided [13].

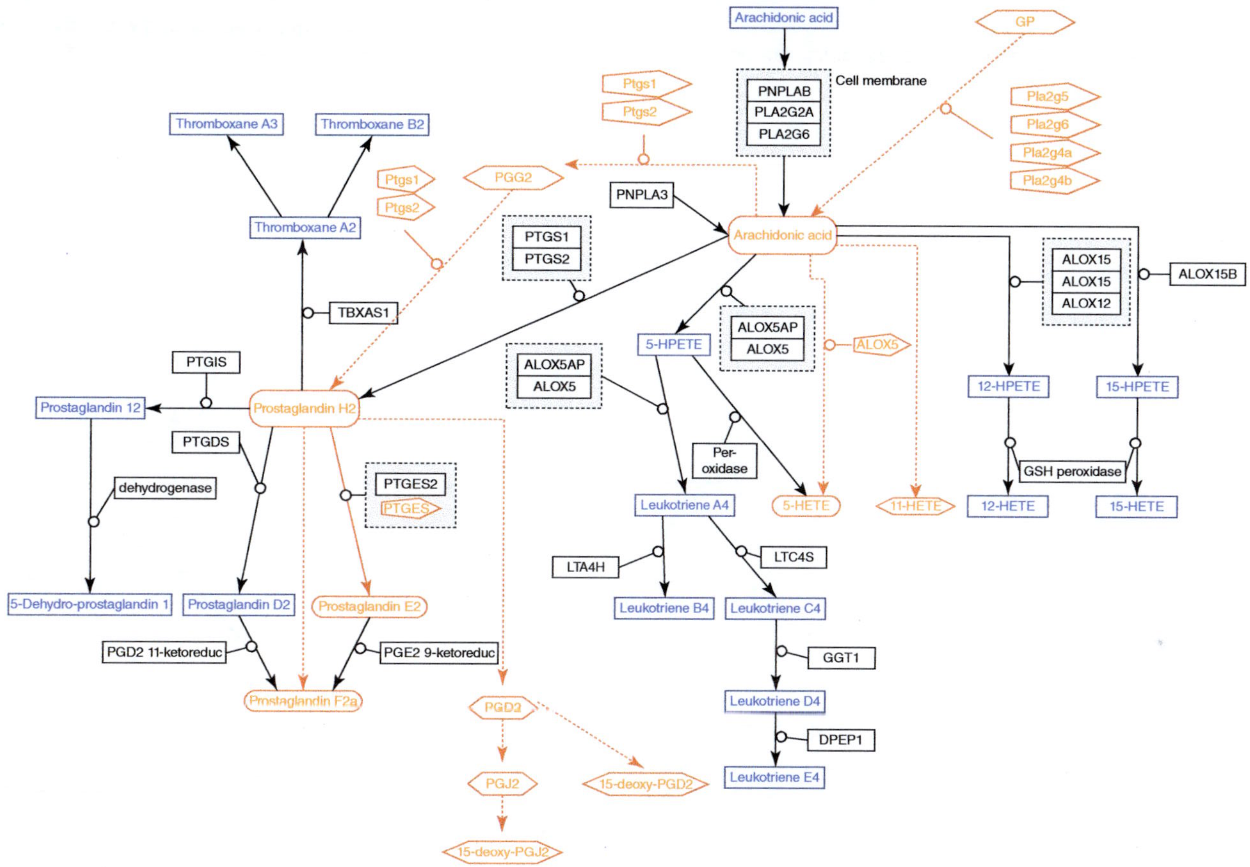

Figure 12.1 Oxylipin generation from arachidonic acid (*Homo sapiens*). A summary pathway showing the main cyclooxygenase and lipoxygenase generation pathways in mammals. Source: Lipid Maps/Public Domain/CC BY 4.0. https://www.wikipathways.org/index.php/Pathway:WP167

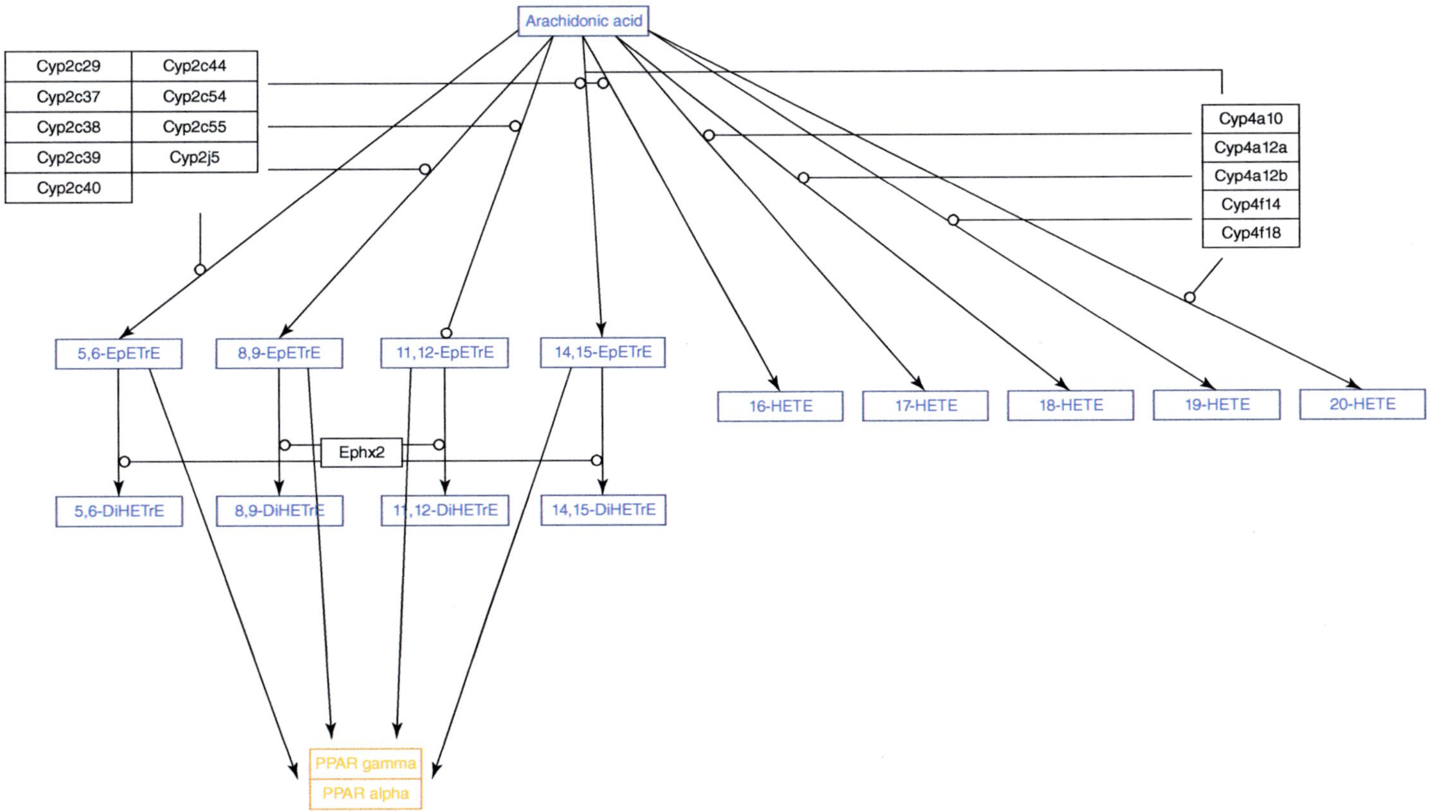

Figure 12.2 Oxylipin generation via the cytochrome P450 pathway (*Mus musculus*). Source: Lipid Maps/Public Domain/CC BY 4.0. https://www.wikipathways.org/index.php/Pathway:WP4349

12.2 Analysis of Oxylipins: Plasma, Tissues, and Cells

While most PGs are detected at low concentrations, others such as hydroxy-linoleic acid derivatives are often 100- to 1000-fold higher concentrated. Furthermore, oxylipins can be formed and degraded after and during sample collection and preparation. Because of this, to ensure a reliable quantitative analysis, all steps from sample collection to MS analysis need to be carefully considered.

12.2.1 Planning of Sample Collection Preparation and Storage

Sample handling needs to be carefully standardized. Inappropriate collection, pretreatment, storage, and analytical sample preparation can strongly influence the oxylipin pattern detected [14–17]. This is because PUFAs, such as arachidonic and docosahexaenoic acids, are prone to artifactual oxidation, which is accelerated by free transition metals from disrupted tissue or cells (Fenton-type chemistry). Here, products include not only enzymatically generated oxylipins but also large numbers of additional regio- and stereoisomers that can be hard to distinguish. Strategies to minimize this include addition of antioxidants, metal chelators that prevent metal ions from redox cycling, and keeping cell and tissue samples cooled on ice during the initial processing. An exception is washed platelets, which need to be isolated at room temperature, to preserve agonist responses.

The test system needs to be carefully considered to decide whether oxylipins are likely to be formed in biologically meaningful amounts. First, asking if biosynthetic enzymes, e.g. COX, LOX, and CYP450s, are expressed in the tissue, compartment, or matrix at sufficient levels. This can be assessed by prior knowledge from the literature or screening transcriptional data. For example, the expression of 12-LOX and COX-1 in platelets has been known for decades. Furthermore, species differences exist, with *ALOX15/Alox15* encoding enzymes with different product profiles in mice/rats/pigs (12-lipoxygenating toward arachidonate primarily and termed 12/15-LOX) versus humans/rabbits (15-lipoxygenating, called 15-LOX) [18, 19]. The impact of activation/stimulation (time, type, and dose of stimulus) of the biological sample needs to be considered as oxylipin patterns change over time, for example, induction and activation of COX-2 by proinflammatory LPS in macrophages [20]. If using cultured cells, oxylipins are generally enriched in the supernatant as they are secreted to act as paracrine stimuli on other cells [21, 22].

12.2.2 Consideration of Experimental System, Focusing on Plasma and Serum

For the analysis of blood cells, one can choose purified cells, such as platelets, monocytes, or neutrophils, either alone or in mixed cell populations [23–26]. All have highly specific expression patterns for oxylipin-generating enzymes, and if combined, then transcellular metabolism to secondary products may occur. Whole blood can be analyzed, before or after the use of activators such as LPS [27]. Anticoagulated whole blood can be centrifuged to remove cells, generating plasma. For this, clotting is prevented using anti-coagulants, such as calcium chelators (Ethylenediaminetetraacetic acid [EDTA] and sodium citrate), or heparin. These have distinct effects on oxylipins [28, 29], with chelation being preferred because of its dual action on both coagulation and platelet/leukocyte activation. Heparin prevents coagulation but does not prevent activation of platelets. When whole blood is not anti-coagulated, or is actively coagulated using activators such as glass particles, the removal of cells along with fibrinogen results in serum. Generation of serum represents an *ex vivo* coagulation assay forming high(er) oxylipin levels because of the activation of white cells and platelets during blood clotting. Thus, oxylipin levels are not representative of what was present *in vivo* before sampling, and great care would also be needed to ensure that non-enzymatic oxidation to form isoprostanes was prevented. Serum oxylipin analysis can be considered an *ex vivo* "capacity assay" where the maximum ability of the blood cells to generate the lipids is being measured. This can be useful, for example, if testing the ability of an *in vivo* administered COX-1 inhibitor to prevent platelet thromboxane generation, but it does not reflect endogenous *in vivo* circulating oxylipins. Notwithstanding this, serum has been successfully used in a large number of published studies [30–32]. Plasma oxylipins can indirectly reflect their formation and action in other tissues such as kidney, liver, lung, etc. The relative contribution of circulating blood cells versus other tissues to measured plasma levels can be difficult to delineate, especially during inflammation.

When isolating plasma or purified cell populations from whole blood, the method of venepuncture needs to be considered. Narrow needles/tubing and vacuum from vacutainers can activate white cells and platelets and sometimes cause hemolysis. This is a well-recognized "sampling artifact" that causes significant elevations in oxylipins. Another consideration relates to which oxylipin pool is of interest experimentally and biologically. These lipids are generally measured as their free acid forms (unesterified), but they are found esterified into complex lipids. Indeed, the majority of hydroxy-PUFA and epoxy-PUFA present in plasma or cells appear to be bound to larger lipids, particularly phospholipids (PLs), and are termed oxidized PL (oxPL) [14]. Furthermore, these can be generated acutely by activated white cells or platelets [33]. OxPL can be detected either as their intact molecular species (see Section 12.5) or as the free acid species generated following hydrolysis. This can be achieved using saponification/base hydrolysis or by using phospholipaseA2. During saponification, prostanoids such as PGE_2 are destroyed and cannot be recovered [34]. For these, enzymatic hydrolysis is preferred [35]. Notably, with this, structural information on the precursor lipids is lost. A detailed protocol for the quantification of "total" oxylipins in blood can be found here [16, 36], and their analysis as intact oxPL is detailed later.

12.2.3 Obtaining and Handling Plasma for Oxylipin Analysis

Many cohorts have stored plasma and serum; thus, these tend to be the most commonly analyzed samples in population studies. Oxylipin measurements in the plasma/serum from cohorts can be fraught with difficulty as sometimes the sampling methods are not well described, and oxidation can increase in long-term storage even at −80 °C [37]. Each step, such as transitory storage, transport, or long-term storage, can influence the resultant levels of oxylipins detected [14, 15, 17, 29]. The key for successful quantitation is to define all parameters that could cause variability or interferences in advance of starting sample collection. A list of these is provided below, and these have been extensively tested in Refs. [14, 15, 17, 29].

(1) Itemize all plasticware (brand and specific type) and test for possible interferences caused, for example, by plasticizers. A list of suitable materials is provided here [16].
(2) Collect blood at a defined time (±2 hours) of the day, ideally following overnight fasting. Use a needle with a sufficient gauge (20 G or larger) and EDTA collection tubes.
   - Fasting state and/or sampling at different times can lead to high inter-individual variability, while inter-day variability of plasma oxylipins of fasted at the same time of day subjects is low [14, 38].
(3) During transport, keep the blood cool – but not on ice – and store for maximum two hours, avoid shaking and minimize transport.
   - Freezing (including freeze/thaw cycles) or shaking can cause hemolysis and/or cell activation, distorting oxylipin levels detected [17].
(4) To generate plasma, centrifuge at 1000 g for 15 minutes at 4 °C and remove plasma from the tube within one hour.
(5) Generate aliquots (e.g. 500 μl for free [39, 40] or 100 μl for total oxylipins [36]). Whenever possible, generate back-up samples. Pipette the exact volume needed and mix directly with an equal volume of methanol or other organic solvents.
   - Methanol stabilizes the biological samples during long-term storage [29]
(6) Freeze the sample immediately, e.g. by snap freezing in liquid nitrogen. Record the time from blood draw till freezing for each sample (batch). Time to freezing should not exceed three hours [15, 17].
(7) Store at −80 °C to analyze the samples as soon as possible but within at most two years. If the samples need to be shipped, send them on dry ice with a temperature logger.
   - Oxylipin levels can change during storage. At temperatures over −30 °C, significant changes occur, particularly if no methanol is added. In particular, non-enzymatically generated oxylipins, including F2-isoprostanes (F2-IsoPs), can form plasma PUFAs at these temperatures. However, at −80 °C, levels of samples prepared as outlined can be regarded as stable for about two years and maybe longer [14, 29].

12.2.4 Extraction of Oxylipins from Plasma

Before LC-MS/MS, the plasma needs to be processed to remove proteins and other interfering compounds as ESI and MS detection are strongly influenced by matrix composition. Processing generally uses solid-phase extraction (SPE) columns that contain reverse-phase LC solid phase, allowing selective elution of the column of oxylipins following removal of polar compounds and neutral lipids. SPE can also enable the concentration of samples, e.g. by factor 10 [39, 40], facilitating detection of low levels of oxylipins. However, several quantitative oxylipin methods have been described, leading to an overall dilution of the samples using rapid sample preparation, without SPE [41] or online SPE following dilution [42]. Before sample clean up, stable isotope-labeled internal standards are added, which correct for losses during extraction and allow quantitation, as described here [36, 39, 40]. Moreover, proteins can be precipitated by using an excess of organic solvent such as methanol, if this has not been added during sample collection (see above) [36, 39].

Use IS recovery as a quality marker for the sample preparation in each sample: Following SPE, the extract is reconstituted in an organic solvent. "Internal standard 2" (IS2) is then added, for example, 1-(1-(ethylsulfonyl)piperidin-4-yl)-3-(4-(trifluoromethoxy)phenyl)urea, 12-(3-adamantan-1-yl-ureido)-dodecanoicacid, 12-oxo-phytodienoic acid, and aleuritic acid [15, 43]. With that, the extraction recovery of the internal standards themselves can be quantified in each sample after extraction. Using that method, only extracted samples with an internal standard recovery of at least 60% would be used for quantification. If recovery appears lower than expected, adding internal standards after the SPE step and comparing the results allow us to pinpoint whether the extraction efficacy of the oxylipins is not sufficient or the signal is reduced because of ion suppression (a matrix effect). This is described in the below paragraph in more detail. Finally, an ion suppression analysis (by post-column addition of an internal standard (IS) solution using a syringe pump and connected via a T-piece between column and MS while monitoring its transition during the injection of a sample) allows a detailed analysis of eluting ion suppressing matrix compounds [36, 44]. For oxylipins, several successful protocols for SPE extraction have been described in the literature, and we refer the reader to detailed procedures here [16, 36, 45].

12.2.5 Setup of LC-MS/MS Analytical Method

Analysis of oxylipins using the newer scheduled MRM methods that can detect and quantify 100–200 analytes is technically demanding and requires significant time, expertise, and training. A dedicated postdoctoral scientist, overseeing all analyses, e.g. supervising, training staff and students, and monitoring instrument performance is essential. A number of these methods have been described by several laboratories, as reviewed here [14, 46]. Most often, the LC system uses sub-2 μm particle size filled columns that generate a very high backpressure of >500 bar, ESI in the negative-ion mode, and a triple-quadrupole (QqQ) analyzer operated in the SRM/MRM mode (see earlier). Optimized methods for a large number of oxylipins including isoprostanes or so-called "specialist resolving mediators" are described here [15, 39, 40].

The following parameters need to be defined for instrument/method qualification:

- Which oxylipins should be included? A quantitative method requires standards for each analyte, and an increasing number of oxylipins are commercially available. However, most are not fully characterized regarding purity. If desired, concentrations can be corrected by concentration factors determined based on UV spectroscopy and LC-ESI(−)MS analysis in selected ion monitoring. The detailed procedures can be found here [47].

- Efficient chromatographic separation: For UHPLC, column and mobile-phase gradient selection should lead to narrow peaks (full width at half-maximum [FWHM] height about three to four seconds [39, 40]) to allow sufficient separation of the large number of oxylipins. Several structurally different oxylipins show identical fragment spectra in mass spectrometry [46] and need to be separated by chromatography. Examples of these, which elute as critical separation pairs, are PGE2/PGD2, PGB2/PGJ2 12(13)-DiHODE/15,16-DiHODE, w-hydroxy-PUFA/(w-1)-hydroxy-PUFA, 8(9)-EpETE/11(12)-EpETE, 11(12)-EpETE/14(15)-EpETE, 8(9)-EpETrE/11(12)-EpETrE, and 7-HDHA/11-HDHA [16]. Their efficient separation needs to be visually confirmed using standards.
- Instability of some lipids: Most oxylipins, including multiply oxygenated PUFA called SPM, are relatively stable under the analytical conditions outlined here. Exceptions include lipids with hydroperoxide functional groups. These, are highly unstable in disrupted tissue samples because of their reactivity toward transitional metal ions. Other lipids that are not possible to analyze are prostaglandin H2 and thromboxane A2, which are unstable in aqueous environments generating more stable end products that can be measured instead.
- Co-elution problems: Because of the large number of oxylipins generated in a biological system, co-elution with stereoisomers can occur, and this is particularly a problem with multiply oxygenated PUFA. Enantiomers (mirror image) do not separate on reverse-phase LC, while diastereomers (non-mirror image) do. This issue is exemplified with SPM, including PD1 [48]. The specific problem with assigning names where there may be co-eluting enantiomers is covered in more detail later.
- Sensitive and selective mass spectrometric detection: The LC-MS/MS parameters (source setting, gas flows, temperature, and electronical potentials) for the MRM transitions need to be optimized [39, 40]. Each primary standard should be optimized for compound-specific parameters such as collision energy.
- Scheduled MRM: This approach allows for detection of large numbers of oxylipins as individual lipids are only monitored around the time when they elute. Here, detection windows need to be narrow but also wide enough to allow a sufficient background trace to be recorded. This requires a stable retention time and a cycle time that records sufficient data points per peak (varying from six upward, with some laboratories using <20) balanced with sufficiently long dwell times (the time during which the instrument is analyzing a transition, ideally 10–50 ms). A starting point, which may need adapting based on the platform and number of analytes, would be 30–60 seconds window and a total cycle time of 0.4 seconds peaks with an FWHM of around 3–4 seconds [15, 39, 40]. However, what is critical is that there is sufficient baseline measured to enable robust estimation of background noise, which is used for LOD/quantitation (LOD/LOQ) determination as detailed below. Thus, the windows must be optimized appropriate to the specific instrument conditions. When measuring very large numbers of lipids, it is generally feasible to only monitor one transition per analyte because of the

analytical capacity of instruments. However, if needed, to ensure higher confidence in identification, an additional strategy can involve the monitoring of two per analyte, one being the quantifier and the other the qualifier. The ratio of response of qualifier to quantifier being consistent with a standard would be considered as further evidence, in particular the existence of a chromatographic peak with identical retention time for both transitions. If this is not routinely possible, then it can be built in as part of structure confirmation. MS/MS spectra can also be acquired; however, when lipids are present in low amounts in biological samples, spectra obtained at the expected elution time are generally extremely noisy and are then not suitable for identification purposes. Only spectra that show a strong visual match to the pure standard provide good evidence for a proposed structure.

- Define lower LOD: Using diluted solutions of standards, the instrument LOD needs to be defined based on a signal-to-noise ratio (s/n) of ≥3 for each oxylipin with the developed method. With the most sensitive of the current instrumentation, a LOD of >0.1–1 pg on column, which means with a typical injection of 10 μl, a concentration in an injected solution of 10–100 pg/ml is feasible for most oxylipins [39, 40].
- Internal standards: Isotopically labeled internal standards should be chosen to represent the sub-categories of oxylipins being measured, including as many as feasible. These should be added to samples at about 10- to 30-folds the concentration of the LOD of the analyte (about 50–100 ng/ml). Prepare sufficient amount of IS solution for samples in advance and add to samples before extraction.
- Standard curves: A calibration series is generated by sequential dilution of mixtures of primary standards. A detailed pipetting scheme can be found in Ref. [15]. Include at least two concentrations of lipids per order of magnitude from the LOD to the highest concentration present in samples or slightly above the expected upper limit of quantification (about 1 μg/ml). This should remain below the upper limits of dynamic range for the instrument (based on linear response to serial dilutions). Standard solutions are stable for several months at −80 °C, and freeze/thawing should be avoided by storing in appropriate glass vials or ampules. Unstable oxylipins, such as hydroperoxy-PUFA, should not be included in calibration series.
- Limits of quantification and calibration: The LOQ for standards and analytes measured in research laboratory assays is generally defined as a peak exceeding an s/n ratio of ~5–10, in line with recommendations from external agencies, such as FDA, WHO, and others. A linear calibration function can be generated by weighting the concentration $1/x$ or $1/x^2$ (because random variation is relative and thus absolutely smaller for lower concentrations). For all calibrators, the accuracy can be calculated and needs to be within ±15 and (±20% for the calibrator of the ultraviolet (ultraviolet), lower limit of quantitation [LLOQ]), which limits the LLOQ. Matrix-matched calibration can be helpful in specific cases, which is

however not required because of the used isotopically labeled standards (the so-called isotope dilution analysis), and also feasible when dealing with many different types of tissues on a routine basis.

- System qualification: It can be helpful, in addition to using the calibration solutions from the MS manufacturer, to make use of a defined solution of oxylipins (e.g. one of the standards) for the purposes of ensuring good system performance before analyzing the samples. Using this approach, a user can confirm specific performance criteria, relating to minimum chromatographic and mass spectrometric performance. These can include background intensity, retention times, peak widths, resolution of critical separation pairs, absolute intensity (peak height and area) of IS, and accuracy.

12.2.6 Quality Assessment and Control

It is important to closely monitor method performance so that if problems occur, they can be quickly identified and rectified. Variations in instrument performance (sensitivity) and sample preparation (IS recovery) can be routinely monitored. Where laboratories are conducting large-scale analyses of cohort samples, it is critical to demonstrate assay stability between batches over time. For this, we recommend a quality control (QS) sample, such as a large volume of (pooled) plasma, which is stored in appropriate aliquots. With every analytical batch, such as the 20 samples co-extracted on a SPE manifold, one of the QC samples would be prepared in parallel. Documenting the detected concentration and comparing it with historical controls allow determination of accuracy and precision of the analysis, which is typically ±15% for most oxylipins.

12.3 Challenges Presented by Oxylipin Isomers

12.3.1 Analytical Challenges of Isomers

A significant challenge in oxylipin analysis lies in discriminating specific constitutional isomers and in particular stereoisomers, especially when analyzing complex biological mixtures. Importantly, the bioactivity and (stereo-)chemical structure of oxylipins are inherently intertwined. For example, LTB_4 and its diastereomer 5*S*,12*S*-diHETE not only arise from different biochemical pathways but also vastly differ in their bioactivities [49]. LTB_4 is a very strong chemotactic agent, predominantly produced in polymorphonuclear cells by the actions of 5-lipoxygenase (LO) and leukotriene (LT) A_4 hydrolase, and its diastereomer 5*S*,12*S*-diHETE on the other hand is a platelet neutrophil interaction product involving 5- and 12-LOX for its production [50, 51]. An oxylipin's stereochemical definition is of great importance to our understanding of its origin and bioactivity. Of note, the actions of the major mammalian enzymes involved in oxylipin production in the immune system are stereochemically *S*-configuration specific while some skin isomers generate R-enantiomers. However, tandem mass spectra of stereoisomers and in some cases even constitutional isomers are identical. A recent example illustrates co-elution and highly similar MS/MS fragmentation of the constitutional isomers LXA_4 and a 5D2-isoprostane. Using biological interpretation and MS^3 analysis, the authors were able to dissect the two isomers [52]. Nevertheless, this illustrates that even with powerful LC-MS/MS approaches and monitoring presumably analyte-specific mass transitions (m/z 351 → 115), likely well-separated constitutional isomers can still be mistaken for each other.

The situation becomes even more complicated when taking stereochemistry into consideration. Two major forms of stereoisomers contribute to the stereochemical complexity of oxylipins, geometric double-bond isomers and configurational isomers including enantiomers as well as diastereomers (being non-mirror image stereoisomers). Each chiral center or geometric double bond in an oxylipin can contribute to two stereoisomers, and where there are multiples of these, the theoretical number of isomers increases consecutively by a factor of 2. In practice, geometric double-bond isomers will likely contribute less as their configurations are fixed and are relatively stable for specific oxylipin substrates, although this is not always the case. As one example, LTB_4 possesses two stereocenters at the hydroxylated positions 5 and 12 of which each can be *R*- or *S*-configured; additionally, LTB_4 possesses four double bonds that can either be *E*- or *Z*-configured; hence, $2^6 = 64$ stereoisomers can theoretically be formed. Fortunately, this is a theoretical number, and considering "enantiomerically pure biochemical starting materials," such as arachidonic acid (all *Z*-configured), this number will be much lower. Nevertheless, a significant number of enzymatic and non-enzymatic (stereo-)isomers can be present in biological samples and consequently several analytical approaches have investigated the stereochemical classification of oxylipins. Most important and routinely applied is the separation of diastereomers using reversed-phase chromatography. As the physiochemical characteristics of diastereomers differ (unlike enantiomers), many of these species are readily separable using routine C18-based chromatographic systems. For example, the non-enzymatic hydrolysis products of LTA_4, 6-*trans*-LTB_4, and 6-*trans*-12-*epi*-LTB_4 are readily separable on C18-reversed-phase columns. In addition, the use of chiral phases has proven to be highly useful for the separation of not only enantiomers (*R* versus *S*) but also diastereomers [29]. Just recently, two reports have shown the applicability of chiral separations for oxylipin analysis [53, 54]. Notably, these have traditionally been restricted to normal phase systems; however, novel column chemistries (e.g. the Chiralpak AD-RH and IA-U) allow operation under reversed-phase conditions, rendering chiral separations more practical in combination with ESI tandem mass spectrometry [29, 53]. Nevertheless, the sheer number of possible stereoisomers combined with the complexity of biological sample materials has led to the development and application of additional technologies for the successful stereochemical resolution of oxylipins. Recently, ion mobility separations have been introduced to the field [55]. Originally differential mobility separation was introduced by Kapron et al. for the separation of prostanoids [56] and later adapted by Jonasdottir et al. [50] for the separation of LTB_4 isomers. Lately also, drift tube-based ion mobility has entered the field of oxylipin analysis [57]. All these approaches rely on the availability of sufficiently pure standard materials with stereochemical assignments. Many platforms initially assign identity and hence stereochemistry of a specific oxylipin based on reversed-phase LC-MS/MS analysis. Particularly for well-described diastereomers, this is to some extent possible and considering time and budgetary constraints is also understandable. However, enantiomers cannot be separated using reversed-phase systems; in turn, the assignment of a stereocenter being *R* or *S* can only be accomplished by chiral chromatography with enantiomerically pure standards at hand. Furthermore, even though diastereomers are, in most cases, well separatable using reversed-phase systems, orthogonal chromatographic resolution is key for the unambiguous assignment of a stereochemically defined diastereomer. In other words, even though LC-MS/MS analysis results between a standard and a specific analyte closely match, there is no guarantee that all geometric and configurational descriptors do so as well. Thus, if unambiguous stereochemical description of an oxylipin is of importance, matching between the standard and the analyte, applying one or more of the above-discussed analytical solutions should be considered. Here, ideally, reversed-phase LC-MS/MS results would be confirmed by chiral chromatography or ion mobility separation. An example of this is provided in Figure 12.3. Here, we see that in peritoneal cells either with or without a 24 hours challenge with zymosan, a peak around 9.3 minutes is visible (Figure 12.3a). However, this peak comprises both LTB_4 (5*S*,12*R*-diHETE) and its diastereomer 5*S*,12*S*-diHETE, which do not separate sufficiently on reverse-phase LC, and thus, only one overall peak is seen for both lipids. Analyzing the same extract using differential ion mobility (differential mobility spectrometry [DMS]) shows that there are different lipids in the samples, with LTB_4 predominating after inflammatory challenge and 5*S*,12*S*-diHETE predominating before (Figure 12.3b). If sensitivity is an issue, which cannot be met by chiral separation or ion mobility, "orthogonal" eluent systems can be applied to reversed-phase columns as an intermediate measure, i.e. swapping methanol for acetonitrile, adding MS-compatible buffers, and changing the pH. Finally, it is important to realize that all types of LC-MS/MS-based stereochemical assignments are of relative nature, and a final absolute assignment of stereochemistry can only be made by complimentary technologies such as nuclear magnetic resonance (NMR). In cases where stereochemistry is ambiguous, non-chiral names would be more chemically correct. This is discussed in more detail below.

12.3.2 Biological Considerations of Isomers

The biology of the tissue can provide important clues to which lipids are present. For example, in the case of PGE_2, if the MRM/SRM channel shows a large number of closely eluting peaks, of fairly similar size, then one would need to consider

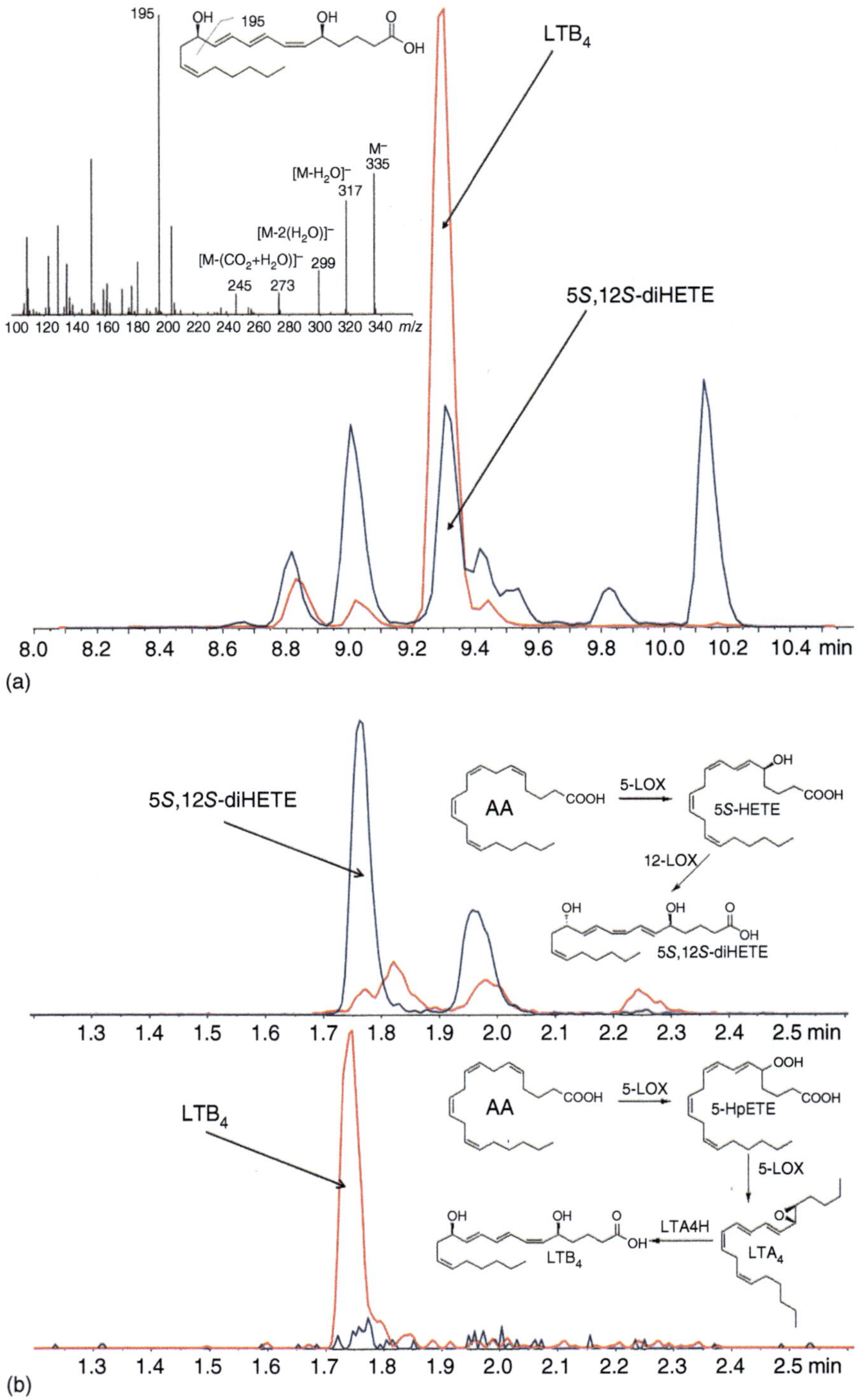

195
OH
195
OH
O
OH
M⁻
335
[M-H2O]⁻
317
[M-2(H2O)]⁻
[M-(CO2+H2O)]⁻ 299
245
273
100 120 140 160 180 200 220 240 260 280 300 320 340 m/z
LTB4
5S,12S-diHETE
8.0 8.2 8.4 8.6 8.8 9.0 9.2 9.4 9.6 9.8 10.0 10.2 10.4 min
(a)
5S,12S-diHETE
AA
COOH
5-LOX
OH
COOH
5S-HETE
12-LOX
OH
OH
O
OH
5S,12S-diHETE
1.3 1.4 1.5 1.6 1.7 1.8 1.9 2.0 2.1 2.2 2.3 2.4 2.5 min
LTB4
AA
COOH
5-LOX
OOH
COOH
5-HpETE
5-LOX
O
LTA4H
LTA4
OH
OH
O
OH
LTB4
1.3 1.4 1.5 1.6 1.7 1.8 1.9 2.0 2.1 2.2 2.3 2.4 2.5 min
(b)

whether they all originate from non-enzymatic oxidation. On the other hand, if the sample is from inflammatory activated cells, and a prominent signal at the retention time for PGE_2 is seen, with only small peaks for the other related lipids, then it is expected that PGE_2 is the lipid measured and that COX-2 and PGES are involved. Further confirmation can be provided by showing inhibition by indomethacin. For many lipids, unless chiral chromatography is used to verify the structure of the lipid made, or the isomeric composition of the system is well known, stereochemistry can be omitted when reporting structures. In the case of platelets, 12(*S*)-HETE predominates because of the well-known high activity of platelet 12-LOX [58]. However, if measuring this lipid elsewhere, it can be called simply 12-HETE. Some oxylipins are named based on their function rather than their structure with these names inferring stereochemistry. These include the docosahexaenoic acid (DHA) and eicosapentaenoic acid (EPA) oxygenation products: lipoxins, resolvins, protectins, and maresins, commonly called specialist pro-resolving mediators (SPMs). Here, the number of stereoisomers that would form via non-enzymatic oxidation of arachidonic acid (AA), DHA, or EPA, considering chirality alone, can be either 4 (with two chiral centers, e.g. PD1) or 8 (with three chiral centers, e.g. RvD1, D2, and D4). As these lipids have more than one chiral center, they form groups of non-mirror image stereoisomers that separate on reverse phase (diastereomers) when one or more (but not all) of the chiral centers have different configurations. However, when all the centers have opposite configurations, they are enantiomers (mirror-image stereoisomers) which do co-elute on reverse-phase LC-MS/MS. In summary, for all oxylipins, two or more peaks eluting closely together of relatively similar size, rather than a single isomer on its own, should always be seen as suspicion of non-enzymatic oxidation either in a biological system or during sample preparation. Unfortunately, the narrow windows used for scheduled MRM/SRM can potentially lead to these peaks being missed. Even if a single peak is seen, without chiral chromatography on that sample type, one cannot be sure which stereoisomer(s) are present. Labeling the lipid based on what is known, e.g. 7,8,17-triHDOHE, omitting

Figure 12.3 LTB4 and its enantiomer, 5*S*,12*S*-diHETE, are not separated on reverse-phase LC-MS/MS but can be separated using differential ion mobility analysis. (a) LC-MS/MS analysis of an ethanol extract from peritoneal cells monitoring the SRM transition *m/z* 335 → 195. Red trace, 24 hours after zymosan A challenge; blue trace, 2 hours after PBS injection (control group). Upper left corner, MS/MS spectrum of LTB_4 and its isomers, showing the typical fragment ion *m/z* 195. (b) Analysis of murine peritoneal cell ethanol extracts using μLC–DMS-MS/MS. Upper panel, control mice (PBS injection). Lower panel, zymosan A challenged mice. Red traces, SV 4500 V and COV 17.9 V; blue traces, SV 4500 V and COV 20.3 V. Additionally, the biochemical pathways leading to LTB_4 (lower panel) and 5*S*,12*S*-diHETE (upper panel) are shown. Abbreviations: AA, arachidonic acid; 5*S*-HETE, 5(*S*)-hydroxyeicosatetraenoic acid; 12-LOX, 12-lipoxygenase; 5HpETE, 5-hydroperoxy-eicosatetraenoic acid; and LTA4, leukotriene A4. Unknown: undefined isomer in the trace of 5*S*,12*S*-diHETE. Source: Republished with permission of American Chemical Society, from Jonasdottir et al. [50], permission conveyed through Copyright Clearance Center, Inc.

the stereochemistry inferred by the trivial name RvD1, avoids this error. Unfortunately, many studies have used reverse-phase chromatography along with stereochemical naming of lipids in complex biological mixtures without the confirmation of enatiomeric structure.

12.4 Analysis of Urine Oxylipin Metabolites

While plasma is more frequently collected in human clinical trials and studies, urine is an alternative biological fluid with distinct advantages for the quantitation of oxylipin biomarkers. Urine collection is non-invasive and can be used in large-scale trials with high participant agreement. Samples can be collected by study participants at home, as many oxylipin urinary metabolites are stable at 4°C (a standard home refrigerator) for a few days. Further, urine can be collected in a sufficient volume that allows for multiple analyses and can be stored for future studies. Important for the analysis of oxylipin metabolites, urine is a filtered fluid with low levels of cell contamination and metals; thus, this matrix is less susceptible than plasma to ex vivo oxidation during sample collection and storage [59, 60]. MS methods for oxylipin urinary metabolites are similar to those used for plasma, for example, the use of pooled urine QC samples. Analysis of oxylipin metabolites, however, has its own set of considerations regarding sample collection and preparation. First, and most importantly, because of analyte stability and extraction, all urinary metabolites cannot be analyzed in a single method. Herein, we will first discuss general considerations for urine analysis and then focus on factors specific to individual classes of urine metabolites.

12.4.1 General Considerations

- Determination of experimental system: First, when planning to measure oxylipins and/or their urinary metabolites, it should be determined if urine is an appropriate matrix that will answer the scientific question in the population being studied. Urinary eicosanoid metabolites have often been measured in large-scale clinical trials as biomarkers of disease risk. Additionally, they have been used to assess therapeutic interventions [61–65]. Metabolic enzymes and metabolite formation differ between humans and animals, particularly rodents, and between males and females in both humans and animals [66, 67]. Additionally, there are differences in oxylipin metabolite patterns between pre- and post-menopausal women [68, 69]. Finally, specific pathophysiological conditions can alter urinary metabolic profiles. For example, kidney dysfunction increases excretion of unmetabolized oxylipins in urine [70].
- Planning of sample collection preparation and storage: Urine samples should be collected at a consistent time of day (i.e. first morning void after overnight fasting). If following a therapeutic intervention or procedure, collection should be timed such that metabolites have sufficient time to be excreted. On collection, the

sample should be placed on ice or in a refrigerator (4 °C) to prevent degradation of metabolites. Samples should remain at room temperature no longer than one hour. Multiple aliquots of at least 2 ml should be prepped for storage. The samples should be vortexed before aliquoting. We recommend that samples should not be spun to remove particulates as it can lead to unequal distribution of oxylipin metabolite patterns between aliquots [60].

- Normalization: The concentration of a urinary metabolite is dependent on its excretion rate and the urinary flow rate. Urinary biomarkers are commonly reported as a ratio of analyte concentration to urinary creatinine excretion, which is typically assumed to be linear to biomarker excretion. Specific gravity and total urine excretion volume are alternative normalization factors. Particular attention to normalization should be paid when renal function is changing rapidly or in the case of acute kidney injury [59, 60, 71].

12.4.2 Prostaglandins (PGs)

For decades, it was considered that the most accurate index of endogenous PG production in humans was the measurement of excreted urinary metabolites using MS [66, 72, 73]. The advent of modern, highly sensitive MS has improved our ability to explore local PG formation (i.e. Cerebrospinal fluid [CSF] and nasal fluid) and enabled broad spectrum, multi-class oxylipin analysis [74–76]. However, urine still provides a robust option for the evaluation of systemic (whole body) PG production [60]. Quantification of PG urinary metabolites (Figure 12.4) has allowed for the assessment of PG production in various diseases and intervention strategies (non-steroidal anti-inflammatory drugs [NSAIDs], COXIBs, and nutritional supplements) [70, 74, 77–83]. In fact, in a recent meta-analysis of urinary biomarkers of colorectal cancer (CRC), measurement of the major urinary metabolite of PGE_2 was shown to be the most clinically promising for detection of CRC risk [84].

- PGE_2: PGE-M is the most abundant PG metabolite in human urine. The first method to quantify PGE-M using GC-MS was published in 1976 [85]. A second method using GC-MS/MS was published in 1990 [86]. However, both are extremely labor intensive, involving multiple purification and derivatization steps. PGE-M is quite unstable, with the two carbonyl groups being susceptible to dehydration and other reactions, as is the C-11 hydroxyl group. PGE-M begins to degrade in urine left at room temperature for 1.5 hours (Milne and Morrow, unpublished results). The development of an LC-MS/MS method for accurate PGE-M quantification was accomplished following the synthesis of analytical standards derivatized with methyloxime HCl or [$^2H_3$]-methyloxime HCl (internal standard) [87]. There are thus several considerations that must be taken into account. First, urine should be derivatized with methyloxime HCl immediately upon thawing. Extraction using a C18 SPE cartridge is required following derivatization to remove unreacted methyloxime HCl. The internal standard has been derivatized with [$^2H_3$]-methyloxime HCl during synthesis and purified; thus, it cannot be added to the sample until after SPE. (Note: This internal standard is stable at −80 °C for 15+ years, but it is not commercially available in the derivatized form) [66]. Methyloxime HCl reacts

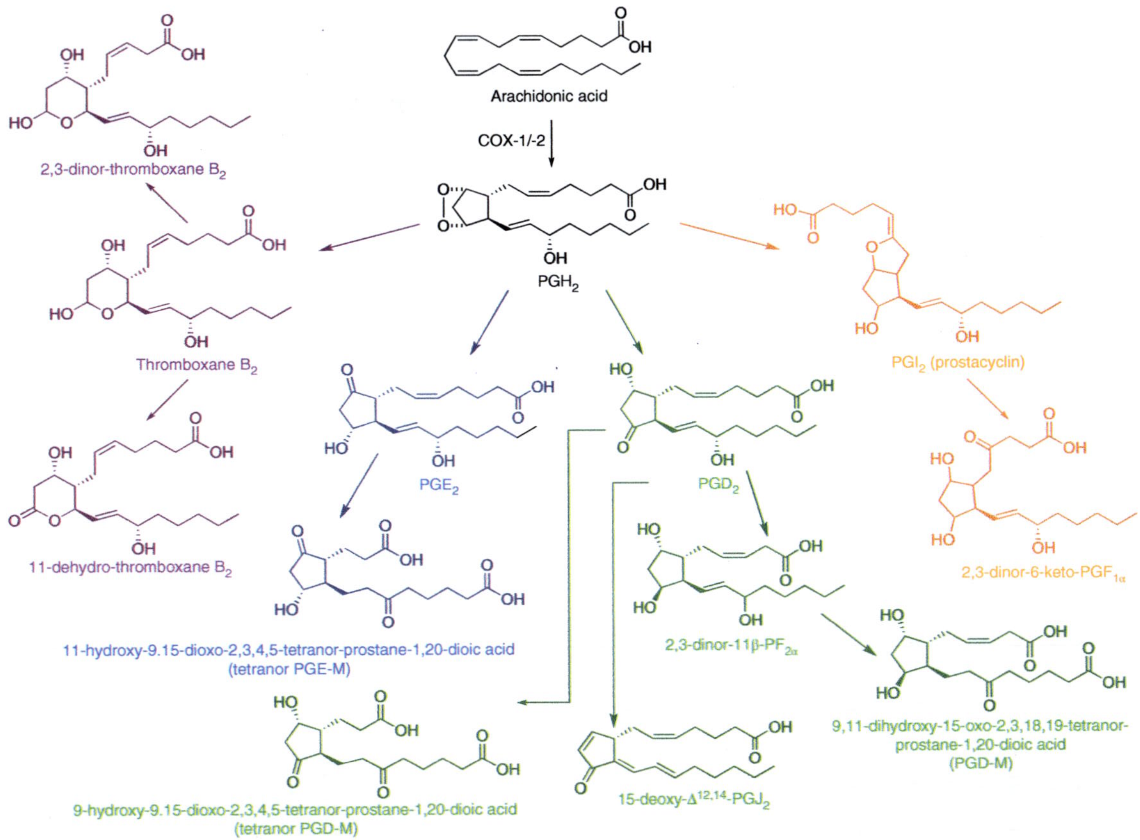

Figure 12.4 Urinary metabolites cyclooxygenase-derived prostaglandins.

with the two keto groups (C-9 and C-15) to form four (syn/anti) methoxime isomers (Figure 12.5a). These separate into two distinct peaks when using a shallow gradient on reversed-phase LC (C18). The ratio of the peak height of endogenous PGE-M to the peak height of the internal standard is calculated for each methoxime isomeric peak. The ratios of the two PGE-M peaks are averaged for quantification [66].

- PGD_2: PGD_2 is isomeric to PGE_2, differing only in the location of the keto and hydroxyl groups on the prostane ring. Despite their similarity, the metabolic profile of PGD_2 is more varied than PGE_2. The two most abundant urinary metabolites of PGD_2 are PGD-M and tetranor PGD-M (Figure 12.4) [88]. PGD-M was identified in

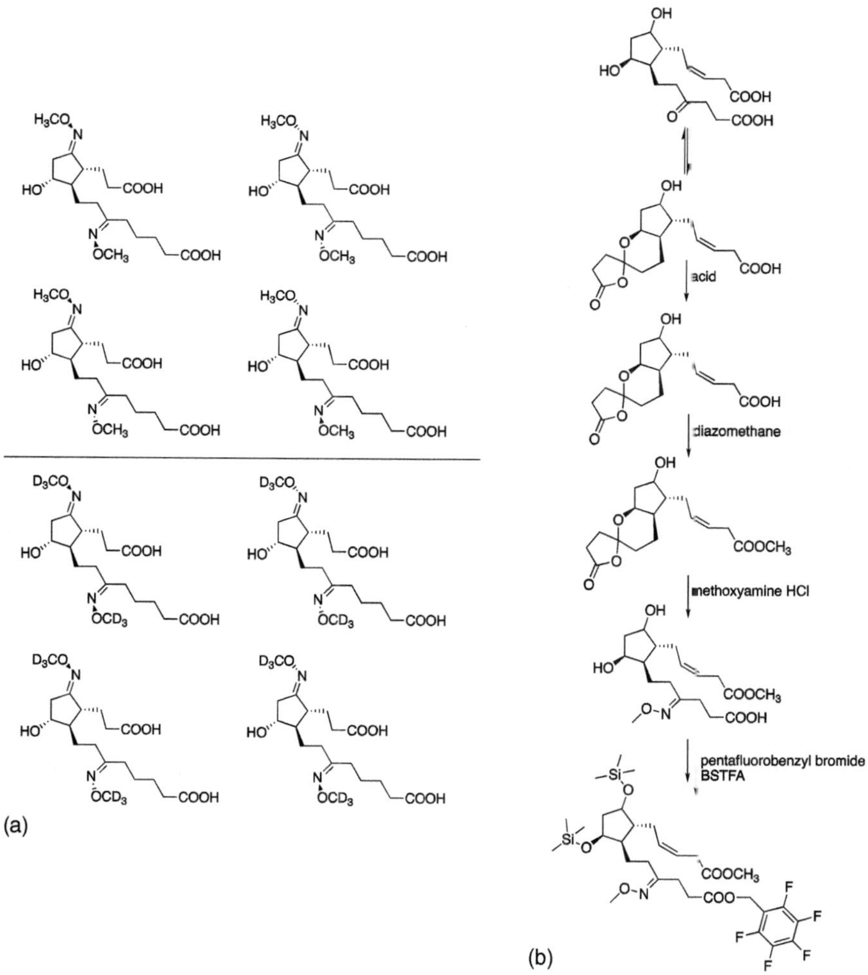

Figure 12.5 Derivatization strategies are used to stabilize urinary metabolites of PGE_2 and PGD_2 for MS analysis. (a) *Syn-* and *anti*-methoxime isomers of PGE-M formed during derivatized with methyloxime HCl or [$^2H_3$]-methyloxime HCl. (b) Cyclization of PGD-M and stabilization by derivatization for GC-MS analysis.

1985 as the major urinary metabolite of PGD_2 [89]. Because of the proximity of the C-11 hydroxyl group to the C-15 keto group, PGD-M readily undergoes cyclization, existing in equilibrium as shown in Figure 12.5b. PGD-M has been quantified using GC-MS following multiple derivatization and purification steps that open up the cyclized form of PGD-M (Figure 12.5b) [90]. No LC-MS method for quantification of PGD-M is currently available. However, 2,3-dinor-11β-$PGF_{2\alpha}$, an intermediate metabolite between PGD_2 and PGD-M, has been detected in urine using LC-MS/MS [91]. Song et al. identified tetranor PGD-M as a urinary metabolite of PGD_2 in 2008 [92]. This metabolite is analogous to PGE-M and formed in concentrations comparable to PGD-M. Tetranor PGD-M is susceptible to the same stability issues as PGE-M. Likewise, the reaction of tetranor PGD-M with methyloxime HCl results in the formation of four (*syn/anti*) methoxime isomers. These isomers are separable from those of PGE-M [93, 94].

- PGJ_2 and 15dPGJ_2: PGD_2 can also undergo metabolism by dehydration to yield PGJ_2, which is further metabolized to 15-deoxy-$\Delta^{12,13}$-PGJ_2 (15dPGJ_2). Very little 15dPGJ_2 is detected in urine [95]. However, 15dPGJ_2 can be metabolized by a human liver cell line (HepG2) to a glutathione conjugate. This is metabolized analogous to leukotriene C_4, a glutathione conjugate of leukotriene A_4, yielding an end metabolite that is a 15dPGJ_2 cysteine conjugate. This may explain why 15dPGJ_2 levels in urine are low [96, 97]. When assessing the formation of endogenous PGD_2, it is important to consider all routes of metabolism. The biological relevance of these metabolic pathways is incompletely understood.
- Thromboxane and prostacyclin: For over 40 years, the production of prostacyclin (PGI_2) and TxA_2 has been central to the understanding of cardiovascular health [98]. PGI_2 is a product of endothelial cells in the vasculature and is antithrombotic and a vasodilator. TxA_2 is generated in platelets and is prothrombotic and a vasoconstrictor [99]. PGI_2 and TxA_2 are rapidly metabolized. The major urinary metabolite of PGI_2 is 2,3-dinor-6-keto-$PGF_{1\alpha}$ (PGI-M). The major urinary metabolites of TxA_2 are 11-dehydro-thromboxane B_2 (11dTxB_2) and 2,3-dinor-TxB_2 [100, 101]. Infusion studies that alter the levels of PGI_2 and TxA_2 in the circulation have shown that urinary PGI-M and 11dTxB_2 accurately reflect these changes [102]. These urinary metabolites have been extremely useful in the assessment of therapeutic interventions that modulate PGI_2 and TxA_2 formation. For example, measurement of PGI-M and 11dTxB_2 contributed to the development of low-dose aspirin for cardioprotection [65]. Further, quantification of these metabolites was found to be essential to assessing the role of PGI_2 and TxA_2 in the cardiovascular hazards associated with NSAID usage (i.e. rofecoxib) [103]. Interestingly, Nakashima and Schneider have recently reported that 11dTxB_2 can be formed from PGD_2 via a Baeyer–Village oxidation reaction [104]. The oxidation of PGD_2 to 11dTxB_2 was accomplished *in vitro* by oxidation with hydrogen peroxide (H_2O_2), which can be formed endogenously during the settings of oxidative stress. These findings provide a potentially interesting link between the metabolism of PGD_2 and TxA_2.
- Leukotrienes (LTs): LTs are generated from AA, which is first released from cells by the action of cPLA2. The 5-LO enzyme associates with 5-LO activating protein

(FLAP) to convert AA into 5-HETE and subsequently leukotriene A4, the 5,6-epoxide of AA. LTA hydrolase converts LTA_4 to LTB_4, a dihydroxy-AA metabolite. Alternatively, LTA_4 can be acted upon by LTC_4 synthase to form the glutathione conjugate LTC_4. LTC_4 is converted to the bioactive LTD_4 and the end product LTE_4 by peptidases. LTs are important in the inflammatory response, particularly in the pathophysiology of asthma and mast cell activation [105]. LTE_4 is readily detectable in the urine. Because of its chemical structure and requirements for extraction and purification, LTE_4 cannot be measured simultaneously with other oxylipin metabolites. LTB_4 metabolism is more complex. Eleven different metabolites of LTB_4 are excreted in human urine following infusion [105]. The major urinary metabolites are LTB_4-glucuronide and 20-carboxy-LTB_4 [106, 107]. These were detected in low levels but not in the urine of subjects not infused with LTB_4 [107]. Interestingly, Morita et al. recently reported the detection of LTB_4-ethanolamide in the urine of healthy humans as well as in patients with diabetes mellitus [108]. The authors describe this as a metabolite of arachidonoyl-ethanolamine rather than AA. The levels of this metabolite were lower in patients with diabetic neuropathy (stage 3–4) than in healthy subjects or in patients with diabetic neuropathy (stage 1–2).

- Cytochrome P450 oxylipins: Cytochrome P450 (CYP) products of AA oxidation include the EETs (CYP2C and CYP2J), which are metabolized by soluble epoxide hydrolase (sEH) to dihydroxy molecules (DHETs), and 20-HETE (CYP4A, CYP4F). The specific CYP enzymes as well as the generated oxidation products vary by species, sex, and organ [109]. While EETs have been detected in urine (primarily kidney derived), they are generally excreted as the DHET metabolite conjugated with glucuronide. 20-HETE is also excreted as a glucuronide conjugate. Thus, urine must be treated with glucuronidase enzymes before LC-MS analysis to ensure complete capture of total DHET or 20-HETE [110, 111]. Importantly, unlike PGs and LTs, urinary DHETs and 20-HETE do not necessarily reflect the systemic levels of these molecules. Intravenous administration of radiolabeled 14,15-EET to dogs found a significant increase in 14,15-DHET in the plasma, but little radioactivity was detected in the urine [112], thus implying that urinary CYP oxylipins originate in the kidney. Urinary DHETs, however, are increased in diabetic kidney disease, diet-induced hypertension, and pregnancy-induced hypertension [111–114].
- Isoprostanes and related metabolites: F_2-IsoPs are a well-studied class of non-enzymatically generated oxylipins. F_2-IsoPs are formed via autoxidation by different types of free radicals [75] and have been shown to be reliable biomarkers of endogenous lipid peroxidation and oxidative stress. These molecules are chemically stable in biological fluids when stored correctly [115]. F_2-IsoPs are isomeric to $PGF_{2\alpha}$ [115]. F_2-IsoPs can form four classes of regioisomers depending on the position where the oxygen molecule is inserted on the arachidonic acid carbon backbone. The four sub-families consist of 16 diastereoisomers as the hydroxy group on the cyclopentane ring can be arranged in eight different configurations, resulting in total 64 F_2-IsoPs [115]. Multiple nomenclature systems for F_2-IsoPs exist. Despite the potential existence of 64 isomers, the most widely measured

F_2-IsoP is 15-F_{2t}-IsoP (also known as 8-iso-PG$F_{2\alpha}$ or iP$F_{2\alpha}$-III) as it is one of the most abundant of the 64 stereoisomers *in vivo* and was the first to be synthesized [88]. Even though 15-F_{2t}-IsoP is commonly measured in urine in its free form, Yan et al. (2007) demonstrated a 40% increase in the IsoP concentration in urine after treating samples with glucuronidase before MS analysis [71]. Moreover, H. Li, J. A. Lawson, M. Reilly, M. Adiyaman, S. W. Hwang, J. Rokach, G. A. FitzGerald (1999). Quantitative high performance liquid chromatography/tandem mass spectrometric analysis of the four classes of F(2)-isoprostanes in human urine. Proc Natl Acad Sci U S A. Nov 9;96(23):13381-6. Demonstrated that 15-F_{2t}-IsoP is not the most abundant isomer in urine, with a range of 1.11 ± 0.45 pg/mg creatinine for 15-F_{2t}-IsoP but a range of 8.33 ± 3.17 pg/mg creatinine for 5-F_2-IsoPs (5-*epi*-5-F_2-IsoP). This has also been observed by others [70] (Milne and Yang, 2021, unpublished results). Although 5-F_2-IsoPs might be found in higher concentration in human urine, little is known about their biological activities [116]. Two major urinary metabolites of 15-F_{2t}-IsoP have been identified, 2,3-dinor-15-F_{2t}-IsoP and 2,3-dinor-5,6-dihydro-15-F_{2t}-IsoP (Figure 12.6). In addition to 2,3-dinor- and 2,3-dinor-5,6-dihydro- metabolites, 13,14-dihydro-15-keto- and 2,3,4,5-tetranor- derivatives of 15-F_{2t}-IsoP have been identified [115]. 2,3-dinor-5,6-dihydro-15-F_{2t}-IsoP is not subject to autoxidation nor renal production and may be a more sensitive marker of endogenous oxidative stress in urine than intact F_2-IsoPs [68]. It is possible that glucuronide conjugates of 15-F_{2t}-IsoP-M could exist as they have a similar structure to 15-F_{2t}-IsoP; however, these compounds have not been studied to date. Since their discovery over 30 years ago, GC-MS has been the method most commonly used to quantify F_2-IsoPs. The use of LC-MS/MS for their analysis is still in its infancy, hindered until recently by technology and the availability of internal standards. Durand and colleagues have synthesized many isomers; thus, the field is ripe for exploration of their biological activity [117, 118]. The ability to quantify individual isomers and metabolites will certainly expand our understanding of IsoP formation in human physiology and pathophysiology.

- Non-AA-derived IsoPs and isomers of other PGs: It is worth mentioning that PUFA other than AA can generate IsoPs. Adrenic acid (AdA), EPA, docosapentaenoic acid, and DHA generate F_{2t}-dihomo-isoprostanes (F_{2t}-dihomo-IsoP), F_{3t}-isoprostanes (F_{3t}-IsoP), and F_{4t}-neuroprostanes (F_{4t}-NeuroP). These IsoPs have been proposed as biomarkers of neurological diseases such as Alzheimer's, Rett Syndrome, epilepsy, and age-related macular degeneration (AMD) [119–122] and were proposed to predict neonatal morbidity in preterm infants [123, 124]. When conducting EPA and DHA supplementation studies, the analysis of urinary F_3-IsoPs and F_4-NeuroPs could be considered as supplementation of these fatty acids can alter the fatty acid composition of plasma, cells, and tissues in humans [125, 126]. In addition to F-ring molecules, compounds isomeric to PGE_2, PGD_2, and TxB_2 can be generated via non-enzymatic lipid peroxidation. 15-E_{2t}-IsoP (also referred to as 8-iso-PGE_2 or iPE_2-III) was found to have both vasoconstrictive and vasodilatory effects, suggesting a biological activity of this molecule in the cardiovascular system. 15-E_{2t}-IsoP has been detected in the urine of triathlon elite athletes [80].

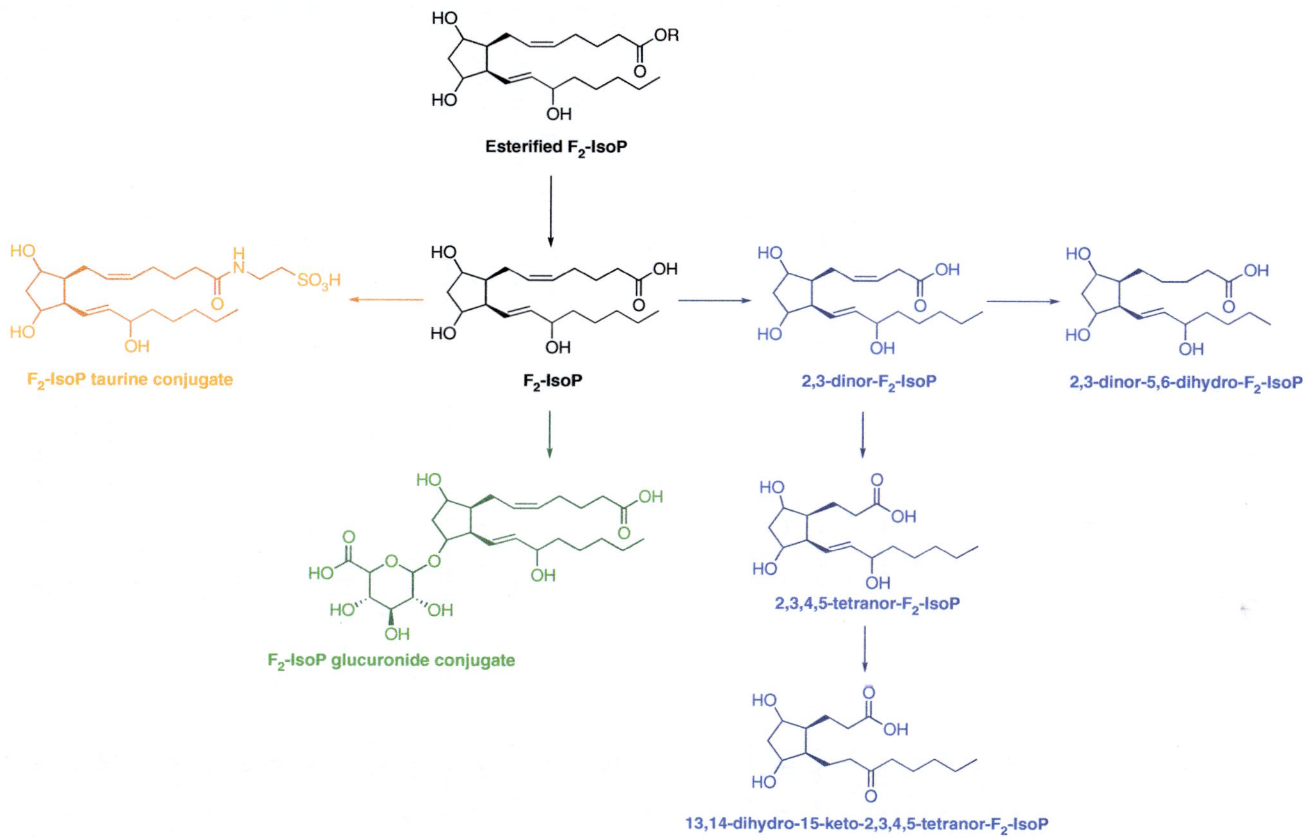

Figure 12.6 Potential routes of metabolism of F_2-Isoprostanes include conjugation, β-oxidation, dehydrogenation, and reduction.

12.5 Analysis of Oxylipins Attached to Phospholipids

There is increasing evidence that considerable amounts of oxylipins are rapidly generated attached to complex lipids such as phospholipids, sterol esters, and glycerides [127]. This generally involves initial enzymatic generation of oxylipins from free FA, followed by their fast esterification into lysoPL via Lands cycle enzymes; although in the case of 15-LOX1, direct oxygenation of the PL occurs [128]. LC-MS/MS analysis of esterified oxylipins is hampered by the relative lack of synthetic standards; thus, structures are often presented as incompletely annotated forms based on the information available, and changes expressed as fold change differences. This pragmatic approach allows biology to be characterized without overstatement of findings in relation to amounts or structures. The exception to this is HETE-PE/phosphatidylcholine (PC), where a limited number of synthetic standards are available including commercially [129]. Furthermore, standards for other monohydroxy-oxylipins such as hydroxydocosahexaenoic acid (HDOHE) or HODEs attached to complex lipids can be generated using air or LOX-mediated oxygenation [129]. In biological samples and cells, often many different molecular species are seen, co-eluting closely together on reverse-phase LC-MS/MS. One example is platelets where the prominent forms are phosphatidylethanolamines (PE) or -cholines (PC) with 12*S*-HETE attached [130]. While these lipids can be quantified, many others formed in the same cells can only be expressed as fold-changes using comparison with an internal standard [131]. Relating to this, early work used dimyristoyl-PE/PC as it was absent in isolated cells; however for plasma, as there is a low endogenous signal, we instead use 15:0/18:1-*d7*-PE/PC. As discussed earlier, a different approach to analyzing oxidized PL was taken by Dennis and colleagues who hydrolyzed chemically and measured the total released oxylipins instead of the intact PL [34]. The advantage is that it allows quantitation of a higher number of oxylipins as there are synthetic standards for the free acid forms. However, the precise esterified structures are missed. As analysis of esterified oxylipins is complex, a number of informatics tools were recently developed to assist, e.g. a computational database on LIPID MAPS (https://www.lipidmaps.org/tools/ms/gp_ox_form.php) LPPTiger, an informatics tool from the Fedorova group that predicts structures from MS/MS data [132], and a library of structures [133]. Analysis of PL-esterified oxylipins generally relies on precursor-to-product ion MRM transitions in the negative-ion mode, where the carboxylate anion of the FA generated from *Sn2* is used as the product ion. While this works very well for phospholipids, where the precursor generates a strong negative ion, it is less useful for glycerides or sterol esters which rely on the positive-ion mode. As oxylipins are rather unstable during collision-induced dissociation, they readily generate internal daughter ions that can inform on positional isomers. In the case of HETEs, a strong ion at *m/z* 319.2 is seen for the intact FA, along with smaller ions that indicate the position of oxygenation, e.g. *m/z* 115 for 5-HETE. Using reverse-phase HPLC/UPLC, positional isomers of HETEs separate slightly, allowing quantitation of the individual isomers, when using their distinct internal product ions [129]. A useful mode for "fishing" for esterified oxylipins in new sample types is precursor scanning LC-MS/MS in the negative-ion mode. This approach allowed initial identification of many HETE-PE and -PC in blood cells such as neutrophils, monocytes, and platelets, as well as esterified PGE_2, and could easily be applied to additional structures in other cell types [35, 128, 130]. It is relatively insensitive however and would not detect many of the lower abundant multiply oxygenates species present. There, prediction of structures using informatics or manual approaches followed by scanning predicted MRM transitions can work well. For general methods for the preparation of oxPL standards and analysis of these lipids in tissues using LC-MS/MS, specifically mono-oxygenated forms, see Morgan et al. [129].

12.6 Conclusions

Methods that allow the quantitation of large numbers of oxylipins and their metabolites in a single run are increasing in popularity. This is in part driven by instrumentation advances such as the ability to conduct chromatography at ever-increasing pressures, improving peak resolution, and reducing analysis time, combined with MS instruments that scan extremely fast and are highly sensitive. However, a note of caution is that these assays are technically demanding to set up and run, and their routine implementation requires highly trained technical support and expert oversight. It is also important to keep biology and biochemistry in mind when interpreting data, with the pattern of products and their isomers guiding the interpretation of findings, e.g. whether the lipids were generated via enzymes, which enzymes were involved, etc. Consideration of sampling: blood or urine, how and where it will be obtained, transported, and stored is essential, and this should ideally happen before the initiation of the study. This review summarizes all these issues and should be a useful guide to those new to the field as well as others seeking guidance around detailed aspects of interpretation of complex data from analysis of biological samples.

Funding Acknowledgment

Funding to LIPID MAPS from the Wellcome Trust is gratefully acknowledged (203 014/Z/16/Z). NHS is supported by a grant (SCHE 1801) of the German Research Foundation (DFG). GLM is supported by the Vanderbilt Diabetes Research and Training Center (United States National Institutes of Health grant DK020593).

References

1 https://lipidmaps.org/databases/lmsd/browse.

2 Moncada, S. and Vane, J.R. (1978). Pharmacology and endogenous roles of prostaglandin endoperoxides, thromboxane A2, and prostacyclin. *Pharmacol Rev.* 30: 293–331.

3 Research CfDEa and Research CfBEa (2015). https://www.fda.gov/regulatory-information/search-fda-guidance-documents/analytical-procedures-and-methods-validation-drugs-and-biologics.

4 Araujo, P. (2009). Key aspects of analytical method validation and linearity evaluation. *J. Chromatogr. B Analyt. Technol. Biomed. Life Sci.* 877: 2224–2234.

5 European Commission (2002). 2002/657/EC: Commission Decision of 12 August 2002 implementing Council Directive 96/23/EC concerning the performance of analytical methods and the interpretation of results (Text with EEA relevance) (notified under document number C(2002) 3044) https://opeuropaeu/en/publication-detail/-/publication/ed928116-a955-4a84-b10a-cf7a82bad858/language-en.

6 Desimoni, E. and Brunetti, B. (2015). About estimating the limit of detection by the signal to noise approach. *Pharm. Anal. Acta* 6: 4.

7 FDA U. Bioanalytical Method Validation Guidance for Industry. *FDA-2013-D-1020.*

8 Boqué, R. and Heyden, Y.H. (2009). The limit of detection. https://www.researchgate.net/profile/Stefano_Polesello/post/How_to_calculate_LOD_and_LOQ_of_analyte_by_hplc/attachment/59d6356379197b8077992f7e/AS:384058174525440@1468578012222/download/The+Limit+of+Detection-LCGC.doc: LCGC Europe.

9 Wenzl, T., Haedrich, J., Schaechtele, A., et al. (2016). Guidance Document on the Estimation of LOD and LOQ for Measurements in the Field of Contaminants in Feed and Food. *EUR 28099; Publications Office of the European Union: Luxembourg*. https://publications.jrc.ec.europa.eu/repository/bitstream/JRC102946/eur%2028099%20en_lod%20loq%20guidance%20document.pdf.

10 http://www.pharmacopeia.cn/v29240/usp29nf24s0_c1225.html.

11 https://www.ema.europa.eu/en/ich-q2-r1-validation-analytical-procedures-text-methodology.

12 https://www.gmp-compliance.org/guidelines/gmp-guideline/who-supplementary-guidelines-on-good-manufacturing-practices-validation.

13 O'Donnell, V., Schebb, N.H., Milne, G.L., et al. (2021). Failure to apply standard limit-of-detection or limit-of-quantitation criteria to specialized pro-resolving mediator analysis incorrectly characterizes their presence in biological samples. Zenodo.

14 Gladine, C., Ostermann, A.I., Newman, J.W., and Schebb, N.H. (2019). MS-based targeted metabolomics of eicosanoids and other oxylipins: analytical and inter-individual variabilities. *Free Radic. Biol. Med.* 144: 72–89.

15 Koch, E., Mainka, M., Dalle, C. et al. (2020). Stability of oxylipins during plasma generation and long-term storage. *Talanta* 217: 121074.

16 Mainka, M., Dalle, C., Petera, M. et al. (2020). Harmonized procedures lead to comparable quantification of total oxylipins across laboratories. *J. Lipid Res.* 61: 1424–1436.

17 Rund, K.M., Nolte, F., Doricic, J. et al. (2020). Clinical blood sampling for oxylipin analysis – effect of storage and pneumatic tube transport of blood on free and total oxylipin profile in human plasma and serum. *Analyst* 145: 2378–2388.

18 Kutzner, L., Goloshchapova, K., Heydeck, D. et al. (2017). Mammalian ALOX15 orthologs exhibit pronounced dual positional specificity with docosahexaenoic acid. *Biochim. Biophys. Acta, Mol. Cell. Biol. Lipids* 1862: 666–675.

19 Ivanov, I., Kuhn, H., and Heydeck, D. (2015). Structural and functional biology of arachidonic acid 15-lipoxygenase-1 (ALOX15). *Gene* 573: 1–32.

20 Dennis, E.A., Deems, R.A., Harkewicz, R. et al. (2010). A mouse macrophage lipidome. *J. Biol. Chem.* 285: 39976–39985.

21 Ebert, R., Cumbana, R., Lehmann, C. et al. (2020). Long-term stimulation of toll-like receptor-2 and -4 upregulates 5-LO and 15-LO-2 expression thereby inducing a lipid mediator shift in human monocyte-derived macrophages. *Biochim. Biophys. Acta, Mol. Cell. Biol. Lipids* 1865: 158702.

22 Willenberg, I., Meschede, A.K., Gueler, F. et al. (2015). Food polyphenols fail to cause a biologically relevant reduction of COX-2 activity. *PLoS One* 10: e0139147.

23 Cooper, B.A. and Rosenblatt, D. (1976). Folate coenzyme forms in fibroblasts from patients deficient in 5,10-methylenetetrahydrofolate reductase. *Biochem. Soc. Trans.* 4: 921–922.

24 Crescente, M., Menke, L., Chan, M.V. et al. (2019). Eicosanoids in platelets and the effect of their modulation by aspirin in the cardiovascular system (and beyond). *Br. J. Pharmacol.* 176: 988–999.

25 Gijon, M.A., Zarini, S., and Murphy, R.C. (2007). Biosynthesis of eicosanoids and transcellular metabolism of leukotrienes in murine bone marrow cells. *J. Lipid Res.* 48: 716–725.

26 McAdam, B.F., Mardini, I.A., Habib, A. et al. (2000). Effect of regulated expression of human cyclooxygenase isoforms on eicosanoid and isoeicosanoid production in inflammation. *J. Clin. Invest.* 105: 1473–1482.

27 Schuchardt, J.P., Ostermann, A.I., Stork, L. et al. (2017). Effect of DHA supplementation on oxylipin levels in plasma and immune cell stimulated blood. *Prostaglandins Leukot. Essent. Fatty Acids* 121: 76–87.

28 Goodfriend, T.L., Pedersen, T.L., Grekin, R.J. et al. (2007). Heparin, lipoproteins, and oxygenated fatty acids in blood: a cautionary note. *Prostaglandins Leukot. Essent. Fatty Acids* 77: 363–366.

29 Jonasdottir, H.S., Brouwers, H., Toes, R.E.M. et al. (2018). Effects of anticoagulants and storage conditions on clinical oxylipid levels in human plasma. *Biochim. Biophys. Acta, Mol. Cell. Biol. Lipids* 1863: 1511–1522.

30 Lundstrom, S.L., Yang, J., Brannan, J.D. et al. (2013). Lipid mediator serum profiles in asthmatics significantly shift following dietary supplementation with omega-3 fatty acids. *Mol. Nutr. Food Res.* 57: 1378–1389.

31 Schuchardt, J.P., Schmidt, S., Kressel, G. et al. (2013). Comparison of free serum oxylipin concentrations in hyper- vs. normolipidemic men. *Prostaglandins Leukot. Essent. Fatty Acids* 89: 19–29.

32 Schuchardt, J.P., Schmidt, S., Kressel, G. et al. (2014). Modulation of blood oxylipin levels by long-chain omega-3 fatty acid supplementation in hyper- and normolipidemic men. *Prostaglandins Leukot. Essent. Fatty Acids* 90: 27–37.

33 O'Donnell, V.B. and Murphy, R.C. (2012). New families of bioactive oxidized phospholipids generated by immune cells: identification and signaling actions. *Blood* 120: 1985–1992.

34 Quehenberger, O., Dahlberg-Wright, S., Jiang, J. et al. (2018). Quantitative determination of esterified eicosanoids and related oxygenated metabolites after base hydrolysis. *J. Lipid Res.* 59: 2436–2445.

35 Aldrovandi, M., Hammond, V.J., Podmore, H. et al. (2013). Human platelets generate phospholipid-esterified prostaglandins via cyclooxygenase-1 that are inhibited by low dose aspirin supplementation. *J. Lipid Res.* 54: 3085–3097.

36 Ostermann, A.I., Koch, E., Rund, K.M. et al. (2020). Targeting esterified oxylipins by LC-MS – effect of sample preparation on oxylipin pattern. *Prostaglandins Other Lipid Mediat.* 146: 106384.

37 Meckelmann, S.W., Hawksworth, J.I., White, D. et al. (2020). Metabolic dysregulation of the lysophospholipid/autotaxin axis in the chromosome 9p21 gene SNP rs10757274. *Circ. Genom. Precis. Med.* 13: e002806.

38 Ostermann, A.I., Greupner, T., Kutzner, L. et al. (2018). Intra-individual variance of the human plasma oxylipin pattern: low inter-day variability in fasting blood samples versus high variability during the day. *Anal. Methods* 10: 4935–4944.

39 Kutzner, L., Rund, K.M., Ostermann, A.I. et al. (2019). Development of an optimized LC-MS method for the detection of specialized pro-resolving mediators in biological samples. *Front. Pharmacol.* 10: 169.

40 Rund, K.M., Ostermann, A.I., Kutzner, L. et al. (2018). Development of an LC-ESI(−)-MS/MS method for the simultaneous quantification of 35 isoprostanes and isofurans derived from the major n3- and n6-PUFAs. *Anal. Chim. Acta* 1037: 63–74.

41 Pedersen, T.L., Gray, I.J., and Newman, J.W. (2021). Plasma and serum oxylipin, endocannabinoid, bile acid, steroid, fatty acid and nonsteroidal anti-inflammatory drug quantification in a 96-well plate format. *Anal. Chim. Acta* 1143: 189–200.

42 Jonasdottir, H., Ioan-Facsinay, A., Kwekkeboom, J. et al. (2014). An advanced LC-MS/MS platform for the analysis of specialized pro-resolving lipid mediators. *Chromatographia* 78: 1–11.

43 Yang, J., Schmelzer, K., Georgi, K., and Hammock, B.D. (2009). Quantitative profiling method for oxylipin metabolome by liquid chromatography electrospray ionization tandem mass spectrometry. *Anal. Chem.* 81: 8085–8093.

44 Ostermann, A.I., Willenberg, I., and Schebb, N.H. (2015). Comparison of sample preparation methods for the quantitative analysis of eicosanoids and other oxylipins in plasma by means of LC-MS/MS. *Anal. Bioanal. Chem.* 407: 1403–1414.

45 Giera, M. (2018). Clinical Metabolomics Methods and Protocols.

46 Willenberg, I., Ostermann, A.I., and Schebb, N.H. (2015). Targeted metabolomics of the arachidonic acid cascade: current state and challenges of LC-MS analysis of oxylipins. *Anal. Bioanal. Chem.* 407: 2675–2683.

47 Hartung, N.M., Mainka, M., Kampschulte, N. et al. (2019). A strategy for validating concentrations of oxylipin standards for external calibration. *Prostaglandins Other Lipid Mediat.* 141: 22–24.

48 Rund, K.M., Peng, S., Greite, R. et al. (2020). Dietary omega-3 PUFA improved tubular function after ischemia induced acute kidney injury in mice but did not attenuate impairment of renal function. *Prostaglandins Other Lipid Mediat.* 146: 106386.

49 Lee, T.H., Menica-Huerta, J.M., Shih, C. et al. (1984). Characterization and biologic properties of 5,12-dihydroxy derivatives of eicosapentaenoic acid, including leukotriene B5 and the double lipoxygenase product. *J. Biol. Chem.* 259: 2383–2389.

50 Jonasdottir, H.S., Papan, C., Fabritz, S. et al. (2015). Differential mobility separation of leukotrienes and protectins. *Anal. Chem.* 87: 5036–5040.

51 von Schacky, C., Marcus, A.J., Safier, L.B. et al. (1990). Platelet-neutrophil interactions. 12*S*,20- and 5*S*,12*S*-dihydroxyeicosapentaenoic acids: two novel neutrophil metabolites from platelet-derived 12*S*-hydroxyeicosapentaenoic acid. *J. Lipid Res.* 31: 801–810.

52 Chakraborty, T., Thuer, E., Heijink, M. et al. (2018). Eicosanoid biosynthesis influences the virulence of *Candida parapsilosis*. *Virulence* 9: 1019–1035.

53 Cebo, M., Fu, X., Gawaz, M. et al. (2020). Enantioselective ultra-high performance liquid chromatography-tandem mass spectrometry method based on sub-2microm particle polysaccharide column for chiral separation of oxylipins and its application for the analysis of autoxidized fatty acids and platelet releasates. *J. Chromatogr. A* 1624: 461206.

54 Blum, M., Dogan, I., Karber, M. et al. (2019). Chiral lipidomics of monoepoxy and monohydroxy metabolites derived from long-chain polyunsaturated fatty acids. *J. Lipid Res.* 60: 135–148.

55 Liakh, I., Pakiet, A., Sledzinski, T., and Mika, A. (2020). Methods of the analysis of oxylipins in biological samples. *Molecules* 25: 349.

56 Kapron, J., Wu, J., Mauriala, T. et al. (2006). Simultaneous analysis of prostanoids using liquid chromatography/high-field asymmetric waveform ion mobility spectrometry/tandem mass spectrometry. *Rapid Commun. Mass Spectrom.* 20: 1504–1510.

57 Hinz, C., Liggi, S., Mocciaro, G. et al. (2019). A comprehensive UHPLC ion mobility quadrupole time-of-flight method for profiling and quantification of eicosanoids, other oxylipins, and fatty acids. *Anal. Chem.* 91: 8025–8035.

58 Hamberg, M. and Samuelsson, B. (1974). Prostaglandin endoperoxides. Novel transformations of arachidonic acid in human platelets. *Proc. Natl. Acad. Sci. U.S.A.* 71: 3400–3404.

59 Il'yasova, D., Scarbrough, P., and Spasojevic, I. (2012). Urinary biomarkers of oxidative status. *Clin. Chim. Acta* 413: 1446–1453.

60 Khamis, M.M., Adamko, D.J., and El-Aneed, A. (2017). Mass spectrometric based approaches in urine metabolomics and biomarker discovery. *Mass Spectrom. Rev.* 36: 115–134.

61 Bain, G., King, C.D., Rewolinski, M. et al. (2010). Pharmacodynamics and pharmacokinetics of AM103, a novel inhibitor of 5-lipoxygenase-activating protein (FLAP). *Clin. Pharmacol. Ther.* 87: 437–444.

62 Balgoma, D., Larsson, J., Rokach, J. et al. (2013). Quantification of lipid mediator metabolites in human urine from asthma patients by electrospray ionization mass spectrometry: controlling matrix effects. *Anal. Chem.* 85: 7866–7874.

63 Drew, D.A., Chin, S.M., Gilpin, K.K. et al. (2017). ASPirin intervention for the REDuction of colorectal cancer risk (ASPIRED): a study protocol for a randomized controlled trial. *Trials* 18: 50.

64 Duffield-Lillico, A.J., Boyle, J.O., Zhou, X.K. et al. (2009). Levels of prostaglandin E metabolite and leukotriene E_4 are increased in the urine of smokers: evidence that celecoxib shunts arachidonic acid into the 5-lipoxygenase pathway. *Cancer Prev. Res. (Phila).* 2: 322–329.

65 FitzGerald, G.A., Oates, J.A., Hawiger, J. et al. (1983). Endogenous biosynthesis of prostacyclin and thromboxane and platelet function during chronic administration of aspirin in man. *J. Clin. Invest.* 71: 676–688.

66 Murphey, L.J., Williams, M.K., Sanchez, S.C. et al. (2004). Quantification of the major urinary metabolite of PGE2 by a liquid chromatographic/mass spectrometric assay: determination of cyclooxygenase-specific PGE2 synthesis in healthy humans and those with lung cancer. *Anal. Biochem.* 334: 266–275.

67 Saude, E.J., Adamko, D., Rowe, B.H. et al. (2007). Variation of metabolites in normal human urine. *Metabolomics* 3: 439–451.

68 Dorjgochoo, T., Gao, Y.T., Chow, W.H. et al. (2011). Obesity, age, and oxidative stress in middle-aged and older women. *Antioxid. Redox Signal.* 14: 2453–2460.

69 Helmersson, J., Mattsson, P., and Basu, S. (2002). Prostaglandin $F_{2\alpha}$ metabolite and F_2-isoprostane excretion rates in migraine. *Clin. Sci. (Lond).* 102: 39–43.

70 Medina, S., De Las, H.-G.I., Casas-Pina, T. et al. (2020). Urinary oxylipin signature as biomarkers to monitor the allograft function during the first six months post-renal transplantation. *Free Radic. Biol. Med.* 146: 340–349.

71 Yan, W., Byrd, G.D., and Ogden, M.W. (2007). Quantitation of isoprostane isomers in human urine from smokers and nonsmokers by LC-MS/MS. *J. Lipid Res.* 48: 1607–1617.

72 Catella, F., Nowak, J., and Fitzgerald, G.A. (1986). Measurement of renal and non-renal eicosanoid synthesis. *Am. J. Med.* 81: 23–29.

73 Nugteren, D.H. (1975). The determination of prostaglandin metabolites in human urine. *J. Biol. Chem.* 250: 2808–2812.

74 Gomes, J.A., Milne, G., Kallianpur, A., and Shriver, L. (2021). Isofurans and isoprostanes as potential markers of delayed cerebral ischemia following aneurysmal subarachnoid hemorrhage: a prospective observational study. *Neurocrit. Care* 36 (1): 202–207.

75 Liakh, I., Pakiet, A., Sledzinski, T., and Mika, A. (2019). Modern methods of sample preparation for the analysis of oxylipins in biological samples. *Molecules* 24.

76 Scott, W.C., Cahill, K.N., Milne, G.L. et al. (2021). Inflammatory heterogeneity in aspirin-exacerbated respiratory disease. *J. Allergy Clin. Immunol.* 147: 1318–1328.e5.

77 Chan, M.V., Hayman, M.A., Sivapalaratnam, S. et al. (2021). Identification of a homozygous recessive variant in PTGS1 resulting in a congenital aspirin-like defect in platelet function. *Haematologica* 106: 1423–1432.

78 Drew, D.A., Schuck, M.M., Magicheva-Gupta, M.V. et al. (2020). Effect of low-dose and standard-dose aspirin on PGE2 biosynthesis among individuals with colorectal adenomas: a randomized clinical trial. *Cancer Prev. Res. (Phila).* 13: 877–888.

79 Edelman, M.J., Wang, X., Hodgson, L. et al. (2017). Phase III randomized, placebo-controlled, double-blind trial of celecoxib in addition to standard chemotherapy for advanced non-small-cell lung cancer with cyclooxygenase-2 overexpression: CALGB 30801 (Alliance). *J. Clin. Oncol.* 35: 2184–2192.

80 Garcia-Flores, L.A., Medina, S., Gomez, C. et al. (2018). Aronia-citrus juice (polyphenol-rich juice) intake and elite triathlon training: a lipidomic approach using representative oxylipins in urine. *Food Funct.* 9: 463–475.

81 Lara-Guzman, O.J., Medina, S., Alvarez, R. et al. (2020). Oxylipin regulation by phenolic compounds from coffee beverage: positive outcomes from a randomized controlled trial in healthy adults and macrophage derived foam cells. *Free Radic. Biol. Med.* 160: 604–617.

82 Montine, T.J., Sonnen, J.A., Milne, G. et al. (2010). Elevated ratio of urinary metabolites of thromboxane and prostacyclin is associated with adverse cardiovascular events in ADAPT. *PLoS One* 5: e9340.

83 Reyes-Uribe, L., Wu, W., Gelincik, O. et al. (2021). Naproxen chemoprevention promotes immune activation in Lynch syndrome colorectal mucosa. *Gut* 70: 555–566.

84 Altobelli, E., Angeletti, P.M., and Latella, G. (2016). Role of urinary biomarkers in the diagnosis of adenoma and colorectal cancer: a systematic review and meta-analysis. *J. Cancer* 7: 1984–2004.

85 Seyberth, H.W., Sweetman, B.J., Frolich, J.C., and Oates, J.A. (1976). Quantifications of the major urinary metabolite of the E prostaglandins by mass spectrometry: evaluation of the method's application to clinical studies. *Prostaglandins* 11: 381–397.

86 Schweer, H., Meese, C.O., and Seyberth, H.W. (1990). Determination of 11 alpha-hydroxy-9,15-dioxo-2,3,4,5,20-pentanor-19-carboxyprostan oic acid and 9 alpha,11 alpha-dihydroxy-15-oxo-2,3,4,5,20-pentanor-19-carboxyprostanoi c acid by gas chromatography/negative ion chemical ionization triple-stage quadrupole mass spectrometry. *Anal. Biochem.* 189: 54–58.

87 Taber, D.F. and Teng, D. (2002). Total synthesis of the ethyl ester of the major urinary metabolite of prostaglandin E_2. *J. Org. Chem.* 67: 1607–1612.

88 Milne, G.L. (2016). Lipid biomarkers of inflammation. In: *Chronic Inflammation: Molecular Pathophysiology, Nutritional and Therapeutic Interventions* (ed. S. Roy, D. Bagchi and S.P. Raychaudhuri), 275. CRC Press.

89 Liston, T.E. and Roberts, L.J. 2nd (1985). Transformation of prostaglandin D2 to 9 alpha, 11 beta-(15*S*)-trihydroxyprosta-(5*Z*,13*E*)-dien-1-oic acid (9 alpha, 11 beta-prostaglandin F_2): a unique biologically active prostaglandin produced enzymatically in vivo in humans. *Proc. Natl. Acad. Sci. U.S.A.* 82: 6030–6034.

90 Morrow, J.D., Prakash, C., Awad, J.A. et al. (1991). Quantification of the major urinary metabolite of prostaglandin D2 by a stable isotope dilution mass spectrometric assay. *Anal. Biochem.* 193: 142–148.

91 Kolmert, J., Gomez, C., Balgoma, D. et al. (2021). Wheelock CE and U-biopred study group obotUBSG. Urinary leukotriene E4 and prostaglandin D2 metabolites increase in adult and childhood severe asthma characterized by type 2 inflammation. A clinical observational study. *Am. J. Respir. Crit. Care Med.* 203: 37–53.

92 Song, W.L., Wang, M., Ricciotti, E. et al. (2008). Tetranor PGDM, an abundant urinary metabolite reflects biosynthesis of prostaglandin D2 in mice and humans. *J. Biol. Chem.* 283: 1179–1188.

93 Mohebati, A., Milne, G.L., Zhou, X.K. et al. (2013). Effect of zileuton and celecoxib on urinary LTE4 and PGE-M levels in smokers. *Cancer Prev. Res. (Phila).* 6: 646–655.

94 Zhang, Y., Zhang, G., Clarke, P.A. et al. (2011). Simultaneous and high-throughput quantitation of urinary tetranor PGDM and tetranor PGEM by online SPE-LC-MS/MS as inflammatory biomarkers. *J. Mass Spectrom.* 46: 705–711.

95 Bell-Parikh, L.C., Ide, T., Lawson, J.A. et al. (2003). Biosynthesis of 15-deoxy-$\Delta^{12,14}$-PGJ2 and the ligation of PPARγ. *J. Clin. Invest.* 112: 945–955.

96 Brunoldi, E.M., Zanoni, G., Vidari, G. et al. (2007). Cyclopentenone prostaglandin, 15-deoxy-$\Delta^{12,14}$-PGJ2, is metabolized by HepG2 cells via conjugation with glutathione. *Chem. Res. Toxicol.* 20: 1528–1535.

97 Hardy, K.D., Cox, B.E., Milne, G.L. et al. (2011). Nonenzymatic free radical-catalyzed generation of 15-deoxy-$\Delta^{12,14}$-prostaglandin J_2-like compounds (deoxy-J_2-isoprostanes) in vivo. *J. Lipid Res.* 52: 113–124.

98 Lands, W.E. (1979). The biosynthesis and metabolism of prostaglandins. *Annu. Rev. Physiol.* 41: 633–652.

99 Imig, J.D. (2020). Eicosanoid blood vessel regulation in physiological and pathological states. *Clin. Sci. (Lond).* 134: 2707–2727.

100 Mitchell, J.A., Knowles, R.B., Kirkby, N.S. et al. (2018). Kidney transplantation in a patient lacking cytosolic phospholipase A2 proves renal origins of urinary PGI-M and TX-M. *Circ. Res.* 122: 555–559.

101 Morrow, J.D. and Minton, T.A. (1993). Improved assay for the quantification of 11-dehydrothromboxane B2 by gas chromatography-mass spectrometry. *J. Chromatogr.* 612: 179–185.

102 Grosser, T., Naji, A., and FitzGerald, G.A. (2018). Urinary prostaglandin metabolites: an incomplete reckoning and a flush to judgment. *Circ Res.* 122: 537–539.

103 McAdam, B.F., Byrne, D., Morrow, J.D., and Oates, J.A. (2005). Contribution of cyclooxygenase-2 to elevated biosynthesis of thromboxane A2 and prostacyclin in cigarette smokers. *Circulation* 112: 1024–1029.

104 Nakashima, F. and Schneider, C. (2020). Transformation of prostaglandin D2 to 11-dehydro thromboxane B2 by Baeyer–Villiger oxidation. *Lipids* 55: 73–78.

105 Murphy, R.C. and Gijon, M.A. (2007). Biosynthesis and metabolism of leukotrienes. *Biochem. J.* 405: 379–395.

106 Berry, K.A., Borgeat, P., Gosselin, J. et al. (2003). Urinary metabolites of leukotriene B4 in the human subject. *J. Biol. Chem.* 278: 24449–24460.

107 Mita, H., Turikisawa, N., Yamada, T., and Taniguchi, M. (2007). Quantification of leukotriene B4 glucuronide in human urine. *Prostaglandins Other Lipid Mediat.* 83: 42–49.

108 Morita, Y., Kurano, M., Sakai, E. et al. (2021). Simultaneous analyses of urinary eicosanoids and related mediators identified tetranor-prostaglandin E metabolite as a novel biomarker of diabetic nephropathy. *J. Lipid Res.* 62: 100120.

109 Capdevila, J.H. and Falck, J.R. (2001). The CYP P450 arachidonic acid monooxygenases: from cell signaling to blood pressure regulation. *Biochem. Biophys. Res. Commun.* 285: 571–576.

110 Kirchheiner, J., Meineke, I., Fuhr, U. et al. (2008). Impact of genetic polymorphisms in CYP2C8 and rosiglitazone intake on the urinary excretion of dihydroxyeicosatrienoic acids. *Pharmacogenomics* 9: 277–288.

111 Santos, J.M., Park, J.A., Joiakim, A. et al. (2017). The role of soluble epoxide hydrolase in preeclampsia. *Med. Hypotheses* 108: 81–85.

112 Catella, F., Lawson, J.A., Fitzgerald, D.J., and FitzGerald, G.A. (1990). Endogenous biosynthesis of arachidonic acid epoxides in humans: increased formation in pregnancy-induced hypertension. *Proc. Natl. Acad. Sci. U.S.A.* 87: 5893–5897.

113 Mota-Zamorano, S., Robles, N.R., Lopez-Gomez, J. et al. (2021). Plasma and urinary concentrations of arachidonic acid-derived eicosanoids are associated with diabetic kidney disease. *EXCLI J.* 20: 698–708.

114 Wang, M.H., Smith, A., Zhou, Y. et al. (2003). Downregulation of renal CYP-derived eicosanoid synthesis in rats with diet-induced hypertension. *Hypertension* 42: 594–599.

115 Milne, G.L., Dai, Q., and Roberts, L.J. 2nd (2015). The isoprostanes – 25 years later. *Biochim. Biophys. Acta* 1851: 433–445.

116 Pandya, B.A. and Snapper, M.L. (2008). A cross-metathesis route to the 5-F_2-isoprostanes. *J. Org. Chem.* 73: 3754–3758.

117 Cuyamendous, C., de la Torre, A., Lee, Y.Y. et al. (2016). The novelty of phytofurans, isofurans, dihomo-isofurans and neurofurans: discovery, synthesis and potential application. *Biochimie* 130: 49–62.

118 Galano, J.M., Lee, Y.Y., Oger, C. et al. (2017). Isoprostanes, neuroprostanes and phytoprostanes: an overview of 25 years of research in chemistry and biology. *Prog. Lipid Res.* 68: 83–108.

119 Leung, H.H., Leung, K.S., Durand, T. et al. (2020). Measurement of enzymatic and nonenzymatic polyunsaturated fatty acid oxidation products in plasma and urine of macular degeneration using LC-QTOF-MS/MS. *Lipids* 55: 693–706.

120 Miller, E., Morel, A., Saso, L., and Saluk, J. (2014). Isoprostanes and neuroprostanes as biomarkers of oxidative stress in neurodegenerative diseases. *Oxid. Med. Cell Longev.* 2014: 572491.

121 Pena-Bautista, C., Vigor, C., Galano, J.M. et al. (2019). New screening approach for Alzheimer's disease risk assessment from urine lipid peroxidation compounds. *Sci. Rep.* 9: 14244.

122 Signorini, C., De Felice, C., Galano, J.M. et al. (2018). Isoprostanoids in clinical and experimental neurological disease models. *Antioxidants (Basel).* 7.

123 Kuligowski, J., Aguar, M., Rook, D. et al. (2015). Urinary lipid peroxidation byproducts: are they relevant for predicting neonatal morbidity in preterm infants? *Antioxid. Redox Signal.* 23: 178–184.

124 Pena-Bautista, C., Durand, T., Vigor, C. et al. (2019). Non-invasive assessment of oxidative stress in preterm infants. *Free Radic. Biol. Med.* 142: 73–81.

125 Calder, P.C. and Yaqoob, P. (2009). Omega-3 polyunsaturated fatty acids and human health outcomes. *Biofactors* 35: 266–272.

126 Skarke, C., Alamuddin, N., Lawson, J.A. et al. (2015). Bioactive products formed in humans from fish oils. *J. Lipid Res.* 56: 1808–1820.

127 O'Donnell, V.B., Aldrovandi, M., Murphy, R.C., and Kronke, G. (2019). Enzymatically oxidized phospholipids assume center stage as essential regulators of innate immunity and cell death. *Sci. Signal.* 12.

128 Maskrey, B.H., Bermudez-Fajardo, A., Morgan, A.H. et al. (2007). Activated platelets and monocytes generate four hydroxyphosphatidylethanolamines via lipoxygenase. *J. Biol. Chem.* 282: 20151–20163.

129 Morgan, A.H., Hammond, V.J., Morgan, L. et al. (2010). Quantitative assays for esterified oxylipins generated by immune cells. *Nat. Protoc.* 5: 1919–1931.

130 Thomas, C.P., Morgan, L.T., Maskrey, B.H. et al. (2010). Phospholipid-esterified eicosanoids are generated in agonist-activated human platelets and enhance tissue factor-dependent thrombin generation. *J. Biol. Chem.* 285: 6891–6903.

131 Slatter, D.A., Aldrovandi, M., O'Connor, A. et al. (2016). Mapping the human platelet lipidome reveals cytosolic phospholipase A2 as a regulator of mitochondrial bioenergetics during activation. *Cell Metab.* 23: 930–944.

132 Ni, Z., Angelidou, G., Hoffmann, R., and Fedorova, M. (2017). LPPtiger software for lipidome-specific prediction and identification of oxidized phospholipids from LC-MS datasets. *Sci. Rep.* 7: 15138.

133 Matsuoka, Y., Takahashi, M., Sugiura, Y. et al. (2021). Structural library and visualization of endogenously oxidized phosphatidylcholines using mass spectrometry-based techniques. *Nat. Commun.* 12: 6339.

13

Mass Spectrometry for Analysis of Glycerolipids

Wm. Craig Byrdwell

U.S. Department of Agriculture, Agricultural Research Service, Methods and Application of Food Composition Lab, 10300 Baltimore Avenue, Building 161, Beltsville, MD 20705, USA

13.1 Introduction

Glycerolipids (GLs), or acylglycerols (AGs), also simply called as glycerides, are molecules made by connecting fatty acyl chains via ester linkages to a three-carbon glycerol moiety that makes up the "backbone" of the molecules. As the glycerol backbone has three positions that can be esterified, acylglycerols come in three forms, monoacylglycerols (MAGs) or monoglycerides (MGs), diacylglycerols (DAGs) or diglycerides (DGs), and triacylglycerols (TAGs) or triglycerides (TGs), having one, two, or three acyl moieties, respectively, attached to the glycerol backbone, as shown in Figures 13.1 and 13.2. Because of the increasing need for standardization in referring to lipid species arising from lipidomics, we have adopted the http://LipidMAPS.org nomenclature here.

Although MGs, DGs, and TGs are each a class of molecules unto themselves, they share so much structural homology that they produce many identical fragments between classes, when analyzed using mass spectrometry (MS). DGs are the core of other lipid classes, such as phospholipids (PLs) and glycolipids (saccharolipids, SLs). Therefore, becoming familiar with the fragments from TGs provides insights into not only MGs and DGs but also other lipid classes. Other chapters in this volume expertly address the analysis of PLs, so these are not discussed in detail here. As most PLs are zwitterionic, PLs are usually charged lipids, while TGs, MGs, and DGs are called neutral lipids. Other neutral lipids include glycolipids, steryl and wax esters, and others. Only glycerol-containing neutral AGs are discussed in this chapter.

Using the IUPAC nomenclature for glycerolipids, when the Fischer projection of the glycerol backbone is arranged such that the hydroxy group on the middle carbon

Mass Spectrometry for Lipidomics: Methods and Applications, First Edition.
Edited by Michal Holčapek and Kim Ekroos.

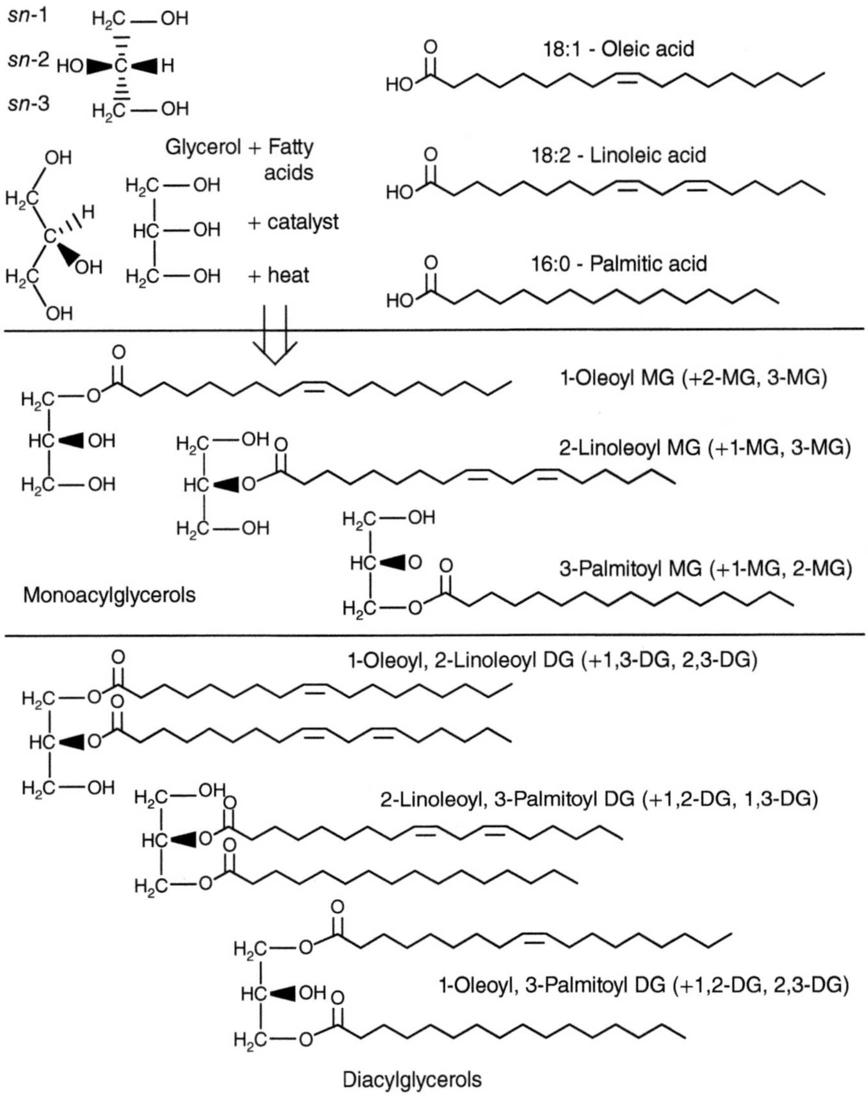

Figure 13.1 Acylglycerol structures, part 1: monoacylglycerols and diacylglycerols. Every combination of fatty acids and glycerol positions may be formed. The *sn*-2 carbon is a chiral center if the fatty acids at *sn*-1 and *sn*-3 are different.

is to the left, then carbon 1 is on top and is stereospecifically numbered, *sn*, as *sn*-1, as can be seen in Figure 13.1. GLs have a pro-chiral center at the center carbon (*sn*-2) of the glycerol backbone, which is chiral if the substituents at *sn*-1 and *sn*-3 are not identical. For example, Figure 13.2 shows three different fatty acids (FAs) attaching to the glycerol backbone as fatty acyl chains, giving a TG with a chiral center at *sn*-2. TGs that are stereoisomers, or optical isomers, are mirror images of each other, such as oleoyl, linoleoyl, palmitoyl glycerol (OLP) and palmitoyl,

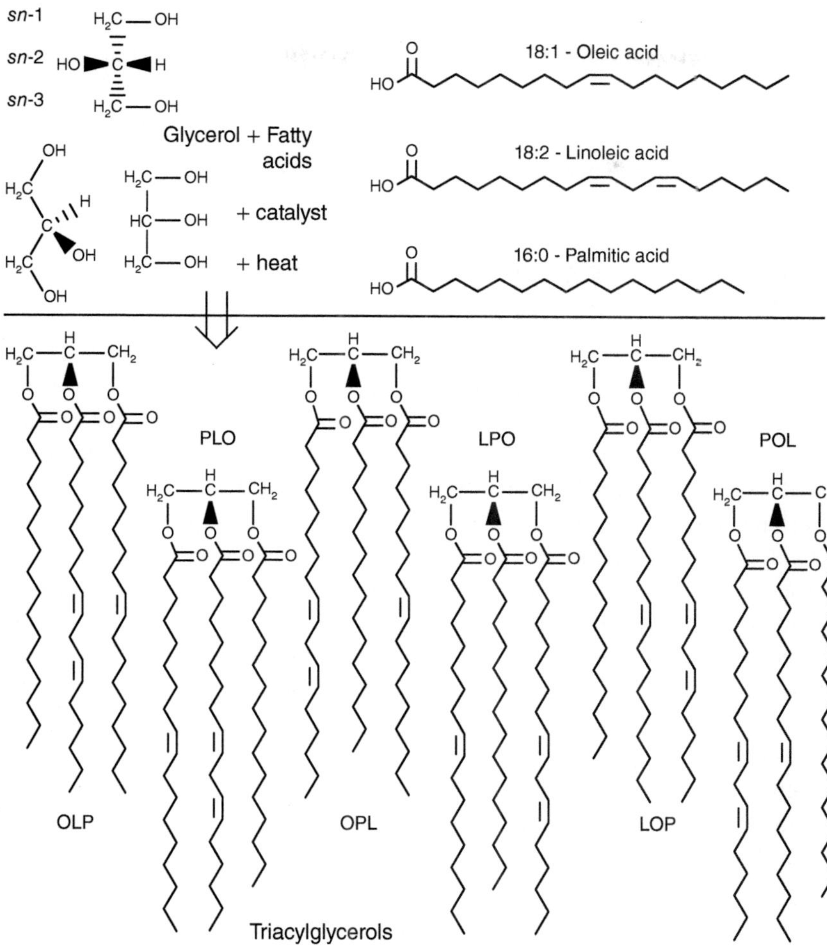

Figure 13.2 Acylglycerols, part 2: triacylglycerols. Regioisomers are formed by the different positions of the FAs on the glycerol backbone *sn*-1, *sn*-2, and *sn*-3, and stereoisomers are present from the chiral center at *sn*-2 (e.g. OLP versus PLO), if *sn*-1 versus *sn*-3 are different FAs. Structural isomers (not shown) have different FAs with the same C:U, for example, PL = PO (34:2 = 16:0_18:2 = 16:1_18:1), so PLO = OPoO = *m/z* 857.7593 for $[M+H]^+$.

linoleoyl, oleoyl glycerol (PLO), Figure 13.2, and can only be physically resolved using lengthy chiral chromatography to be discussed later. Regioisomers, such as OLP, OPL, and linoleoyl, oleoyl, palmitoyl glycerol (LOP), which have the FAs arranged in a different order on the glycerol backbone may be separable using silver ion chromatography (depending on the location of unsaturation), chiral chromatography, or other but may be partially resolved during non-aqueous reversed-phase (NARP) liquid chromatography (LC).

Analysis of TGs sometimes requires the synthesis of a known set of TG molecular species. The number of TGs that come from the synthesis of glycerol with FAs depends on the number of FAs used and the degree of isomeric specificity desired. The number of TGs possible, with and without isomers, from n FAs is [1] as follows:

Total, all isomers (regioisomers and stereoisomers): n^3
Regioisomers (stereoisomers not defined): $(n^3 + n^2)/2$
TGs (regioisomers and stereoisomers not defined): $(n^3 + 3n^2 + 2n)/6$

13.1.1 Gas Chromatography with Flame Ionization Detection for Fatty Acid Analysis

FA analysis is an important complement to intact molecule analysis of AGs by MS. The online tutorial by Christie on the topic of fatty acid methyl esters (FAMEs) is required reading for those seeking expert knowledge on the topic of derivatization options for GC analysis of FAs [2], as well as the type of information that can be derived from GC-MS analysis of different derivatives [3]. For analytical purposes, perhaps the most common and useful fat analysis technique is the conversion of the FAs from TGs directly to FAMEs, which are volatile derivatives suitable for analysis by gas chromatography (GC). The most popular cheap and easy detection method for FAMEs is the flame ionization detector (FID), which gives signal proportional to weights of FAMEs. It has long been known to be reliable for quantitation over many orders of magnitude using response factors (RFs) determined by analysis of FAME standard mixtures. A modern capillary GC separation of FAMEs can provide excellent resolution and allow separation of most double-bond isomers and branched FAs. For additional structural information, FA derivatives such as 4,4-dimethyloxazoline (DMOX) derivatives or 3-pyridylcarbinol ("picolinyl") esters are valuable for double-bond localization by GC-MS [4] and so represent an important complement to intact molecule analysis of AGs. Because of their widespread use and value for identifying double-bond locations and branch points, the GC-MS mass spectra of pyridylcarbinol ester and DMOX derivative from Christie [5] are shown in Figure 13.3.

MGs, DGs, and TGs have been separated and quantified as classes in a wide range of samples using only high-temperature (HT) GC-FID [6], but such methods are not applicable to all samples as TGs often degrade at the temperatures used to elute them in high-temperature gas chromatography (HTGC). Nevertheless, there is an Association of Official Analytical Chemists (AOAC) method for HTGC of TGs [7]. HTGC methods for TG analysis have recently been reviewed [8, 9].

The FID produces a signal on a weight % basis. Atmospheric pressure ionization (API) techniques produce a signal that is a molar response. GC-FID data for FAME need to be converted to a mole % basis before they are used for operations on LC-MS data. The first important use of FA mole % data is calculating the statistically expected TG composition from the FA composition using the following equations [10]:

$$\text{Type I TAG\%} = \frac{\text{FA\%} \times \text{FA\%} \times FA\%}{10000}$$

$$\text{Type II TAG\%} = \frac{(\text{FA}_1\% \times \text{FA}_1\% \times \text{FA}_2\%) \times 3}{10000}$$

$$\text{Type III TAG\%} = \frac{(\text{FA}_1\% \times \text{FA}_2\% \times \text{FA}_3\%) \times 6}{10000}$$

These equations should be used to calculate what the statistically expected TG composition is that would result from a particular FA mole% composition determined by GC-FID, if all FAs were distributed randomly. Differences to the statistically expected composition point to the non-random behavior, such as preferences for specific combinations of FAs.

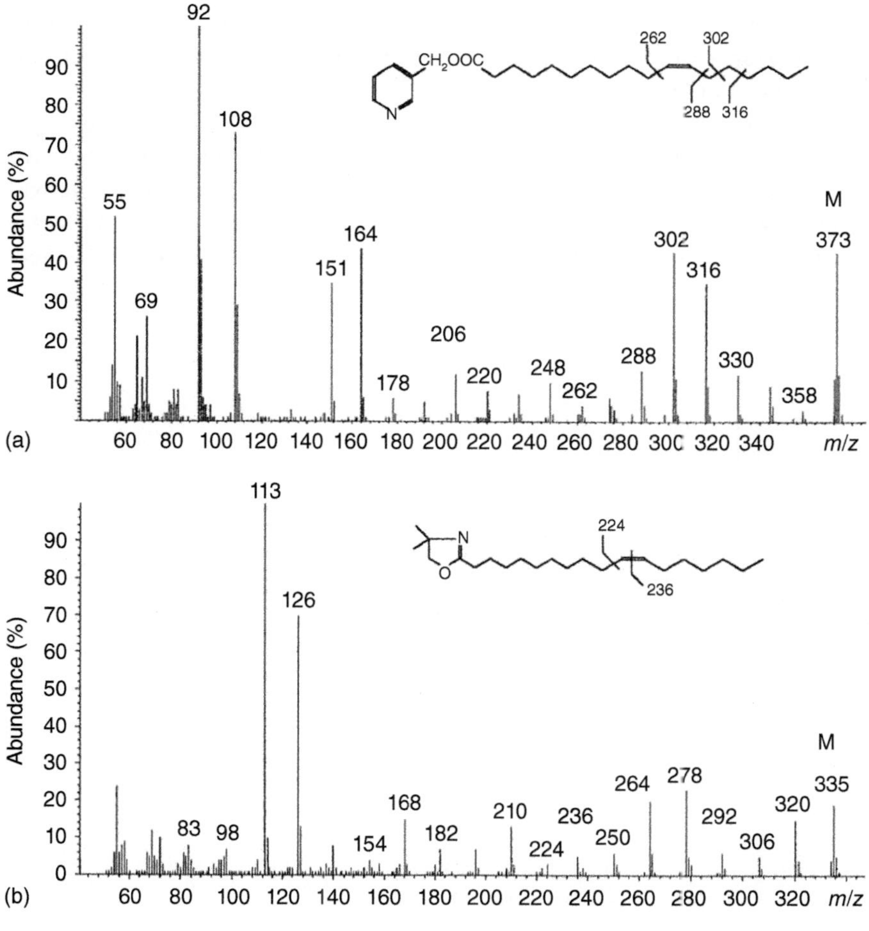

Figure 13.3 GC-electron ionization (EI)-MS mass spectra of 5,9,12-octadecatrienoic acid as (a) the pyridylcarbinol ("picolinyl") ester derivative and (b) the dimethyloxazoline (DMOX) derivative. Source: Reproduced from Christie [5] with permission from John Wiley and Sons.

13.2 Monoacylglycerols (MAGs)

Monoacylglycerols have only one fatty acyl chain and two free hydroxyl groups on the glycerol backbone, Figure 13.1. Therefore, MGs are the most polar (least non-polar) of the acylglycerols and elute first by reversed-phase (RP)-high-performance liquid chromatography (HPLC). *sn*-1-MGs cannot readily be resolved from *sn*-3-MGs chromatographically, but these can be separated from *sn*-2-MGs, so the *sn*-1 and *sn*-3 MGs are often grouped together as α-MGs, while *sn*-2-MGs are β-MGs. α-MGs are more thermodynamically stable than β-MGs, such that a mixture that is rich in β-MGs quickly rearranges to become more balanced in isomers [11].

Although this chapter is focused more on LC coupled to modern API techniques, MGs are very amenable to GC analysis, so a few important articles that are commonly referenced are cited here, to provide a starting point for further investigation of the GC-MS approaches that have proved most useful. From an API mass spectrometry standpoint, the first thing to be aware of is that hydroxyl groups are very labile and so are easily lost by dehydration, abstracting a neighboring hydrogen and leaving a site of unsaturation.

MGs can be made volatile for GC in a way similar to FAMEs as MGs are only a little larger. TMS ethers of MGs are perhaps the most popular derivatives and have continued to be analyzed by GC-MS for decades. The mass spectra of di-TMS derivatives of MGs from butter oil [12] serve as good examples of typical spectra and so are reproduced here as Figure 13.4. More recent examples show that TMS ethers

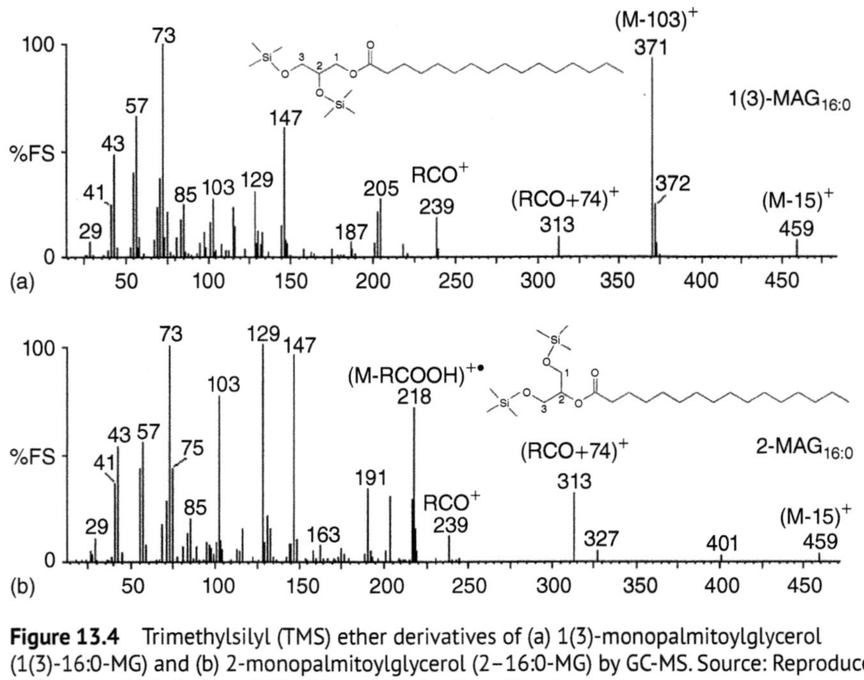

Figure 13.4 Trimethylsilyl (TMS) ether derivatives of (a) 1(3)-monopalmitoylglycerol (1(3)-16:0-MG) and (b) 2-monopalmitoylglycerol (2–16:0-MG) by GC-MS. Source: Reproduced from Liu and Kinderlerer [12] with permission from Elsevier.

continue to be valuable tools for MG analysis. Destaillats et al. [13] used GC-MS for analysis of the TMS derivative of 16 : 0-MG as a model to differentiate α- and β-DGs and then applied the approach to TMS derivatives of a commercial distilled MG mixture. Cao et al. [14] have recently used GC-MS analysis of TMS-MGs to identify six MGs that served as markers for cooking oil quality. There are official GC methods for analysis of TMS derivatives of MGs and DGs by GC. The American Oil Chemists' Society (AOCS) official method is Cd 11b-91: "Mono- and diglycerides by capillary gas chromatography" [15]. The AOAC official method is 993.18, "Mono- and diglycerides in fats and oils. Gas chromatographic method" [16].

As LC-MS methods and softer ionization methods were developed, they were used to identify intact MGs without the need for derivatization. Holčapek et al. [17] analyzed the FAMEs and the intact, underivatized AGs in biodiesel made by transesterification of rapeseed oil using LC-MS with atmospheric pressure chemical ionization (APCI) MS. Mass spectra from that report are reproduced here because they represent an example of the common fragments shared across all classes of AGs. Figure 13.5 shows the HPLC-APCI-MS mass spectrum of the MGs 1-oleoylmonoacylglycerol (1-O) and 1-linoleoylmonoacylglycerol (1-L). The important ions in Figure 13.5 are listed in Table 13.1. Some of the most common fragments from all AGs are the $[RCOO+58]^+$, or $[RCO+74]^+$, fragment and its dehydrated fragment $[RCOO+58-H_2O]^+$, the acylium ion $[RCO]^+$, and its dehydrated fragment $[RCO-H_2O]^+$.

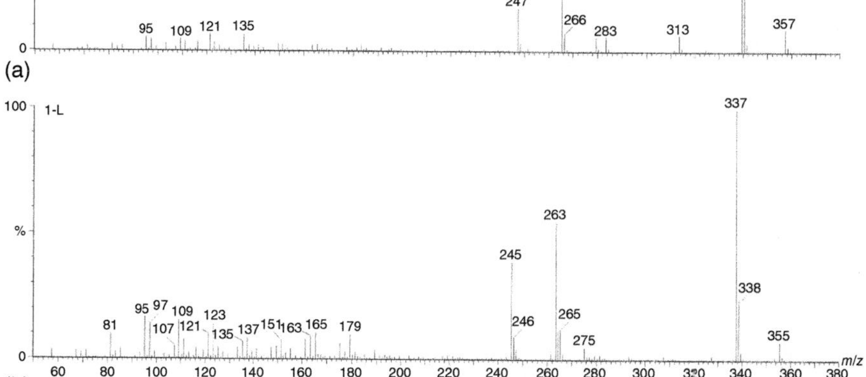

Figure 13.5 HPLC-APCI-MS mass spectrum of (a) 1-oleoylmonoacylglycerol (1-O) and (b) 1-linoleoylmonoacylglycerol (1-L). Source: Reproduced from Holčapek et al. [17] with permission from Elsevier.

Table 13.1 Common fatty acids (FAs) and the common fragments seen in the mass spectra of acylglycerols (AGs) containing the fatty acyl chains.

Fatty	Systematic (IUPAC)			Fatty Acid	$[RCOO+58]^+$	$[RCOO+58-H_2O]^+$			
Acyl group	**Name**	**Common name**	**Abbreviation**	**RCOOH**	$[RCO+74]^+$	$[RCO+74-H_2O]^+$	$[RCO]^+$	$[RCO-H_2O]^+$	$[RCOO]^-$
4:0	Butanoic	Butyric	Bu	88.0524	145.0859	127.0754	71.0491	53.0386	87.0452
5:0	Pentanoic	Valeric	V	102.0681	159.1016	141.0910	85.0648	67.0542	101.0608
6:0	Hexanoic	Caproic	Co	116.0837	173.1172	155.1067	99.0804	81.0699	115.0765
7:0	Heptanoic	Enanthic	En	130.0994	187.1329	169.1223	113.0961	95.0855	129.0921
8:0	Octanoic	Caprylic	Cy	144.1150	201.1485	183.1380	127.1117	109.1012	143.1078
9:0	Nonanoic	Pelargonic	Pg	158.1307	215.1642	197.1536	141.1274	123.1168	157.1234
10:0	Decanoic	Capric	Ca	172.1463	229.1798	211.1693	155.1430	137.1325	171.1391
11:0	Undecanoic	Undecylic	Un	186.1620	243.1955	225.1849	169.1587	151.1481	185.1547
12:0	Dodecanoic	Lauric	La	200.1776	257.2111	239.2006	183.1743	165.1638	199.1704
13:0	Tridecanoic	Tridecylic	Tr	214.1933	271.2268	253.2162	197.1900	179.1794	213.1860
14:0	**Tetradecanoic**	**Myristic**	**M**	**228.2089**	**285.2424**	**267.2319**	**211.2056**	**193.1951**	**227.2017**
15:0	Pentadecanoic	Pentadecylic	Pn	242.2246	299.2581	281.2475	225.2213	207.2107	241.2173
16:0	**Hexadecanoic**	**Palmitic**	**P**	**256.2402**	**313.2737**	**295.2632**	**239.2369**	**221.2264**	**255.2330**
16:1	**(9*Z*)-Hexadec-9-enoic**	**Palmitoleic (*n*-7)**	**Po**	**254.2246**	**311.2581**	**293.2475**	**237.2213**	**219.2107**	**253.2173**

17:0	Heptadecanoic	Margaric	Ma	270.2559	327.2894	309.2788	253.2526	235.2420	269.2486
18:0	**Octadecanoic**	**Stearic**	**S**	**284.2715**	**341.3050**	**323.2945**	**267.2682**	**249.2577**	**283.2643**
18:1	**(9*Z*)-Octadec-9-enoic**	**Oleic (*n*-9)**	**O**	**282.2559**	**339.2894**	**321.2788**	**265.2526**	**247.2420**	**281.2486**
18:2	**(9*Z*,12*Z*)-Octadeca-9,12-dienoic**	**Linoleic (ω-6)**	**L**	**280.2402**	**337.2737**	**319.2632**	**263.2369**	**245.2264**	**279.2330**
18:3	**(9*Z*,12*Z*,15*Z*)-Octadeca-9,12,15-trienoic**	**α-Linolenic (ω-3)**	**Ln**	**278.2246**	**335.2581**	**317.2475**	**261.2213**	**243.2107**	**277.2173**
18:4	(6*Z*,9*Z*,12*Z*,15*Z*)-Octadeca-6,9,12,15-tetraenoic	Stearidonic (ω-3)	Sd	276.2089	333.2424	315.2319	259.2056	241.1951	275.2017
19:0	Nonadecanoic	Nonadecylic	No	298.2872	355.3207	337.3101	281.2839	263.2733	297.2799
20:0	**Eicosanoic = Icosanoic**	**Arachidic**	**A**	**312.3028**	**369.3363**	**351.3258**	**295.2995**	**277.2890**	**311.2956**
20:1	**(9*Z*)-Icos-9-enoic**	**Gadoleic (*n*-11)**	**G**	**310.2872**	**367.3207**	**349.3101**	**293.2839**	**275.2733**	**309.2799**
20:2	(11*Z*,14*Z*)-Icosa-11,14-dienoic	Dihomo Linoleic (ω-6)	Ed	308.2715	365.3050	347.2945	291.2682	273.2577	307.2643
20:3	(11*Z*,14*Z*,17*Z*)-Icosa-11,14,17-trienoic	Dihomo Linolenic (ω-3) = eicosatrienoic acid (ETA)	Etr	306.2559	363.2894	345.2788	289.2526	271.2420	305.2486
20:4	(5*Z*,8*Z*,11*Z*,14*Z*)-Icosa-5,8,11,14-tetraenoic	Arachidonic (ω-6)	Ad	304.2402	361.2737	343.2632	287.2369	269.2264	303.2330
20:5	(5*Z*,8*Z*,11*Z*,14*Z*,17*Z*)-Icosa-5,8,11,14,17-pentaenoic	EPA (ω-3)	EPA	302.2246	359.2581	341.2475	285.2213	267.2107	301.2173
21:0	Heneicosanoic	Heneicosylic	He	326.3185	383.3520	365.3414	309.3152	291.3046	325.3112
22:0	**Docosanoic**	**Behenic**	**B**	**340.3341**	**397.3676**	**379.3571**	**323.3308**	**305.3203**	**339.3269**

(Continued)

Table 13.1 (Continued)

Fatty	Systematic (IUPAC)			Fatty Acid	[RCOO+58]$^+$	[RCOO+58−H_2O]$^+$			
Acyl group	Name	Common name	Abbreviation	RCOOH	[RCO+74]$^+$	[RCO+74−H_2O]$^+$	[RCO]$^+$	[RCO−H_2O]$^+$	[RCOO]$^-$
22:1	**(13*Z*)-Docos-13-enoic**	**Erucic (*n*-9)**	**E**	**338.3185**	**395.3520**	**377.3414**	**321.3152**	**303.3046**	**337.3112**
22:2	(13*Z*,16*Z*)-Docosa-13,16-dienoic	Docosadienoic (ω-6)	Dde	336.3028	393.3363	375.3258	319.2995	301.2890	335.2956
22:3	(5*Z*,13*Z*,16*Z*)-docosa-5,13,16-trienoic	Eranthic (ω-6)	Dtr	334.2872	391.3207	373.3101	317.2839	299.2733	333.2799
22:4	(7*Z*,10*Z*,13*Z*,16*Z*)-docosa-7,10,13,16-tetraenoic	Adrenic (ω-6)	Dte	332.2715	389.3050	371.2945	315.2682	297.2577	331.2643
22:5	**(7*Z*,10*Z*,13*Z*,16*Z*,19*Z*)-docosa-7,10,13,16,19-pentaenoic**	**DPA (ω-3)**	**DPA**	**330.2559**	**387.2894**	**369.2788**	**313.2526**	**295.2420**	**329.2486**
22:6	**(4*Z*,7*Z*,10*Z*,13*Z*,16*Z*,19*Z*)-docosa-4,7,10,13,16,19-hexaenoic**	**DHA (ω-3)**	**DHA**	**328.2402**	**385.2737**	**367.2632**	**311.2369**	**293.2264**	**327.2330**
23:0	Tricosanoic	Tricosylic	Tc	354.3498	411.3833	393.3727	337.3465	319.3359	353.3425
24:0	**Tetracosanoic**	**Lignoceric**	**Lg**	**368.3654**	**425.3989**	**407.3884**	**351.3621**	**333.3516**	**367.3582**
24:1	**(15*Z*)-tetracos-15-enoic**	**Nervonic (*n*-9)**	**N**	**366.3498**	**423.3833**	**405.3727**	**349.3465**	**331.3359**	**365.3425**
25:0	Pentacosanoic	Pentacosylic	Pc	382.3811	439.4146	421.4040	365.3778	347.3672	381.3738
26:0	**Hexacosanoic**	**Cerotic**	**Ce**	**396.3967**	**453.4302**	**435.4197**	**379.3934**	**361.3829**	**395.3895**
26:1	**(17*Z*)-hexacos-17-enoic**	**Ximinic (*n*-9)**	**Xi**	**394.3811**	**451.4146**	**433.4040**	**377.3778**	**359.3672**	**393.3738**
27:0	Heptacosanoic	Heptacosylic	Hc	410.4124	467.4459	449.4353	393.4091	375.3985	409.4051

28:0	**Octacosanoic**	**Montanic**	**Mo**	**424.4280**	**481.4615**	**463.4510**	**407.4247**	**389.4142**	**423.4208**
28:1	Octacosenoic	Octacosenylic	Oce	422.4124	479.4459	461.4353	405.4091	387.3985	421.4051
29:0	Nonacosanoic	Nonacosylic	Nc	438.4437	495.4772	477.4666	421.4404	403.4298	437.4364
30:0	Triacontanoic	Melissic	Me	452.4593	509.4928	491.4823	435.4560	417.4455	451.4521
31:0	Hena triacontanoic	Henatriacontylic	Ht	466.4750	523.5085	505.4979	449.4717	431.4611	465.4677
32:0	Dotriacontanoic	Lacceroic	Lc	480.4906	537.5241	519.5136	463.4873	445.4768	479.4834
33:0	Triatriacontanoic	Psyllic	Tt	494.5063	551.5398	533.5292	477.5030	459.4924	493.4990
34:0	Tetratriacontanoic	Gheddic	Gh	508.5219	565.5554	547.5449	491.5186	473.5081	507.5147
35:0	Pentatriacontanoic	Ceroplastic	C35:0	522.5376	579.5711	561.5605	505.5343	487.5237	521.5303
36:0	Hexatriacontanoic	Hexatriacontylic	C36:0	536.5532	593.5867	575.5762	519.5499	501.5394	535.5460

Masses are exact masses expected in high-resolution accurate-mass mass spectra, including loss or addition of a mole of electrons per monoisotopic mass for ions. Bold FAs represent those most commonly encountered during of AGs in common edible oils. See The Lipid Library for systematic and trivial FA names, at: https://lipidlibrary.aocs.org/resource-material/trivial-names-of-fatty-acids-part-1

MGs have been analyzed as polar impurities formed during the production of biodiesel as their TMS derivatives by Yang et al. [18] using GC-MS. The target molecules for biodiesel production are FAME, but incomplete reaction and side reactions yield a variety of by-products. Free fatty acids (FFAs) are the most detrimental to biodiesel quality, so their amounts are an important indicator of fuel quality. Also present, although in lower amounts, are unreacted or interesterified TGs and partial glycerolysis products such as MGs and DGs. More recently, MGs have been recognized as important signaling molecules, especially the endocannabinoid 2-arachidonylglycerol [19]. Also, 2-oleoylglycerol has been found to activate GPR119 (G-coupled protein receptor) and stimulate GLP-1 (glucagon-like peptide-1), making GPR119 a fat sensor [20]. A very recent report showed that 2-arachidonoyl glycerol stimulated cholecystokinin secretion (a peptide hormone that regulates gastric motility, appetite, and other functions) [21].

There are far too many reports of MG analysis by MS to list (>500 articles in Scopus). Therefore, we can only cite a few reviews and more recent unique reports to provide a starting point for those undertaking MG analysis. Given that GC is a mature, reliable, and well-proven analytical technique, the original works that demonstrated the utility of specific derivatives (e.g. TMS) remain relevant and useful to this day. However, lipids remain ever-popular objects of investigation because of their biological and industrial importance; so as new MS methods are developed, they are inevitably used for AG analysis. For example, 2D-GC is a relatively new approach that Indrasti et al. [22] applied to MG and DG analysis. Another recent variation on a classical GC approach was the report of Nina Naquiah et al. [23], who used the AOCS official method Cd-11c-93 to separate MGs and DGs from a glycerolysis reaction, followed by FAME analysis and elemental analysis–isotope ratio (EA-IR) MS. They found that that the $\delta^{13}C$ values for MGs and DGs from lard could be easily distinguished from those for chicken fat, beef fat, and mutton fat.

Three API techniques were compared by Cai et al. [24, 25]. Atmospheric pressure photoionization (APPI)-MS was compared to APCI-MS and electrospray ionization (ESI)-MS for analysis of neutral lipids including FAME, MGs, DGs, and TGs, as well as FFA and fatty acid ethyl esters (FAEEs). Figure 13.6 shows the mass spectra of MGs by APPI-MS, APCI-MS, and ESI-MS, demonstrating several common observations related to AGs. First, both APPI-MS and APCI-MS exhibited a protonated molecule, $[M+H]^+$, and a base peak representing the very common loss of a hydroxyl group from the protonated molecule by dehydration, $[M+H-H_2O]^+$, which for MGs gives an ion equivalent to the $[RCOO+58]^+$ ion seen in Table 13.1. The mass spectra in Figure 13.6a,c demonstrate the now commonly recognized observation that APPI-MS and APCI-MS give mass spectra that are essentially indistinguishable. Nowadays, it is common to supply acetone or toluene as a dopant to promote ionization as these have ionization energies below the energies provided by the 10.0 and 10.6 eV lines of the Krypton lamp [26].

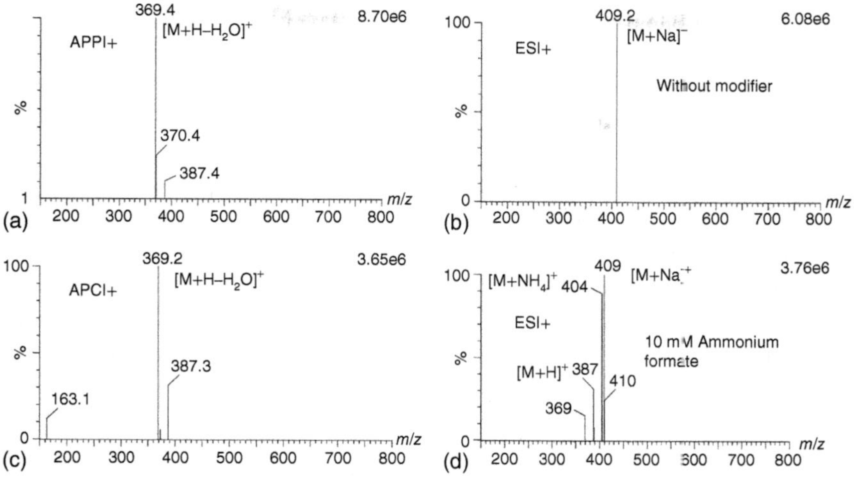

Figure 13.6 (a) Atmospheric pressure photoionization (APPI) mass spectrometry (MS) spectrum of monoarachidin (20:0-MAG) by infusion in isooctane/isopropyl alcohol (1:1) with no dopant; (b) electrospray ionization (ESI)-MS without an added electrolyte; (c) atmospheric pressure chemical ionization (APCI)-MS; and (d) ESI-MS mass spectrum of 20:0-MAG with 10 mM ammonium formate electrolyte added to the infusion mobile phase. Source: Reprinted with permission from Cai and Syage [24]. Copyright 2006 American Chemical Society.

13.3 Diacylglycerols (DAGs)

Similar to MGs, DGs are also more polar than TGs and also serve as emulsifiers and surfactants, so DGs are often measured in combination with MGs [27]. Furthermore, the FDA designation as generally recognized as safe (GRAS) applied to DGs as well as MGs [28]. DGs are a product of partial glycerolysis and can also be formed synthetically, and conditions can be chosen to favor DG production. As with other AG classes, William Christie's The Lipid Web provides an excellent primer on DG structures, biosynthesis, and metabolism [29].

While DGs are by-products of partial enzymatic glycerolysis of TGs, they are metabolized differently and have different effects in the body, particularly 1,3-DGs. For example, 1,3-DG-rich oil suppressed both body weight and regional fat deposition including visceral and hepatic fat in healthy men [30] and thus has been seen as a potentially valuable tool to reduce the incidence of obesity. A DG oil rich in α-linolenic acid similarly reduced body fat in men [31]. However, a review from 2020 [32] indicated that some concerns had been raised about the presence of glycidol esters of FAs as probable carcinogens by the German Federal Institute for Risk Assessment. Because of this, DG oils were removed from the market, and as of the date of that review, they were still off the market.

Early reports (i.e. 1970s) of DGs were similar to those of MGs because DGs were often derivatized to acetyl, trimethylsilyl (TMS) ethers [33–35], *t*-butyl dimethyl silyl (TBDMS) ethers [36, 37], or other derivative [38–40] and analyzed by GC-MS. Often, PLs were hydrolyzed, and the resultant DGs and related backbone structures were analyzed in the same way as normal DGs [36, 37]. Derivatization and GC analysis of DGs remains a perfectly valid approach to analysis to this day [41–43].

As with MGs, DGs are most commonly analyzed as part of a broader analysis of AGs or lipids, with few articles specifically targeted at DGs. However, Itabashi, Kuksis, Myher, Marai, and coworkers have published numerous reports of the analysis of DGs that were derivatized to their dinitrophenylurethane (DNPU) derivatives to allow a chiral separation of the DG isomers [44, 45]. As ESI-MS became more common, it was used for detection of the DNPU-MGs, other derivatives, and DGs from PLs [46, 47] (Japanese language). By application of phospholipase C to phosphatidyl choline (PC), the PC species were hydrolyzed to DG isomers (e.g. PO-DG) and reverse isomers (e.g. OP-DG), which were analyzed as their sodium adducts by ESI-MS [47]. ESI-MS and MS/MS spectra were shown for DGs, and 3,5-dinitrobenzoate (DNB), nicotinate, and 2-anthrylurethane (2-AU) derivatives of DGs, as pairs of isomers and reverse isomers in positive-ion mode, while mass spectra of 3,5-DNPU derivatives in negative-ion mode were shown [47]. Later, Honda et al. [48] applied negative-ion ESI-MS to analysis of DNPU derivatives of DGs released from digalactosyl diacylglycerols (DGDGs) using chiral chromatography and reversed-phase HPLC.

Matrix-assisted laser desorption ionization (MALDI)-MS was first reported in 1987 [49] and by the mid to late 1990s, MALDI was used with both Fourier transform ion cyclotron resonance (FT-ICR) mass detection for PLs [50] and time-of-flight (TOF) detection for TGs, PLs, and galactosyl-DGs [51], and other lipid analysis [52], and was described for quantification of model DGs [53]. Those experiments showed trends common to other types of mass spectrometry, specifically that the carbon chain length affected the MALDI-MS signal, with shorter chains generally giving more signal than longer chains, such that SA-DG < SL-DG < PO-DG, although dipalmitoyl (PP)-DG < PO-DG. A similar inverse correlation between carbon chain length and signal was observed for TGs by MALDI-MS [54]. The use of MALDI-MS for lipid analysis has been reviewed [55–61], as well as mentioned in other reviews of MS for lipids overall [62].

In their report on APPI-MS for lipid analysis, Cai and Syage [24] demonstrated spectra for diarachidin (20:0,20:0-DG) that are very representative of spectra still observed by the API techniques, so they are reproduced here as Figure 13.7. APPI-MS and APCI-MS mass spectra of diarachidin, di-20:0-DG, Figure 13.7a,c, respectively, show no protonated molecule at all for the DGs, but only the common DG-like fragment, $[DAG]^+$, that represents $[M+H-H_2O]^+$. The same $[DAG]^+$ fragment is commonly observed from TGs. One example of LC-APCI-MS for targeted DG analysis was reported by Lee et al. [63], who used HPLC on a C6-phenyl column for analysis of DGs in mouse muscle tissue.

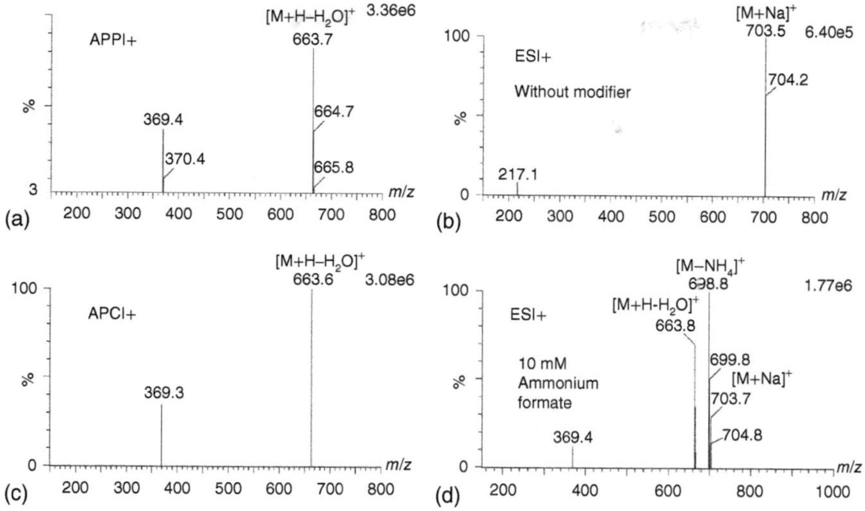

Figure 13.7 (a) Atmospheric pressure photoionization (APPI) mass spectrometry (MS) spectrum of diarachidin (20:0-MAG) by infusion in isooctane/isopropyl alcohol (1:1) with no dopant; (b) electrospray ionization (ESI)-MS without an added electrolyte; (c) atmospheric pressure chemical ionization (APCI)-MS; and (d) ESI-MS mass spectrum of 20:0/20:0-DAG with 10 mM ammonium formate electrolyte added to the infusion mobile phase. Source: Reprinted with permission from Cai and Syage [24]. Copyright 2006 American Chemical Society.

13.3.1 Electrospray Ionization (ESI) for DGs

In the 1990s, ESI-MS also started being applied to lipids, with the first application of ESI-MS to AGs apparently being by Duffin et al. [64]. Other approaches for MS of AGs have been employed, such as particle–beam interface, but are not included here, so we can focus on more modern methods. Those older techniques have been included in earlier reviews [62, 65, 66].

One of the more popular applications of ESI-MS remains the analysis of DGs that are enzymatically cleaved from PLs to be measured as derivatives using ESI-MS. For instance, Palyzová and Řezanka [67] recently demonstrated the use of phospholipase C to cleave the head groups from PLs (predominantly phosphatidyl glycerols, PGs) in Gram-positive bacteria, which contained substantial amounts of branched chain FAs. The lipid extracts were first separated into classes using hydrophilic interaction liquid chromatography (HILIC). Then, PGs were collected and treated with phospholipase C to yield DGs, which were then acetylated to acetyl-TGs and analyzed by RP-HPLC-ESI-MS. Nowadays, it is more common to determine the presence and/or quantity of DGs as a part of an overall lipidomic analysis using ESI-MS. For instance, He et al. [68] recently used ESI-MS lipidomics for analysis of FFAs, MGs, DGs, TGs, glycolipids, PLs, and sphingolipids, so a variety of DGs were included in their analysis.

13.4 Triacylglycerols (TAGs)

TAGs have long been subjects of nutritional and analytical research because of their importance in the diet and in biological systems. Because of this, practically every new MS technique has been applied to TGs soon after its development. Reviews of MS for lipid analysis have appeared over the years [69–73], including the recent reviews of HTGC of TGs mentioned above [8, 9]. We have provided multiple reviews of APCI-MS and ESI-MS of TGs over the years [74–80]. For those undertaking TG analysis, the reviews above are a valuable starting point, and it is also useful to be aware of steps that should be taken to maximize the stability of samples, as well as some best practices for analysis, discussed by Ulmer et al. [81].

13.4.1 Early Reports Described Structural Information that Comes from APCI-MS of TGs

Original early reports are mentioned here because many articles are appearing in the literature that cite only recent examples and not the original sources. Tyrefors et al. demonstrated a custom-made APCI interface to supercritical fluid chromatography in 1993 [82] and used supercritical fluid chromatography (SFC)-APCI-MS for several classes of molecules, including trilaurin (TG 12:0/12:0/12:0) and stearic FAME. In 1995, Byrdwell and coworkers [83–85] reported the first applications of HPLC-APCI-MS, first to a mixture of TG standards [83] and then to soybean oil [85] and unusual seed oils [84]. Shortly thereafter, quantification of TAGs using response factors derived from several different approaches was demonstrated [86], discussed further below. The first trend that was immediately apparent was the fact that the amount of protonated molecule, $[M+H]^+$, produced by APCI-MS was inversely dependent on the degree of unsaturation in the TGs [83, 86]. Saturated TGs gave little or no $[M+H]^+$, while polyunsaturated (PU)-TGs gave an $[M+H]^+$ base peak. Soon, other trends were reported that demonstrated that the ion abundances of $[DAG]^+$ fragments depended on the positions of FAs on the glycerol backbone [87–89]. Mottram and Evershed [88] used mixtures of di-acid TGs, AAB(BAA)/ABA, to demonstrate that the ratio of $[AA]^+/[AB]^+$ was smaller when "B" was in the *sn*-2 position. They also showed that for tri-acid TGs, ABC/BAC/ACB/CBA/CAB/BCA TGs, the $[M-RCOO]^+$ fragment from loss of the *sn*-2 FA was the smallest of the $[DAG]^+$ fragments. Those reports led to the conclusion that loss of the *sn*-2 FA was energetically disfavored and could be used to identify TG regioisomers. If there was no non-statistical preference for FA loss from TGs, the ratio of $[AA]^+$ fragments to $[AB]^+$ from di-acid TGs (AAB/BAA/ABA) would always be 1:2, or ½, or 0.5.

In 2003, Jakab et al. [90] used those trends for quantitative analysis of TG regioisomers. They analyzed mixtures of 1,3-dilinoleoyl, oleoyl glycerol (LOL) + 1,2-dilinoleoyl, oleoyl glycerol (LLO) in known concentrations, thereby creating a calibration line based on the $[LL]^+/[OL]^+$ ratios. Then, they analyzed several seed oils, including grape seed oil, olive oil, and others, to quantify the amount of LOL versus LLO. In 2004, Fauconnot et al. [91] analyzed seven TG regioisomer pairs and found that mixtures of known composition of AAB(BAA)/ABA TGs produced linear calibration lines from the $[AA]^+/[AB]^+$ ratios. They then reported the compositions of several fats and oils, including palm oil, cocoa butter, and beef, pork, and chicken fat.

Soon after, Byrdwell [79] gave the equations necessary to get the regioisomer information (e.g. %LOL) from just the $[AA]^+/[AB]^+$ ratios of the two pure regioisomers for di-acid TGs (AAB/BAA/ABA). Building on that work, Byrdwell [92] developed "The Bottom Up Solution to the triacylglycerol lipidome" and identified three Critical Ratios (CRs) that could be used to (i) derive the structural information desired about regioisomers and the degree of unsaturation from TGs and (ii) reproduce the original mass spectrum using fewer CRs than raw abundances, constituting a compact library, or reduced data set. Byrdwell introduced the nomenclature of Type I TGs for monoacid (AAA) TGs, Type II for di-acid (AAB/BAA/ABA) TGs, and Type III for tri-acid (ABC/ACB/BAC/CBA/BCA/CAB) TGs.

In 2010, Holčapek et al. [93] added to the APCI-MS knowledge base by determining the $[AA]^+$ and $[AB]^+$ abundances for a variety of synthesized TGs on five different instruments all employing APCI. Sets of three Type I TGs at a time were randomized to produce combinations of Type II TGs ("doublets") and Type III TGs ("triplets"). The data for Type III TGs were converted to CR2 and CR3 for comparison to previously established trends [94]. Holčapek et al. [93] results confirmed several trends widely observed in APCI-MS of TGs. First, all ABA Type II TGs gave CR2 less than the statistically expected value of 0.5, without exception, while all AAB/BAA Type II TGs gave CR2 greater than 0.5. From those data, one can conclude that a CR2 in the range 0.2–0.3 is a good indicator of the predominance of ABA TGs, whereas a CR2 in the range 0.7–0.95 is a good indicator of AAB/BAA TGs. Values in between need to be compared to tabulated results or TG regioisomer standards. Also, Byrdwell showed [95] that as APPI-MS spectra are so similar to APCI-MS mass spectra, the APPI-MS data can be put in the form of CRs and processed through the Updated Bottom Up Solution and that the CRs from APPI-MS are closer to APCI-MS than those from ESI-MS. The CRs were also constructed for ESI-MS/MS [95] and were found to agree pretty well with APCI-MS, but not as well as APPI-MS agreed.

13.4.2 Quantification of TGs by APCI-MS and APPI-MS

ESI-MS, APCI-MS, and APPI-MS all give different responses to TGs based on their carbon chain length and degree of unsaturation. The signal generally decreases with increasing carbon chain length. For APCI-MS and APPI-MS, saturates give more signal than polyunsaturates (PUs) [86]. On the other hand, ESI-MS produces $[M+NH_4]^+$ or other adduct ions from all TGs, and PU-TGs give more response than saturated TGs [96].

There are multiple approaches that have been used for quantification of TGs, and most of them are targeted at a percent-relative composition of TG molecular species rather than absolute quantification. Absolute quantification is best when a calibration line can be constructed for each TG to be quantified. However, unfortunately, there are not nearly enough TG standards available to provide coverage of most TGs to be quantified.

In our early report of quantification of TGs, we compared quantification using calibration lines for individual TG standards, response factors from a synthetic 35-TG mixture, response factors produced from an interesterified version of the analyzed oil, and response factors from GC-FID applied to LC-APCI-MS. Calibration lines gave good linearity ($r^2 > 0.99$), but there were too few standards available to allow quantification of all TGs. By far the most effective approach, and the one that continues to provide the best results, is the use of response factors from GC-FID compared to LC-APCI-MS. This is accomplished by calculating the FA composition from the TG composition determined by LC-APCI-MS. Then, this is compared to the mole % FA composition (converted from the weight % composition) determined by calibrated GC-FID, which is known to be very reliable over several orders of magnitude. A response factor for each FA is made from the ratio of the GC-FID FA composition to the LC-APCI-MS FA composition. Then, the FA RFs are multiplied together for each TG to get a set of TG RFs. These were multiplied times the raw TG areas to get RF-adjusted areas, and the complete TG composition was recalculated. Then, the FA composition from the RF-adjusted TG composition was calculated. Not only did the FA composition from TGs agree better with the GC-FID FA composition than the agreement in FA composition from all other approaches but also the TG composition for randomly distributed TGs in synthetic or randomized mixtures (randomization confirmed by lipase analysis) agreed better with the statistically predicted compositions than other approaches. Thus, both the TG compositions and the FA compositions agreed better using the GC-FID-derived RF approach.

One very different approach to quantification of TGs has been described in multiple reports by the group of Kallio et al. [97, 98]. They originally used ammonia negative chemical ionization (NCI) by direct insertion probe (DIP) tandem sector quadrupole (TSQ) MS to identify TGs in human milk using a combination of $[M-H]^-$ precursor ions and $[M-H-RCOOH-100]^-$ and $[RCOO]^-$ product ions. Later, the authors adapted their NCI method for APCI-MS with ammonia as the nebulizer (sheath) gas [99], again using the $[M-H]^-$ ions, $[M-H-RCOOH-100]^-$, and $[RCOO]^-$ ions. They constructed calibration curves for ABA/AAB/BAA (Type II) TGs from the ratios of $[M-H-R_ACOOH-100]^-/([M-H-R_ACOOH-100]^- + [M-H-R_BCOOH-100]^-)$ for reference TGs as well as several seed/kernel oils (rapeseed, sunflower seed, palm, and other oils).

Holčapek et al. [100] developed a different approach to quantification based on APCI-MS of 23 monoacid (Type I) TGs. First, the response factors (RFs) for each TG were calculated relative to triolein, OOO, 18:1_18:1_18:1, by dividing the area for OOO by the area of each TG. Then, the three RFs of the monoacid TGs for each FA in a TG were averaged. These RFs were applied to TGs in nut, seed, and kernel oils (i.e. walnut, hazelnut, cashew, almond, poppy seed, yellow melon, mango stone, fig seed, and date seed) separated by NARP-HPLC detected by APCI-MS. The FA composition was calculated from the TG composition and was compared to the FA composition determined by GC-FID.

Fasciotti and Netto [101] quantified triolein in extra virgin olive oil (EVOO) using an external standard approach and the method of standard addition, as well as percent relative quantification of select TG species from multiple reaction monitoring (MRM) transitions of $[M+H]^+ \rightarrow [DAG]^+$ with no RFs applied. Other reports in the literature apply no response factors but simply report unadjusted area percentages. For instance, Řezanka et al. [102] recently reported the use of both NARP-HPLC-APCI-MS and chiral chromatography APCI-MS for analysis of TG enantiomers in ximenia oil, including nine synthesized enantiomers. Řezanka et al. [102, 103] are the most prolific and authoritative authors on chiral separation of TGs, often with detection by APCI-MS. Although separations can be quite long, up to hundreds of minutes, the chiral chromatography reported by these authors is the most effective approach for separating and quantifying TG enantiomers.

A report by Baiocchi et al. [104] on tuna oil and algal oil TGs analyzed by APCI-MS, MS/MS, and MS^3 published a possible exception to the rule that the loss of the *sn*-2 FA from TGs is energetically unfavorable. They showed that TGs having two polyunsaturated fatty acids (PUFAs) next to each other, as in PoEpDh (16:1_20:5_22:6), gave the smallest fragment for the $[EpDh]^+$ fragment, instead of the $[PoDh]^+$, in full-scan mass spectra, but in MS/MS spectra, $[PoDh]^+$ was the smallest $[DAG]^+$ fragment. The TG compositions of tuna oil and algal oil showed numerous TG species in which a lone docosahexaenoic acid (DHA) was located in the 1,3-positions and was only in the *sn*-2 position when two DHA FAs were present. In contrast, Zhang et al. [105] analyzed tuna, salmon, and anchovy oils by RP-HPLC-APCI-MS, which included enzymatic degradation for regioisomer determination, and found that DHA was mainly found in the *sn*-2 position, whereas eicosapentaenoic acid (EPA) was more evenly distributed among the backbone positions.

13.4.3 Covalent Adduct Chemical Ionization (CACI)

Xu and Brenna [106] have reported adaptation of their covalent adduct chemical ionization (CACI) to APCI for atmospheric pressure covalent adduct chemical ionization (APCACI). Helium was used as the nebulizing gas to optimize the $[M+54]^+$ (1-methyleneimino)-1-ethenylium (MIE, $CH_2{=}C{=}N{+}{=}CH_2$) ion, although ions at $[M+40]^+$, $[M+54]^+$, $[M+81]^+$, and $[M+95]^+$ were all tested and found to provide fragments that allowed double bond locations to be deduced for monoenoic and dienoic TGs. The $[M+40]^+$ ion was deemed best for diagnostic purposes because of its larger abundance and slightly simpler spectra. Fragmentation of either ketene adduct $[RCH{=}C{=}O+54]^+$ or $[RCH{=}C{=}O+40]^+$ in APCI-MS^3 mass spectra yielded two strong diagnostic ions that allowed localization of the positions of double bonds in fatty acyl chains, as depicted in Figure 13.8.

13.4.4 Quantification of TGs by ESI-MS Using Shotgun Lipidomics

In 2001, Han and Gross [107] demonstrated quantification of TGs as their lithium adducts in unseparated mixtures by infusion of standards and a rat myocardium extract. They used neutral loss scanning (NLS) for the mass of each FA that could be

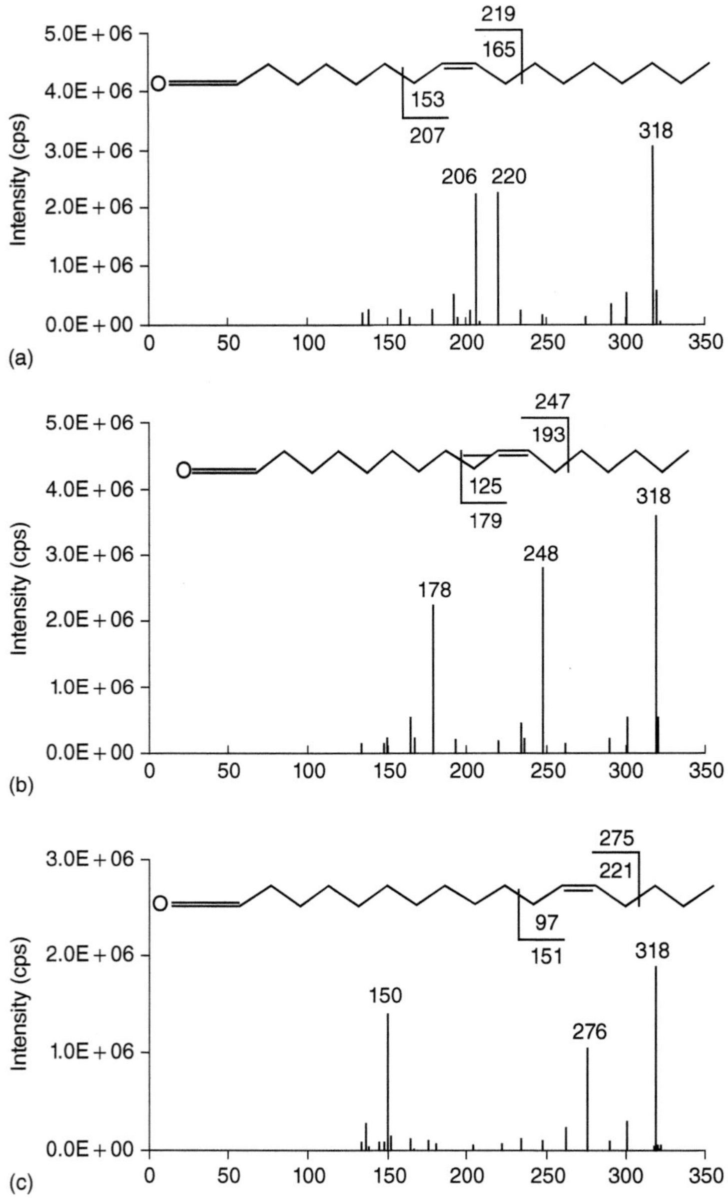

Figure 13.8 Atmospheric pressure covalent adduct chemical ionization (APCACI)-MS/MS/MS of TGs containing 18:1 isomers. (a) (16:0/Δ9-18:1/Δ9-18:1)-TG; (b) (16:0/Δ11-18:1/Δ11-18:1)-TG; and (c) (16:0/Δ13-18:1/Δ13-18:1)-TG. The [ketene + 54]$^+$ ions at *m/z* 318 came from MS/MS of [M+54]$^+$ of TGs were selected and collisionally dissociated (*m/z* 913 → *m/z* 318 → products) to yield a different pair of diagnostic ions for each isomer. Source: Reprinted with permission from Xu and Brenna [106]. Copyright 2007 American Chemical Society.

lost from a TG. They applied two ^{13}C correction factors, one accounting for the difference in carbon chain length between the internal standard (IS) and the analytes (since a different number of carbons means a different amount of isotope peaks) and a second accounting for the $2\times{}^{13}C$ variants of an M-2 ion contributed to the M ion, which can also be referred to as the A and the A+2 ions, respectively. Note that most isotope corrections only account for the ^{13}C atoms as these are the major contributors and do not include the smaller contributions from ^{18}O and ^{2}H in TGs. They also applied response factors calculated from a least-squares regression of linear calibration lines for a variety of TG molecular species. They confirmed the observation that had earlier been reported by Duffin et al. [64] for $[M+Na]^+$ and $[M+NH_4]^+$ adducts that the signal decreased with increasing carbon chain length (the behenic and erucic labels seem to have been reversed in that report) and increased with increasing unsaturation.

The report by Han and Gross [107] and the report soon after [108] signaled the start of the era of "shotgun lipidomics" [109]. Since then, many other researchers have employed this and similar shotgun approaches. Later, their work was extended to shotgun lipidomics of TGs in Arabidopsis seeds using NLS with ammonium adducts, $[M+NH_4]^+$ [110]. They employed 10 neutral loss scans, corresponding to the 10 FAs identified by GC-FID, and used tri-17:1 as an IS. As has become normal, they used isotope correction for the "A+2" overlap. "Adjustment factors" were calculated from 15 commercial TG species relative to the IS. Interestingly, the adjustment factors were at a minimum for TGs with three sites of unsaturation but higher for TGs with lower and higher numbers of DBs. The results obtained by ESI-MS were compared to the statistically expected amounts of TGs calculated from the equations given in the section on GC-FID above.

Recent examples of shotgun lipidomics for quantitative analysis of AGs include the 2020 report by Yeo and Parrish [111], who used NLS of masses of FAs lost from TG ammonium adducts, $[M+NH_4]^+$, for quantification of TGs in salmon muscle. The authors showed an increase in signal going from 3 to 6 double bonds and then a decrease going from 6 to 12 DBs, followed by an increase again at 15 DBs. That work was similar to the analysis of the lipids in fish heads of big eye tuna, Atlantic salmon, and bighead carp by Zhang et al. in 2019 [112].

13.4.5 Quantification of TGs by ESI-MS with HPLC/UHPLC Separation

A very recent example of lipidomic analysis of tilapia muscle, head, and viscera following a 15 minute ultra-high-performance liquid chromatography (UHPLC) separation was reported by He et al. [68]. They used UHPLC-ESI-high-resolution, accurate-mass (HRAM)-MS of $[M+NH_4]^+$ to identify multiple classes of lipids, including TGs, DGs, MGs, FFAs, SLs, and PLs. Classes were also separated using solid-phase extraction (SPE) and then derivatized to FAMEs for analysis by GC-FID. Stereospecific analysis of TGs was done by hydrolyzing them with porcine pancreas lipase, followed by separation of the FFAs from 2-MAGs by SPE and then GC-FID of the FAME.

Another recent representative example is that of Eum et al. [113] who used lipidomics with 18 internal standards across lipid classes, with d_5-17:0/17:0 for DGs and d_5-17:0/17:1/17:0 for TGs. They analyzed mouse serum, kidney, and heart tissue extracts using two different nanoflow UHPLC-ESI-MS/MS systems, in the positive- and negative-ion modes, for both untargeted lipidomics and then targeted quantification. TGs were quantified at the brutto structure level (carbon number [CN]:DB) for tissue extracts from young versus old mice.

13.4.6 Quantification of Regioisomers by ESI-MS

In addition to the overall TG composition, one of the most commonly sought pieces of information is the identity of the FA in the *sn*-2 position of TGs due to the relationship to lipase digestion and bioavailability of the FAs in the different positions. ESI-MS has long been used for quantification of regioisomers, as demonstrated by Hsu and Turk in 1999 [114], who used low-energy collisionally activated dissociation (CAD) of lithium adducts of TGs for analysis of regioisomers. They showed that loss of FAs from the $[M+Li]^+$ produced two sets of fragments for each FA, as $[M+Li-RCOOH]^+$ and $[M+Li-RCOOLi]^+$, with the latter being equivalent to a typical $[DAG]^+$ fragment. Typical ESI-MS/MS spectra of $[M+Li]^+$ precursors of two Type III TG regioisomers are shown in Figure 13.9. Using the TG isomer pair PSO and POS, they showed that the neutral loss of the *sn*-2 FA produced a smaller abundance than losses from the 1,3-positions. The authors also used deuterium-labeled TG standards to propose fragmentation mechanisms.

Hvattum [115] reported NARP-HPLC-ESI-MS of TG standards and showed that the smallest $[DAG]^+$ fragment from SOL (18:0/18:1/18:2) came from loss of the *sn*-2 FA, while the lower mass acylium ions, $[RCO]^+$, were more dependent on the degree of unsaturation than the FA's position on the glycerol backbone. Byrdwell and Neff [96] showed that low-energy collision-induced dissociation (CID) MS/MS of $[M+NH_4]^+$ adducts gave different $[DAG]^+$ fragment ratios, $[AA]^+/[AB]^+$, from Type II regioisomer standards, such as OPO/OOP, POP/PPO, and OOS. They also confirmed that in ESI-MS^3 mass spectra, the four most common ions were $[RCO]^+$, $[RCO-H_2O]^+$, $[RCOO+58]^+$, and $[RCOO+58-H_2O]^+$, as presented in Table 13.1, and that the identity of the FA chain had a larger effect on ion abundances than their positions. Marzilli et al. [116] confirmed all of the aforementioned trends and also pointed out a $[DAG-74]^+$ fragment formed from $[DAG]^+$-containing unsaturated FAs. These reports laid the foundation for the numerous reports that have been published since that time that have used the above-mentioned fragments for identification and quantification of TG regioisomers.

Lin and Arcinas [117] fractionated olive oil and used ESI-MS^3 of the $[M+Li]^+ \rightarrow [M+Li-RCOOH]^+ \rightarrow$ fragments of TGs in the fractions for the identification of regioisomers. They showed a non-linear calibration curve from the ESI-MS^3 data for OOP/(OOP+OPO). In 2010, Hsu and Turk [118] used ESI-MS^3 of the $[M+Li]^+ \rightarrow [M+Li-RCOOH]^+ \rightarrow$ fragments from TG standards to localize the positions of double bonds in unsaturated FA chains (using very low abundance fragment ions). Soon after, Cubero Herrara et al. [119] applied ESI-MS^3 of $[M+Li]^+$

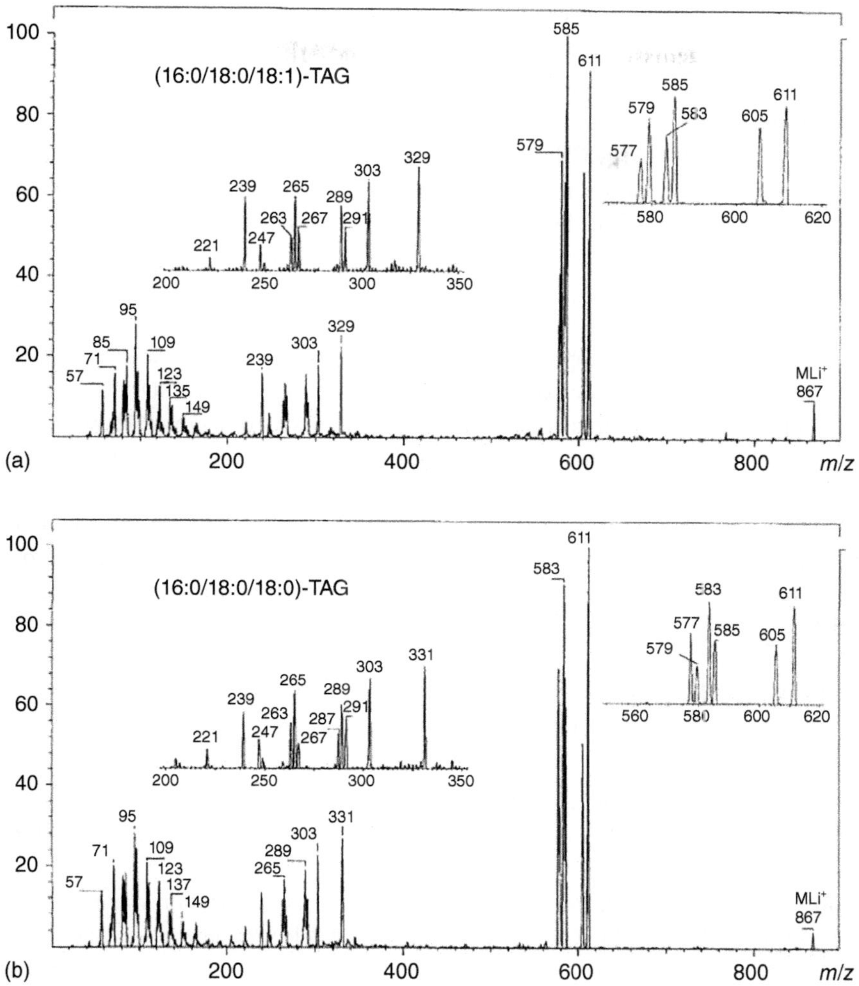

Figure 13.9 ESI-MS/MS mass spectra of lithiated adducts of two TG regioisomers with $[M+Li]^+$ at *m/z* 867.8 that contain three different fatty acid substituents. (a) ESI-MS/MS of lithiated adduct of (16:0/18:0/18:1)-TG and (b) ESI-MS/MS of lithiated adduct of (16:0/18:1/18:0)-TG. Source: Reprinted with permission from Hsu and Turk [114]. Copyright 1999 American Chemical Society.

adducts to regiospecific analysis of ABA/AAB pairs in DHA- and EPA-rich TGs from fish oils. Note that in the ESI-MS^3 spectra of $[M+Li]^+$, loss of the *sn*-2 group is preferred (having already lost a 1(3)-FA in MS/MS) [114, 117–119].

One application that serves as a mini-review up to that point and presents a different outlook on TG regioisomer quantification by ESI-MS^3 of $[M+Li]^+$ was reported by Ramaley et al. [120] in 2013. They presented a model that indicated that the calibration curves from ESI-MS^3 fragments from $[M+Li]^+$ are inherently non-linear, so calibration curves should be fitted using a second-order polynomial. In 2007, Leskinen et al. [121] summarized and reproduced previous work that showed linear calibration lines from HPLC-APCI-MS, UHPLC-APCI-MS, and UHPLC-ESI-MS/MS. $[DAG]^+$ fragments in HPLC-APCI-MS/MS and UHPLC-APCI-MS/MS mass spectra did not reflect the identity of the FA at the *sn*-2 position but instead were more affected by unsaturation and other factors.

In contrast to high-energy CID, which was found to not provide an indication of the positions of FA chains on the glycerol backbone of TGs [122], low-energy CID did give the lowest abundance for the loss of the *sn*-2 component [115] similar to the trends observed by APCI-MS [87, 88]. Kalo et al. [123] used low-energy CID of $[M+NH_4]^+$ adducts for regiospecific analysis of TGs in butter fat separated using three normal-phase columns in series. Even up to very recently, the principles mentioned above continue to be applied to samples. For instance, Liu and Rochfort [124] employed ESI-MS/MS to make calibration lines for OPO/OOP and OSO/OOS regioisomer pairs to determine their presence in Holstein and Jersey cows. They found that Jersey milk contained more OPO, but a lower amount of OSO compared to Holstein milk.

An interesting report by Lévêque et al. [125] used a different approach than most other researchers by adding $AgNO_3$ post-column as the electrolyte to promote ionization in ESI-MS, followed by MS^5. The fragmentation scheme was thus as follows:

$$\begin{aligned}&[M+Ag+AgNO_3]^+ \rightarrow [M+Ag]^+ \rightarrow [M-RCOOH+Ag]^+\\&\quad\rightarrow [M-RCOOH-H]^+ \rightarrow [RCO+74]^+\end{aligned}$$

Calibration lines based on MS^5 with $r^2 > 0.99$ were made for the regioisomer pairs POO/OPO, SOO/OSO, and PPO/POP, which were then applied to soybean oil and rapeseed oil.

Another interesting and novel approach to double-bond localization was reported by the group at SCIEX [126], who described the use of "electron impact excitation of ions from organics" (EIEIO) combined with differential mobility spectrometry (DMS). While the authors did demonstrate that it was possible to deduce such structural information and applied the approach to several oils (olive, avocado, coconut, etc.), it depended on fragment abundances that were very low, often fractions of 0.1%, used a somewhat complex workflow, and required specialized instrumentation.

13.4.7 Ion Mobility MS for TGs

One of the more recent developments in the commercially available MS options is ion mobility spectrometry (IMS). There are multiple types of IMS, including classic drift-time IMS, traveling wave ion mobility spectrometry (TWIMS), field asymmetric waveform ion mobility spectrometry (FAIMS) (also called DMS), trapped IMS, and several others, so a few reviews of those are helpful to differentiate them [127–131]. Applications to lipid analysis were reviewed up to 2015 by Paglia et al. [132].

Schvartsburg et al. [133] used the newly developed FAIMS (a.k.a., DMS) for lipid analysis using a planar FAIMS unit coupled to a linear ion trap. They found that *sn*-1,2-PLa (16:0/12:0) could be separated from 1,3-PLa using 74% helium at a dispersion voltage of 4.0 V. Trend lines for saturated species differed from the unsaturated ones. Šala et al. [134] used a Sciex SelexION differential mobility separation device to optimize parameters such as separation voltage, chemical modifier, pressure of throttle gas, and others and to demonstrate separation of commercially available TG regioisomers and a tissue extract. The authors employed silver ions, $[M+Ag]^+$, in which case the Ag^+ ions coordinated with sites of unsaturation, just as they do during Ag-ion LC. They used the OSO/SOO isomer pair to optimize parameters and then applied those optimized parameters to SOS/SSO, POP/OPP, OPO/OOP, and an extract of porcine adipose tissue.

13.4.8 Oz-ID for TGs

In 2007, Mitchell and coworkers [135] extended their ozone-induced dissociation (Oz-ID) method to ozone ESI-MS for analysis of multiple classes of PLs and a TG, depicted in Figure 13.10. The authors showed that an ozone generator attached to the desolvation gas inlet of ESI-MS provided sufficient ozone gas for in-source ozonolysis. Initially demonstrated for the ammonium adduct of the monounsaturated TG POP, ozone attacked the single site of unsaturation and produced two diagnostic ions containing an aldehyde functional group and an α-methoxy hydroperoxide. Later [136], this technique was applied to OOO, as its $[M+Na]^+$, and again two ozonolysis products (the aldehyde and the "Criegee ion") identified the double bond position. Then [137], the ozone generator was attached to the collision cell of a linear ion trap instrument, and ESI-MS^3 with CID/OzID of $[M+Na]^+$ precursors was used for characterization of additional TGs with monounsaturated FAs and for TGs containing monounsaturated FAs from *Drosophila* extracts. OzESI-MS has recently [138] been applied to isomer-resolved mass spectral imaging using MALDI-Q-ToF on an instrument with traveling wave (T-Wave) IMS technology.

13.4.9 Paternò–Büchi Reactions

Franklin and Xia [139] demonstrated another new approach for double bond localization in TGs using the Paternò–Büchi (PB) reaction, which is the 2+2 photochemical cycloaddition of a carbonyl (from acetone) across an olefin double bond to give an oxetane 4-member ring. As each PB derivative gave multiple ions for each derivatized TG DB, the results were demonstrated for TG with few sites of unsaturation, such as SOO (18:0_18:1_18:1), which showed diagnostic fragments indicating Δ9 and Δ11 DBs.

13.4.10 Lipidomics

Lipidomic analysis is quickly becoming the method of choice for analysis of TGs and other lipid classes. However, unfortunately, the current state of lipidomics is likened to "the wild west" of American history. As of the time of this writing, there are >1700 references in Scopus for "lipidomics AND (triacylglycerol* OR triglyceride*)." Lipidomics software seems to take the burden of extensive knowledge off the

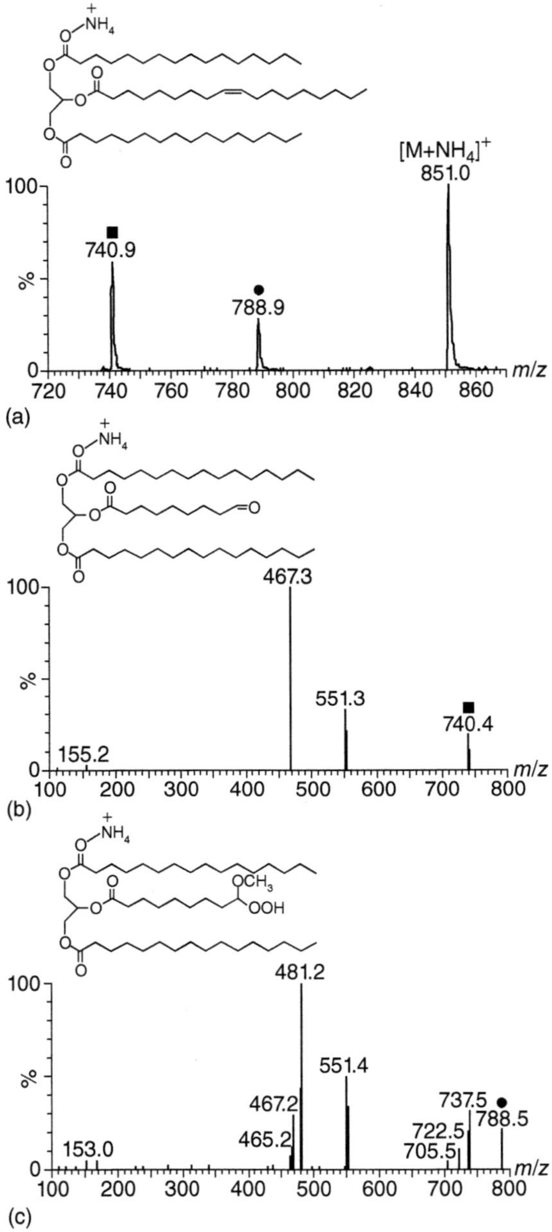

Figure 13.10 (a) OzESI-MS spectrum of a 1 μM methanolic solution of (16:0/9*Z*-18:1/16:0)-TG with 5 mM ammonium acetate recorded as a precursor ion scan of *m/z* 551. (b) The MS/MS spectrum of the OzESI-MS product ion at *m/z* 740.9. (c) The MS/MS spectrum of the OzESI-MS product ion at *m/z* 788.9. The ● and ■ symbols identify the ozonolysis product ions as α-methoxyhydroperoxides and aldehydes, respectively. Source: Reprinted with permission from Thomas et al. [135]. Copyright 2007 American Chemical Society.

researchers employing the software and allows those with little to no experience to quickly produce and report results of lipid compositional analysis. Unfortunately, users trust the software to provide definitive answers when there is much more uncertainty present than many users realize. Kofeler and a large group of lipidomics researchers [140] recently published an article demonstrating the need for improvements in standardization in lipidomics data.

The need for standardization of lipidomics data is becoming a popular topic of discussion, and those associated with the LIPID MAPS consortium have recently updated their guidance for nomenclature to be used for lipid annotation [141]. Table 13.2 shows the latest nomenclature suggested for glycerolipids. One interlaboratory study [142] was designed to provide data regarding the lipidomic analysis of National Institute of Standards and Technology (NIST) standard reference material (SRM) 1950 – human plasma.

Lipidomics can be divided into two main categories: (i) shotgun lipidomics, or lipidomics by infusion without prior separation, and (ii) lipidomics after separation. As mentioned in the above section, Dr. Xianlin Han pioneered the use of shotgun lipidomics and has continued to pursue and advance the field. Numerous reviews and tutorials on shotgun lipidomics by Han et al. [143–152] have summarized the lessoned learned from years of infusion-based lipid analysis.

Lipidomics has now become so widespread that there are reviews for lipidomics applied to numerous specialized areas of biology and tissue or sample types. In fact, the number of reviews of lipidomics is so large that they cannot be listed here because the citations take many pages, so only a few representative recent reviews can be cited, such as those that have focused on pediatric non-alcoholic fatty liver disease [153]; skin and sebum [154]; semen analysis [155]; musculoskeletal disorders [156]; brain lipidomics [157]; Alzheimer's disease [158, 159]; schizophrenia [160, 161]; drug development [162, 163]; adverse drug reactions [164]; cardiovascular disease [165]; obesity, diabetes, and liver function [166]; fish and aquaculture [167, 168]; environmental analysis [169]; foods and "foodomics" [170]; and other areas.

Among the many issues, those in the field of lipidomics are grappling with is the wide range of software packages, both proprietary and open source, available for lipidomics. Most MS instrument manufacturers have developed their own lipidomics platforms, including (i) ThermoScientific – LipidSearch and Compound Discoverer; (ii) AB Sciex – Lipidizer workflow manager and LipidView; and (iii) Agilent – MassHunter with Lipid Anotator. Multiple third-party and open-source software programs are also available, including (i) MS-DIAL [171]; (ii) Lipid Data Analyzer (LDA) [172]; (iii) Lipid Miner; (iv) Lipid Xplorer [173]; (v) LIQUID (Lipid Quantification and Identification) [174]; (vi) Lipid Hunter [175]; (vii) LipidQA [176]; and others, such as those listed at The Feihn Lab website on lipid analysis: https://fiehnlab.ucdavis.edu/staff/kind/metabolomics/lipidanalysis. It is hard to determine which software is most effective for lipid analysis. Often, it is difficult to tell in advance which software package is most effective and user-friendly. It is a process of trial-and-error, with substantial learning curves involved in the initial trials to see if the software is useful.

Table 13.2 Examples of shorthand notation of glycerolipids from the LIPID MAPS consortium.

Bond type	Species level[a]	Molecular species level[b]	*sn*-Position level[c]	Full structure level[d]
Acyl	MG 18:0	MG 18:0	MG 0:0/18:0/0:0	
Alkyl	MG O-18:0	MG O-18:0	MG 0:0/O-18:0/0:0	
Diacyl	DG 34:1	DG 16:0_18:1	DG 16:0/18:1/ 0:0	DG 16:0/ 18:1(9*Z*)/0:0
Acyl-alkyl	DG O-34:1	DG O-16:0_18:1	DG O-16:0/18:1/0:0	DG O-16:0/ 18:1(9*Z*)/0:0
Dialkyl	DG dO-32:1	DG O-16:0_O-16:1	DG O-16:0/ O-16:1/0:0	DG O-16:0/ O-16:1(9*Z*)/0:0
	DG 30:1[e]			
Triacyl	TG 52:2	TG 16:0_18:1_18:1	TG 16:0/18:1/ 18:1	TG 16:0/18:1(9*Z*)/ 18:1(11*Z*)
		TG 16:0_36:2 (only one acyl chain identified)	TG 16: 0_18:1(*sn*-2)_18:1[f]	
Acyl-alkyl	TG O-52:2	TG O-16:0_18:1_18:1	TG O-16:0/ 18:1/18:1	TG O-16:0/18:1 (9*Z*)/18:1(11*Z*)
	TG 51:2[e]			
Acyl-dialkyl	TG dO-52:2	TG O-18:1_O-16:0_18:1	TG O-18:1/O-16:0/ 18:1	TG O-18:1(9*Z*)/ O-16:0/18:1(9*Z*)
	TG 50:2[e]			
Trialkyl	TG tO-52:2	TG O-18:1_O-16:0_O-18:1	TG O-18:1/ O-16:0/O-18:1	TG O-18:1(9*Z*)/O-16:0/O-18:1(9*Z*)
	TG 49:2[e]			
TG-Estolide	TG 68:3;O_2	TG 18:1_18:1_32:1;O_2	TG 16:0;O(FA16:0)/ 18:1/18:1	TG 16:0;5O(FA16:0)/ 18:1(9i)/18:1(9*Z*)

a) Annotation based on exact mass measurements using a high-resolution mass spectrometer, which allows differentiation of isobaric acyl and alkyl species.
b) Annotation requires MS/MS and detection of FA chain-specific fragments.
c) *sn*-Positions determined by specific analysis such as differential mobility spectrometry or LC separation of isomeric species using silver ions.
d) DB positions determined by independent techniques such as ozonolysis or photochemical derivatization.
e) Annotation using low-resolution MS including the assumption of acyl chains only.
f) Only the acyl chain at the *sn*-2-position is defined.

13.4.11 TG Quantification Using Lipidomics Software

One strength of lipidomics software is the ability to locate and integrate small peaks that might otherwise be overlooked. Lipidomics software can be used to quantify TGs in samples, but some general guidelines should be kept in mind, such as proper selection of internal standards, correction for isotope effects, etc., which have been outlined by Yang and Han [177].

For example, Bird et al. [178] used RP-HPLC-ESI-HRAM-MS for lipidomics with searching of the Metlin, Human Metabolome Database (HMDB), and LIPID MAPS databases to identify 86 TGs in rat serum. In 69 of the TGs, they were able to identify the specific FAs of the TG isomers. A recent review of lipidomic analysis of TGs and PLs in marine species was reported by Yeo and Parrish [167].

13.4.12 Future Directions

One indicator of coming developments in the LC-MS analysis of AGs is what has previously happened in GC analysis of lipids. The same kinds of multi-dimensional analyses that have been reported by GC are being reported for LC analysis of lipids. In addition to GC×GC, parallel second dimensions have been used [179–183], referred to as GC×2GC. Other researchers used parallel multiplexed GC first dimensions that went to a single second dimension [184], referred to as 2GC×GC. Recently, a unique approach has been described that used two parallel first dimensions and two parallel second dimensions [185], referred to as 2GC×2GC. Most experiments employed the FID [179–182, 185], but MS was also used in GC×2GC for detection of one of the parallel second dimensions (GC×(GC-FID+GC-MS)) [183]. MS was also used in the experiments for 2GC×GC-MS [184]. Obviously, LC×LC has already been used extensively for AGs. A recent report by Byrdwell et al. [186] described 3D-LC experiments that may be referred to as LC×2LC as they used parallel second dimensions analogous to those in GC×2GC. This trend is expected to continue, as more complex multi-dimensional LC*x*MS*y* separations are reported.

It is clear that lipidomics will continue its path toward dominance of the field of lipid analysis because of its software-driven ease of use. On the other hand, complex and detailed analysis that does not lend itself to automation has been used extensively to address the remaining challenges regarding lipid structure determination [187]. In this chapter, we have cited research that allows increasing amounts of information to be derived from TGs. There is little doubt that with the right amount of effort and instrumentation, we are at the point where complete structural elucidation of GLs is possible. However, many of these approaches are demanding, in terms of time, expertise, and equipment. Thus, there are two tracks that continue to evolve in lipid analysis: automated lipidomics and detailed manual analysis.

A serious shortcoming for lipidomic analysis remains the inability to analyze APCI- and APPI-MS data. Lipidomics software is designed for use with ESI-MS, in which single adducts are expected for each GL. The $[DAG]^+$ fragments in TG mass spectra are not correctly recognized by most lipidomics software packages. This is unfortunate as APCI- and APPI-MS can offer more reliable options for determining the regioisomeric configurations of AGs from fragment ratios than ESI-MS/MS data. However, their potential cannot be realized without software that allows both adducts and fragments in full-scan spectra.

Acknowledgments

This work was supported by the United States Department of Agriculture (USDA) Agricultural Research Service. Mention or use of specific products or brands do not represent or imply endorsement by the USDA.

References

1 Daubert, B. Lectures of the 1949 short course. *J. Am. Oil Chem. Soc.* 26 (1949): 547–590. https://doi.org/10.1007/BF02822067.
2 Christie, W.W. (2021). A beginner's guide to mass spectrometry of fatty acids. Part 1. Methyl esters, Lipid Web. https://www.lipidmaps.org/resources/lipidweb/lipidweb_html/ms/basics/ms-begin-1/index.htm.
3 Christie, W.W. (2021). A beginner's guide to mass spectrometry of fatty acids. Part 2. 3-Pyridylcarbinol, 4,4-dimethyloxazoline and pyrrolidine derivatives, Lipid Web. https://lipidmaps.org/resources/lipidweb/lipidweb_html/ms/basics/ms-begin-2/index.htm.
4 Christie, W.W. (2021). Introduction to mass spectrometry of fatty acids and these web pages. https://www.lipidmaps.org/resources/lipidweb/lipidweb_html/ms/basics/introduction/index.htm.
5 Christie, W.W. (1998). Gas chromatography-mass spectrometry methods for structural analysis of fatty acids. *Lipids* 33: 343–353. https://doi.org/10.1007/s11745-998-0214-x.
6 Ruiz-Samblás, C., González-Casado, A., and Cuadros-Rodríguez, L. (2015). Triacylglycerols determination by high-temperature gas chromatography in the analysis of vegetable oils and foods: a review of the past 10 years. *Crit. Rev. Food Sci. Nutr.* 55: 1618–1631. https://doi.org/10.1080/10408398.2012.713045.
7 AOAC (1992). Official Method 986.19-1992: Triglycerides in fats and oils. Gas chromatographic method. http://www.aoacofficialmethod.org/index.php?main_page=product_info&cPath=1&products_id=2331.
8 Waktola, H.D., Zeng, A.X., Chin, S.-T., and Marriott, P.J. (2020). Advanced gas chromatography and mass spectrometry technologies for fatty acids and triacylglycerols analysis. *TrAC, Trends Anal. Chem.* 129: https://doi.org/10.1016/j.trac.2020.115957.

9 Mota, M.F.S., Waktola, H.D., Nolvachai, Y., and Marriott, P.J. (2021). Gas chromatography – mass spectrometry for characterisation, assessment of quality and authentication of seed and vegetable oils. *TrAC, Trends Anal. Chem.* 138: https://doi.org/10.1016/j.trac.2021.116238.

10 Bailey, A.E. and Swern, D. (1964). *Bailey's Industrial Oil and Fat Products*, 3e. New York, NY: Wiley.

11 Serdarevich, B. (1967). Glyceride isomerizations in lipid chemistry. *J. Am. Oil Chem. Soc.* 44: 381–393. https://doi.org/10.1007/BF02666775.

12 Liu, Q.-T. and Kinderlerer, J.L. (1999). Preparative thin-layer chromatographic separation and subsequent gas chromatographic-mass spectrometric analysis of monoacylglycerols derived from butter oil by fungal degradation. *J. Chromatogr. A* 855: 617–624. https://doi.org/10.1016/S0021-9673(99)00726-8.

13 Destaillats, F., Cruz-Hernandez, C., Nagy, K., and Dionisi, F. (2010). Identification of monoacylglycerol regio-isomers by gas chromatography-mass spectrometry. *J. Chromatogr. A* 1217: 1543–1548. https://doi.org/10.1016/j.chroma.2010.01.016.

14 Cao, G., Ding, C., Ruan, D. et al. (2019). Gas chromatography-mass spectrometry based profiling reveals six monoglycerides as markers of used cooking oil. *Food Control* 96: 494–498. https://doi.org/10.1016/j.foodcont.2018.10.013.

15 AOCS (2017). Official Method Cd 11b-91: Mono- and diglycerides by capillary gas chromatography. https://www.aocs.org/attain-lab-services/methods/methods/search-results?method=111521.

16 AOAC (1996). Official Method 993.18: Mono- and diglycerides in fats and oils. Gas chromatographic method. http://www.aoacofficialmethod.org/index.php?main_page=product_info&cPath=1&products_id=353.

17 Holčapek, M., Jandera, P., Fischer, J., and Prokeš, B. (1999). Analytical monitoring of the production of biodiesel by high-performance liquid chromatography with various detection methods. *J. Chromatogr. A* 858: 13–31. https://doi.org/10.1016/S0021-9673(99)00790-6.

18 Yang, Z., Hollebone, B.P., Wang, Z. et al. (2011). Determination of polar impurities in biodiesels using solid-phase extraction and gas chromatography-mass spectrometry. *J. Sep. Sci.* 34: 409–421. https://doi.org/10.1002/jssc.201000533.

19 Poursharifi, P., Madiraju, S.R.M., and Prentki, M. (2017). Monoacylglycerol signalling and ABHD6 in health and disease. *Diabetes Obes. Metab.* 19: 76–89. https://doi.org/10.1111/dom.13008.

20 Hansen, H.S., Rosenkilde, M.M., Holst, J.J., and Schwartz, T.W. (2012). GPR119 as a fat sensor. *Trends Pharmacol. Sci.* 33: 374–381. https://doi.org/10.1016/j.tips.2012.03.014.

21 Ochiai, K., Hirooka, R., Sakaino, M. et al. (2021). 2-Arachidonoyl glycerol potently induces cholecystokinin secretion in murine enteroendocrine STC-1 cells via cannabinoid receptor CB1. *Lipids* 1–9. https://doi.org/10.1002/lipd.12323.

22 Indrasti, D., Che Man, Y.B., Chin, S.T. et al. (2010). Regiospecific analysis of mono- and diglycerides in glycerolysis products by GC × GC-TOF-MS, JAOCS. *J. Am. Oil Chem. Soc.* 87: 1255–1262. https://doi.org/10.1007/s11746-010-1614-x.

23 Nina Naquiah, A.N., Marikkar, J.M.N., and Shuhaimi, M. (2016). Differentiation of partial acylglycerols derived from different animal fats by EA-IRMS and GCMS techniques. *Grasas Aceites* 67: https://doi.org/10.3989/gya.0757152.

24 Cai, S.S. and Syage, J.A. (2006). Comparison of atmospheric pressure photoionization, atmospheric pressure chemical ionization, and electrospray ionization mass spectrometry for analysis of lipids. *Anal. Chem.* 78: 1191–1199. https://doi.org/10.1021/ac0515834.

25 Cai, S.S. and Syage, J.A. (2006). Atmospheric pressure photoionization mass spectrometry for analysis of fatty acid and acylglycerol lipids. *J. Chromatogr. A* 1110: 15–26. https://doi.org/10.1016/j.chroma.2006.01.050.

26 Cai, S.S., Short, L.C., Syage, J.A. et al. (2007). Liquid chromatography-atmospheric pressure photoionization-mass spectrometry analysis of triacylglycerol lipids – effects of mobile phases on sensitivity. *J. Chromatogr. A* 1173: 88–97. https://doi.org/10.1016/j.chroma.2007.10.008.

27 Subroto, E. (2020). Monoacylglycerols and diacylglycerols for fat-based food products: a review. *Food Res.* 4: 932–943. https://doi.org/10.26656/fr.2017.4(4).398.

28 FDA 2020. 21 CFR 184.1505 Mono- and diglycerides, Washington DC, USA. https://www.accessdata.fda.gov/scripts/cdrh/cfdocs/cfcfr/CFRSearch.cfm?fr=184.1505.

29 Christie, W.W. (2021). Diacylglycerols, Lipid Web. https://lipidmaps.org/resources/lipidweb/lipidweb_html/lipids/simple/dg/index.htm (accessed 21 June 2021).

30 Nagao, T., Watanabe, H., Goto, N. et al. (2000). Dietary diacylglycerol suppresses accumulation of body fat compared to triacylglycerol in men in a double-blind controlled trial. *J. Nutr.* 130: 792–797. https://doi.org/10.1093/jn/130.4.792.

31 Takei, A., Katsuragi, Y., Abe, C. et al. (2001). Effects of dietary α-linolenic acid-rich diacylglycerol on body fat in man (1): lowering effect on body fat. *J. Oleo Sci.* 50: 735–746. https://doi.org/10.5650/jos.50.735.

32 Lee, Y.-Y., Tang, T.-K., Phuah, E.-T. et al. (2020). Production, safety, health effects and applications of diacylglycerol functional oil in food systems: a review. *Crit. Rev. Food Sci. Nutr.* 60: 2509–2525. https://doi.org/10.1080/10408398.2019.1650001.

33 Satouchi, K. and Saito, K. (1977). Studies on trimethylsilyl derivatives of 1-alk-1-enyl-2-acyl glycerols by gas-liquid chromatography mass spectrometry. *Biol. Mass Spectrom.* 4: 107–112. https://doi.org/10.1002/bms.1200040210.

34 Curstedt, T. (1977). Analysis of molecular species of ether analogues of phosphatidylcholines from biological samples. *Biochim. Biophys. Acta* 489: 79–88. https://doi.org/10.1016/0005-2760(77)90234-X.

35 Myher, J.J. and Kuksis, A. (1989). Relative gas-liquid chromatographic retention factors of trimethylsilyl ethers of diradylglycerols on polar capillary columns. *J. Chromatogr. A* 471: 187–204. https://doi.org/10.1016/S0021-9673(00)94167-0.

36 Myher, J.J., Kuksis, A., Shepherd, J. et al. (1981). Effect of saturated and unsaturated fat diets on molecular species of phosphatidylcholine and sphingomyelin of human plasma lipoproteins. *Biochim. Biophys. Acta, Lipids Lipid Metab.* 666: 110–119. https://doi.org/10.1016/0005-2760(81)90096-5.

37 Dickens, B.F., Ramesha, C.S., and Thompson, G.A. Jr. (1982). Quantification of phospholipid molecular species by coupled gas chromatography-mass spectrometry of deuterated samples. *Anal. Biochem.* 127: 37–48. https://doi.org/10.1016/0003-2697(82)90141-5.

38 Harvey, D.J. (1991). Nicotinylidene derivatives for the structural elucidation of glycerol mono-ethers and mono-esters by gas chromatography/mass spectrometry. *Biol. Mass Spectrom.* 20: 87–93. https://doi.org/10.1002/bms.1200200207.

39 Falardeau, P., Robillard, M., and Hui, R. (1993). Quantification of diacylglycerols by capillary gas chromatography-negative ion chemical ionization-mass spectrometry. *Anal. Biochem.* 208: 311–316. https://doi.org/10.1006/abio.1993.1052.

40 Dasgupta, A. and Banerjee, P. (1993). Microwave induced rapid preparation of acetyl, trifluoroacetyl and *tert*-butyl dimethylsilyl derivatives of fatty alcohols and diacylglycerols for gas chromatography-mass spectrometric analysis. *Chem. Phys. Lipids* 65: 217–224. https://doi.org/10.1016/0009-3084(93)90019-Y.

41 Zhu, H., Clegg, M.S., Shoemaker, C.F., and Wang, S.C. (2013). Characterization of diacylglycerol isomers in edible oils using gas chromatography-ion trap electron ionization mass spectrometry. *J. Chromatogr. A* 1304: 194–202. https://doi.org/10.1016/j.chroma.2013.06.058.

42 Cai, T., Shu, Q., Hou, J. et al. (2015). Profiling and relative quantitation of phosphoinositides by multiple precursor ion scanning based on phosphate methylation and isotopic labeling. *Anal. Chem.* 87: 513–521. https://doi.org/10.1021/ac503224j.

43 Lee, J.Y., Park, N.H., Lee, W. et al. (2016). Comprehensive chemical profiling of Pinellia species tuber and processed *Pinellia tuber* by gas chromatography–mass spectrometry and liquid chromatography–atmospheric pressure chemical ionization–tandem mass spectrometry. *J. Chromatogr. A* 1471: 164–177. https://doi.org/10.1016/j.chroma.2016.10.033.

44 Itabashi, Y., Marai, L., and Kuksis, A. (1991). Identification of natural diacylglycerols as the 3,5-dinitrophenylurethanes by chiral phase liquid chromatography with mass spectrometry. *Lipids* 26: 951–956. https://doi.org/10.1007/BF02535983.

45 Marai, L., Kuksis, A., Myher, J.J., and Itabashi, Y. (1992). Liquid chromatography chloride attachment negative chemical ionization mass spectrometry of diacylglycerol dinitrophenylurethanes. *Biol. Mass Spectrom.* 21: 541–547. https://doi.org/10.1002/bms.1200211104.

46 Takano, H. and Itabashi, Y. (2002). Molecular species analysis of 1,3-diacylglycerols in edible oils by HPLC/ESI-MS. *Bunseki Kagaku* 51: 437–442. https://doi.org/10.2116/bunsekikagaku.51.437.

47 Hashidate, T. and Itabashi, Y. (2005). Electrospray ionization mass spectrometry of regioisomeric 1,2-diacylglycerols. *Bunseki Kagaku* 54: 807–816. https://doi.org/10.2116/bunsekikagaku.54.807.

48 Honda, M., Kashima, A., Takahashi, K., and Itabashi, Y. (2009). Molecular species of digalactosyldiacylglycerols in the brown alga *Sargassum yezoense*. *Nippon Suisan Gakkaishi* 75: 1061–1069. https://doi.org/10.2331/suisan.75.1061.

49 Karas, M., Bachmann, D., Bahr, U., and Hillenkamp, F. (1987). Matrix-assisted ultraviolet laser desorption of non-volatile compounds. *Int. J. Mass Spectrom. Ion Processes* 78: 53–68. https://doi.org/10.1016/0168-1176(87)87041-6.

50 Marto, J.A., White, F.M., Seldomrldge, S., and Marshall, A.G. (1995). Structural characterization of phospholipids by matrix-assisted laser desorption/ionization Fourier transform ion cyclotron resonance mass spectrometry. *Anal. Chem.* 67: 3979–3984. https://doi.org/10.1021/ac00117a025.

51 Zöllner, P., Schmid, E.R., and Allmaier, G. (1996). K4[Fe(CN)6]/glycerol – a new liquid matrix system for matrix-assisted laser desorption/ionization mass spectrometry of hydrophobic compounds. *Rapid Commun. Mass Spectrom.* 10: 1278–1282. https://doi.org/10.1002/(SICI)1097-0231(19960731)10:10<1278::AID-RCM661>3.0.CO;2-L.

52 Schiller, J., Arnhold, J., Benard, S. et al. (1999). Lipid analysis by matrix-assisted laser desorption and ionization mass spectrometry: a methodological approach. *Anal. Biochem.* 267: 46–56. https://doi.org/10.1006/abio.1998.3001.

53 Benard, S., Arnhold, J., Lehnert, M. et al. (1999). Experiments towards quantification of saturated and polyunsaturated diacylglycerols by matrix-assisted laser desorption and ionization time-of-flight mass spectrometry. *Chem. Phys. Lipids* 100: 115–125. https://doi.org/10.1016/S0009-3084(99)00045-6.

54 Asbury, G.R., Al-Saad, K., Siems, W.F. et al. (1999). Analysis of triacylglycerols and whole oils by matrix-assisted laser desorption/ionization time of flight mass spectrometry. *J. Am. Soc. Mass Spectrom.* 10: 983–991. https://doi.org/10.1016/S1044-0305(99)00063-X.

55 Schiller, J., Süß, R., Arnhold, J. et al. (2004). Matrix-assisted laser desorption and ionization time-of-flight (MALDI-TOF) mass spectrometry in lipid and phospholipid research. *Prog. Lipid Res.* 43: 449–488. https://doi.org/10.1016/j.plipres.2004.08.001.

56 Schiller, J., Suss, R., Fuchs, B. et al. (2007). MALDI-TOF MS in lipidomics. *Front. Biosci.* 12: 2568–2579. https://doi.org/10.2741/2255.

57 Fuchs, B., Süß, R., and Schiller, J. (2010). An update of MALDI-TOF mass spectrometry in lipid research. *Prog. Lipid Res.* 49: 450–475. https://doi.org/10.1016/j.plipres.2010.07.001.

58 Zemski Berry, K.A., Hankin, J.A., Barkley, R.M. et al. (2011). MALDI imaging of lipid biochemistry in tissues by mass spectrometry. *Chem. Rev.* 111: 6491–6512. https://doi.org/10.1021/cr200280p.

59 Fuchs, B., Bresler, K., and Schiller, J. (2011). Oxidative changes of lipids monitored by MALDI MS. *Chem. Phys. Lipids* 164: 782–795. https://doi.org/10.1016/j.chemphyslip.2011.09.006.

60 Fuchs, B. (2012). Analysis of phospolipids and glycolipids by thin-layer chromatography-matrix-assisted laser desorption and ionization mass spectrometry. *J. Chromatogr. A* 1259: 62–73. https://doi.org/10.1016/j.chroma.2012.03.068.

61 Leopold, J., Popkova, Y., Engel, K.M., and Schiller, J. (2018). Recent developments of useful MALDI matrices for the mass spectrometric characterization of lipids. *Biomolecules* 8: https://doi.org/10.3390/biom8040173.

62 Murphy, R.C., Fiedler, J., and Hevko, J. (2001). Analysis of nonvolatile lipids by mass spectrometry. *Chem. Rev.* 101: 479–526. https://doi.org/10.1021/cr9900883.

63 Lee, S.-Y., Kim, J.R., Ha, M.-Y. et al. (2013). Measurements of diacylglycerols in skeletal muscle by atmospheric pressure chemical ionization mass spectrometry. *Lipids* 48: 287–296. https://doi.org/10.1007/s11745-013-3766-6.

64 Duffin, K.L., Henion, J.D., and Shieh, J.J. (1991). Electrospray and tandem mass spectrometric characterization of acylglycerol mixtures that are dissolved in nonpolar solvents. *Anal. Chem.* 63: 1781–1788. https://doi.org/10.1021/ac00017a023.

65 Rohwedder, W.K. (1985). Mass spectrometry of lipids labeled with stable isotopes. *Prog. Lipid Res.* 24: 1–18. https://doi.org/10.1016/0163-7827(85)90006-2.

66 Myher, J.J. and Kuksis, A. (1995). General strategies in chromatographic analysis of lipids. *J. Chromatogr. B Biomed. Sci. Appl.* 671: 3–33. https://doi.org/10.1016/0378-4347(95)00178-L.

67 Palyzová, A. and Řezanka, T. (2021). Separation and identification of diacylglycerols containing branched chain fatty acids by liquid chromatography-mass spectrometry. *J. Chromatogr. A* 1635: https://doi.org/10.1016/j.chroma.2020.461708.

68 He, C., Cao, J., Bao, Y. et al. (2021). Characterization of lipid profiling in three parts (muscle, head and viscera) of tilapia (*Oreochromis niloticus*) using lipidomics with UPLC-ESI-Q-TOF-MS. *Food Chem.* 347: https://doi.org/10.1016/j.foodchem.2021.129057.

69 Fuchs, B. and Schiller, J. (2009). Application of MALDI-TOF mass spectrometry in lipidomics. *Eur. J. Lipid Sci. Technol.* 111: 83–98. https://doi.org/10.1002/ejlt.200800223.

70 Murphy, R.C., Leiker, T.J., and Barkley, R.M. (2011). Glycerolipid and cholesterol ester analyses in biological samples by mass spectrometry. *Biochim. Biophys. Acta, Mol. Cell. Biol. Lipids* 1811: 776–783. https://doi.org/10.1016/j.bbalip.2011.06.019.

71 Cacciola, F., Donato, P., Beccaria, M. et al. (2012). Advances in LC-MS for food analysis. *LC GC Eur.* 25: 15–24.

72 Bosque-Sendra, J.M., Cuadros-Rodríguez, L., Ruiz-Samblás, C., and de la Mata, A.P. (2012). Combining chromatography and chemometrics for the characterization and authentication of fats and oils from triacylglycerol compositional data – a review. *Anal. Chim. Acta* 724: 1–11. https://doi.org/10.1016/j.aca.2012.02.041.

73 Xu, S.-L., Wei, F., Xie, Y. et al. (2018). Research advances based on mass spectrometry for profiling of triacylglycerols in oils and fats and their applications. *Electrophoresis* 39: 1558–1568. https://doi.org/10.1002/elps.201700481.

74 Byrdwell, W.C. and Neff, W.E. (1997). Qualitative and quantitative analysis of triacylglycerols using atmospheric pressure chemical ionization mass spectrometry. In: *New Techniques and Applications in Lipid Analysis* (ed. R.E. McDonald and M.M. Mossoba), 45–80. Champaign, IL: AOCS Press.

75 Byrdwell, W.C. (2001). Atmospheric pressure chemical ionization mass spectrometry for analysis of lipids. *Lipids* 36: 327–346. https://doi.org/10.1007/s11745-001-0725-5.

76 Byrdwell, W.C. (2003). APCI-MS in lipid analysis. In: *Advances in Lipid Methodology – Five* (ed. R.O. Adlof), 171–253. Bridgwater, England: The Oily Press https://doi.org/10.1016/B978-0-9531949-6-4.50010-6.

77 Byrdwell, W.C. (2005). Liquid chromatography/atmospheric pressure chemical ionization mass spectrometry for lipid analysis. In: *HPLC Acyl Lipids* (ed. J.-T. Lin and T.A. McKeon), 269–314. New York, NY: HNB Publishing.

78 Byrdwell, W.C. (2005). Liquid chromatography with electrospray ionization mass spectrometry for lipid analysis. In: *HPLC Acyl Lipids* (ed. J.-T. Lin and T.A. McKeon), 315–371. New York, NY: HNB Publishing.

79 Byrdwell, W.C. (2005). Qualitative and quantitative analysis of triacylglycerols by atmospheric pressure ionization (APCI and ESI) mass spectrometry techniques. In: *Modern Methods for Lipid Analysis by Liquid Chromatography/Mass Spectrometry and Related Techniques* (ed. W.C. Byrdwell), 298–412. Champaign, IL: AOCS Press.

80 Byrdwell, W.C. (2020). Analysis of lipids: triacylglycerols, phospholipids, fatty acids and others. In: *Bailey's Industrial Oil and Fat Products*, 7e (ed. F. Shahidi), 293–336. Wiley.

81 Ulmer, C.Z., Koelmel, J.P., Jones, C.M. et al. (2021). A review of efforts to improve lipid stability during sample preparation and standardization efforts to ensure accuracy in the reporting of lipid measurements. *Lipids* 56: 3–16. https://doi.org/10.1002/lipd.12263.

82 Tyrefors, L.N., Moulder, R.X., and Markides, K.E. (1993). Interface for open tubular column supercritical fluid chromatography/atmospheric pressure chemical ionization mass spectrometry. *Anal. Chem.* 65: 2835–2840. https://doi.org/10.1021/ac00068a021.

83 Byrdwell, W.C. and Emken, E.A. (1995). Analysis of triglycerides using atmospheric pressure chemical ionization mass spectrometry. *Lipids* 30: 173–175. https://doi.org/10.1007/BF02538272.

84 Neff, W.E. and Byrdwell, W.C. (1995). Triacylglycerol analysis by high performance liquid chromatography-atmospheric pressure chemical ionization mass spectrometry: *Crepis alpina* and *Vernonia galamensis* seed oils. *J. Liq. Chromatogr.* 18: 4165–4181. https://doi.org/10.1080/10826079508013753.

85 Neff, W.E. and Byrdwell, W.C. (1995). Soybean oil triacylglycerol analysis by reversed-phase high-performance liquid chromatography coupled with atmospheric pressure chemical ionization mass spectrometry. *J. Am. Oil Chem. Soc.* 72: 1185–1191. https://doi.org/10.1007/BF02540986.

86 Byrdwell, W.C., Emken, E.A., Neff, W.E., and Adlof, R.O. (1996). Quantitative analysis of triglycerides using atmospheric pressure chemical ionization-mass spectrometry. *Lipids* 31: 919–935. https://doi.org/10.1007/BF02522685.

87 Laakso, P. and Voutilainen, P. (1996). Analysis of triacylglycerols by silver-ion high-performance liquid chromatography-atmospheric pressure chemical ionization mass spectrometry. *Lipids* 31: 1311–1322. https://doi.org/10.1007/BF02587918.

88 Mottram, H.R. and Evershed, R.P. (1996). Structure analysis of triacylglycerol positional isomers using atmospheric pressure chemical ionisation mass spectrometry. *Tetrahedron Lett.* 37: 8593–8596. https://doi.org/10.1016/0040-4039(96)01964-8.

89 Mottram, H.R., Woodbury, S.E., and Evershed, R.P. (1997). Identification of triacylglycerol positional isomers present in vegetable oils by high performance liquid chromatography/atmospheric pressure chemical ionization mass spectrometry. *Rapid Commun. Mass Spectrom.* 11: 1240–1252. https://doi.org/10.1002/(SICI)1097-0231(199708)11:12<1240::AID-RCM990>3.0.CO;2-5.

90 Jakab, A., Jablonkai, I., and Forgács, E. (2003). Quantification of the ratio of positional isomer dilinoleoyl-oleoyl glycerols in vegetable oils. *Rapid Commun. Mass Spectrom.* 17: 2295–2302. https://doi.org/10.1002/rcm.1193.

91 Fauconnot, L., Hau, J., Aeschlimann, J.M. et al. (2004). Quantitative analysis of triacylglycerol regioisomers in fats and oils using reversed-phase high-performance liquid chromatography and atmospheric pressure chemical ionization mass spectrometry. *Rapid Commun. Mass Spectrom.* 18: 218–224. https://doi.org/10.1002/rcm.1317.

92 Byrdwell, W.C. (2005). The Bottom-Up Solution to the triacylglycerol lipidome using atmospheric pressure chemical ionization mass spectrometry. *Lipids* 40: 383–417. https://doi.org/10.1007/s11745-006-1398-9.

93 Holčapek, M., Dvorakova, H., Lísa, M. et al. (2010). Regioisomeric analysis of triacylglycerols using silver-ion liquid chromatography-atmospheric pressure chemical ionization mass spectrometry: comparison of five different mass analyzers. *J. Chromatogr. A* 1217: 8186–8194. https://doi.org/10.1016/j.chroma.2010.10.064.

94 Byrdwell, W.C. (2015). The Updated Bottom Up Solution applied to mass spectrometry of soybean oil in a dietary supplement gelcap. *Anal. Bioanal. Chem.* 407: 5143–5160. https://doi.org/10.1007/s00216-015-8590-9.

95 Byrdwell, W.C. (2015). The Updated Bottom Up Solution applied to atmospheric pressure photoionization and electrospray ionization mass spectrometry. *J. Am. Oil Chem. Soc.* 92: 1533–1547. https://doi.org/10.1007/s11746-015-2735-z.

96 Byrdwell, W.C. and Neff, W.E. (2002). Dual parallel electrospray ionization and atmospheric pressure chemical ionization mass spectrometry (MS), MS/MS and MS/MS/MS for the analysis of triacylglycerols and triacylglycerol oxidation products. *Rapid Commun. Mass Spectrom.* 16: 300–319. https://doi.org/10.1002/rcm.581.

97 Currie, G.J. and Kallio, H. (1993). Triacylglycerols of human milk: rapid analysis by ammonia negative ion tandem mass spectrometry. *Lipids* 28: 217–222. https://doi.org/10.1007/BF02536642.

98 Kallio, H. and Rua, P. (1994). Distribution of the major fatty acids of human milk between *sn*-2 and *sn*-1,3 positions of triacylglycerols. *J. Am. Oil Chem. Soc.* 71: 985–992. https://doi.org/10.1007/BF02542266.

99 Leskinen, H.M., Suomela, J.P., and Kallio, H.P. (2010). Quantification of triacylglycerol regioisomers by ultra-high-performance liquid chromatography and ammonia negative ion atmospheric pressure chemical ionization tandem mass spectrometry. *Rapid Commun. Mass Spectrom.* 24: 1–5. https://doi.org/10.1002/rcm.4346.

100 Holčapek, M., Lísa, M., Jandera, P., and Kabátová, N. (2005). Quantitation of triacylglycerols in plant oils using HPLC with APCI-MS, evaporative light-scattering, and UV detection. *J. Sep. Sci.* 28: 1315–1333. https://doi.org/10.1002/jssc.200500088.

101 Fasciotti, M. and Pereira Netto, A.D. (2010). Optimization and application of methods of triacylglycerol evaluation for characterization of olive oil adulteration by soybean oil with HPLC-APCI-MS-MS. *Talanta* 81: 1116–1125. https://doi.org/10.1016/j.talanta.2010.02.006.

102 Řezanka, T., Kolouchová, I., Nedbalová, L., and Sigler, K. (2018). Enantiomeric separation of triacylglycerols containing very long chain fatty acids. *J. Chromatogr. A* 1557: 9–19. https://doi.org/10.1016/j.chroma.2018.04.064.

103 Řezanka, T. and Sigler, K. (2014). Separation of enantiomeric triacylglycerols by chiral-phase HPLC. *Lipids* 49: 1251–1260. https://doi.org/10.1007/s11745-014-3959-7.

104 Baiocchi, C., Medana, C., Dal Bello, F. et al. (2015). Analysis of regioisomers of polyunsaturated triacylglycerols in marine matrices by HPLC/HRMS. *Food Chem.* 166: 551–560. https://doi.org/10.1016/j.foodchem.2014.06.067.

105 Zhang, H., Zhao, H., Zhang, Y. et al. (2018). Characterization of positional distribution of fatty acids and triacylglycerol molecular compositions of marine fish oils rich in omega-3 polyunsaturated fatty acids. *Biomed Res. Int.* 2018: https://doi.org/10.1155/2018/3529682.

106 Xu, Y. and Brenna, J.T. (2007). Atmospheric pressure covalent adduct chemical ionization tandem mass spectrometry for double bond localization in monoene- and diene-containing triacylglycerols. *Anal. Chem.* 79: 2525–2536. https://doi.org/10.1021/ac062055a.

107 Han, X. and Gross, R.W. (2001). Quantitative analysis and molecular species fingerprinting of triacylglyceride molecular species directly from lipid extracts of biological samples by electrospray ionization tandem mass spectrometry. *Anal. Biochem.* 295: 88–100. https://doi.org/10.1006/abio.2001.5178.

108 Han, X. and Gross, R.W. (2003). Global analyses of cellular lipidomes directly from crude extracts of biological samples by ESI mass spectrometry: a bridge to lipidomics. *J. Lipid Res.* 44: 1071–1079. https://doi.org/10.1194/jlr.R300004-JLR200.

109 Han, X. and Gross, R.W. (2005). Shotgun lipidomics: electrospray ionization mass spectrometric analysis and quantitation of cellular lipidomes directly from crude extracts of biological samples. *Mass Spectrom. Rev.* 24: 367–412. https://doi.org/10.1002/mas.20023.

110 Li, M., Baughman, E., Roth, M.R. et al. (2014). Quantitative profiling and pattern analysis of triacylglycerol species in Arabidopsis seeds by electrospray ionization mass spectrometry. *Plant J.* 77: 160–172. https://doi.org/10.1111/tpj.12365.

111 Yeo, J. and Parrish, C.C. (2020). Evaluation of triacylglycerol (TAG) profiles and their contents in salmon muscle tissue using ESI-MS/MS spectrometry with multiple neutral loss scans. *Food Chem.* 324: https://doi.org/10.1016/j.foodchem.2020.126816.

112 Zhang, J., Tao, N., Zhao, Y. et al. (2019). Comparison of the fatty acid and triglyceride profiles of big eye tuna (*Thunnus obesus*), Atlantic salmon (*Salmo salar*) and bighead carp (*Aristichthysnobilis*) heads. *Molecules* 24: https://doi.org/10.3390/molecules24213983.

113 Eum, J.Y., Lee, J.C., Yi, S.S. et al. (2020). Aging-related lipidomic changes in mouse serum, kidney, and heart by nanoflow ultrahigh-performance liquid chromatography-tandem mass spectrometry. *J. Chromatogr. A* 1618: https://doi.org/10.1016/j.chroma.2020.460849.

114 Hsu, F.F. and Turk, J. (1999). Structural characterization of triacylglycerols as lithiated adducts by electrospray ionization mass spectrometry using low-energy collisionally activated dissociation on a triple stage quadrupole instrument. *J. Am. Soc. Mass Spectrom.* 10: 587–599. https://doi.org/10.1016/S1044-0305(99)00035-5.

115 Hvattum, E. (2001). Analysis of triacylglycerols with non-aqueous reversed-phase liquid chromatography and positive ion electrospray tandem mass spectrometry. *Rapid Commun. Mass Spectrom.* 15: 187–190. https://doi.org/10.1002/1097-0231(20010215)15:3<187::AID-RCM211>3.0.CO;2-T.

116 Marzilli, L.A., Fay, L.B., Dionisi, F., and Vouros, P. (2003). Structural characterization of triacylglycerols using electrospray ionization-MSn ion-trap MS. *J.Am.Oil Chem.Soc.* 80: 195–202. https://doi.org/10.1007/s11746-003-0676-2.

117 Lin, J.T. and Arcinas, A. (2008). Analysis of regiospecific triacylglycerols by electrospray ionization-mass spectrometry(3) of lithiated adducts. *J. Agric. Food Chem.* 56: 4909–4915. https://doi.org/10.1021/jf072837k.

118 Hsu, F.F. and Turk, J. (2010). Electrospray ionization multiple-stage linear ion-trap mass spectrometry for structural elucidation of triacylglycerols: assignment of fatty acyl groups on the glycerol backbone and location of double bonds. *J. Am. Soc. Mass Spectrom.* 21: 657–669. https://doi.org/10.1016/j.jasms.2010.01.007.

119 Cubero Herrera, L., Ramaley, L., Potvin, M.A., and Melanson, J.E. (2013). A method for determining regioisomer abundances of polyunsaturated triacylglycerols in omega-3 enriched fish oils using reversed-phase liquid chromatography and triple-stage mass spectrometry. *Food Chem.* 139: 655–662. https://doi.org/10.1016/j.foodchem.2012.12.059.

120 Ramaley, L., Herrera, L.C., and Melanson, J.E. (2013). Applicability of non-linear versus linear fractional abundance calibration plots for the quantitative determination of triacylglycerol regioisomers by tandem mass spectrometry. *Rapid Commun. Mass Spectrom.* 27: 1251–1259. https://doi.org/10.1002/rcm.6569.

121 Leskinen, H., Suomela, J.P., and Kallio, H. (2007). Quantification of triacylglycerol regioisomers in oils and fat using different mass spectrometric and liquid chromatographic methods. *Rapid Commun. Mass Spectrom.* 21: 2361–2373. https://doi.org/10.1002/rcm.3090.

122 Cheng, C., Gross, M.L., and Pittenauer, E. (1998). Complete structural elucidation of triacylglycerols by tandem sector mass spectrometry. *Anal. Chem.* 70: 4417–4426. https://doi.org/10.1021/ac9805192.

123 Kalo, P., Kemppinen, A., Ollilainen, V., and Kuksis, A. (2004). Regiospecific determination of short-chain triacylglycerols in butterfat by normal-phase HPLC with on-line electrospray - tandem mass spectrometry. *Lipids* 39: 915–928. https://doi.org/10.1007/s11745-004-1314-3.

124 Liu, Z. and Rochfort, S. (2021). Bovine milk triacylglycerol regioisomer ratio shows remarkable inter-breed and inter-cow variation. *Molecules* 26: https://doi.org/10.3390/molecules26133938.

125 Lévêque, N.L., Héron, S., and Tchapla, A. (2010). Regioisomer characterization of triacylglycerols by non-aqueous reversed-phase liquid chromatography/electrospray ionization mass spectrometry using silver nitrate as a post column reagent. *J. Mass Spectrom.* 45: 284–296. https://doi.org/10.1002/jms.1713.

126 Baba, T., Campbell, J.L., Le Blanc, J.C.Y., and Baker, P.R.S. (2016). Structural identification of triacylglycerol isomers using electron impact excitation of ions from organics (EIEIO). *J. Lipid Res.* 57: 2015–2027. https://doi.org/10.1194/jlr.M070177.

127 Creaser, C.S., Griffiths, J.M.R., Bramwell, C.J. et al. (2004). Ion mobility spectrometry: a review. Part 1. Structural analysis by mobility measurement. *Analyst* 129: 984–994. https://doi.org/10.1039/b404531a.

128 Cumeras, R., Figueras, E., Davis, C.E. et al. (2015). Review on ion mobility spectrometry. Part 1: current instrumentation. *Analyst* 140: 1376–1390. https://doi.org/10.1039/c4an01100g.

129 Cumeras, R., Figueras, E., Davis, C.E. et al. (2015). Review on ion mobility spectrometry. Part 2: hyphenated methods and effects of experimental parameters. *Analyst* 140: 1391–1410. https://doi.org/10.1039/c4an01101e.

130 Kafle, G.K., Khot, L.R., Sankaran, S. et al. (2016). State of ion mobility spectrometry and applications in agriculture: a review. *Eng. Agric. Environ. Food* 9: 346–357. https://doi.org/10.1016/j.eaef.2016.05.004.

131 Alikord, M., Mohammadi, A., Kamankesh, M., and Shariatifar, N. (2021). Food safety and quality assessment: comprehensive review and recent trends in the applications of ion mobility spectrometry (IMS). *Crit. Rev. Food Sci. Nutr.* https://doi.org/10.1080/10408398.2021.1879003.

132 Paglia, G., Kliman, M., Claude, E. et al. (2015). Applications of ion-mobility mass spectrometry for lipid analysis. *Anal. Bioanal. Chem.* 407: 4995–5007. https://doi.org/10.1007/s00216-015-8664-8.

133 Shvartsburg, A.A., Isaac, G., Leveque, N. et al. (2011). Separation and classification of lipids using differential ion mobility spectrometry. *J. Am. Soc. Mass Spectrom.* 22: 1146–1155. https://doi.org/10.1007/s13361-011-0114-z.

134 Šala, M., Lísa, M., Campbell, J.L., and Holčapek, M. (2016). Determination of triacylglycerol regioisomers using differential mobility spectrometry. *Rapid Commun. Mass Spectrom.* 30: 256–264. https://doi.org/10.1002/rcm.7430.

135 Thomas, M.C., Mitchell, T.W., Harman, D.G. et al. (2007). Elucidation of double bond position in unsaturated lipids by ozone electrospray ionization mass spectrometry. *Anal. Chem.* 79: 5013–5022. https://doi.org/10.1021/ac0702185.

136 Thomas, M.C., Mitchell, T.W., Harman, D.G. et al. (2008). Ozone-induced dissociation: elucidation of double bond position within mass-selected lipid ions. *Anal. Chem.* 80: 303–311. https://doi.org/10.1021/ac7017684.

137 Marshall, D.L., Pham, H.T., Bhujel, M. et al. (2016). Sequential collision- and ozone-induced dissociation enables assignment of relative acyl chain position in triacylglycerols. *Anal. Chem.* 88: 2685–2692. https://doi.org/10.1021/acs.analchem.5b04001.

138 Claes, B.S.R., Bowman, A.P., Poad, B.L.J. et al. (2021). Mass spectrometry imaging of lipids with isomer resolution using high-pressure ozone-induced dissociation. *Anal. Chem.* 93: 9826–9834. https://doi.org/10.1021/acs.analchem.1c01377.

139 Franklin, E.T. and Xia, Y. (2020). Structural elucidation of triacylglycerol using online acetone Paternò–Büchi reaction coupled with reversed-phase liquid chromatography mass spectrometry. *Analyst* 145: 6532–6540. https://doi.org/10.1039/d0an01353f.

140 Köfeler, H.C., Eichmann, T.O., Ahrends, R. et al. (2021). Quality control requirements for the correct annotation of lipidomics data. *Nat. Commun.* 12: https://doi.org/10.1038/s41467-021-24984-y.

141 Liebisch, G., Fahy, E., Aoki, J. et al. (2020). Update on LIPID MAPS classification, nomenclature, and shorthand notation for MS-derived lipid structures. *J. Lipid Res.* 61: 1539–1555. https://doi.org/10.1194/jlr.S120001025.

142 Bowden, J.A., Heckert, A., Ulmer, C.Z. et al. (2017). Harmonizing lipidomics: NIST interlaboratory comparison exercise for lipidomics using SRM 1950-metabolites in frozen human plasma. *J. Lipid Res.* 58: 2275–2288. https://doi.org/10.1194/jlr.M079012.

143 Han, X., Yang, K., and Gross, R.W. (2012). Multi-dimensional mass spectrometry-based shotgun lipidomics and novel strategies for lipidomic analyses. *Mass Spectrom. Rev.* 31: 134–178. https://doi.org/10.1002/mas.20342.

144 Han, X. (2016). Lipidomics for studying metabolism. *Nat. Rev. Endocrinol.* 12: 668–679. https://doi.org/10.1038/nrendo.2016.98.

145 Wang, M., Wang, C., Han, R.H., and Han, X. (2016). Novel advances in shotgun lipidomics for biology and medicine. *Prog. Lipid Res.* 61: 83–108. https://doi.org/10.1016/j.plipres.2015.12.002.

146 Yang, K. and Han, X. (2016). Lipidomics: techniques, applications, and outcomes related to biomedical sciences. *Trends Biochem. Sci* 41: 954–969. https://doi.org/10.1016/j.tibs.2016.08.010.

147 Hu, C., Wang, C., He, L., and Han, X. (2019). Novel strategies for enhancing shotgun lipidomics for comprehensive analysis of cellular lipidomes. *TrAC, Trends Anal. Chem.* 120: https://doi.org/10.1016/j.trac.2018.11.028.

148 Wang, J. and Han, X. (2019). Analytical challenges of shotgun lipidomics at different resolution of measurements. *TrAC, Trends Anal. Chem.* 121: https://doi.org/10.1016/j.trac.2019.115697.

149 Wang, J., Wang, C., and Han, X. (2019). Tutorial on lipidomics. *Anal. Chim. Acta* 1061: 28–41. https://doi.org/10.1016/j.aca.2019.01.043.

150 Hu, C., Duan, Q., and Han, X. (2020). Strategies to improve/eliminate the limitations in shotgun lipidomics. *Proteomics* 20: https://doi.org/10.1002/pmic.201900070.

151 Hu, C., Luo, W., Xu, J., and Han, X. (2020). Recognition and avoidance of ion source-generated artifacts in lipidomics analysis. *Mass Spectrom. Rev.* https://doi.org/10.1002/mas.21659.

152 Han, X. and Ye, H. (2021). Overview of lipidomic analysis of triglyceride molecular species in biological lipid extracts. *J. Agric. Food Chem.* 69: 8895–8909. https://doi.org/10.1021/acs.jafc.0c07175.

153 Sessa, A.D., Riccio, S., Pirozzi, E. et al. (2021). Advances in paediatric nonalcoholic fatty liver disease: role of lipidomics. *World J. Gastroenterol.* 27: 3815–3824. https://doi.org/10.3748/wjg.v27.i25.3815.

154 Kendall, A.C., Koszyczarek, M.M., Jones, E.A. et al. (2018). Lipidomics for translational skin research: a primer for the uninitiated. *Exp. Dermatol.* 27: 721–728. https://doi.org/10.1111/exd.13558.

155 Evans, H.C., Dinh, T.T.N., Hardcastle, M.L. et al. (2021). Advancing semen evaluation using lipidomics. *Front. Vet. Sci.* 8: https://doi.org/10.3389/fvets.2021.601794.

156 Mo, C., Du, Y., and O'Connell, T.M. (2021). Applications of lipidomics to age-related musculoskeletal disorders. *Curr. Osteoporos. Rep.* 19: 151–157. https://doi.org/10.1007/s11914-021-00656-0.

157 Chakraborty, A., Kumar Praharaj, S., Krishnananda Prabhu, R.V., and Mukhyaprana Prabhu, M. (2020). Lipidomics and cognitive dysfunction – a narrative review. *Turkish J. Biochem.* 45: https://doi.org/10.1515/tjb-2020-0134.

158 Cuperlovic-Culf, M. and Badhwar, A. (2020). Recent advances from metabolomics and lipidomics application in Alzheimer's disease inspiring drug discovery. *Expert Opin. Drug Discovery* 15: 319–331. https://doi.org/10.1080/17460441.2020.1674808.

159 Wong, M.W., Braidy, N., Poljak, A., and Sachdev, P.S. (2017). The application of lipidomics to biomarker research and pathomechanisms in Alzheimer's disease. *Curr. Opin. Psychiatry.* 30: 136–144. https://doi.org/10.1097/YCO.0000000000000303.

160 Zhuo, C., Hou, W., Tian, H. et al. (2020). Lipidomics of the brain, retina, and biofluids: from the biological landscape to potential clinical application in schizophrenia. *Transl. Psychiatry* 10: https://doi.org/10.1038/s41398-020-01080-1.

161 Sethi, S. and Brietzke, E. (2017). Recent advances in lipidomics: analytical and clinical perspectives. *Prostaglandins Other Lipid Mediat.* 128, 129: 8–16. https://doi.org/10.1016/j.prostaglandins.2016.12.002.

162 Dehairs, J., Derua, R., Rueda-Rincon, N., and Swinnen, J.V. (2015). Lipidomics in drug development. *Drug Discov. Today Technol.* 13: 33–38. https://doi.org/10.1016/j.ddtec.2015.03.002.

163 Koeberle, A. (2016). Target identification and lead discovery by functional lipidomics. *Fut. Med. Chem.* 8: 2169–2171. https://doi.org/10.4155/fmc-2016-0182.

164 Saito, K. (2021). Application of comprehensive lipidomics to biomarker research on adverse drug reactions. *Drug Metab. Pharmacokinet.* 37: https://doi.org/10.1016/j.dmpk.2020.100377.

165 Tabassum, R. and Ripatti, S. (2021). Integrating lipidomics and genomics: emerging tools to understand cardiovascular diseases. *Cell. Mol. Life Sci.* 78: 2565–2584. https://doi.org/10.1007/s00018-020-03715-4.

166 Masoodi, M., Gastaldelli, A., Hyötyläinen, T. et al. (2021). Metabolomics and lipidomics in NAFLD: biomarkers and non-invasive diagnostic tests. *Nat. Rev. Gastroenterol. Hepatol.* https://doi.org/10.1038/s41575-021-00502-9.

167 Yeo, J. and Parrish, C.C. (2021). Mass spectrometry-based lipidomics in the characterization of individual triacylglycerol (TAG) and phospholipid (PL) species from marine sources and their beneficial health effects. *Rev. Fish. Sci. Aquac.* https://doi.org/10.1080/23308249.2021.1897968.

168 Laudicella, V.A., Whitfield, P.D., Carboni, S. et al. (2020). Application of lipidomics in bivalve aquaculture, a review. *Rev. Aquac.* 12: 678–702. https://doi.org/10.1111/raq.12346.

169 Aristizabal-Henao, J.J., Ahmadireskety, A., Griffin, E.K. et al. (2020). Lipidomics and environmental toxicology: recent trends. *Curr. Opin. Environ. Sci. Heal.* 15: 26–31. https://doi.org/10.1016/j.coesh.2020.04.004.

170 Wu, B., Wei, F., Xu, S. et al. (2021). Mass spectrometry-based lipidomics as a powerful platform in foodomics research. *Trends Food Sci. Technol.* 107: 358–376. https://doi.org/10.1016/j.tifs.2020.10.045.

171 Tsugawa, H., Ikeda, K., Takahashi, M. et al. (2020). A lipidome atlas in MS-DIAL 4. *Nat. Biotechnol.* 38: 1159–1163. https://doi.org/10.1038/s41587-020-0531-2.

172 Hartler, J., Triebl, A., Ziegl, A. et al. (2017). Deciphering lipid structures based on platform-independent decision rules. *Nat. Methods* 14: 1171–1174. https://doi.org/10.1038/nmeth.4470.

173 Herzog, R., Schuhmann, K., Schwudke, D. et al. (2012). Lipidxplorer: a software for consensual cross-platform lipidomics. *PLoS One* 7: https://doi.org/10.1371/journal.pone.0029851.

174 Kyle, J.E., Crowell, K.L., Casey, C.P. et al. (2017). LIQUID: an-open source software for identifying lipids in LC-MS/MS-based lipidomics data. *Bioinformatics* 33: 1744–1746. https://doi.org/10.1093/bioinformatics/btx046.

175 Ni, Z., Angelidou, G., Lange, M. et al. (2017). LipidHunter identifies phospholipids by high-throughput processing of LC-MS and shotgun lipidomics datasets. *Anal. Chem.* 89: 8800–8807. https://doi.org/10.1021/acs.analchem.7b01126.

176 Song, H., Hsu, F.-F., Ladenson, J., and Turk, J. (2007). Algorithm for processing raw mass spectrometric data to identify and quantitate complex lipid molecular species in mixtures by data-dependent scanning and fragment ion database searching. *J. Am. Soc. Mass Spectrom.* 18: 1848–1858. https://doi.org/10.1016/j.jasms.2007.07.023.

177 Yang, K. and Han, X. (2011). Accurate quantification of lipid species by electrospray ionization mass spectrometry – meets a key challenge in lipidomics. *Metabolites* 1: 21–40. https://doi.org/10.3390/metabo1010021.

178 Bird, S.S., Marur, V.R., Sniatynski, M.J. et al. (2011). Serum lipidomics profiling using LC-MS and high-energy collisional dissociation fragmentation: focus on triglyceride detection and characterization. *Anal. Chem.* 83: 6648–6657. https://doi.org/10.1021/ac201195d.

179 Seeley, J.V., Kramp, F.J., and Sharpe, K.S. (2001). A dual-secondary column comprehensive two-dimensional gas chromatograph for the analysis of volatile organic compound mixtures. *J. Sep. Sci.* 24: 444–450. https://doi.org/10.1002/1615-9314(20010601)24:6<444::AID-JSSC444>3.0.CO;2-5.

180 Seeley, J.V., Kramp, F.J., Sharpe, K.S., and Seeley, S.K. (2002). Characterization of gaseous mixtures of organic compounds with dual-secondary column comprehensive two-dimensional gas chromatography (GC × 2 GC). *J. Sep. Sci.* 25: 53–59. https://doi.org/10.1002/1615-9314(20020101)25:1/2<53::AID-JSSC53>3.0.CO;2-V.

181 Savareear, B. and Shellie, R.A. (2013). Multiplexed dual second-dimension column comprehensive two-dimensional gas chromatography (GC × 2GC) using thermal modulation and contra-directional second-dimension columns. *Anal. Chim. Acta* 803: 160–165. https://doi.org/10.1016/j.aca.2013.05.013.

182 Reichenbach, S.E., Rempe, D.W., Tao, Q. et al. (2015). Alignment for comprehensive two-dimensional gas chromatography with dual secondary columns and detectors. *Anal. Chem.* 87: 10056–10063. https://doi.org/10.1021/acs.analchem.5b02718.

183 Sgorbini, B., Cagliero, C., Boggia, L. et al. (2015). Parallel dual secondary-column-dual detection comprehensive two-dimensional gas chromatography: a flexible and reliable analytical tool for essential oils quantitative profiling. *Flavour Fragance J.* 30: 366–380. https://doi.org/10.1002/ffj.3255.

184 Savareear, B., Jacobs, M.R., and Shellie, R.A. (2014). Multiplexed dual first-dimension comprehensive two-dimensional gas chromatography-mass spectrometry with contra-directional thermal modulation. *J. Chromatogr. A* 1365: 183–190. https://doi.org/10.1016/j.chroma.2014.09.014.

185 Yan, D., Tedone, L., Koutoulis, A. et al. (2017). Parallel comprehensive two-dimensional gas chromatography. *J. Chromatogr. A* 1524: 202–209. https://doi.org/10.1016/j.chroma.2017.09.063.

186 Byrdwell, W.C., Kotapati, H., Goldschmidt, R. et al. (2022). Three-dimensional liquid chromatography with parallel second dimensions, LC × 2LC, and quadruple parallel mass spectrometry, LC3MS4, for infant formula analysis. *J. Chromatogr. A* 1661: 462682. https://doi.org/10.1016/j.chroma.2021.462682.

187 Hancock, S.E., Poad, B.L.J., Batarseh, A. et al. (2017). Advances and unresolved challenges in the structural characterization of isomeric lipids. *Anal. Biochem.* 524: 45–55. https://doi.org/10.1016/j.ab.2016.09.014.

14

Lipidomic Analysis of Glycerophospholipid Molecular Species in Biological Samples

Xianlin Han

University of Texas Health Science Center at San Antonio, Barshop Institute for Longevity and Aging Studies, 4939 Charles Katz Dr., San Antonio, TX, 78229, USA

14.1 Introduction

14.1.1 Diverse Functions and Structures of Glycerophospholipids

Glycerophospholipid (GPL) molecular species are ubiquitously present in nature, the essential components of cellular membranes and lipoproteins, and greatly involved in cellular metabolism, homeostasis, trafficking, and signaling [1]. For example, cardiolipin (CL) serves as the glue of electron respiratory complex and is crucial for mitochondrial function [2]; translocation of serine GPL is the indicator of cell apoptosis; choline GPL is the dominant component of monolayer in lipoproteins [1]; accumulation of bis(monoacylglycero)phosphate (BMP) is involved in lysosomal disease [3]; *N*-acyl ethanolamine GPL serves as the precursor of endocannabinoid production; polyphosphoinositol GPL actively involves signal transduction and vesicle transport; platelet activation factor and lysoGPL play important roles in biological systems serving as lipid second messengers; etc. The majority of the GPL classes are the major components of cellular membrane, which dynamically change as environment changes to provide appropriate matrices for optimal functions of membrane proteins and to serve as substrates for releasing signaling lipids. Aberrant GPL metabolism and homeostasis are associated with a variety of diseases and metabolic disorders (e.g. [4–6]).

The category of GPL is defined as those molecular species carrying a glycerol as the backbone and containing at least one group of phosphate (or phosphonate to a less degree) which is commonly esterified to the *sn*-3 hydroxyl group in the glycerol. The category of cellular GPL is very complex, and further divided into different classes, subclasses, and individual molecular species (Figure 14.1a) (see the next paragraph for detail description). As illustrated and defined by its name, individual

Mass Spectrometry for Lipidomics: Methods and Applications, First Edition.
Edited by Michal Holčapek and Kim Ekroos.

molecular species in this category of lipids contain three key components. These include "glycero-" which indicates the presence of at least one glycerol molecule; "phospho-" which indicates the existence of at least one phosphate or phosphodiester; and "lipid" which implies the presence of one or two aliphatic chains connected with the hydroxyl groups of glycerol.

The GPL classes are defined by the moieties (i.e. small molecules) esterified with the phosphate (i.e. X in Figure 14.1a). The commonly seen moieties include hydrogen, choline, ethanolamine, glycerol, inositol, serine, monomethyl ethanolamine, dimethyl ethanolamine, *N*-acyl ethanolamine, monoacyl glycerol, (mono-, di-, and tri-)phosphoinositol, glycerol GPL, etc. corresponding to phosphatidic acid (PA), choline GPL (PC), ethanolamine GPL (PE), glycerol GPL (PG), inositol GPL (PI), serine GPL (PS), monomethyl PE (MMPE), dimethyl PE (DMPE), NAPE, BMP, phosphatidylinositol phosphate (PIP), PIP_2, PIP_3, CL, etc. respectively. Over 10 different GPL classes can be readily counted. GPL subclasses are defined with the different connections of the aliphatic chain with the hydroxyl group of glycerol at the *sn*-1 position (Figure 14.1a). These connections include the linkage of an ester, an ether, and a vinyl ether, corresponding to phosphatidyl, plasmanyl, or plasmenyl, respectively, according to International Union of Pure and Applied Chemistry (IUPAC) nomenclature. These subclasses are commonly annotated as "aa," "ea or ae," and "pa," respectively. There exist a huge number of aliphatic chains in biological systems, considering different numbers of carbon atoms (i.e. chain length), different degrees of unsaturation, different locations of these

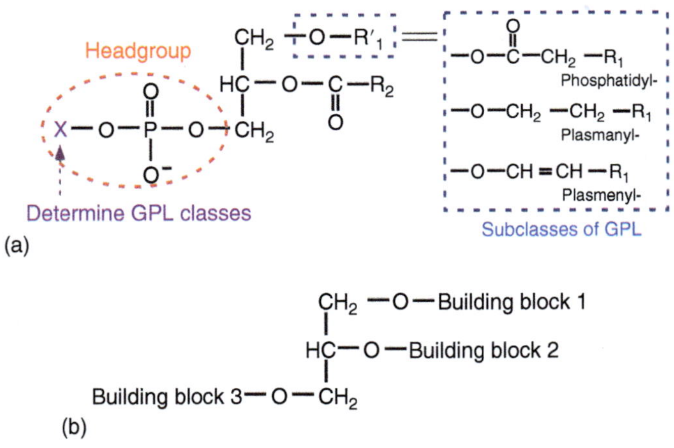

Figure 14.1 Summary of classes, subclasses, and molecular species in glycerophospholipid and a matched schematic illustration of building blocks of glycerophospholipids. (a) The polar moiety (X), which is connected to the phosphate group, defines the individual class of glycerophospholipid (GPL) as described in the text; the linkage (i.e. ester, ether, and vinyl ether) of the aliphatic chain to the hydroxyl group at *sn*-1 position of glycerol defines the structure of an individual subclass of phosphatidyl, plasmanyl, or plasmenyl, respectively; and the identities of R'_1 and R_2 which vary with different number of carbon atoms, number of double bonds, and the location of the double bonds define the individual molecular species of each lipid class. (b) A schematic illustration of building block model of glycerophospholipids as discussed in the text.

double bonds, the existence of branch chain(s), and different modifications. For example, we can easily count for over 30 aliphatic chains without consideration of branch chains and modifications. Accordingly, we can easily estimate that the possible numbers of individual molecular species in the category of GPL should be approximately 30 000 (10 classes × 3 subclasses × 30 chains at *sn*-1 × 30 chains at *sn*-2). In practice, mass spectrometric (MS) measurement has already demonstrated the presence of large numbers of individual plasmenyl ethanolamine GPL and CL molecular species, respectively [7, 8].

14.1.2 Pattern Recognition in Analysis of GPL

14.1.2.1 Recognition of a Building Block Pattern

In fact, GPL (including lysoGPL) molecular species present in biological systems are the combinations of some building blocks connected to the glycerol molecule. These building blocks represent some kinds of hydrolysis products or their analogs. The commonly recognized building blocks include fatty acyls as categorized by LIPID MAPS classification [9] and a variety of polar head groups which classify the GPL classes as described in the last subsection. With this concept, the molecular species of an entire GPL category could be represented by a common chemical structure as shown in Figure 14.1b.

Specifically, molecular species of all GPL lipid classes are the combination of three different building blocks. In this general structure, the building blocks I and II can be a hydrogen (i.e. lysoGPL molecular species) or aliphatic chains (including fatty acyls) which are linked to the *sn*-1 and 2 positions of glycerol with an ester, ether, or vinyl ether bond. Building block III at the *sn*-3 position of glycerol is the phospho(di)esters in GPL and lysoGPL. In GPL molecular species, the fatty acyl chains typically contain 12–24 carbon atoms with variable degrees of unsaturation or modifications. It should be emphasized that the degree of unsaturation must obey the biological principles, and not only the chemical formulas.

The significance of this building block pattern is twofold: (i) ready to construct theoretical lipid databases that are expandable and (ii) effectively identify a large number of individual lipid molecular species through identification of the relatively smaller number of building blocks. Luckily, these building blocks can be readily identified with their corresponding characteristic fragments (see discussion in the following subsection) by using two powerful tandem MS techniques (i.e. neutral loss scan (NLS) and precursor-ion scan (PIS)) [10].

14.1.2.2 Recognition of Fragmentation Patterns of GPL Classes

Identification of individual GPL molecular species should generally cover all the information about GPL class, subclass, and fatty acyl(s) (which should include the chain length, unsaturation, and the location of each double bond), as well as regiospecificity if present. Such a fragmentation pattern providing the information matches well with the building block concept as described above. In fact, the majority of the elucidated fragmentation patterns allow scientists to extract the GPL head groups, the linkages for subclasses, the mass corresponding to a fatty acyl chain, and usually the intensity ratio of the ions carrying the fatty acyl chain information. This suggests that the GPL classes, subclasses, and regioisomers (from the intensity ratios of the ions related to fatty acyl chains) can be well determined. It should be mentioned that determining the distribution of double bonds in a fatty acyl chain still remains to be a challenge in most of the cases although great efforts have been made to solve this issue [11, 12]. Luckily, the majority of the unsaturated fatty acids produced in nature follow the metabolic rule, which is governed by the existing desaturases. Hence, the location of the double bonds in fatty acids is not present randomly. There are numerous publications in the literature on structural elucidation of cellular GPL molecular species by using tandem MS (e.g. [13, 14]). Therefore, fragmentation patterns of the majority of the lipid classes from a variety of ion forms can be summarized from the previous studies.

Recognizing and remembering fragmentation patterns of GPL classes are very useful and important. First, remembering one pattern from one ion form of a GPL class is much easier than to memorize all the tandem MS spectra yielding from individual species of the class with this type of ion form. In fact, all these tandem MS spectra can be readily derived from the fragmentation pattern if one is familiar with it. Moreover, the appearance of a product ion mass spectrum of a GPL molecular species may vary with experimental conditions, whereas the fragmentation pattern of a GPL class is minimally changed with these conditions as previously described [15].

It should be clarified that the fragmentation pattern described in this chapter does not include those determining the double-bond location(s) in fatty acyl chains. Any reader interested in this area of research is advised to read an invaluable review article newly published in *Journal of Lipid Research* by Dr. Yu Xia. Moreover, it should also be recognized that with advances in mass spectrometry with novel technologies such as electron-activated dissociation, the fragmentation patterns of GPL classes might be revised.

14.1.2.3 Molecular Mechanisms Underlying Fragmentation Patterns of GPL Classes

Characterization of GPL molecular structures has been widely conducted in both positive- and negative-ion modes under a variety of experimental conditions [14]. These experimental conditions can lead to generating a variety of quasimolecular ions including different adducts depending on the availability of the small cations or anions in the matrices [16]. Hence, very different fragment ions can be produced from these different quasimolecular ions, all of which are very interesting, particularly from an ion chemistry perspective view. After extensive characterization, Hsu and Turk concluded that the fragmentation processes are essentially identical for those resulting from positive quasimolecular ions, whereas the fragmentation of all negative quasimolecular ions also follows a general rule [13].

In summary, charge-remote fragmentation processes [17] play a major role in fragmentation of GPL classes in the positive-ion mode. This charge-remote fragmentation generally leads to the generation of intense fragment ions yielding from the head groups. In these fragmentation processes, the loss of the fatty acyl chain from the *sn*-1 position is a more favorable pathway than the counterpart loss of the fatty acyl from the *sn*-2 position of the glycerol moiety. Through stable isotope labeling and/or MS^n

analysis, Hsu and Turk revealed that these differential losses of *sn*-1 and -2 fatty acyl chains are due to the involvement of the α-hydrogen atoms at the fatty acyl chain in the elimination of the adjacent fatty acyl chain as an acid, and due to that the α-hydrogen atoms of the *sn*-2 fatty acyl chain are more labile than their counterparts at *sn*-1 [13]. These differential losses of *sn*-1 and 2 fatty acyl chains are critical for identification of regioisomers of GPL species, for example as their alkaline adducts [18, 19].

In the negative-ion mode, the deprotonated ions of GPL molecular species after low-energy collision-induced dissociation (CID) yield the predominant ion peak(s) corresponding to the fatty acyl carboxylates in the mass region of *m/z* 200–350 generated from the fatty acyl chains at the *sn*-1 and -2 positions of glycerol. In most cases, these carboxylates are the base peaks in the tandem MS mass spectra. In the mass spectra, another set of fragment ions around *m/z* 400, corresponding to the $[M\text{-}H\text{-}R_xCH_2CO_2H]^-$ and $[M\text{-}H\text{-}R_xCH{\cdot}C{\cdot}O]^-$ (where $x = 1, 2$) through the loss of the fatty acyl chains as acids or ketenes, respectively, is also present in modest intensities. Charge-driven fragmentation processes [20] are the major mechanism leading to the generation of product ions in the negative-ion mode. In this process, the gas-phase basicity of the deprotonated molecular ions determines the loss of the fatty acyl chain as an acid or as a ketene [20, 21]. The gas-phase basicity defines the affinity of a proton to a molecule or a negatively charged molecular ion according to IUPAC Gold Book [22], whereas the polar head group of individual GPL class determines the gas-phase basicity of their deprotonated ions. Therefore, different molecular species of a GPL class show essentially an identical fragmentation pattern, but distinct mass spectra of different GPL classes are obtained. Generally, the loss of the fatty acyl chain at *sn*-2 as an acid or as a ketene is more favorable than that at *sn*-1 because the loss of the *sn*-2 fatty acyl chain is relatively more favored sterically. This leads to the ion corresponding to $[M\text{-}H\text{-}R_2CH_2CO_2H]^-$ more abundant than that of $[M\text{-}H\text{-}R_1CH_2CO_2H]^-$ and the ion corresponding to $[M\text{-}H\text{-}R_2CH{\cdot}CO]^-$ more intense than that of $[M\text{-}H\text{-}R_1CH{\cdot}CO]^-$. This difference between the losses of the fatty acyl chains at *sn*-1 and -2 positions allows for identification of regioisomers [20, 21, 23–26], particularly with the trap-type instrumentation.

14.1.2.4 Practical Usage of Fragmentation Patterns of GPL Classes in Lipidomics

There exist at least four variables in identification of individual GPL molecular species. These variables include the polar head group (i.e. lipid class), the connection of aliphatic chains at the *sn*-1 position (i.e. lipid subclass), the identity of at least one fatty acyl chain (i.e. the molecular species since the other chain can be derived from the information mentioned), and the regiospecificity (i.e. the positions of individual fatty acyl chains). Therefore, acquisition of a full product ion mass spectrum of a GPL molecular ion would be preferred for thorough characterization of the species.

Although elution time from a liquid chromatography (LC) separation provides valuable information for a particular species, definitive identification of the lipid species to the molecular level needs at least three additional variables. For those researchers who are using LC-MS for identification and quantification of lipid species, getting familiar with the aforementioned fragmentation patterns of individual GPL classes; therefore, deriving the product ion mass spectra of individual lipid species is very important. It should be pointed out that although a few databases and/or libraries can be used to aid the identification of lipid species, manual identification of these species is inevitable at the current stage of lipidomics development. This is partially due to the presence of unresolved isomeric/isobaric species. It is difficult to search the currently available databases/libraries to definitively identify an individual molecular species from a product ion mass spectrum of mixed species [27]. Therefore, getting familiar with the fragmentation patterns of GPL classes would greatly aid the identification of the GPL molecular species in the case. Finally, getting familiar with the fragmentation patterns of individual GPL lipid classes would also be essential for establishment of multiple reaction monitoring (MRM) methods for quantification of those individual molecular species of GPL classes after LC-MS.

For those who would like to employ shotgun lipidomics (particularly multi-dimensional mass spectrometry-based shotgun lipidomics [MDMS-SL]) for identification and quantification of individual species of a GPL class, getting familiar with the fragmentation patterns of individual GPL classes could enable them to design different tandem MS scans for building block analysis in the PIS, NLS, or both modes to selectively identify individual molecular species of a GPL class of interest, or to identify individual molecular species after product ion analysis.

14.2 Fragmentation Patterns of GPL Classes

In this section, some of the representative fragmentation patterns of individual GPL classes under some common experimental conditions are summarized because the majority of these experimental conditions can be readily transferred to a practical usage for a large-scale analysis of lipids in lipidomics. It is advised that any advanced reader should always look for the details from the original studies on the topic and/or study several invaluable review articles as well as a newly published book on the topic [13, 14, 28, 29]. The review article by Hsu and Turk [13] and the papers cited should always consider as a resource for a detailed understanding of structural characterization of specific GPL lipid classes.

14.2.1 Choline Glycerophospholipid

14.2.1.1 Positive-Ion Mode

14.2.1.1.1 Protonated Species

Under an acidic condition or in the presence of an ammonium salt as a modifier in the matrix, PC molecular species can readily form protonated adducts in the positive-ion mode since these species contain a quaternary amine in the form of zwitterion with the phosphate. Fragmentation of protonated PC molecular species including all subclasses after low-energy CID yields an ion at *m/z* 184, corresponding to a phosphocholine ion. The structurally informative fragment ions corresponding to the fatty acyl chain(s) are generally in low abundance. These ions are essentially buried in baselines when product ion analysis is conducted with a QqQ-type instrument although these fragment ions can be well visualized by a QqToF-type mass spectrometer. Therefore, structural characterization from protonated PC molecular species is less applicable such as for identification of regioisomers although achievable [29, 30]. Owing to its high abundance with some specificity, the fragment ion at *m/z* 184 is widely used for quantification of PC molecular species by LC-MRM.

Extensive studies have shown that formation of the fragment ion at *m/z* 184 mainly involves the α-hydrogen atoms of the fatty acyl chain at the *sn*-2 position of glycerol due to it being more labile than its counterpart at *sn*-1 [30]. This leads to the more favorable formation of the $[M+H-R_2CH=C=O]^+$ ion than the $[M+H-R_1CH=C=O]^+$ ion, resulting from the losses of the fatty acyl chains at the *sn*-2 and -1 positions as ketenes, respectively. Therefore, the position of the fatty acyl moieties at the glycerol backbone can be assigned from these ketene-type fragment ions to a certain degree.

14.2.1.1.2 Alkaline Adducts

PC molecular species can readily form adducts with alkaline (i.e. Li, Na, K as examined). The formation of the adduct(s) largely depends on the availability or the concentration of the alkaline ions in the matrices. Therefore, if lithium hydroxide or a lithium salt is introduced to the matrix, then lithium adducts can be formed for the analysis of PC molecular species. It should be emphasized that sodium adducts are always displayed in a mass spectrum if there are no any other modifier(s) added to the matrix since sodium ion essentially exists "everywhere."

A common fragmentation pattern yielded from alkaline (Alk) adducts of PC molecular species after low-energy CID has been shown in many studies (e.g. [19, 29–32]). The fragment ions include the one corresponding to the loss of trimethylamine (i.e. $[M+Alk\text{-}59]^+$), the one corresponding to the loss of phosphocholine (i.e. $[M+Alk\text{-}183]^+$), and a product ion corresponding to the loss of alkaline cholinephosphate (i.e. $[M+Alk\text{-}(Alk+182)]^+$).

In addition to these abundant fragment ions corresponding to the phosphocholine head group, fragment ions related to fatty acyl chains of PC molecular species are also present, particularly from those of lithiated PC molecular species. These fragment ions are usually present in relatively low abundance in tandem MS mass spectra of alkaline PC adducts. There exist three pairs of fragment ions in this category. They are the pair of product ions that resulted from the neutral loss of fatty acids (i.e. $[M+Alk\text{-}R_xCO_2H]^+$); the pair of product ions yielded from the neutral loss of alkaline fatty acyl salts (i.e. $[M+Alk\text{-}R_xCO_2Alk]^+$); and the pair of product ions corresponding to the continuous neutral loss of fatty acids after neutral loss of trimethylamine (59 amu) (i.e. $[M+Alk\text{-}(R_xCO_2H+59)]^+$); where $x=1$ and 2, respectively. The loss of the fatty acyl chain at *sn*-1 position is more favored than the counterpart loss of the fatty acyl chain from the *sn*-2 position of glycerol as discussed in the introduction. Finally, the pair of $[M+Alk\text{-}(R_xCO_2H+59)]^+$ ions is usually the most abundant product ions among these fatty acyl-related ions and commonly used for determination of regiosomers of phosphatidylcholine (i.e. PC(aa)) molecular species [13, 30, 32, 33].

Discrimination of PC Subclasses in the Positive-Ion Mode

The fragmentation patterns of alkaline adducts of plasmenylcholine (i.e. PC(ap)) and plasmanylcholine (i.e. PC(ae)) molecular species contain all the fragment ions related to phosphocholine head group as those resulted from PC(aa) species, including $[M+Alk-59]^+$, $[M+Alk-183]^+$, and $[M+Alk-(Alk+182)]^+$. However, the fragmentation patterns of alkaline adducts of PC(ap) and PC(ae) molecular species are also distinct from that of PC(aa) molecular species in three aspects. First, the $[M+Alk-59]^+$ ion is dominated in the product ion spectra of the alkaline adducts of both PC(ap) and PC(ae) molecular species. Second, the product ions corresponding to the neutral losses of fatty acids and fatty acyl salts (i.e. $M+Alk-R_2CO_2H]^+$ and $[M+Alk-R_2CO_2Alk]^+$) are not present. Lastly, in contrast to that of PC(aa) molecular species, the fragment ion corresponding to $[M+Alk-(R_2CO_2H+59)]^+$ is in low abundance. These differences of the fragmentation patterns between ether-containing PC subclasses and PC(aa) subclass are widely used to distinguish ether-containing PC molecular species from diacyl ester-containing PC molecular species [19, 30, 33]. The difference of the fragmentation patterns of PC subclasses can be well explained with the absence of labile α-hydrogen atoms of the *sn*-1 aliphatic chains in the ether-containing PC subclasses. Specifically, these labile α-hydrogen atoms are required for elimination of the adjacent fatty acyl chain as an acid (i.e. R_2CO_2H) as discussed in the introduction. Hence, the formation of $[M+Alk-R_2CO_2H]^+$ and $[M+Alk-(R_2CO_2H+59)]^+$ fragments is not favored from the alkaline adducts of both PC(ap) and PC(ae) molecular species. Moreover, the lack of continuous neutral loss of fatty acyl chain as a fatty acid from the resultant $[M+Alk-59]^+$ due to the absence of labile α-hydrogen atoms explains this fragment ion as a predominant fragment ion displayed in the product ion mass spectra of ether-containing PC subclasses [33].

The difference of the fragmentation patterns between ether-containing PC alkaline adducts is the presence of an abundant fragment ion corresponding to $[M+Alk-(182+Alk)-R_2CO_2H]^+$ in the product ion ESI-MS spectra of PC(ap) alkaline adducts, but not in those of both PC(ae) and PC(aa) counterparts [19]. This prominent ion arises from the further neutral loss of the *sn*-2 fatty acyl from the ion corresponding to ($[M+Alk-(182+Alk)]^+$).

14.2.1.2 Negative-Ion Mode

Molecular species of all PC subclasses can readily form various adduct ions with those small anion(s) present in the matrix (i.e. $[M+X]^-$) (where X = Cl, CH_3CO_2, HCO_2, CF_3CO_2, etc.) in the negative-ion mode of ESI-MS [31, 34–36]. These adducts yield an abundant $[M-15]^-$ ion (i.e. $[M+X-CH_3X]^-$) or even showing as the predominant quasimolecular ions of PC species in many cases [34, 37, 38]. However, this quasimolecular ion can be minimized through tuning the ionization conditions [38, 39], indicating that the neutral loss of CH_3X from the anion adduct is very facile.

The fragmentation pattern of anion adducts of PC molecular species after low-energy CID shows the following fragment ions: (i) a predominant $[M-15]^-$ ion arising from the neutral loss of CH_3X; (ii) one or two very abundant fragment ions around m/z 300 corresponding to fatty acyl carboxylates generated from fatty acyl chain(s) of PC molecular species; and (iii) a cluster of low-to-modest abundance fragment ions around m/z 450 arising from the neutral losses of fatty acids of PC molecular species.

Many studies have shown that the peak intensity of the carboxylate yielded from the *sn*-2 fatty acyl chain is approximately three times more intense than that resulting from the *sn*-1 counterpart of PC(aa) molecular species [29, 31, 40, 41]. This fact has been used to identify the location of fatty acyls and thus the regioisomers of PC(aa) molecular species, as well as the composition of mixed PC(aa) regioisomers to a certain degree of accuracy. It should be remembered that this ratio gets lower than three if the *sn*-2 fatty acyl chain is a polyunsaturated fatty acyl [31]. It gets clear now that this reduced ratio is arising from the continuous fragmentation of polyunsaturated fatty acyl carboxylate to generate an ion corresponding to the neutral loss of carbon dioxide (i.e. [carboxylate-44]$^-$) [8, 42, 43]. This continuous loss of carbon dioxide has been exploited to identify the location of double bond(s) of fatty acyls [43]. With the recognition of carbon dioxide loss, the combined peak intensity of the carboxylate and [carboxylate-44] ion arising from *sn*-2 polyunsaturated fatty acyl chain is still in approximately three times intense than that of the *sn*-1 counterpart [8].

In addition to utilizing the ratios of fatty acyl carboxylates for identification of regiospecificity, the paired ions arising from the neutral losses of fatty acyl chains as either fatty acids (i.e. [M-15-R_xCH_2COOH]$^-$) or ketenes (i.e. [M-15-$R_xCH{\cdot}C{\cdot}O$]$^-$), respectively, can also be applied for determination of the regioisomers of PC(aa) molecular species [31, 37]. In this case, 1-acyl demethylated lysoPC fragment ions arising from the loss of *sn*-2 fatty acyl chain are more intense than those of 2-acyl counterparts yielded from the loss of *sn*-1 counterpart [31, 37]. This is due to the fact that the neutral loss of the *sn*-2 fatty acyl chain is more sterically favored as mentioned in the introduction. Moreover, the [M-15-$R_2CH{\cdot}C{\cdot}O$]$^-$ ion is more abundant than the [M-15-R_2CH_2COOH]$^-$ ion, while the [M-15-$R_1CH{\cdot}C{\cdot}O$]$^-$ ion is less intense than the [M-15-R_1CH_2COOH]$^-$ ion [31, 37].

14.2.1.2.1 *Discrimination of PC Subclasses in the Negative-Ion Mode*

The fragmentation patterns of small anion adducts of ether-containing subclasses are essentially identical to those of PC(aa) molecular species containing two identical fatty acyl chains, i.e. only one each of the carboxylate, [M-15-RCH_2COOH]$^-$, and [M-15-$RCH{\cdot}C{\cdot}O$]$^-$ is present in their product ion mass spectra after low-energy CID. However, the pathways leading to this identical pattern are different. The pattern of ether-containing PC subclasses is due to the absence of the *sn*-1 fatty acyl chain, whereas the pattern from PC(aa) quasimolecular ions containing two identical fatty acyl chains is due to the resultant fragment ions from *sn*-1 and 2 fatty acyls that are identical. Multiple approaches can be employed to distinguish PC(ap) molecular species from PC(ae) molecules present in biological samples. These include the treatment of acid vapor [44, 45], utilization of the paired rule [8], a mass shift with iodine in methanol [46], or detection of a "Boolean" ion [47].

14.2.1.3 Choline Lysoglycerophospholipids

Choline lysoglycerophospholipids (lysoPC) are a subgroup of lipid molecular species in the choline GPL group, in which one of the aliphatic chains is replaced with a hydroxy group. These lysoPC molecular species are the precursors of PC molecular species biosynthesis in the remodeling process with an acyltransferase activity and the product of PC molecular species through a phospholipase activity or oxidative degradation in the case of PC(ap) molecular species. As their parent PC molecular species, lysoPC molecular species can be readily ionized in both positive- and negative-ion modes as proton/alkaline or small anion adducts, respectively.

Characterization of lysoPC molecular species as alkaline adducts is well studied in the positive-ion mode after low-energy CID [18, 19, 33, 35]. The fragmentation pattern of alkaline adducts of lysoPC molecular species contains three groups of informative fragments. Firstly, the abundant fragment ions correspond to the neutral losses of polar head group fragments, including $[M+Alk-59]^+$, $[M+Alk-183]^+$, and $[M+Alk-(Alk+182)]^+$. These fragment ions are essentially identical to those arising from PC molecular species. Secondly, an additional ion at $[M+Alk-103]^+$ corresponding to the loss of choline-like molecule is also present in low abundance in the fragmentation pattern. In contrast to the presence of abundant ions reflecting the losses of fatty acyl chains (i.e. $[M+Alk-R_xCO_2H]^+$ and $[M+Alk-59-R_xCO_2H]^+$) from their parent molecular species, these fragment ions are in relatively low abundance or even not present in the product ion ESI-MS spectra after low-energy CID. This is due to the fact that the lack of a fatty acyl chain leads to missing the initiation of the α-hydrogen at the fatty acyl chain in the elimination of the adjacent fatty acyl as an acid. Lastly, the fragment ions resulting from the phosphocholine head group are present at *m/z* 124+Alk and 104 corresponding to the alkaline adduct of a five-membered ethylene PA and choline ions in the product ion spectra of lysoPC molecular species after low-energy CID. The peak intensity ratio of these two fragment ions has been used to determine the location of the aliphatic chain linked to glycerol [18, 33].

Similar to PC(ae) and PC(ap) molecular species as discussed in Subsection 14.2.1.1, due to the lack of α-hydrogen in both the *sn*-1 and 2 moieties, further dissociation from $[M+Alk-59]^+$ to form $[M+Alk-183]^+$ and $[M+Alk-(Alk+182)]^+$ is significantly reduced [13]. Accordingly, the presence of a prominent $[M+Alk-59]^+$ ion relative to $[M+Alk-183]^+$ and $[M+Alk-(Alk+182)]^+$ ions can be used to discriminate ether-containing lysoPC subclasses from lysoPC(aa) subclass.

The fragmentation pattern of protonated lysoPC molecular species after low-energy CID is very different from that of the alkaline adducts [35]. Fragmentation of all protonated molecular species shows a prominent and distinct ion at *m/z* 184 corresponding to phosphocholine. Product ion ESI-MS spectra of protonated lysoPC molecular species also display another very abundant fragment ion corresponding to the loss of H_2O (18 amu) (i.e. $[M+H-18]^+$). This fragment ion is absent or present in very low abundance in ether-containing lysoPC subclasses. This feature can be applied for distinguishing lysoPC subclasses.

Characterization of anionic adducts of lysoPC molecular species after low-energy CID yields a prominent ion corresponding to the loss of methyl along with the anion adduct from quasimolecular ions giving rise to an $[M\text{-}15]^-$ ion [18, 35]. Another abundant fragment ion from lysoPC species is the fatty acyl carboxylate corresponding to the fatty acyl chain. A few other low-abundance fragment ions could also be present in the product ion ESI-MS spectra of lysoPC anion adducts corresponding to the further loss of fatty acyl chain as a fatty acid or fatty acyl ketene from the $[M\text{-}15]^-$ ion (i.e. $[M\text{-}15\text{-}RCH_2CO_2H]^-$ and $[M\text{-}15\text{-}RCH{\cdot}C{\cdot}O]^-$, respectively). These fragment ions are absent in ether-linked lysoPC subclasses, in which a fragment ion corresponding to the loss of long chain aliphatic alcohol (i.e. $[M\text{-}15\text{-}RCOH]^-$) is present in low abundance [18, 35].

14.2.2 Ethanolamine Glycerophospholipid

14.2.2.1 Positive-Ion Mode

14.2.2.1.1 Protonated Species

PE molecular species under acidic conditions or in the presence of ammonium ions can be readily ionized as proton adducts (i.e. $[M+H]^+$) in the positive-ion mode [34, 48, 49]. It should be recognized that the ionization efficiency of PE molecular species as proton adducts is relatively lower than that of PC molecular species because the quaternary amine of protonated PC species as a positive charge site is much more stable than the primary amine as the charge site of protonated PE molecular species as previously discussed [50].

The fragmentation pattern of protonated PE molecular species contains two types of fragment ions as previously demonstrated [13, 34, 48, 49, 51]. First, an intense fragment ion corresponding to the neutral loss of phosphoethanolamine (i.e. $[M+H\text{-}141]^+$). Second, one or two low-abundance fragment ions corresponding to fatty acylium ions (i.e. $R_xCO^+, x = 1, 2$).

This pattern is very different from that of protonated PC molecular species (see the last subsection). The resultant $[M+H\text{-}141]^+$ ion, rather than a protonated phosphoethanolamine ion at *m/z* 142 (equivalent to the *m/z* 184 ion resultant from protonated PC molecular species), indicates that the phosphoethanolamine is less competitive for a proton, but easy to lose a proton from the primary amine. Extensive mechanistic studies have demonstrated that the fragmentation process leading to the production of $[M+H\text{-}141]^+$ ion also involves the participation of the α-hydrogen atoms of the fatty acyl chains, mainly those at the *sn*-2 position [13, 21, 30].

14.2.2.1.2 Alkaline Adducts

PE molecular species can yield adducts with any alkaline (e.g. Li, Na, and K) in the positive-ion mode when a modifier carrying the alkaline ion is added in the matrix. Sodium adduct is formed if no other alkaline ion is added as mentioned in the subsection of PC. Similar to the formation of proton adducts, the ionization efficiency of PE alkaline adducts is much lower relative to that of PC counterparts. Hence, alkaline adducts of PE molecular species are suppressed by alkaline-adducted PC molecular species in shotgun lipidomics analysis of lipid extracts. Application of these adducts for analysis of PE molecular species after LC-MS separation is also rare.

The fragmentation pattern of alkaline adducts of PE molecular species has been well characterized [31, 52]. Similar to that of PC alkaline adducts, the fragmentation pattern of PE alkaline adducts mainly contains three types of fragment ions. First, a fragment ion corresponds to an alkaline adduct of PA after the neutral loss of aziridine (43 amu) (i.e. $[M+Alk-43]^+$) from the quasimolecular ions. This fragment is equivalent to $[M+Alk-59]^+$ in the fragmentation pattern of PC molecular species. Second, there exist two fragment ions corresponding to alkaline adducts of diacylglycerol (DAG) and DAG-like ion arising from the neutral losses of phosphoethanolamine (141 amu) and alkaline ethanolamine phosphate salt (140+Alk amu), respectively, from PE alkaline adducts. These two fragment ions are equivalent to the $[M+Alk-183]^+$ and $[M+Alk-(182+Alk)]^+$ fragment ions in the fragmentation pattern of PC adducts. These fragment ions along with the presence of low-abundance alkaline ethanolamine phosphate ($[(HO)_2PO_2(CH_2)_2NH_2Alk]^+$) and phosphoric acid alkaline adduct ($[(HO)_3POAlk]^+$) are characteristic of PE molecular species. Third, a type of fragment ions arises from the neutral losses of the fatty acyl chains from either quasimolecular ion or alkaline PA adduct (i.e. $[M+Alk-R_xCO_2H]^+$ and $[M+Li-43-R_xCO_2H]^+$, where $x = 1, 2$). The loss of the fatty acyl chain at *sn*-1 position is more favored than the counterpart loss of the fatty acyl chain at *sn*-2 position of the glycerol moiety as discussed in the introduction of the chapter.

14.2.2.1.3 Discrimination of PE Subclasses in the Positive-Ion Mode

Similar to the case of PC class, product ion ESI-MS analysis of PE alkaline adducts after low-energy CID enables us to assign individual molecular species in different PE subclasses [13]. For example, in the case of lithium adducts, the ions corresponding to $[M+Li-43]^+$, $[M+Li-141]^+$, $[M+Li-(140+Li)]^+$, and the ion at *m/z* 148 ($[(HO)_2PO_2(CH_2)_2NH_2Li]^+$) commonly yielded from lithiated PE(aa) molecular species are also observed in the product ion spectra of the $[M+Li]^+$ ions of PE(ap) and PE(ae) molecular species. However, fragmentation of lithiated PE(ap) molecular species contains unique fragment ions. For instance, there exist two unique fragment ions from quasimolecular ions of lithiated PE(ap) species, i.e. an ion arising from the combined losses of aziridine and the alk-1′-enyl residue at the position *sn*-1 as an alcohol ($C_{16}H_{33}CH{\cdot}CHOH$), and an ion yielding from further loss of fatty acyl chain at *sn*-2 position as a free acid [13]. These two ions identify the vinyl ether linkage at position *sn*-1 and the fatty acyl at *sn*-2, respectively. In contrast, these two counterpart fragment ions are absent in the fragmentation of lithiated PE(aa) [13].

14.2.2.1.4 Analysis of PE Molecular Species after Methylation

Because of a superior ionization efficiency of PC molecular species than that of PE molecular species in the positive-ion mode, methylation of PE with a variety of reagents to form corresponding PC counterparts has been exploited for quantitative analysis or relative comparison of PE molecular species and its analogs [53–57]. These methods could be used for lipidomics in certain applications.

14.2.2.2 Negative-Ion Mode

14.2.2.2.1 Deprotonated Species

PE molecular species can be readily ionized and form deprotonated ions in the negative-ion mode [34, 49, 58]. The fragmentation pattern of deprotonated PE molecular species after low-energy CID contains three types of fragment ions. First, abundant one or two intense fragment ions around *m/z* 300 corresponding to fatty acyl carboxylates, depending on whether the fatty acyl chains are identical or not, respectively. The peak intensity of the carboxylate resulting from the *sn*-2 fatty acyl is approximately three times intense due to its sterically favorable loss than that from *sn*-1 fatty acyl of deprotonated PE(aa) molecular species. Second, there exist a cluster of low-to-modest abundance fragment ions around *m/z* 450 arising from the neutral losses of fatty acyl chain(s) as fatty acid(s) (i.e. $[M\text{-}H\text{-}R_xCH_2COOH]^-$) and fatty acyl ketenes (i.e. $[M\text{-}H\text{-}R_xCH{\cdot}C{\cdot}O]^-$) (where $x = 1, 2$ if present) of PE(aa) molecular species. Similar to the fragmentation pattern of PC(aa) molecular species in the negative-ion mode, 1-acyl lysoPE fragment ion arising from the loss of the *sn*-2 fatty acyl is more intense than that of 2-acyl lysoPE fragment ion yielding from the loss of the *sn*-1 fatty acyl [21, 31, 37]. Again, the $[M\text{-}H\text{-}R_2CH{\cdot}C{\cdot}O]^-$ ion is generally more intense than the $[M\text{-}H\text{-}R_2CH_2COOH]^-$ ion, while the $[M\text{-}H\text{-}R_1CH{\cdot}C{\cdot}O]^-$ ion is less abundant than the $[M\text{-}H\text{-}R_1CH_2COOH]^-$ ion [8, 31]. The intensity ratios of these pairs of ions can be used to determine the regioisomers of PE(aa) species. The third type of fragment ion is that at *m/z* 196, corresponding to a glycerophosphoethanolamine anion derivative, which is usually present in low abundance and characteristic of the phosphoethanolamine head group.

It should be remembered that the significant loss of CO_2 from polyunsaturated fatty acyl carboxylate is always present as previously described [8, 43]. Hence, fragmentation of deprotonated PE molecular species consisting of polyunsaturated fatty acyl chains is readily distinguishable from that with saturated ones [59, 60]. Therefore, the peak intensity of the polyunsaturated fatty acyl carboxylate is lower than that expected.

14.2.2.2.2 Discrimination of PE Subclasses in the Negative-Ion Mode

Fragmentation of both PE(ap) and PE(ae) molecular species after CID is dominated with a single set of fragment ions corresponding to fatty acyl carboxylate from *sn*-2 fatty acyl, $[M\text{-}H\text{-}R_2CH_2COOH]^-$, and $[M\text{-}H\text{-}R_2CH{\cdot}C{\cdot}O]^-$. These fragment ions can be used to identify the fatty acyl identity. An ion corresponding to the alkenyl moiety can be detected in low abundance to determine the plasmenyl- or plasmanyl-identity [8, 61]. Confirmation of the structural assignments using a comparison of the two mass spectra of PE molecular species obtained before and after destructive removal of PE(ap) by acid treatment [8, 44, 45] or iodine treatment [46] could be employed. It should be recognized that acid treatment may result in severe sample losses as previously discussed [45].

14.2.2.2.3 Analyzing PE Molecular Species After Derivatization

PE molecular species contain a relatively specific primary amine moiety. By exploiting this specific reactivity, multiple derivative methods have been developed for effective and enhanced analyses of PE molecular species. These methods include the ones with Fmoc chloride and 4-(dimethylamino)benzoic acid [62–64]. Another derivative method was also employed for the purpose [46]. The enhancement of ionization efficiency in the negative-ion mode is achieved after derivatization with a charge-neutral moiety through turning weakly anionic lipids (or weakly zwitterionic lipids) into anionic lipids. Moreover, with the tagged group, PE molecular species can be shifted out from an overlapped mass region. Finally, the tagged group usually forms a specific and facile neutral loss in fragmentation of PE molecular species after low-energy CID so that enhanced analysis of these species can be achieved by exploiting this specific neutral loss as previously demonstrated [63].

Fragmentation of 4-(dimethylamino)benzoic acid-derivatized PE molecular species in the negative-ion mode yielded predominant fragments corresponding to fatty acyl carboxylate(s) and low-abundance fragment ion(s) corresponding to the separate neutral loss of individual fatty acyl chains as fatty acyl ketenes [64]. The derivative of PE molecule species with 4-(dimethylamino)benzoic acid can also readily form a stable positive charge site at dimethylamino moiety under acidic conditions. This charge site can lead to charge-remote fragmentation of the derivatives similar to those previously described [65]. The pattern of charge-remote fragmentation of the derivatized PE molecular species includes two intense fragment ions containing the charge site and a fragment ion yielded from the neutral loss of the entire head group (including the derivatized moiety) and corresponding to DAG-like ion [64].

14.2.2.2.4 Ethanolamine Lysoglycerophospholipids

Negative-Ion Mode Fragmentation of lysoPE (a) molecular species after low-energy CID yields two fragment ions at *m/z* 214 and 196, arising from the losses of the fatty acyl chains as a fatty acyl ketene and an acid (i.e. $[M\text{-}H\text{-}RCH{\cdot}C{\cdot}O]^-$ and $[M\text{-}H\text{-}RCH_2COOH]^-$), respectively. It should be recognized that the intensity of the ion at *m/z* 214 is lower than that of *m/z* 196 ion in fragmentation of deprotonated 1-acyl lysoPE, whereas the abundances of these two ions are reversed in fragmentation of 2-acy lysoPE counterparts. This is consistent with the notion that the gas-phase $[M-H]^-$ ions of 1- and of 2-acyl lysoPE species are weakly basic ions and undergo more facile loss of the fatty acyl chain as a ketene at *sn*-2 than that at *sn*-1. Thus, regioisomers of 1- and 2-acyl lysoPE molecular species can be differentiated in the negative-ion mode.

The fragmentation patterns between deprotonated lysophosphatidylethanolamine and ether-containing lysoPE molecular species are very different [13, 18]. The former contains a predominant ion around *m/z* 300 corresponding to the sole fatty acyl carboxylate that identifies the fatty acyl identity of the molecules. Ether-linked lysoPE molecular species can be readily recognized in the absence of a fatty acyl carboxylate which is replaced by a low-abundance ion corresponding to the *sn*-1-alkenyl or -alkyl chain. An additional product ion (i.e. $[M-62]^-$, corresponding to the neutral loss of ethanolamine from the precursor ion) is also present in the spectra of ether-containing lysoPE molecular species [18]. Moreover, the fragment ion at *m/z* 196 is usually the most abundant one in the fragmentation pattern of ether-containing lysoPE molecular species.

Positive-Ion Mode

The fragmentation pattern of lysoPE species as alkaline adducts contains four types of fragment ions [18]. First, there is an abundant ion at $[M-Alk-43]^+$ (where Alk = alkaline) arising from the neutral loss of vinylamine. Second, a less abundant ion at $[M+Alk-61]^+$ generating from the neutral loss of ethanolamine is present in the product-ion ESI-MS mass spectra. Third, a low-abundance ion at $[M+Alk-141]^+$ resulting from the neutral loss of phosphoethanolamine exists. Finally, a modest ion at $[M+Alk-(140+Alk)]^+$ corresponding to the neutral loss of alkaline ethanolamine phosphate is displayed in the spectra.

Fragmentation of protonated acyl lysoPE species displays two fragment ions at $[M+H-141]^+$ and $[M+H-18]^+$ corresponding to the losses of phosphoethanolamine and a water molecule, respectively [66]. Their alkenyl lysoPE counterparts yield three product ions at $[M+H-172]^+$, $[M+H-154]^+$, and $[M+H-18]^+$ [66]. The first two ions correspond to the losses of phosphoethanolamine derivatives and the third one is the loss of a water molecule. Obviously, the fragment ion corresponding to the loss of phosphoethanolamine from lysoPE(a) molecular species does not occur in the fragmentation pathway of alkenyl lysoPE molecular species. This feature can be readily used for discrimination of the molecular species between subclasses.

LysoPE molecular species can form alkaline adducts which can be used for identification of regioisomers of lysoPE. Specifically, these isomers can be identified by the significant difference of the ratio between the fragment ion pair at $[M+Alk-(140+Alk)]^+$ and $[M+Alk-61]^+$ as previously discussed [18]. Finally, a few fragment ions resulting from the phosphoethanolamine head group are present, including the one at *m/z* 148 corresponding to the lithiated phosphoethanolamine [52].

14.2.2.3 Phosphatidylinositol and Polyphosphoinositides

14.2.2.3.1 Positive-Ion Mode

Ionization of anionic GPL molecular species in the positive-ion mode is not favorable. However, these molecules can be ionized to a certain degree as alkaline adducts in the presence of alkaline ion(s) in the matrix or protonated ions under acidic conditions. Generally, ionization efficacy as alkaline adducts is markedly lower than what is observed as the $[M+H]^+$ ions in the positive-ion mode. Characterization of lithium and dilithium adducts of PI molecular species (i.e. $[M+Li]^+$ and $[M-H+2Li]^+$) or their protonated species has been performed [13]. It should be pointed out that ionization is suppressed if there co-exists any PC, PE, or many other lipids as in biological lipid extracts.

14.2.2.3.2 Negative-Ion Mode

PI molecular species can be readily ionized as the deprotonated form. The fragmentation pattern of deprotonated PI molecular species (i.e. $[M-H]^-$) is very informative and more complicated than those of other anionic GPL lipid classes [59], which contains four types of fragments. The *first type* is a cluster of fragment ions around *m/z* 550 in low abundance arising from the losses of fatty acyl chains as fatty acids and fatty acyl ketenes (i.e. $[M\text{-}H\text{-}R_xCH_2CO_2H]^-$ and $[M\text{-}H\text{-}R_xCH{\cdot}C{\cdot}O]^-$, where $x = 1, 2$), respectively. The *second type* is a cluster of fragment ions around *m/z* 400 arising from the further loss of either inositol or (inositol-H_2O) from the first type of cluster ions (i.e. $[M\text{-}H\text{-}R_xCH_2CO_2H\text{-}(\text{inositol-}H_2O)]^-$ or $[M\text{-}H\text{-}R_xCH{\cdot}C{\cdot}O\text{-inositol}]^-$). It seems that these fragment ions arising from charge-driven processes are occurring preferentially at the *sn*-2 position [59]. This preference can be used to assign the positions of fatty acyl chains in the molecules. The *third* one is a cluster of abundant fragment ions corresponding to fatty acyl carboxylates. The intensity of the $R_2CO_2^-$ ion is either relatively lower than or nearly equal to that of the $R_1CO_2^-$ ion. Those showing the lower intensity ones are the polyunsaturated fatty acyls due to the secondary loss of CO_2 from the $R_2CO_2^-$ ion yielding an ion corresponding to $[R_2CO_2\text{-}44]^-$. In such a case, the combination of the intensities of both ions is still near to that of the $R_1CO_2^-$ ion. The *last type* is a cluster of fragment ions arising from the PI head group including those at *m/z* 315, 297, 279, 259, 241, and 223. Usually, the ion at *m/z* 241 is the most abundant one. The *m/z* 297 ion yields from consecutive losses of the fatty acyls as fatty acids (i.e. $[M\text{-}H\text{-}R_1CO_2H\text{-}R_2CO_2H]^-$).

The fragmentation patterns of deprotonated polyphosphoinositides (PPIs) (including PIP and diphosphate (PIP_2) molecular species) (i.e. $[M-H]^-$) after low-energy CID are similar to those of PI molecular species [59]. However, the doubly charged (i.e. $[M\text{-}2H]^{2-}$) ions of PIP and PIP_2 species undergo fragmentation pathways that are similar to that of deprotonated PE species, which are basic [59]. These results suggest that the further deprotonated gaseous $[M\text{-}2H]^{2-}$ ions of PIP and PIP_2 are basic precursors.

14.2.2.3.3 Analyzing Polyphosphoinositides after Methylation

PPIs refer to a category of lipid classes with 1 [PI(3)P, PI(4)P, and PI(5)P], 2 [PI(3,4)P_2, PI(3,5)P_2, and PI(4,5)P_2], and 3 [PI(3,4,5)P_3] phosphate groups attached to the inositol group of PI molecules. Due to their low abundance, high polarity, and similar hydrophobicity, analysis of this category of lipid classes to the molecular structural level is very challenging. Recent studies demonstrated that methylated PPI molecular species via trimethylsilyl-diazomethane (TMS-diazomethane) greatly improve the aforementioned difficulties, leading to virtually total analysis of PPI molecular species by LC-MS [67, 68] or shotgun lipidomics [69]. For example, a unique methylation pattern resulting from individual PPI class was recognized and exploited for total identification of individual PPI molecular species including fatty acyl identities and positions, and the number of phosphates and their locations in shotgun lipidomics [69].

14.2.2.4 Phosphatidic Acid

14.2.2.4.1 Positive-Ion Mode

Dissociation of protonated PA molecular species (i.e. $[M+H]^+$) after low-energy CID displays an intense fragment ion arising from the neutral loss of phosphoric acid (i.e. $[M+H-(HO)_2P(O)OH]^+$) [13]. This outcome is consistent with the notion that the gas-phase phosphoric acid is less competitive for a proton to form a protonated $(HO)_2P(O)OH$ ion as discussed in the introduction of the chapter. Other fragment ions present in the fragmentation pattern of protonated PA molecular species include the acylium ions as well as the ions corresponding to the further dissociation of $[M+H-(HO)_2P(O)OH]^+$ through the loss of the fatty acyls as ketenes at positions *sn*-1 and -2, respectively. These fragment ions are generally present in low abundance. Moreover, identification of the fatty acyls from the fragment ions is possible with efforts. However, assignment of fatty acyl chain regioisomers is impractical.

14.2.2.4.2 Negative-Ion Mode

Dissociation of deprotonated PA molecular species (i.e. $[M-H]^-$) after low-energy CID yields three types of fragment ions. *First*, one or two intense ions corresponding to fatty acyl carboxylate(s). *Second*, a cluster of abundant fragment ions arise from the loss of one fatty acyl as either a fatty acid or a fatty acyl ketene. *Third*, an abundant ion at *m/z* 153 corresponding to glycerophosphate derivative is present.

Charge-driven fragmentation processes involve the participation of the exchangeable hydrogen of the phosphate head group as discussed in the introduction of this chapter [20]. Hence, more intense $[M-H-R_2CH_2CO_2H]^-$ and $[M-H-R_2CH{\cdot}C{\cdot}O]^-$ ions than the $[M-H-R_1CH_2CO_2H]^-$ and $[M-H-R'_1CH{\cdot}C{\cdot}O]^-$ ions, respectively, are yielded from deprotonated PA molecular ions after CID. The difference between these pairs of fragment ions along with the finding that the $R_2CO_2^-$ is more intense than the $R_1CO_2^-$ ion that is commonly used to determine the PA regioisomers. However, it should be reminded that the significantly secondary loss of CO_2 from *sn*-2 polyunsaturated fatty acyl carboxylate [8, 43] may change the peak intensity ratios.

14.2.2.5 Phosphatidylserine

14.2.2.5.1 Positive-Ion Mode

Similar to the PI molecular species, ionization of PS molecular species in the positive-ion mode is not favorable. The fragmentation pattern of protonated PS molecular species is similar to that of protonated PE molecular species. This suggests that the fragmentation processes and the gas-phase basicities of the PE and PS head groups are similar. Specifically, fragmentation of protonated PS molecular species yields a dominant fragment ion corresponding to $[M+H-185]^+$, arising from the elimination of the phosphoserine moiety as that of the PE counterpart [13]. The second type of fragment ions from protonated PS molecular species results from further dissociation of $[M+H-185]^+$, through losses of the fatty acyl chains as ketenes at positions *sn*-1 and 2, respectively. Moreover, the acylium ions (R_xCO^+) are also observed, but in low abundance.

Due to the presence of a primary amine in PS head group, methylation of the amine group to enhance ionization sensitivity and change the fragmentation pattern in the positive-ion mode has been similarly explored to that of PE molecular species [54, 57, 70]. Readers interested in this area of work could find the details in the references.

14.2.2.5.2 Negative-Ion Mode

PS molecular species readily form deprotonated molecular ions (i.e. $[M-H]^-$) in the negative-ion mode. In addition to the formation of this ion, two other relevant ions can also be formed under certain experimental conditions. The first one corresponds to a deprotonated PA counterpart (i.e. $[M\text{-}H\text{-}87]^-$) that resulted from the neutral loss of serine (i.e. 87 amu), representing an ion generated in the ion source due to its facile loss. To minimize this source-generated ion, the ionization conditions should be optimized to minimize this fragmentation as recently discussed [38]. Doubly charged molecular ions of PS molecular species (i.e. $[M\text{-}2H]^{2-}$) could also be formed under basic conditions. The fragmentation pattern from this type of PS molecular ion has not been well characterized.

The fragmentation pattern of deprotonated PS molecular ions includes an intense fragment ion (i.e. $[M\text{-}H\text{-}87]^-$) corresponding to a deprotonated PA counterpart arising from the facile loss of serine (i.e. 87 amu). The rest of the fragmentation pattern of deprotonated PS molecular ions is essentially identical to those resulted from the deprotonated PA counterparts as described above. This result suggests that a loss of serine to the formation of $[M\text{-}H\text{-}87]^-$ is the primary fragmentation process that leads to its further fragmentation. This pathway explains the ion-source generated fragment ion and has been confirmed by MS^n studies utilizing an ion-trap instrument [24].

14.2.2.6 Phosphatidylglycerol

14.2.2.6.1 Positive-Ion Mode

Pure PG molecular species can form protonated ions under acidic conditions, which are suppressed by other GPL lipids such as PC and PE in the case of analysis of biological lipid extracts without pre-separation. Fragmentation of protonated PG species after low-energy CID yields a predominant fragment ion corresponding to the loss of phosphoglycerol (i.e. $[M+H\text{-}(HO)_2P(O)OX]^+$, where X = glycerol) [13]. The fragmentation pattern also contains the acylium ions (R_xCO^+) (where $x = 1, 2$) as well as the ions arising from further loss of the fatty acyl chains as ketenes at positions *sn*-1 and -2, respectively, from protonated PG molecular ions (i.e. $[M+H\text{-}(HO)_2P(O)OX]^+$). These acylium ions are generally present in low abundance. The fatty acyl identities can be determined from these ions, but assigning the regioisomers of PG molecular species is impractical. Hsu and Turk made this type of assignment from the MS^3 analysis [24].

14.2.2.6.2 Negative-Ion Mode

PG molecular species readily form deprotonated molecular ions in the negative-ion mode. The fragmentation pattern of deprotonated PG ions after low-energy CID contains three sets of modest-to-abundant, informative fragment ions. The *first* type of fragment ions corresponds to the fatty acyl carboxylates (i.e. $R_xCO_2^-$, where $x = 1, 2$).

These ions are usually the most abundant ions in the product ion mass spectra of deprotonated PG ions. The $R_2CO_2^-$ ion peak is more intense than that of the $R_1CO_2^-$ ion as discussed in the introduction of the chapter. The *second* fragment ions are a cluster of fragment ions arising from the losses of fatty acyl chains as fatty acids and fatty acyl ketenes, corresponding to $[M\text{-}H\text{-}R_xCH_2CO_2H]^-$ and $[M\text{-}H\text{-}R_xCH{\cdot}C{\cdot}O]^-$ ions, where $x = 1, 2$. The $[M\text{-}H\text{-}R_2CH_2CO_2H]^-$ and $[M\text{-}H\text{-}R_2CH{\cdot}C{\cdot}O]^-$ ions are more abundant than the counterpart ions at $[M\text{-}H\text{-}R_1CH_2CO_2H]^-$ and $[M\text{-}H\text{-}R_1CH{\cdot}C{\cdot}O]^-$ after low-energy CID [23]. Moreover, the loss of a fatty acyl ketene yielding the $[M\text{-}H\text{-}R_2CH{\cdot}C{\cdot}O]^-$ ion is more preferential than the loss of a fatty acid yielding $[M\text{-}H\text{-}R_2CH_2CO_2H]^-$ ion at *sn*-2. In contrast, the $[M\text{-}H\text{-}R_1CH_2CO_2H]^-$ ion is more abundant than the $[M\text{-}H\text{-}R_1CH{\cdot}C{\cdot}O]^-$ ion arising from the analogous losses of the fatty acyl chain at position *sn*-1 as a free acid or a fatty acyl ketene, respectively. This is consistent with the notion that PG is a weakly acidic GPL class [23], and the gas-phase basicity of the $[M-H]^-$ ion is between that of PE and PA. These observations are also consistent with the notion that the α-hydrogen of the fatty acyl chain at position *sn*-2 is more labile and undergoes more facile loss of ketene as aforementioned. The presence of the more intense $[M\text{-}H\text{-}R_2CH{\cdot}C{\cdot}O]^-$ ion peak than that of the $[M\text{-}H\text{-}R_1CH{\cdot}C{\cdot}O]^-$ ion along with the finding that the $R_2CO_2^-$ is more abundant than the $R_1CO_2^-$ ion is readily applicable for the structural identification of PG molecular species including regioisomers. The *third* type of fragments is a set of ions at *m/z* 227, 209, 171, and 153 (which are usually in low abundance) corresponding to the combined loss of a fatty acyl ketene and a fatty acid, the loss of fatty acids, glycerol phosphate, and a phosphoglycerol derivative, respectively. These ions indicate the polar head groups [23].

14.2.2.7 Bis(Monoacylglycero)Phosphate

14.2.2.7.1 Positive-Ion Mode

BMP molecular species are isomeric to PG counterparts. In BMP, two fatty acyl chains are acylated to different glycerol molecules, whereas these fatty acyl chains are located in one glycerol in PG molecular species. What is described above for PG molecular species should be applicable for BMP molecular species. Whether PG and BMP isomers after dissociation in the positive-ion mode can be discriminated remains unknown since no studies on characterization of BMP molecular species in the positive-ion mode have been done. However, it could be anticipated that the fragment ions arising from charge-remote processes occurring preferentially at the *sn*-2 position in PG molecular species, but not in BMP (in which the fatty acyl chain at the *sn*-2 position is absent) [59], could yield differential fragmentation patterns between the PG and BMP isomers. This prediction is evidenced in the fragmentation patterns after methylation of PG and BMP isomers (Figure 14.2). These differential fragmentation patterns allow researchers to readily discriminate the PG and BMP isomers [71].

14.2.2.7.2 Negative-Ion Mode

Extensive characterization of BMP molecular species by ESI-MS after low-energy CID has not been fully conducted, particularly in comparison to PG molecular species. We found that the fragmentation pattern of deprotonated BMP molecular

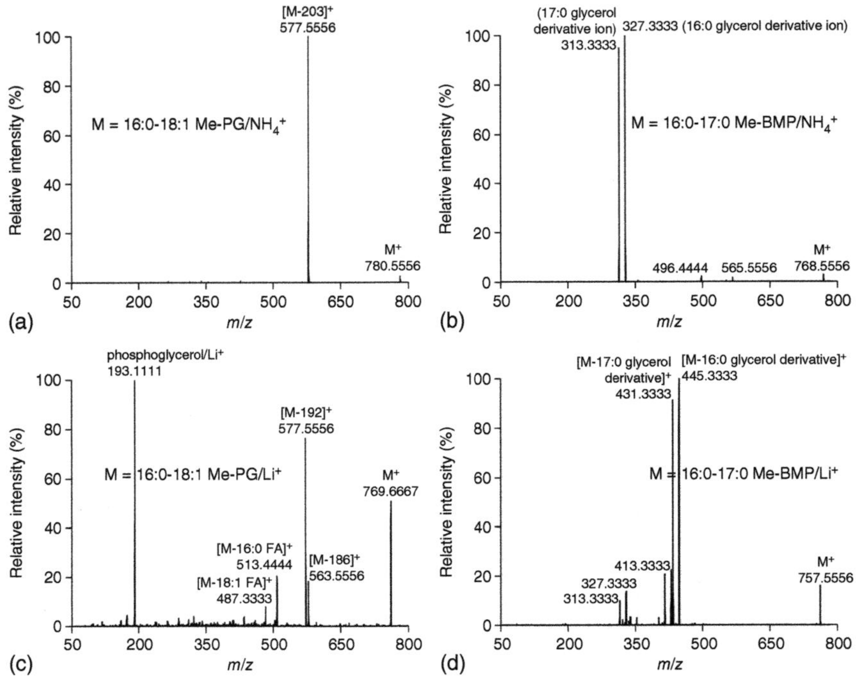

Figure 14.2 Representative product ion analysis of methylated PG and BMP species in the positive-ion mode. Product-ion analyses of ammonium (a and b) or lithium (c and d) adducts of 16:0-18:1 Me-PG (a and c) and 16:0-17:0 Me-BMP (b and d) were performed with a collision energy of 25 eV for ammonium adducts and 40 eV for lithium adducts and gas pressure of 1.0 mT. It should be pointed out that the fragmentation patterns of protonated and sodiated species were identical to those of ammonium and lithium adducts, respectively.

species contains some different fragment features in comparison to those from their isomeric PG species. First, the fatty acyl carboxylates ($R_xCO_2^-$, where $x = 1, 2$) are the most abundant ions in the spectra as in both cases of PG and BMP molecular species, but the intensities of these ions (if two different fatty acyls are present in a BMP species) are essentially identical in the case of BMP, whereas the $R_2CO_2^-$ ion peak is more intense than that of the $R_1CO_2^-$ peak in the case of PG molecular species. It should be remembered that the significantly secondary loss of CO_2 from polyunsaturated fatty acyl carboxylate [8, 43] may change the peak intensity ratios. Moreover, among the cluster of fragment ions at *m/z* 227, 209, 171, and 153 corresponding to the combined loss of a fatty acyl ketene and a fatty acid, the loss of fatty acids, the glycerol phosphate, and a phosphoglycerol derivative, respectively, present in the product ion spectra of PG molecular species, only the phosphoglycerol derivative ion at *m/z* 153 can be detected in low abundance in the case of BMP molecular species. All others are either absent or present in very low intensity relative to the intensity of the ion at *m/z* 153. These distinct features can be used to readily identify PG and BMP isomers to a certain degree [3].

14.2.2.8 Cardiolipin

14.2.2.8.1 Positive-Ion Mode

CL is a unique class of anionic GPL and contains two phosphodiester moieties in each molecule. Thus, ionization of CL molecular species in the positive-ion mode is not sensitive. CL can form alkaline adducts (i.e. $[CL\text{-}2H+3Alk]^+$, Alk = Li and Na) under certain experimental conditions [72]. Dissociation of its alkaline adducts predominantly yields one or a pair of fragment ion(s) from CL molecular species containing identical or different DAG moieties corresponding to the loss of a DAG molecule by using a linear ion trap-type instrument [72]. Further fragmentation of these fragment ions by the instrument yields fragment ions corresponding to the neutral losses of lithium fatty acyl carboxylates [72]. Hsu and Turk used these fragment ions to assign the structure of CL molecular species including fatty acyl positions.

14.2.2.8.2 Negative-Ion Mode

CL molecular species can readily form both singly and doubly charged ions in the negative-ion mode. In most cases by ESI-MS, the peak of $[M-2H]^{2-}$ ion is more intense than that of the $[M-H]^-$ ion. In contrast to ESI-MS, MALDI-MS analysis in the negative-ion mode displays predominant singly charged deprotonated CL molecular species [73, 74].

Dissociation of deprotonated doubly charged CL molecule species containing four identical FA chains after low-energy CID is well documented [31, 75]. Fragmentation of such an ion after low-energy CID yields a predominant carboxylate. The product ion spectra of $[M-2H]^{2-}$ ion also display a doubly charged fragment ion arising from the loss of the fatty acyl ketene. However, the fragment ions corresponding to the loss of fatty acids are not observed, consistent with the notion that the $[M-2H]^{2-}$ ion is a basic precursor ion and undergoes more facile acyl ketene than fatty acid loss [59]. Moreover, the product ion spectra of doubly charged CL molecular species containing four identical fatty acyl chains also give rise to a singly charged fragment ion, corresponding to the residual ion after splitting a carboxylate anion from the $[M-2H]^{2-}$ ion.

The fragmentation pattern of $[M-H]^-$ ion of CL molecular species containing four identical fatty acyl chains is very different from that of their $[M-2H]^{2-}$ ions. Dissociation of the $[M-H]^-$ ion from the molecular species containing four identical fatty acyl chains after low-energy CID gives rise to a fragment ion in modest-to-abundant intensity corresponding to the loss of a free fatty acid and contains ions corresponding to a deprotonated PA anion and a deprotonated dehydrated PG anion (i.e. $[M\text{-}H\text{-}PA]^-$) [20]. The fragment ions arising from the further loss of the fatty acyls as a ketene or a free acid, respectively, corresponding to lysoPA and lysoPG anion derivatives, are also present [31, 75] (Figure 7.5b).

Synthetic CL molecular species containing different fatty acyls are commercially unavailable. Characterization of this type of CL molecular species is usually conducted after direct selection of a CL ion from the full MS analysis of biological lipid extracts. Hence, the selected ions of CL molecular species may represent a mixture of regioisomers and the fragmentation pattern obtained from such studies may not exactly reflect the fragmentation pattern of a single CL molecular species. Han et al. [76] have determined a few CL ions from lipid extracts of mouse myocardium, while Hsu et al. [75] studied some isomeric ions of CL species from different origins including bacteria. Generally, the fragment ions yielded from the synthetic CL molecular species can be displayed in product ion spectra of the natural CL samples. However, the peak intensities of the latter vary from sample to sample, reflecting the presence of different regioisomers. It is intriguing that the ratio of the peak intensities of the ions corresponding to the fatty acyls is essentially equivalent to the ratio of the numbers of each fatty acyl chain [76].

14.2.2.9 Anionic Lysoglycerophospholipids

Characterization of anionic lysoGPL species as deprotonated species after dissociation with low-energy CID has extensively been conducted [29, 66, 77–79]. Fragmentation patterns of deprotonated anionic lysoGPL molecular species are essentially identical to their parent diacyl counterparts. However, the fragment ion at *m/z* 153, corresponding to a glycerophosphate derivative, is much more intense in the fragmentation patterns of anionic lysoGPL molecular species than their diacyl counterparts. It should be recognized that the product ion spectra of deprotonated anionic lysoGPL regioisomers are essentially identical. Hence, discrimination between regioisomers of anionic lysoGPL molecular species is impractical.

14.2.2.10 Other Glycerophospholipids

14.2.2.10.1 N-Acyl Phosphatidylethanolamine

Characterization of the fragmentation patterns of deprotonated *N*-acyl PE molecular species after low-energy CID has been conducted in a few studies [80–83]. Overall, dissociation of these ions after CID yields a few extra features in addition to those of its parent PE counterparts (see above). First, a cluster of fragment ions is present, corresponding to the loss of a fatty acyl as a fatty acid or a fatty acyl ketene. Second, the cluster of fragment ions arising from the losses of fatty acyl as a fatty acid or a fatty acyl ketene becomes more complicated with the combination of the loss of the *N*-acyl as a fatty acyl ketene. However, the fatty acyl carboxylate arising from *N*-acyl is absent. This feature can be used to identify the identity of *N*-acyl.

14.2.2.10.2 N-Acyl Phosphatidylserine

The fragmentation pattern of deprotonated *N*-acyl PS species is essentially identical to that of deprotonated PS species. The only difference is the replacement of a fragment ion corresponding to the loss of serine with *N*-acyl serine [84].

14.2.2.10.3 Acyl Phosphatidylglycerol

The fragmentation pattern of acyl PG is very similar to that of *N*-acyl PE, but is more complicated than that of *N*-acyl PE due to the presence of the additional loss of the third fatty acyl as a fatty acid [60, 81]. The fragmentation pattern of acyl PG after CID includes (i) a cluster of fragment ions arising from the loss of one fatty acyl chain as a fatty acid or a fatty acyl ketene, respectively; (ii) a cluster of fragment ions arising from the losses of two fatty acyls as fatty acids or ketenes or their combinations; and (iii) a cluster of fragment ions corresponding to those of fatty acyl carboxylates, among which the fatty acyl carboxylate arising from the third acyl chain on the head group is usually less intense than those from other two fatty acyl chains, in which a steric effect is present.

14.2.2.10.4 Cyclic Phosphatidic Acid

Fragmentation of deprotonated cyclic PA molecular species by ESI-MS in the negative-ion mode after low-energy CID has been conducted [85]. The fragmentation pattern of deprotonated cyclic PA molecular species contains two prominent fragment ions, one corresponding to the fatty acyl carboxylates of the molecular species and the other at *m/z* 153, corresponding to a glycerophosphate derivative.

Acknowledgments

This work was partially supported by National Institute on Aging RF1 AG061872; National Institute on Aging RF1 AG061729; National Institute on Aging P30 AG013319; National Institute on Aging P30 AG044271; and Methodist Hospital Foundation.

References

1 Vance, D.E. and Vance, J.E. (2008). *Biochemistry of Lipids, Lipoproteins and Membranes*. Amsterdam: Elsevier Science B.V.
2 Zhang, M., Mileykovskaya, E., and Dowhan, W. (2002). Gluing the respiratory chain together. Cardiolipin is required for supercomplex formation in the inner mitochondrial membrane. *J. Biol. Chem.* 277: 43553–43556.
3 Akgoc, Z., Sena-Esteves, M., Martin, D.R. et al. (2015). Bis(monoacylglycero)phosphate: a secondary storage lipid in the gangliosidoses. *J. Lipid Res.* 56: 1006–1013.
4 Hu, C., Zhang, J., Hong, S. et al. (2021). Oxidative stress-induced aberrant lipid metabolism is an important causal factor for dysfunction of immunocytes from patients with systemic lupus erythematosus. *Free Radic. Biol. Med.* 163: 210–219.
5 Palavicini, J.P., Chen, J., Wang, C. et al. (2020). Early disruption of nerve mitochondrial and myelin lipid homeostasis in obesity-induced diabetes. *JCI Insight* 5: e137286.
6 Qiu, S., Palavicini, J.P., Wang, J. et al. (2021). Adult-onset CNS myelin sulfatide deficiency is sufficient to cause Alzheimer's disease-like neuroinflammation and cognitive impairment. *Mol. Neurodegener.* 16: 64.
7 Cheng, H., Mancuso, D.J., Jiang, X. et al. (2008). Shotgun lipidomics reveals the temporally dependent, highly diversified cardiolipin profile in the mammalian brain: temporally coordinated postnatal diversification of cardiolipin molecular species with neuronal remodeling. *Biochemistry* 47: 5869–5880.

8 Yang, K., Zhao, Z., Gross, R.W., and Han, X. (2007). Shotgun lipidomics identifies a paired rule for the presence of isomeric ether phospholipid molecular species. *PLoS One* 2: e1368.

9 Fahy, E., Subramaniam, S., Brown, H.A. et al. (2005). A comprehensive classification system for lipids. *J. Lipid Res.* 46: 839–861.

10 Han, X. and Gross, R.W. (2005). Shotgun lipidomics: electrospray ionization mass spectrometric analysis and quantitation of the cellular lipidomes directly from crude extracts of biological samples. *Mass Spectrom. Rev.* 24: 367–412.

11 Cao, W., Cheng, S., Yang, J. et al. (2020). Large-scale lipid analysis with C=C location and *sn*-position isomer resolving power. *Nat. Commun.* 11: 375.

12 Ma, X. and Xia, Y. (2014). Pinpointing double bonds in lipids by Paternò–Büchi reactions and mass spectrometry. *Angew. Chem. Int. Ed.* 53: 2592–2596.

13 Hsu, F.F. and Turk, J. (2009). Electrospray ionization with low-energy collisionally activated dissociation tandem mass spectrometry of glycerophospholipids: mechanisms of fragmentation and structural characterization. *J. Chromatogr. B* 877: 2673–2695.

14 Murphy, R.C. (2015). *Tandem Mass Spectrometry of Lipids: Molecular Analysis of Complex Lipids*. Cambridge, UK: Royal Society of Chemistry.

15 Han, X. (2016). *Lipidomics: Comprehensive Mass Spectrometry of Lipids*. Hoboken, NJ: Wiley.

16 Han, X., Yang, K., and Gross, R.W. (2012). Multi-dimensional mass spectrometry-based shotgun lipidomics and novel strategies for lipidomic analyses. *Mass Spectrom. Rev.* 31: 134–178.

17 Cheng, C. and Gross, M.L. (2000). Applications and mechanisms of charge-remote fragmentation. *Mass Spectrom. Rev.* 19: 398–420.

18 Han, X. and Gross, R.W. (1996). Structural determination of lysophospholipid regioisomers by electrospray ionization tandem mass spectrometry. *J. Am. Chem. Soc.* 118: 451–457.

19 Hsu, F.-F., Turk, J., Thukkani, A.K. et al. (2003). Characterization of alkylacyl, alk-1-enylacyl and lyso subclasses of glycerophosphocholine by tandem quadrupole mass spectrometry with electrospray ionization. *J. Mass Spectrom.* 38: 752–763.

20 Hsu, F.F. and Turk, J. (2000). Charge-driven fragmentation processes in diacyl glycerophosphatidic acids upon low-energy collisional activation. A mechanistic proposal. *J. Am. Soc. Mass Spectrom.* 11: 797–803.

21 Hsu, F.F. and Turk, J. (2000). Charge-remote and charge-driven fragmentation processes in diacyl glycerophosphoethanolamine upon low-energy collisional activation: a mechanistic proposal. *J. Am. Soc. Mass Spectrom.* 11: 892–899.

22 IUPAC (1994). *Compendium of Chemical Terminology*. Oxford: Blackwell Scientific Publications.

23 Hsu, F.F. and Turk, J. (2001). Studies on phosphatidylglycerol with triple quadrupole tandem mass spectrometry with electrospray ionization: fragmentation processes and structural characterization. *J. Am. Soc. Mass Spectrom.* 12: 1036–1043.

24 Hsu, F.F. and Turk, J. (2005). Studies on phosphatidylserine by tandem quadrupole and multiple stage quadrupole ion-trap mass spectrometry with electrospray ionization: structural characterization and the fragmentation processes. *J. Am. Soc. Mass Spectrom.* 16: 1510–1522.

25 Nakanishi, H., Iida, Y., Shimizu, T., and Taguchi, R. (2010). Separation and quantification of *sn*-1 and *sn*-2 fatty acid positional isomers in phosphatidylcholine by RPLC-ESIMS/MS. *J. Biochem.* 147: 245–256.

26 Taguchi, R. and Ishikawa, M. (2010). Precise and global identification of phospholipid molecular species by an orbitrap mass spectrometer and automated search engine Lipid Search. *J. Chromatogr. A* 1217: 4229–4239.

27 Kind, T., Liu, K.H., Lee Do, Y. et al. (2013). LipidBlast in silico tandem mass spectrometry database for lipid identification. *Nat. Methods* 10: 755–758.

28 Murphy, R.C. and Axelsen, P.H. (2011). Mass spectrometric analysis of long-chain lipids. *Mass Spectrom. Rev.* 30: 579–599.

29 Pulfer, M. and Murphy, R.C. (2003). Electrospray mass spectrometry of phospholipids. *Mass Spectrom. Rev.* 22: 332–364.

30 Hsu, F.-F. and Turk, J. (2003). Electrospray ionization/tandem quadrupole mass spectrometric studies on phosphatidylcholines: the fragmentation processes. *J. Am. Soc. Mass Spectrom.* 14: 352–363.

31 Han, X. and Gross, R.W. (1995). Structural determination of picomole amounts of phospholipids via electrospray ionization tandem mass spectrometry. *J. Am. Soc. Mass Spectrom.* 6: 1202–1210.

32 Hsu, F.-F., Bohrer, A., and Turk, J. (1998). Formation of lithiated adducts of glycerophosphocholine lipids facilitates their identification by electrospray ionization tandem mass spectrometry. *J. Am. Soc. Mass Spectrom.* 9: 516–526.

33 Yang, K., Zhao, Z., Gross, R.W., and Han, X. (2009). Systematic analysis of choline-containing phospholipids using multi-dimensional mass spectrometry-based shotgun lipidomics. *J. Chromatogr. B* 877: 2924–2936.

34 Kerwin, J.L., Tuininga, A.R., and Ericsson, L.H. (1994). Identification of molecular species of glycerophospholipids and sphingomyelin using electrospray mass spectrometry. *J. Lipid Res.* 35: 1102–1114.

35 Khaselev, N. and Murphy, R.C. (2000). Electrospray ionization mass spectrometry of lysoglycerophosphocholine lipid subclasses. *J. Am. Soc. Mass Spectrom.* 11: 283–291.

36 Weintraub, S.T., Pinckard, R.N., and Hail, M. (1991). Electrospray ionization for analysis of platelet-activating factor. *Rapid Commun. Mass Spectrom.* 5: 309–311.

37 Houjou, T., Yamatani, K., Nakanishi, H. et al. (2004). Rapid and selective identification of molecular species in phosphatidylcholine and sphingomyelin by conditional neutral loss scanning and MS^3. *Rapid Commun. Mass Spectrom.* 18: 3123–3130.

38 Hu, C., Luo, W., Xu, J., and Han, X. (2022). Recognition and avoidance of ion source-generated artifacts in lipidomics analysis. *Mass Spectrom. Rev.* 41: 15–31.

39 Han, X., Yang, K., Yang, J. et al. (2006). Factors influencing the electrospray intrasource separation and selective ionization of glycerophospholipids. *J. Am. Soc. Mass Spectrom.* 17: 264–274.

40 Ejsing, C.S., Duchoslav, E., Sampaio, J. et al. (2006). Automated identification and quantification of glycerophospholipid molecular species by multiple precursor ion scanning. *Anal. Chem.* 78: 6202–6214.

41 Ekroos, K., Chernushevich, I.V., Simons, K., and Shevchenko, A. (2002). Quantitative profiling of phospholipids by multiple precursor ion scanning on a hybrid quadrupole time-of-flight mass spectrometer. *Anal. Chem.* 74: 941–949.

42 Ekroos, K., Ejsing, C.S., Bahr, U. et al. (2003). Charting molecular composition of phosphatidylcholines by fatty acid scanning and ion trap MS^3 fragmentation. *J. Lipid Res.* 44: 2181–2192.

43 Yang, K., Zhao, Z., Gross, R.W., and Han, X. (2011). Identification and quantitation of unsaturated fatty acid isomers by electrospray ionization tandem mass spectrometry: a shotgun lipidomics approach. *Anal. Chem.* 83: 4243–4250.

44 Ford, D.A., Rosenbloom, K.B., and Gross, R.W. (1992). The primary determinant of rabbit myocardial ethanolamine phosphotransferase substrate selectivity is the covalent nature of the *sn*-1 aliphatic group of diradyl glycerol acceptors. *J. Biol. Chem.* 267: 11222–11228.

45 Kayganich, K.A. and Murphy, R.C. (1992). Fast atom bombardment tandem mass spectrometric identification of diacyl, alkylacyl, and alk-1-enylacyl molecular species of glycerophosphoethanolamine in human polymorphonuclear leukocytes. *Anal. Chem.* 64: 2965–2971.

46 Fhaner, C.J., Liu, S., Zhou, X., and Reid, G.E. (2013). Functional group selective derivatization and gas-phase fragmentation reactions of plasmalogen glycerophospholipids. *Mass Spectrom. (Tokyo)* 2: S0015.

47 Schwudke, D., Hannich, J.T., Surendranath, V. et al. (2007). Top-down lipidomic screens by multivariate analysis of high-resolution survey mass spectra. *Anal. Chem.* 79: 4083–4093.

48 Brugger, B., Erben, G., Sandhoff, R. et al. (1997). Quantitative analysis of biological membrane lipids at the low picomole level by nano-electrospray ionization tandem mass spectrometry. *Proc. Natl. Acad. Sci. U.S.A.* 94: 2339–2344.

49 Kim, H.Y., Wang, T.C., and Ma, Y.C. (1994). Liquid chromatography/mass spectrometry of phospholipids using electrospray ionization. *Anal. Chem.* 66: 3977–3982.

50 Han, X. and Gross, R.W. (2003). Global analyses of cellular lipidomes directly from crude extracts of biological samples by ESI mass spectrometry: a bridge to lipidomics. *J. Lipid Res.* 44: 1071–1079.

51 Brouwers, J.F., Vernooij, E.A., Tielens, A.G., and Van Golde, L.M. (1999). Rapid separation and identification of phosphatidylethanolamine molecular species. *J. Lipid Res.* 40: 164–169.

52 Hsu, F.F. and Turk, J. (2000). Characterization of phosphatidylethanolamine as a lithiated adduct by triple quadrupole tandem mass spectrometry with electrospray ionization. *J. Mass Spectrom.* 35: 595–606.

53 Bilgin, M., Markgraf, D.F., Duchoslav, E. et al. (2011). Quantitative profiling of PE, MMPE, DMPE, and PC lipid species by multiple precursor ion scanning: a tool for monitoring PE metabolism. *Biochim. Biophys. Acta* 1811: 1081–1089.

54 Shields, S.W.J., Rosales, C.A., Roberts, J.A. et al. (2021). iTrEnDi: In situ trimethylation enhancement using diazomethane: improved and expanded glycerophospholipid and sphingolipid analyses via a microscale autonomous derivatization platform. *Anal. Chem.* 93: 1084–1091.

55 Wang, M., Kim, G.H., Wei, F. et al. (2015). Improved method for quantitative analysis of methylated phosphatidylethanolamine species and its application for analysis of diabetic-mouse liver samples. *Anal. Bioanal. Chem.* 407: 5021–5032.

56 Wang, X., Wei, F., Xu, J.Q. et al. (2016). Profiling and relative quantification of phosphatidylethanolamine based on acetone stable isotope derivatization. *Anal. Chim. Acta* 902: 142–153.

57 Wasslen, K.V., Canez, C.R., Lee, H. et al. (2014). Trimethylation enhancement using diazomethane (TrEnDi) II: rapid in-solution concomitant quaternization of glycerophospholipid amino groups and methylation of phosphate groups via reaction with diazomethane significantly enhances sensitivity in mass spectrometry analyses via a fixed, permanent positive charge. *Anal. Chem.* 86: 9523–9532.

58 Han, X. and Gross, R.W. (1994). Electrospray ionization mass spectroscopic analysis of human erythrocyte plasma membrane phospholipids. *Proc. Natl. Acad Sci. U.S.A.* 91: 10635–10639.

59 Hsu, F.-F. and Turk, J. (2000). Characterization of phosphatidylinositol, phosphatidylinositol-4-phosphate, and phosphatidylinositol-4,5-bisphosphate by electrospray ionization tandem mass spectrometry: a mechanistic study. *J. Am. Soc. Mass Spectrom.* 11: 986–999.

60 Hsu, F.F., Turk, J., Shi, Y., and Groisman, E.A. (2004). Characterization of acylphosphatidylglycerols from *Salmonella typhimurium* by tandem mass spectrometry with electrospray ionization. *J. Am. Soc. Mass Spectrom.* 15: 1–11.

61 Schwudke, D., Oegema, J., Burton, L. et al. (2006). Lipid profiling by multiple precursor and neutral loss scanning driven by the data-dependent acquisition. *Anal. Chem.* 78: 585–595.

62 Berry, K.A. and Murphy, R.C. (2005). Analysis of cell membrane aminophospholipids as isotope-tagged derivatives. *J. Lipid Res.* 46: 1038–1046.

63 Han, X., Yang, K., Cheng, H. et al. (2005). Shotgun lipidomics of phosphoethanolamine-containing lipids in biological samples after one-step in situ derivatization. *J. Lipid Res.* 46: 1548–1560.

64 Zemski Berry, K.A., Turner, W.W., Vannieuwenhze, M.S., and Murphy, R.C. (2009). Stable isotope labeled 4-(dimethylamino)benzoic acid derivatives of glycerophosphoethanolamine lipids. *Anal. Chem.* 81: 6633–6640.

65 Wang, M., Han, R.H., and Han, X. (2013). Fatty acidomics: global analysis of lipid species containing a carboxyl group with a charge-remote fragmentation-assisted approach. *Anal. Chem.* 85: 9312–9320.

66 Chen, S. (1997). Tandem mass spectrometric approach for determining structure of molecular species of aminophospholipids. *Lipids* 32: 85–100.

67 Clark, J., Anderson, K.E., Juvin, V. et al. (2011). Quantification of $PtdInsP_3$ molecular species in cells and tissues by mass spectrometry. *Nat. Methods* 8: 267–272.

68 Kielkowska, A., Niewczas, I., Anderson, K.E. et al. (2014). A new approach to measuring phosphoinositides in cells by mass spectrometry. *Adv. Biol. Regul.* 54: 131–141.

69 Wang, C., Palavicini, J.P., Wang, M. et al. (2016). Comprehensive and quantitative analysis of polyphosphoinositide species by shotgun lipidomics revealed their alterations in db/db mouse brain. *Anal. Chem.* 88: 12137–12144.

70 Betancourt, S.K., Canez, C.R., Shields, S.W.J. et al. (2017). Trimethylation enhancement using ^{13}C-diazomethane: gas-phase charge inversion of modified phospholipid cations for enhanced structural characterization. *Anal. Chem.* 89: 9452–9458.

71 Wang, M., Palavicini, J.P., Cseresznye, A., and Han, X. (2017). Strategy for quantitative analysis of isomeric bis(monoacylglycero)phosphate and phosphatidylglycerol species by shotgun lipidomics after one-step methylation. *Anal. Chem.* 89: 8490–8495.

72 Hsu, F.F. and Turk, J. (2010). Toward total structural analysis of cardiolipins: multiple-stage linear ion-trap mass spectrometry on the $[M\text{-}2H\text{+}3Li]^+$ ions. *J. Am. Soc. Mass Spectrom.* 21: 1863–1869.

73 Sun, G., Yang, K., Zhao, Z. et al. (2008). Matrix-assisted laser desorption/ ionization time-of-flight mass spectrometric analysis of cellular glycerophospholipids enabled by multiplexed solvent dependent analyte-matrix interactions. *Anal. Chem.* 80: 7576–7585.

74 Wang, H.Y., Jackson, S.N., and Woods, A.S. (2007). Direct MALDI-MS analysis of cardiolipin from rat organs sections. *J. Am. Soc. Mass Spectrom.* 18: 567–577.

75 Hsu, F.F., Turk, J., Rhoades, E.R. et al. (2005). Structural characterization of cardiolipin by tandem quadrupole and multiple-stage quadrupole ion-trap mass spectrometry with electrospray ionization. *J. Am. Soc. Mass Spectrom.* 16: 491–504.

76 Han, X., Yang, K., Yang, J. et al. (2006). Shotgun lipidomics of cardiolipin molecular species in lipid extracts of biological samples. *J. Lipid Res.* 47: 864–879.

77 Lee, J.Y., Min, H.K., and Moon, M.H. (2011). Simultaneous profiling of lysophospholipids and phospholipids from human plasma by nanoflow liquid chromatography-tandem mass spectrometry. *Anal. Bioanal. Chem.* 400: 2953–2961.

78 Wang, C., Wang, M., and Han, X. (2015). Comprehensive and quantitative analysis of lysophospholipid molecular species present in obese mouse liver by shotgun lipidomics. *Anal. Chem.* 87: 4879–4887.

79 Xiao, Y., Chen, Y., Kennedy, A.W. et al. (2000). Evaluation of plasma lysophospholipids for diagnostic significance using electrospray ionization mass spectrometry (ESI-MS) analyses. *Ann. N.Y. Acad. Sci.* 905: 242–259.

80 Astarita, G., Ahmed, F., and Piomelli, D. (2008). Identification of biosynthetic precursors for the endocannabinoid anandamide in the rat brain. *J. Lipid Res.* 49: 48–57.

81 Holmback, J., Karlsson, A.A., and Arnoldsson, K.C. (2001). Characterization of *N*-acylphosphatidylethanolamine and acylphosphatidylglycerol in oats. *Lipids* 36: 153–165.

82 Kilaru, A., Tamura, P., Isaac, G. et al. (2012). Lipidomic analysis of *N*-acylphosphatidylethanolamine molecular species in Arabidopsis suggests feedback regulation by *N*-acylethanolamines. *Planta* 236: 809–824.

83 Mileykovskaya, E., Ryan, A.C., Mo, X. et al. (2009). Phosphatidic acid and *N*-acylphosphatidylethanolamine form membrane domains in *Escherichia coli* mutant lacking cardiolipin and phosphatidylglycerol. *J. Biol. Chem.* 284: 2990–3000.

84 Guan, Z., Li, S., Smith, D.C. et al. (2007). Identification of *N*-acylphosphatidylserine molecules in eukaryotic cells. *Biochemistry* 46: 14500–14513.

85 Shan, L., Li, S., Jaffe, K., and Davis, L. (2008). Quantitative determination of cyclic phosphatidic acid in human serum by LC/ESI/MS/MS. *J. Chromatogr. B Analyt. Technol. Biomed. Life Sci.* 862: 161–167.

15

Sphingolipids

Lukas Opalka[1], Lisa Schlicker[2,3], and Roger Sandhoff[4]

[1] Charles University, Skin Barrier Research Group, Department of Organic and Bioorganic Chemistry, Faculty of Pharmacy in Hradec Králové, Akademika Heyrovskeho 1203, 500 05 Hradec Králové, Czech Republic
[2] German Cancer Research Center, Tumor Metabolism and Microenvironment, Im Neuenheimer Feld 581, 69120 Heidelberg, Germany
[3] German Cancer Research Center, Protein analysis, Im Neuenheimer Feld 580, 69120 Heidelberg, Germany
[4] German Cancer Research Center, Lipid Pathobiochemistry Group, Im Neuenheimer Feld 581, 69120 Heidelberg, Germany

15.1 Introduction

The core building block of sphingolipids (SLs) is a long aliphatic chain aminoalcohol. Its most prominent member is C18-sphingosine (C18-So), first discovered by J.L.W. Thudichum in alcoholic brain extracts in 1884 [1] and structurally characterized by H. Carter as 2-aminooctadec-4-ene-1,3-diol [2]. In the following years, the stereochemistry of the chiral compound was identified as (2*S*,3*R*,4*E*)-2-aminooctadec-4-ene-1,3-diol. However, depending on organism, organ, and age, the long chain base structure may vary in chain length, saturation grade, stereochemistry, and degree of hydroxylation giving rise to a broad group of so-called sphingoid bases (SBs, see Figure 15.1, for detailed review see [3]). Sphingoid bases can be metabolized further in cells, which includes N-acylation (ceramides, Cer), 1-O-phosphorylation (SB 1-phosphates like sphingosine 1-phosphate [So1P], ceramide 1-phosphates, Cer1P), 1-*O*-phosphocholine attachment (sphingomyelins, SM), 1-O-acylation (1-*O*-acylceramides, 1-OACs), and 1-O-glycosylation. The latter gives rise to large group of glycosphingolipids (GSLs), which upon addition of sialic acids or sulfate groups turn into gangliosides (GG) or sulfatides (Figure 15.2) [4–6].

To date, more than 40 000 hits come up when searching the literature for either "sphingolipid," "cerebroside," "ganglioside," "sphingomyelin," or "ceramide." In combination with "mass spectrometry" more than 6000 hits remain, emphasizing the important aspect of mass spectrometric analysis in sphingolipid research. As a consequence, this chapter will cover only some aspects of SLs in mass spectrometry.

Mass Spectrometry for Lipidomics: Methods and Applications, First Edition.
Edited by Michal Holčapek and Kim Ekroos.

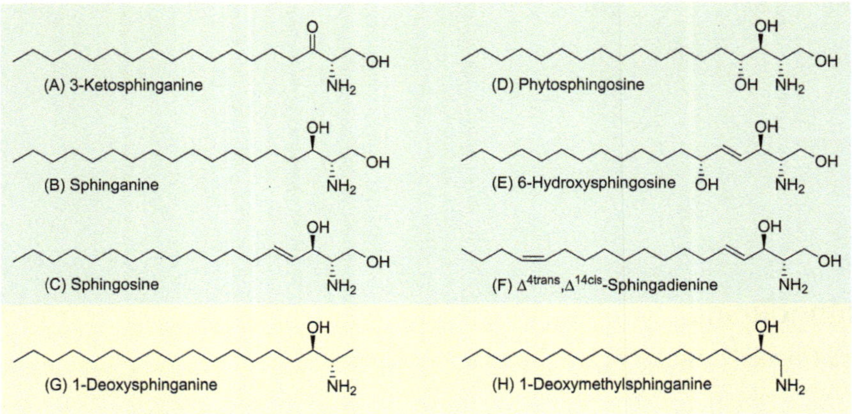

Figure 15.1 Canonical (green background) and non-canonical (yellow background) sphingoid bases of vertebrates: (a) 3-ketosphinganine, (b) sphinganine, (c) sphingosine, (d) phytosphingosine, (e) 6-hydroxy sphingosine, (f) Δ^{4trans},Δ^{14cis}-sphingadiene, (g) 1-deoxysphinganine, and (h) 1-deoxymethylsphinganine. Structures derived by condensation with palmitoyl-CoA are plotted, but chain length may vary.

Starting out with SLs in mammals/vertebrates, we will use the biosynthetic path of SLs to introduce individual groups of compounds together with aspects of mass spectrometric analysis.

15.2 Sphingolipid Nomenclature

Sphingolipid nomenclature defines sphinganine (dihydrosphingosine) and sphing-4-enine (sphingosine) as key structures with specific stereochemistry as D-*erythro* or 2*S*,3*R*,4*E* (for sphingosine). If stereochemistry differs from D-*erythro* configuration, the full systematic name has to be used instead. Shorthand nomenclature is commonly used for sphingolipid description to simplify the text searches, and it is also widely used in sphingolipid research. Backbones, head groups, and sugar moieties are abbreviated according to the sphingolipid major classes as Cer, Cer1P, SM, GlcCer, and the radyl substituents are denoted by an indication of carbon chain length, number of double bonds (separated by a colon) and eventually additional hydroxylation. For example ω-*O*-acylceramide structure systematically named as *N*-(32-(9*Z*,12*Z*-octadecadienoyloxy)dotriacontanoyl)sphing-4-enine is then shortened to Cer(d18:1/h32:0/18:2) meaning that it contains an 18-carbon sphingoid base with one double bond and two ("d" for di) OH groups (= sphingosine); it is N-acylated with 32-carbon acyl chain containing an ω-OH group, which is further esterified with linoleic acid (Figure 15.3) [7]. Hence, sphingoid bases with one, two, or three hydroxyl groups are annotated with the letter m (mono), d, (di), or t (tri) in front of the number of carbon atoms. Complex SLs like gangliosides or sulfatides then use a specific nomenclature corresponding to their carbohydrate composition in position 1 of the sphingoid base which is grouped into families based on the

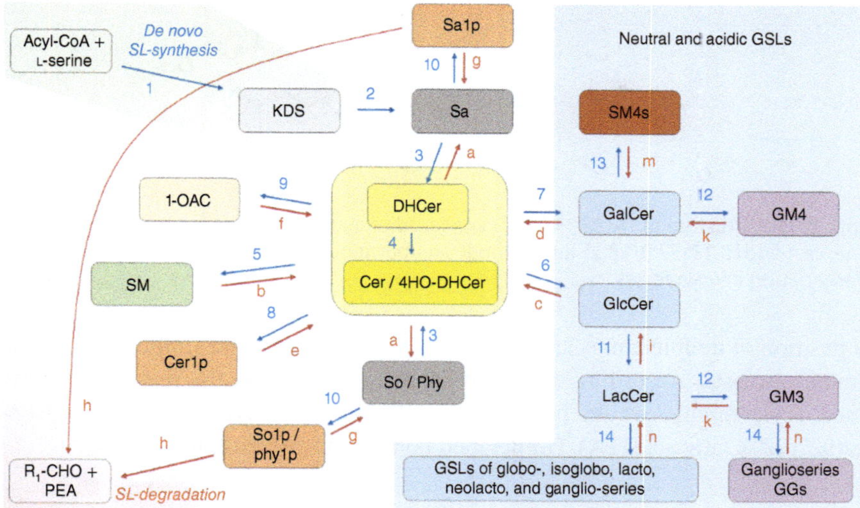

Figure 15.2 Canonical sphingolipid metabolism in vertebrates: Serine palmitoyltransferase complex (SPT, 1) condensates L-serine with acyl CoAs, mainly palmitoyl-CoA to yield 3-ketosphinganines (also called 3-ketodihydrosphingosines, KDS), which are quickly converted to sphinganines (Sa) by ketodihydrosphingosine reductase (KDSR, 2). Subsequently sphinganines are N-acylated by ceramide synthases (3) to yield dihydroceramides (DHCer). In humans and mice there are six ceramide synthases with different preferences for the acyl-CoA chain length. DHCers are substrates for desaturase DEGS1 or hydroxylase DEGS2 (4) yielding ceramides with either a trans-Δ^4-double bond (Cer) or a 4-hydroxy-group (4HO-DHCer/phytoceramide), sphingomyelin synthases (5), ceramide kinases (8), glucosyl- (6) and galactosylceramide synthases (7) synthesizing sphingomyelins (SM), ceramide 1-phosphates (Cer1P), glucosylceramides (GlcCers), and galactosylceramides (GalCers). Diacylglycerol acyl transferases (DGAT1 and DGAT2) and so far unknown acyltransferases in epidermis can also 1-O-acylate ceramides (9) to produce 1-O-acylceramides (1-OAC). GalCers and GlcCers and their downstream products are substrates of glycosyltransferases and sulfotransferases (11, β4-galactosyl transferases, 12, GM4/GM3-synthase, 13, cerebroside sulfotransferase, 14, various glycosyl- and sialyltransferases). Catabolism of sphingolipids (SLs) is also a stepwise process and the sphingoid bases sphingosine (So) and phytosphingosine (Phy) are generated by hydrolysis of corresponding ceramides (a). Free sphingoid bases are further substrates for sphingosine kinases 1 and 2 (10) and corresponding SB 1-phosphates may be cleaved by sphingosine-1-phosphate lyase 1 (SGPL1, h) to yield long chain aldehydes (R_1-CHO, R_1 is an alk-2-enal [from So], an alkanal [from Sa] or a 2 hydroxy alkanal [from Phy]) and phosphoethanolamine (PEA). The latter reaction is the only one for cells to completely degrade SLs. (a) ceramidases (b) sphingomyelinases, (c) glucosylceramidases, (d) β-galactosylceramidase, (e) Cer1P phosphatases and acidic sphingomyelinase, (f) not described, (g) S1P phosphatases, (h) SGPL1, (i) β-galactosidases, (k) neuraminidases, (m) arylsulfatase A, and (n) various glycosidases and neuraminidases.

nature of the glyco-substituents [8]. This shorthand nomenclature will be used throughout this chapter. Such shorthand nomenclature has serious drawbacks, especially when the structure is more complex or contains non-canonical lipid parts. Thus, LIPID MAPS recommend a modified nomenclature for more precise reporting of all lipid species. Such nomenclature includes an annotation of ring double bonds and number of oxygens, allows distinguishing between species where

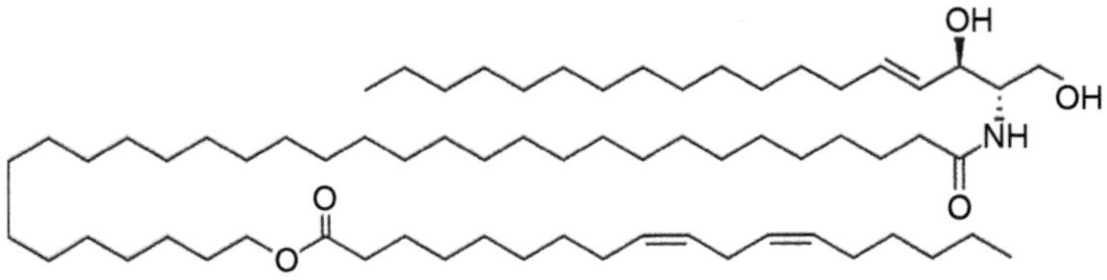

Figure 15.3 Structure of *N*-(32-(9*Z*,12*Z*-octadecadienoyloxy)dotriacontanoyl)sphing-4-enine, i.e. Cer(d18:1/h32:0/18:2), an essential skin barrier ω-esterified ceramide, produced by differentiating keratinocytes.

the position of double bonds and oxygens is known or not, and simplifies the data mining processes. According to this complex nomenclature, the previously mentioned acylceramide would be named as Cer 18:1(4*E*);1OH,3OH/32:0,32O(FA 18:2(9*Z*,12*Z*)), fatty acid (FA). For detailed nomenclature please see [9].

15.3 General Aspects of Sphingolipids in Mass Spectrometry

SLs are amphiphilic lipids which are nowadays most often identified from lipid raw extracts of biological samples using untargeted high-resolution tandem mass spectrometry eventually coupled with liquid chromatography or using targeted liquid chromatography-coupled tandem mass spectrometry, especially when working with low-resolution instruments. Ionization of intact SLs may be achieved with various ionization methods including the commonly used electrospray ionization (ESI) and matrix-associated laser desorption/ionization (MALDI). These methods lead to various molecular adduct ions (positive mode: e.g. $[M+H]^+$, $[M+NH_4]^+$, $[M+\text{Alkali metal}]^+$ and negative mode: $[M-H]^-$, $[M+Cl]^-$, $[M+\text{solvent}-H]^-$). Due to the 2-amino group of the SB, most measurements consider positive mode. For tandem mass spectrometry protonated and deprotonated ions are most important, but other adduct ions may be used for specific purposes as they may lead to different fragmentation patterns. Besides molecular ion adducts, in-source decay (ISD) may sometimes lead to ion signals, which are present due to loss of water or of labile sugar moieties. In certain cases, this may lead to false-positive SL signals, which are best recognized on LC-MS systems due to wrong retention times. For example disialogangliosides of the GD1 group may give rise to ISD of one sialic acid $\Delta(\text{NeuNAc} - H_2O)$, which gives rise to a signal and compound equivalent to monosialoganglioside GM1. Another critical example would be loss of H_2O by ISD of neutral SLs containing a sphingosine backbone and an α-hydroxy fatty acid. This will lead to a mass spectrometric signal equivalent to corresponding neutral sphingolipids with a sphingosine backbone and corresponding non-hydroxylated and monounsaturated fatty acid, e.g. the *m/z* of Cer(d18:1/h24.0) minus loss of H_2O is isobaric to the *m/z* of Cer(d18:1/24:1).

15.4 Sphingolipids in Vertebrates

15.4.1 Sphingoid Bases

Synthesis of sphingoid bases is initialized in eukaryotes on the cytosolic side of the endoplasmic reticulum (ER) by the pyridoxal phosphate-dependent SPT complex composed of two large subunits (in humans SPTLC1 and SPTLC2 or SPTLC1 and SPTLC3) and a small subunit (serine palmitoyl transferase small subunit A [SPTSSA] or serine palmitoyl transferase small subunit B [SPTSSB]). Its activity is regulated by the inhibitory ORMDL1-3 proteins. SPT has a high specificity for L-serine thereby giving rise to 3-ketodihydrosphingosines (KDS) as products. Depending on substrate availability and composition of the SPT complex, KDS with different chain lengths can be obtained. Most mammalian cells however will produce C18-KDS due to a strong preference of the common SPT complex (SPTLC1, SPTLC2, and SPTSSA) toward palmitoyl-CoA. Different sphingoid base chain lengths are observed for example in nervous tissue upon aging (C20-SBs) and in human epidermis (C16 up to C28). Under normal conditions, KDSs are hardly detected as KDS is quickly converted to sphinganine (Sa) by 3-ketodihydrosphingosine reductase (KDSR). However, genetic mutations in the *KDSR* gene causing a rare type of autosomal recessive ichthyosis lead to accumulation of KDS in stratum corneum of the skin correlating with the broad chain length pattern of human skin barrier sphingoid bases (Figure 15.4). Other sphingoid bases than sphinganine are generated after N-acylation from dihydroceramides in mammals (Figure 15.2). Corresponding ceramides then have to be hydrolyzed by ceramidases to generate free sphingosine, phytosphingosine, 6-hydroxy sphingosine, or sphingadiene/sphingadienine [10–13]. To quantify the relative amount of free and bound sphingoid bases in biological samples, strong acid hydrolysis may be performed, which releases functional groups linked in 1-O and N-positions [13].

Due to genetic mutations in SPTLC1 or SPTLC2, the SPT complex may lose its specificity toward L-serine and incorporation of alanine and glycine instead were observed, giving rise to 1-deoxy- and 1-deoxymethyl-sphinganines, respectively [14–17] (Figure 15.5). In mammalian cells, these compounds were initially observed after fumonisin B treatment and are still substrates for ceramide synthases giving rise to 1-deoxy(methyl) ceramides [18]. Missing, however, the 1-*O*-hydroxy group, they are somehow dead-end products that cannot be converted to corresponding Cer1P, SM, 1-OAC, or GSLs [17].

Whereas sphingosines (dX:1), phytosphingosines (tX:0), 6-hydroxy-sphingosines (tX:1), and sphingadienes (dX:2) differ from sphinganines (dX:0) with same carbon number (X) by a nominal mass of −2, +16, +14, −4 amu, 3-KDS are isobaric to sphingosines, i.e. absolutely identical in mass and have to be discriminated by either MS^2, chromatography, or eventually ion mobility. Due to the keto group in 3 position collision-induced dissociation (CID) of protonated KDS will preferentially lead to a McLafferty (McL)-like rearrangement and loss of formaldehyde (Δ30 amu).

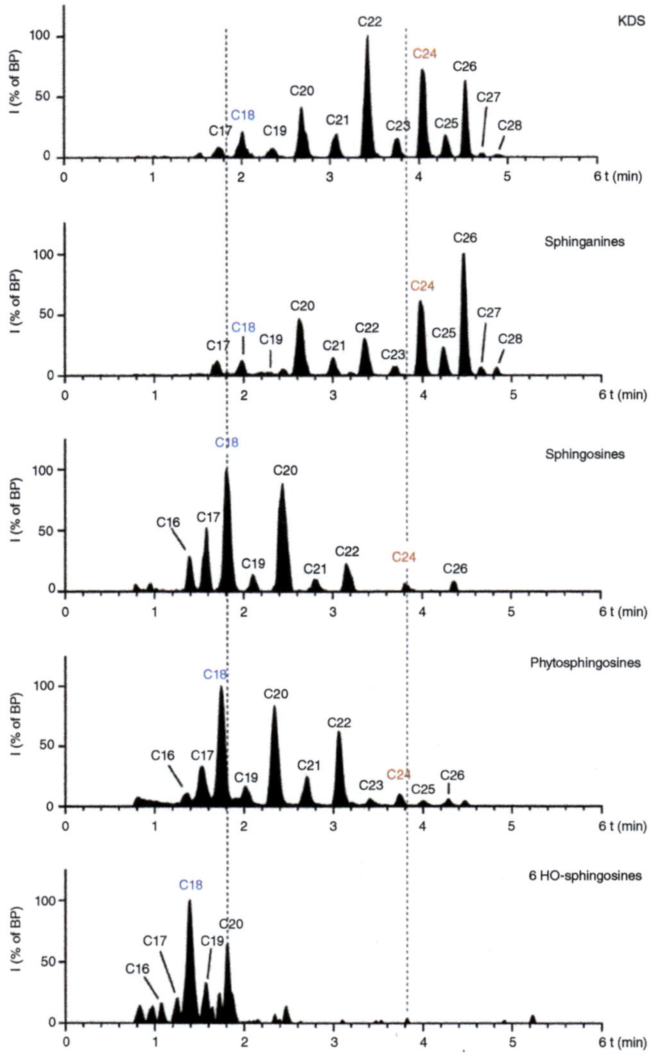

Figure 15.4 C18-ultra-performance liquid chromatography (UPLC)-ESI-(QqQ)MS^2 of sphingoid bases recorded from a lipid raw extract of stratum corneum from a patient with biallelic mutations in the KDSR gene demonstrating the complex SB-composition and broad chain length pattern (C16 up to C28) in human skin barrier (inner forearm). Chromatograms represent MRMs with the transitions of $[M+H]^+$ - $[M+H-H_2CO]^+$ (KDS, 3-ketodihydrosphingosine), $[M+H]^+$ - $[M+H-(H_2O+H_2CO)]^+$ (sphinganines [Sa] and sphingosines [So]), $[M+H]^+$ - $[M+H-(2H_2O)]^+$ (phytosphingosines [Phy]), $[M+H]^+$ - $[M+H-(2H_2O+H_2CO)]^+$ (6-hydroxy sphingosines [6HO-So]). With reversed-phase C18-chromatography, sphingoid bases with a more polar head group, but the same number of carbons elute earlier, i.e. t_R(6HO-So) < t_R(Phy) < t_R(So) < t_R(Sa) < t_R(KDS). Likewise corresponding SBs elute increasingly later with growing carbon chain.

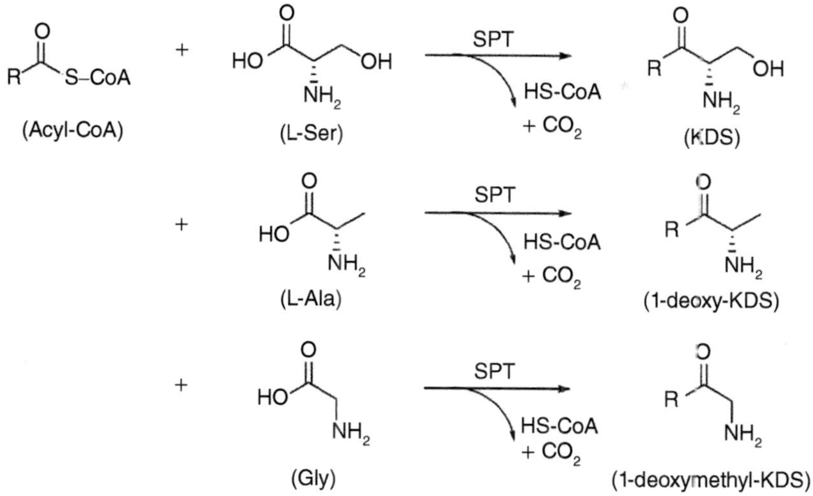

Figure 15.5 The specificity of the serine palmitoyl transferase (SPT) complex toward amino acids determines the ratio of canonical (KDS-derived) and non-canonical (1-deoxy(methyl)-KDS-derived) sphingoid bases. Mutation in the SPTLC1 or SPTLC2 gene may lead to increased synthesis of 1-deoxy-sphinganine and 1-deoxymethyl-sphinganine.

This type of rearrangement is observed in 3-hydroxy-SBs only after primary loss of water, which will establish a C—C double bond in the right position. Therefore, the transition of $[KDS+H]^+$ to $[KDS+H-H_2CO]^+$ was successfully used to specifically quantify KDS from biological samples [13, 19, 20].

Protonated sphingoid bases preferentially undergo loss of water molecules, and to minor extent additional loss of formaldehyde when subjected to CID in tandem mass spectrometry (Figure 15.6) [21–23]. Ultraviolet photodissociation at 193 nm will lead to additional structurally diagnostic product ions [24]. In order to identify and quantify sphingoid bases from complex lipid extracts of biological samples by LC-MS^2 low-resolution instruments like triple quadrupole instruments, these instruments are often used in the most sensitive multi-reaction monitoring (MRM) mode. Although the most intense transitions in MS^2 are preferred for this mode, it is rather the low background noise associated with the respective transitions, which is more important for specific quantifications when dealing with complex biological mixtures. Therefore, we preferred to use rather weak transitions including loss of formaldehyde when analyzing free sphingoid bases from biological samples like human stratum corneum (Figure 15.4).

Missing the 1-hydroxy(methyl) group, 1-deoxy(methyl)sphinganines show only a single water loss upon low-energy CID (Figure 15.6). Corresponding quantification by MRM from complex lipid mixtures, therefore, may rather be associated with higher background signal and taking chromatographic retention time into account is even more important than with canonical sphingoid bases.

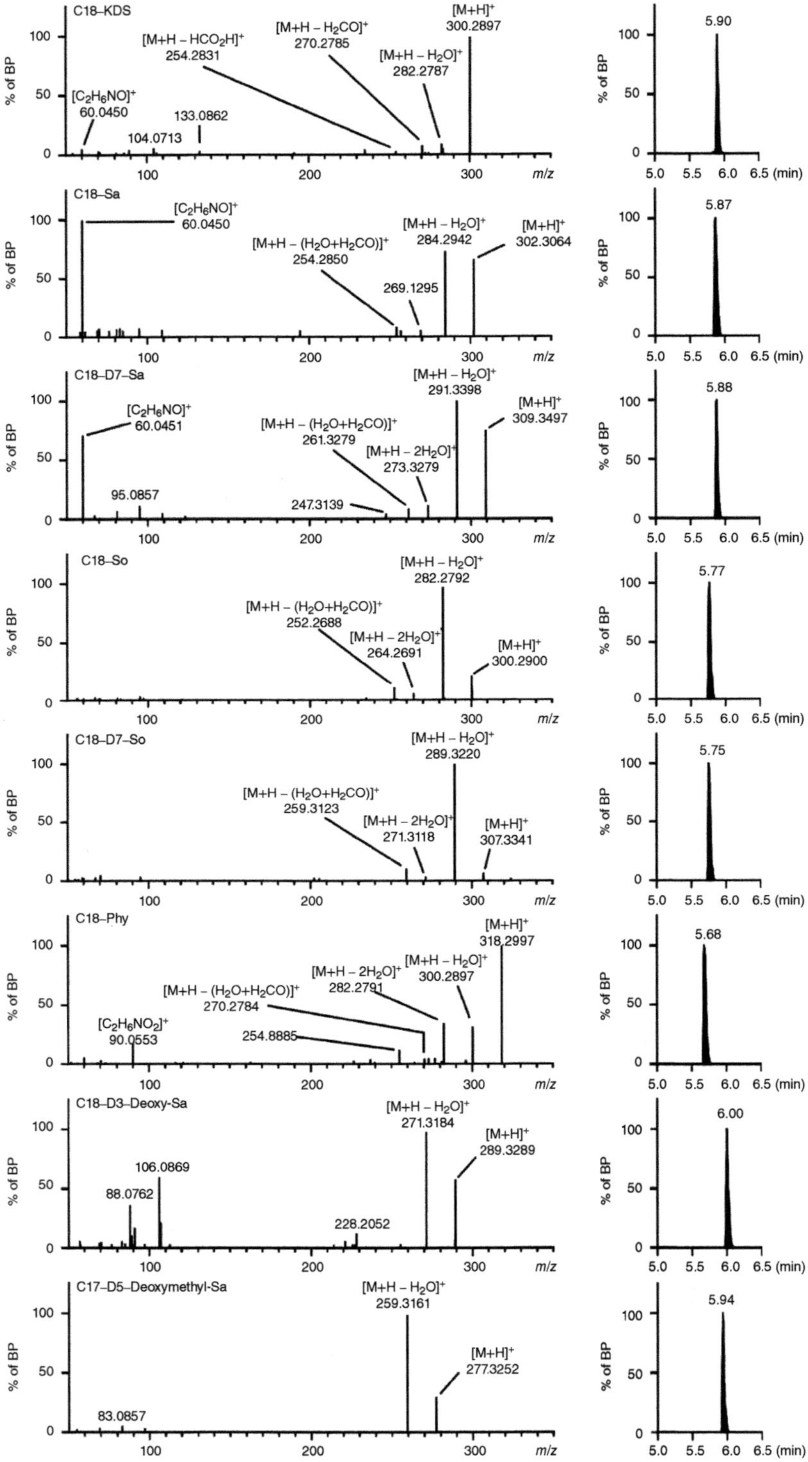
C18–KDS
[M+H]+ 300.2897
[M+H – H2CO]+ 270.2785
[M+H – HCO2H]+ 254.2831
[M+H – H2O]+ 282.2787
[C2H6NO]+ 60.0450
133.0862
104.0713
5.90
C18–Sa
[C2H6NO]+ 60.0450
[M+H – H2O]+ 284.2942
[M+H]+ 302.3064
[M+H – (H2O+H2CO)]+ 254.2850
269.1295
5.87
C18–D7–Sa
[M+H – H2O]+ 291.3398
[M+H]+ 309.3497
[C2H6NO]+ 60.0451
[M+H – (H2O+H2CO)]+ 261.3279
[M+H – 2H2O]+ 273.3279
95.0857
247.3139
5.88
C18–So
[M+H – H2O]+ 282.2792
[M+H – (H2O+H2CO)]+ 252.2688
[M+H – 2H2O]+ 264.2691
[M+H]+ 300.2900
5.77
C18–D7–So
[M+H – H2O]+ 289.3220
[M+H – (H2O+H2CO)]+ 259.3123
[M+H – 2H2O]+ 271.3118
[M+H]+ 307.3341
5.75
C18–Phy
[M+H]+ 318.2997
[M+H – H2O]+ 300.2897
[M+H – 2H2O]+ 282.2791
[M+H – (H2O+H2CO)]+ 270.2784
[C2H6NO2]+ 90.0553
254.8885
5.68
C18–D3–Deoxy-Sa
[M+H – H2O]+ 271.3184
[M+H]+ 289.3289
106.0869
88.0762
228.2052
6.00
C17–D5–Deoxymethyl-Sa
[M+H – H2O]+ 259.3161
[M+H]+ 277.3252
83.0857
5.94
% of BP
m/z
(min)

Figure 15.6 High mass resolution product ion spectra of protonated sphingoid bases. Spectra were obtained on a C8-UPLC-hESI-(Orbitrap Q-Exactive plus) MS^2 system. C18-KDS: ketodihydrosphingosine (m18:0) contains a 3-ketogroup and a corresponding γ-hydroxy proton enabling direct McLafferty (McL)-like β-cleavage leading to loss of formaldehyde besides the possibility of water loss. After rearrangement, a second McL-like β-cleavage may release the $[C_2H_6NO]^+$ ion, corresponding to C_2 and C_3 of the original KDS chain. C18-Sa/C18-D7-Sa: sphinganine (d18:0) initially undergoes loss of one or two water molecules. After loss of one water molecule, a double bond is established that may lead in a McL-like β-cleavage either to further loss of formaldehyde or to release of a $[C_2H_6NO]^+$ ion. C18-So/C18-D7-So: sphingosine(d18:1) very easily loses one or two water molecules by CID (and to some extent already by in-source-decay), which is due to the allylic position of the 3-hydroxy group. Similar to sphinganine, initial loss of water may trigger subsequently β-cleavage and secondary loss of formaldehyde. C18-Phy: phytosphingosine (t18:0) undergoes mainly water loss (up to three times), but subsequent to the first or second water loss neutral release of formaldehyde by β-cleavage may also occur which will lead to loss of formaldehyde or generation of the $[C_2H_6NO]^+$ or the $[C_3H_8NO_2]^+$ ion. C18-D3-Deoxy-Sa and C18-D5-Deoxymethyl-Sa: 1-deoxy-sphinganine (m18:0) and 1-deoxymethyl-sphinganine (m17:0), respectively, will release solely a single water molecule by low CID. Further decomposition by β-cleavage is not observed. Measured *m/z* values fit to calculated *m/z* with a mass error of less than 5 ppm. Similar to C18-LC in Figure 15.4, reversed-phase chromatography with a C8 column leads to elution of sphingoid bases in the order of decreasing polarity: t_R(C18-Phy) < t_R(C18-So) < t_R(C18-Sa) < t_R(C18-KDS) < t_R(C17-Deoxymethyl-Sa) < t_R(C18-Deoxy-Sa).

Alternatively, free deoxy(methyl)- and canonical sphingoid bases had been derivatized with *O*-phthalaldehyde (OPA) in the presence of a thiol (e.g. 2-mercaptoethanol), which is a reaction specific for lipids with primary amino groups [25, 26]. This method, originally developed for fluorescent detection on HPLC, also reduces background and false-positive signals as most of the lipids will not react with this reagent.

15.4.2 Phosphorylated Sphingoid Bases

In mammals, two kinases (in humans abbreviated as SPHK1 and SPHK2) may phosphorylate sphingoid bases. Metabolically, these SB 1-phosphates can then be irreversibly cleaved by the ER-localized S1P lyase (SGPL1), which is the only exit pathway of sphingolipid degradation. Alternatively, SB 1-phosphates are dephosphorylated by S1P phosphatases. Functionally, So1P acts as a second messenger in an auto- and paracrine fashion by activating five specific G protein-coupled receptors termed sphingosine 1-phosphate receptors (S1PR1–5) [27]. So1P signaling is especially essential for vascular development, neurogenesis, and lymphocyte trafficking [28]. Intracellularly, So1P in mitochondria interacts with the electron transport chain, and in the nucleus it was found to regulate histone acetylation [27]. Sa1P may also bind to S1PR1, but may result in different effects [29, 30].

Phosphorylation allows detection of phosphorylated SBs as deprotonated ions and So1P and Sa1P were quantified in biological samples including human plasma with a hydrophilic interaction liquid chromatography (HILIC)-tandem MS system operating in negative ESI (−ESI) mode using the MRM transition of $[M-H]^-$ to

$[PO_3]^-$ [31, 32]. Simultaneous elution of So1P and Sa1P and transition to the identical $[PO_3]^-$ fragment ion result, however, in an isotopic overlap of the $n+2$ isotopic ion of So1P with the monoisotopic signal for Sa1P, which can only be avoided when working with high-resolution mass spectrometers. Similar results for human EDTA plasma were obtained by HILIC-coupled tandem mass spectrometry in positive ESI (+ESI) MRM mode (Transitions: $[\text{Sa1P+H}]^+ \rightarrow [\text{Sa1P+H}-H_3PO_4]^+$ and $[\text{So1P+H}]^+ \rightarrow [\text{So1P+H}-(H_2O+H_3PO_4)]^+$), and Sa1P was detected at about four- to sixfold lower concentration than So1P [33]. However, retention times for So1P and Sa1P were also identical with this method. Recent methods prefer detection of protonated compounds due to higher sensitivity. In positive mode, CID will lead to loss of water, phosphoric acid, or both, when applied to So1P (Figure 15.7). Tang et al.

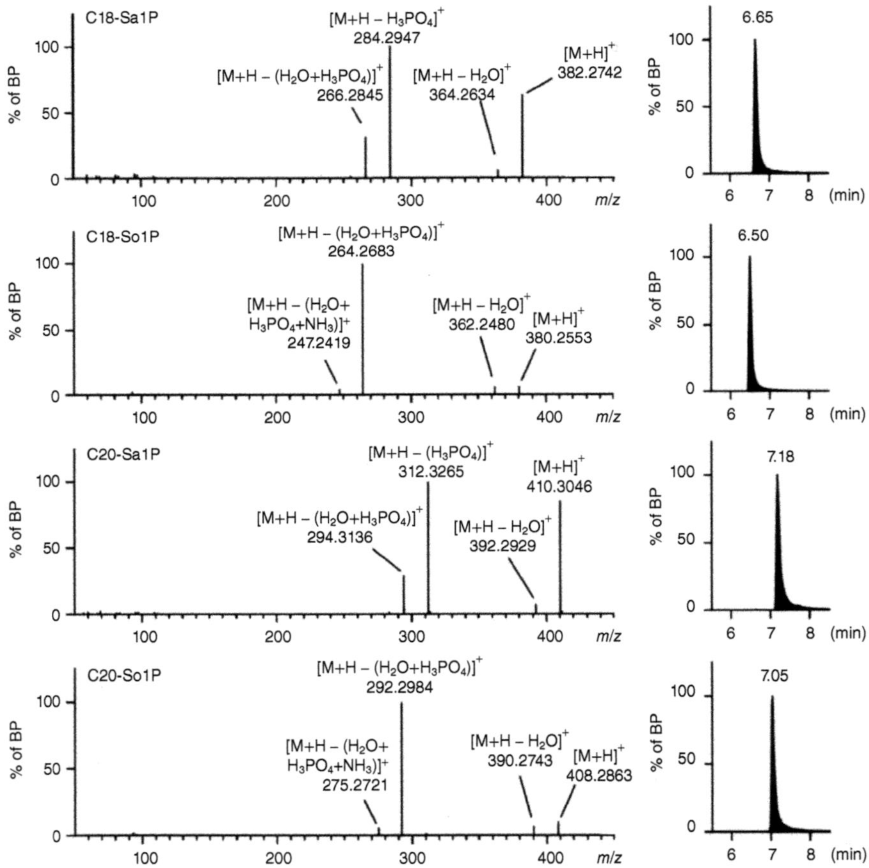

Figure 15.7 Product ion spectra of protonated S1P. High mass resolution product ion spectra of protonated sphinganine 1-phosphates (Sa1P) and sphingosine 1-phosphates (So1P). Spectra were obtained on a C8-UPLC-hESI-(Orbitrap Q-Exactive plus) MS^2 system. With CID loss of water (−18), loss of phosphoric acid (−98, and base peak for Sa1Ps), and loss of water and phosphoric acid (−116, and base peak for So1Ps) are observed. Sa1P elutes subsequently to the corresponding So1P on this system.

also reported the additional loss of NH_3 on top of phosphoric acid and water for CID of So1P [34]. Transition of $[M+H]^+ \rightarrow [M+H-(H_2O+H_3PO_4)]^-$ and $[M+H]^+ \rightarrow [M+H-(H_3PO_4)]^+$ was used to detect So1P and Sa1P in MRM mode after separation by reversed-phase chromatography [35, 36]. With such a method, Sakai et al. achieved a lower limit of quantification (LOQ) of 0.01 ng/ml of cerebrospinal fluid with a modern triple quadrupole instrument [37], while Tang et al. were able to detect down to 25 ng/ml of serum using a C18 column [34]. Using a Q-TOF mass spectrometer, Egom et al. achieved likewise a lower LOQ of 50 fmol/ml [38]. Lipid monophosphate esters like phosphorylated sphingoid bases tend to tail on LC columns resulting in bad peak shape and corresponding low detection sensitivity. Using a metal-free column and the metal chelator EDTA in addition to ammonia and ammonium acetate in the solvent system, Gowda et al. strongly reduced tailing and achieved a limit of detection of 10 fmol/ml for So1P on their quadrupole/TOF MS instrument [39].

Another option to avoid peak tailing and to increase detection sensitivity is derivatization of SB 1-phosphates. Berdyshev et al. achieved this by bisacetylation of SB 1-phosphates at the amino and hydroxyl groups with acetic anhydride dissolved in pyridine by monitoring the transition of a single acetic acid loss in negative ESI mode [40]. Later, Lee et al. performed bismethylation of So1P and Sa1P at the phosphate group by trimethylsilyl-diazomethane resulting in suppressed peak tailing and improved detection sensitivity. Dimethylesters of So1P and Sa1P were subjected to supercritical fluid chromatography and recorded in positive ESI mode with the MRM transition of the $[M+H]^+$ ions to $[So/Sa+H-2H_2O]^+$ fragment ions [41] (Figure 15.8).

15.4.3 Ceramides Including Omega-Esterified Ceramides and 1-*O*-Acylceramides

15.4.3.1 Ceramides with Long and Very Long Acyl Chains

N-acetylation of free sphingoid bases occurs at the ER by so-called ceramide synthases (reaction no. 3 in Figure 15.2). In humans, there are six isoforms CerS1-6, which are expressed in a cell- and tissue-specific manner. Furthermore, cell differentiation and activation as well as pathological signals like in metabolic syndrome or obesity will also regulate the expression pattern of these enzymes [42–46]. In the *de novo* sphingolipid pathway, sphinganines are condensed with coenzyme A-activated acyl chains to yield dihydroceramides (DHCer). Sphingolipid Δ4 desaturases DEGS1 and DEGS2 catalyze the desaturation of DHCer to ceramides or the hydroxylation of DHCer to phytoceramides [47–49]. Mammalian tissue also contains (4*E*,14*Z*)-sphingadiene-based ceramides. The second double bond in the sphingoid base is introduced at the level of ceramide by the ceramide desaturase FADS3, which is expressed in a tissue-specific manner. This sphingoid base changes the biophysical properties of corresponding ceramides as the 14-cis double bond will drive these lipids out of lipid microdomains into the more fluid phase of cell membranes [11, 12]. Likewise, ceramides with a 6-hydoxy sphingosine base are specifically reported for epidermis [50]. So far, however, the molecular mechanism of

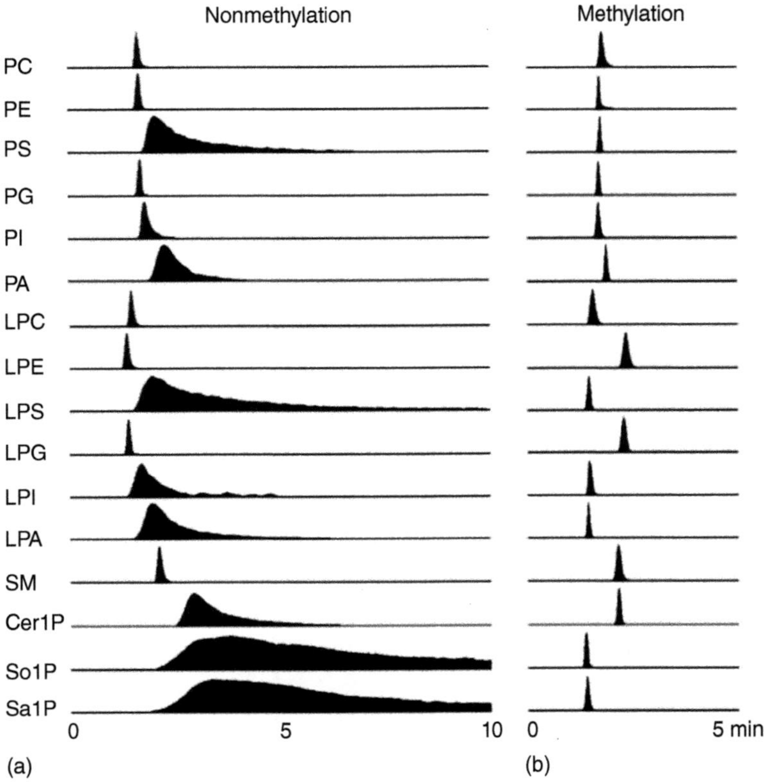

Figure 15.8 Supercritical fluid chromatography of non-methylated (a) and methylated (b) phospholipids. Source: Lee et al. [41]/with permission of Elsevier.

6-OH Cer creation remains unclear [43]. As ceramide synthases are rather promiscuous, they not only integrate various canonical sphingoid bases and acyl chains of different chain lengths (including odd sphingoid bases and odd acyl chains [51]), but also non-canonical sphingoid bases, which are observed under pathological conditions. By that they integrate deoxy- and deoxymethylsphinganine to yield deoxy- and deoxymethyldihydroceramides [18]. Deoxydihydroceramides accumulate upon fumonisin B1 treatment [18] and tumor treatment with paclitaxel [52]; they contribute to the pathogenesis in HSAN1 patients [17, 53, 54] and increase in type 2 diabetes [55] or upon aging [56]. Ceramide synthases also accept the direct product of the SPT reaction, KDS [57, 58]. Under normal conditions, the corresponding product 3-Ketodihydroceramide is hardly detected indicating eventually direct channeling of the SPT product to KDSR. However, lately we found a significant amount of these non-canonical ceramides in stratum corneum and epidermis of patients with mutations in the *KDSR* gene [13].

Using atmospheric pressure ionization methods or MALDI, ceramides can be detected as molecular adduct ions in positive- ($[M+H]^+$, $[M+NH_4]^+$, $[M+Li]^+$, $[M+Na]^+$, $[M+K]^+$, . . .) as well as negative-ion mode ($[M+Cl]^-$, $[M+CH_3CO_2]^-$, $[M+CF_3CO_2]^-$, . . .) and as deprotonated ions in negative mode ($[M-H]^-$). Coupling atmospheric pressure chemical ionization-mass spectrometry to reversed-phase HPLC, ceramides were quantified from as low as 40 000 HeLa cells in the 90th [59]. Depending on the ionization conditions, instrumentation, and molecular ion adduct, ISD due to loss of one or two water molecules may be detected in addition. This is somehow problematic for direct injection analysis as a Cer with a hydroxylated acyl chain (e.g. Cer(d18:1/h24:0)) will turn after ISD-mediated loss of water into a signal reflecting a ceramide with a monounsaturated non-hydroxylated acyl chain (e.g. Cer(d18:1/24:1). Using LC-coupled mass spectrometry, both compounds can be separated in time enabling unambiguous identification. CID of protonated ceramide ions result in fragment ions containing the amino group. Besides the loss of water molecules ($[M+H-(H_2O)_{1-3}]^+$), the additional loss of the acyl chain ($[M+H-((H_2O)_{0-2}+FA)]^+$) and formaldehyde ($[M+H-((H_2O)_{0-1}+FA+H_2CO)]^+$) is observed (Figure 15.9). These product ions enable to reconstruct the sphingoid base and acyl chain composition of naturally occurring ceramides, i.e. assigning the number of carbon atoms, double bonds, and hydroxyl groups present in the sphingoid base and in the N-linked acyl moiety [13, 60, 61].

CID-product ion spectra of protonated 1-deoxydihydroceramides also show loss of water ($[M+H-H_2O]^+$), acyl chain ($[M+H-(FA-H_2O)]^+$), and a combination of both ($[M+H-(FA+H_2O)]^+$), but in addition may release the amidated and protonated acyl chains due to a McL-type fragmentation [62]. Due to loss of the 1-hydroxyl group they experience increased retention on reversed-phase columns as compared to dihydroceramides.

The CID of protonated ketodihydroceramides reveals fragmentation guided by the keto group similar to KDS: $[M+H-(H_2CO+(FA-H_2O))]^+$. The corresponding transition had been used to identify and quantify natural ketodihydroceramides in skin of ichtiotic patients with mutations in the *KDSR* gene. Besides these rather specific product ions, protonated ketodihydroceramides also produce product ions due to the simple loss of the acyl chain ($[M+H-(FA-H_2O)]^+$ and $[M+H-FA]^+$) followed by loss of carbon monoxide ($[M+H-(FA+CO)]^+$) [13].

CID of deprotonated ceramides in negative mode also reveals fragments with structure-important information [63], and further fragmentation by MS^3 and MS^4 allows identifying the location of double bonds and functional groups [64]. By co-injection of LiOH, electrospray ionization with higher-energy CID fragmentation of lithium adducts also enables the characterization of ceramides including the chain length and the modification (α-hydroxy-, β-hydroxy-, ω-hydroxy-FA) of the fatty acyl moiety, and the types of SBs (sphingosine, phytosphingosine, 6-hydroxysphingosine). Such approach permits the differentiation of isomeric structures and by application of sequential MS^3, the double bond position along the fatty acyl chain of the molecule can be located. Importantly, there is apparently no discrimination in the formation of $[M+Li]^+$ ions among the various ceramide subclasses [65]

The $[M+H-(H_2O+FA)]^+$ fragment was used early on to specifically detect and quantify ceramides with a C18-sphingosine backbone ($[M+H]^+ \rightarrow m/z$ 264) or a C18-sphinganine backbone ($[M+H]^+ \rightarrow m/z$ 266) from complex biological lipid extracts [60, 66, 67]. Dihydroceramides miss the allylic double bond and are more resistant to fragmentation. They may also be tracked using the following MRM

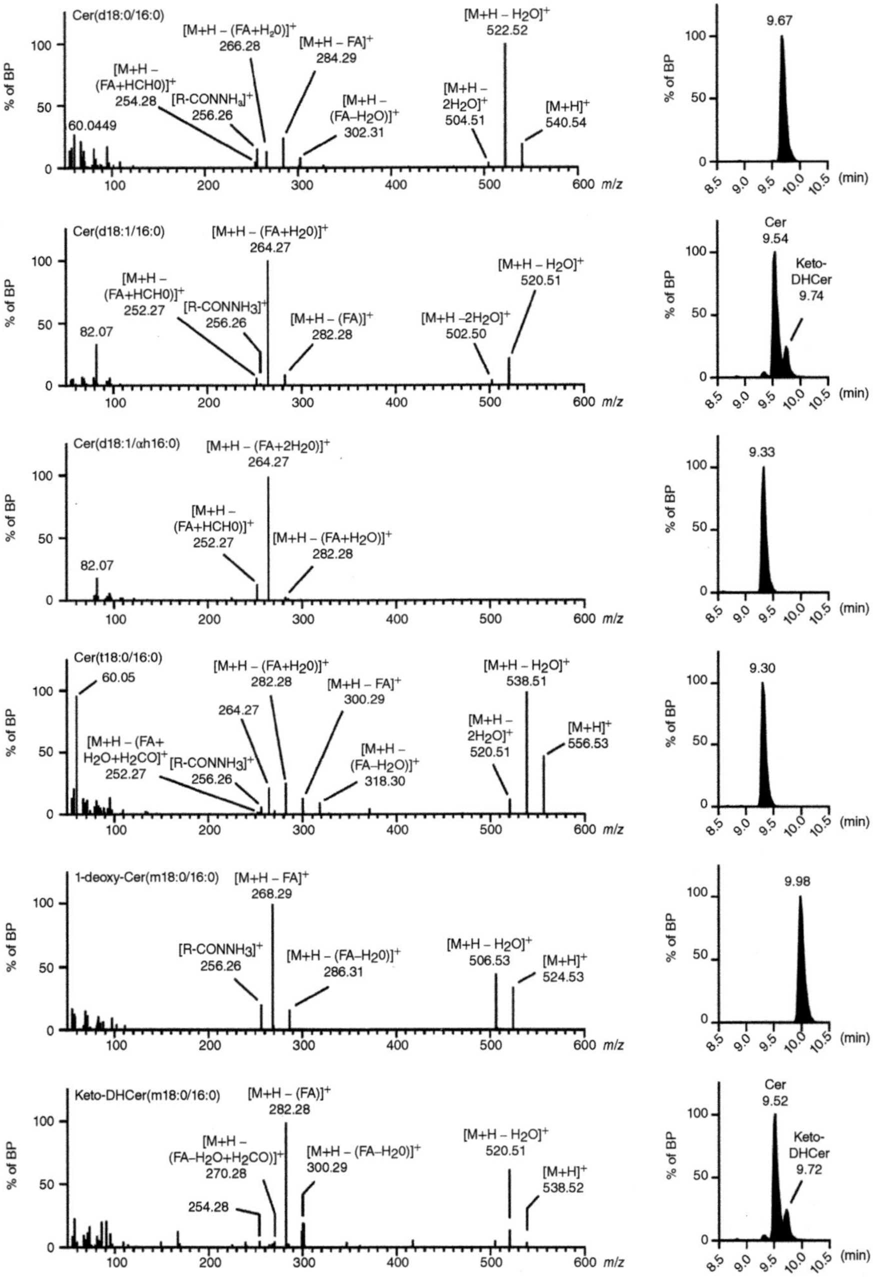

Figure 15.9 High mass resolution product ion spectra of protonated ceramides. Spectra were obtained on a C8-UPLC-hESI-(Orbitrap Q-Exactive plus) MS^2 system. Ceramides containing a Δ4 trans-double bond in the sphingoid base break up easily with *m/z* 264 as major fragment (for C18-SBs). Keto-DHCer reveal a fragment due to loss of dehydrated fatty acid and formaldehyde (for C18-KDS: *m/z* 270) enabling their identity and specific detection. Elution on reversed-phase column appears with decreasing polarity of ceramides in the order observed for sphingoid bases.

transition: $[M+H]^+ \rightarrow [M+H-FA]^+$. Since phytoceramides elute in front of ceramides, which elute slightly earlier than corresponding dihydroceramides on reversed-phase columns, these compounds can be quantified separately from each other by targeted LC-MS^2 due to differences in retention time and *m/z* of both molecular and product ions of the MRM transition. In contrast to MS^1 signals, there is a great disparity between the signal responses of these compounds in the MS^2-approach. We observed about 5 times less signal intensity for the transition of sphinganine-containing dihydroceramides and 12 times less signal intensity for the transition of phytosphingosine-containing phytoceramides as compared to equal concentrations of sphingosine-containing ceramides [68, 69]. When operating with the highly specific MRM mode, these corresponding factors have to be taken into account, if quantification relies on sphingosine-containing internal ceramide standards only. For exact quantification, internal standards corresponding to each ceramide class would be beneficial. Having isotopically labeled internal standards for each endogenous ceramide structure, however, is utopian – human skin, for example, contains probably several thousand different ceramide structures. The latter also shows the limitations of targeted lipid analysis. Modern instruments enable dwell times of about 3 msec for individual MRM transitions. During the period of a single peak elution, such a transition should be recorded at least 10 times to record a good peak shape for quantification. Assuming a peak width of about six seconds would allow then a maximum of 200 different transitions to be recorded in parallel. Even if different transitions are measured for different time sections of the LC gradient, several injections of the same sample would be necessary to cover the recording of all ceramide structures, which could be present in human skin.

Untargeted analysis on the other hand is independent of a transition list. Running it in MS^1 mode, however, requires either high mass resolution [70] or coupling to LC for minimal specificity [71, 72]. Data-dependent fragmentation in MS^2 is limited to peaks with a minimal threshold of intensity and a certain number of peaks that can be fragmented per time. For example t'Kindt et al. estimated to detect with this approach over 1000 ceramides in human stratum corneum and identified unknown ceramides with a novel sphingoid base containing four hydroxyl groups [73]. With LC-independent shotgun lipidomics, a sequential precursor ion fragmentation may be applied to verify the identity of corresponding lipids/ceramides [70]. Unexpected compounds are only recorded by untargeted and not by targeted analysis; however, intelligent data mining is required to recognize these signals [74–76], and it remains difficult for structures not found in databases. Another approach is to screen lipid extracts in a semi-untargeted manner. Classical "triple-quadrupole" instruments offer for this purpose the so-called precursor ion scan and neutral loss scan, which can be used to determine all signals arising from a lipid class out of which all members release a constant fragment ion or a constant neutral loss fragment in tandem mass spectrometry [22, 60, 67, 70, 77]. This approach is often used to evaluate which transitions are relevant for later quantification of a certain sample set in MRM mode. Hence, screening tissue samples with a precursor ion scan of *m/z* 264.3 identifies all ceramides with a C18-sphingosine. It reveals, e.g. the presence of ceramides with α-hydroxy acyl moieties in addition to those with non-hydroxy acyl chains in kidney or brain [78], but would also reveal signals for ceramides with ultra-long and polyunsaturated acyl chains in testes [79] and those with ultra-long ω-hydroxylated acyl chains in the epidermal skin barrier [71, 80].

15.4.3.2 Skin Omega-Hydroxy Ceramides, Free, Esterified, and Protein-Bound

The biosynthesis of ultra-long chain Cer, i.e. ω-hydroxylated, ω-esterified, and covalently bound Cer, contains several additional steps compared to long and very long Cer. The fatty acid acyl chain is prolonged to lengths exceeding 28 carbons by the ELO family elongase of very long chain acids 4 (ELOVL4) [81, 82], and the final lengths of these chains range between 28 and 36 carbons [83]. The ω-hydroxylation then occurs on the cytoplasmic side of the ER by the cytochrome P450 member CYP4F22 [84]. After activation by FATP4 [85, 86], the acyl-CoA is then connected to a sphingoid base using the CerS3 isoform of ceramide synthases [80, 87]. Linoleate is then attached to the ω-OH group by a trans-esterification reaction from triacylglycerols enabled by patatin-like phospholipase domain containing 1 (PNPLA1) stimulated by α/β-hydrolase domain-containing 5 ([88–94]). For the attachment of ω-esterified Cer to the corneocyte envelope and a formation of protein-bound Cer, the linoleate part is subsequently oxidized. First, 12*R*-LOX (lipoxygenase) oxygenates linoleate to a hydroperoxide, which is then isomerized to the form of epoxy-alcohol by eLOX3 [95] and in the last step dehydrogenated by SDR9C7 to an epoxy-enon. The connection to corneocyte surface proteins, mainly involucrin, is then realized by Michael addition or Schiff base formation [96]; the exact mechanism is still under debate (Figure 15.10).

Similar to very long chain Cer, molecular ions of ultra-long chain Cer can be detected in both positive and negative modes after an ionization with atmospheric pressure methods or MALDI. According to the Cer polar head structure, ω-esterified Cer provide (after a CID of their protonated ions) fragments with loss of 1–3 molecules of water ($[M+H-(H_2O)_{1-3}]^+$) or a loss of single water molecule and a molecule of formaldehyde ($[M+H-(CH_2O+H_2O)]^+$). Loss of water due to an ISD could also be observed, especially in ω-esterified Cer based on 6-hydroxysphingosine. Besides these fragments, loss of the linoleic acid and/or loss of a sphingoid base occur, resulting in fragment ions [M+H−FA(18:2)]$^+$, [Linoleoyl-*O*-alk(en)yl-CONH$_2$+H]$^+$ and [HO-alk(en)yl-CONH$_2$+H]$^+$. The latter fragment is also characteristic for

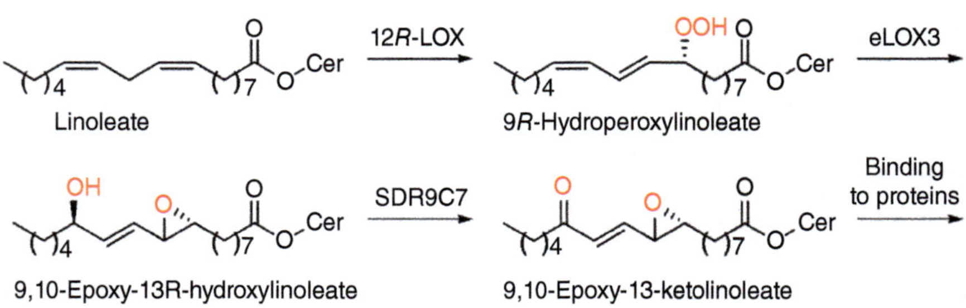

Figure 15.10 Epidermal linoleate oxidation in ω-esterified Cer necessary for the attachment to corneocyte envelope proteins and formation of protein-bound Cer.

ω-hydroxylated Cer and may further provide a loss of water and ammonia molecules. Sphingosine-containing omega esterified ceramides (EOS-Cer) also give rise to a minor fragment ion, which appears due to loss of water, formaldehyde, and dehydrated linoleic acid ($[M+H-(FA(18:2)+H_2CO)]^+$). Although of minor intensity, this transition has been successfully used in an MRM approach to separately quantify intact EOS-Cer containing ω-esterified linoleic acid and EOS-Cer containing ω-esterified oleic acid ($[M+H-(FA(18:1)+CH_2O)]^+$) in murine epidermis on a LC-coupled ESI-(QqQ) tandem mass spectrometer (R. Sandhoff, unpublished results). Although the latter is rather a minor fraction of esterified Cer, it may increase compared to the other under pathological conditions. Fragments characteristic for the loss of water and/or formaldehyde from a sphingoid base $[SB+H-(H_2O)_{1-3}]^+$ or $[SB+H-(H_2O+CH_2O)]^+$ may as well be used for a detection and quantification of ω-esterified and ω-hydroxylated Cer, including the characteristic fragment with *m/z* of 264.3 originating from C18 sphingosine (Figure 15.11). Hsu et al. studied the fragmentation of lithiated adducts of ω-hydroxylated Cer after an ESI ionization providing the fragment $[HO\text{-alk(en)yl-}CONH_2+Li]^+$, which may further lose a water molecule: $[HO\text{-alk(en)yl-}CONH_2+Li-H_2O]^+$. The latter does not appear in corresponding ceramides with an α-hydroxy fatty acid and may be used for diagnostic approaches [97]. Using ESI ionization in a negative mode, the fragmentation pattern is different: The pseudomolecular ion of ω-esterified Cer $[M-H]^-$ mostly fragments on the ester group providing two fragments with the negative charge being allocated either on the linoleate part ($[FA(18:2)-H]^-$) or on the ω-hydroxyl group of ω-hydroxylated Cer ($[M-(FA18:2)]^-$) [51, 98]. This fragment may further lose a molecule of water and/or formaldehyde (in both ω-esterified and ω-hydroxylated Cer). Loss of the sphingoid base further provides fragments $[HO\text{-alk(en)yl-}COO]^-$ and $[HO\text{-alk(en)yl-}CONH]^-$ [64].

The oxidized ω-esterified Cer were analyzed by atmospheric-pressure chemical ionization (APCI) ionization in positive mode after a separation using a normal-phase HPLC. The product of 12*R*-LOX, the 9*R*-hydroperoxide turned out to be unstable in APCI conditions; thus, it was reduced to an alcohol by an excess of triphenylphosphine prior to the analysis. 9*R*-hydroperoxide, 9,10-epoxy-13*R*-alcohol (eLOX3 product), and 9,10-epoxy-13-keton (SDR9C7 product) were identified as $[M+H]^+$ or $[M+H-(H_2O)_{1-2}]^+$ ions. For a selective analysis of the oxidized linoleate part, the reduction of the keto group (if present) and transesterification using sodium methanolate could be used [96, 99]. Protein-bound Cer are released from corneocytes by an alkaline hydrolysis (mostly potassium hydroxide in methanol) providing ω-hydroxylated Cer (which may then be analyzed in the same manner as free ω-hydroxylated Cer) or ω-hydroxylated fatty acid [100].

15.4.3.3 1-*O*-Acylceramides in Skin and Other Tissues

Mammalian cells possess enzyme activity to acylate ceramides in 1-O-position [101]. Based on findings in yeast [102], Dgat2 and Dgat1 had been identified to catalyze this reaction in mammalian cells [103, 104]. Normal 1-OAC levels are lower by powers of 10 compared to ceramide levels. However, decent amounts have been found in epidermis [77], and they increase upon high-fat diet in liver cells [103].

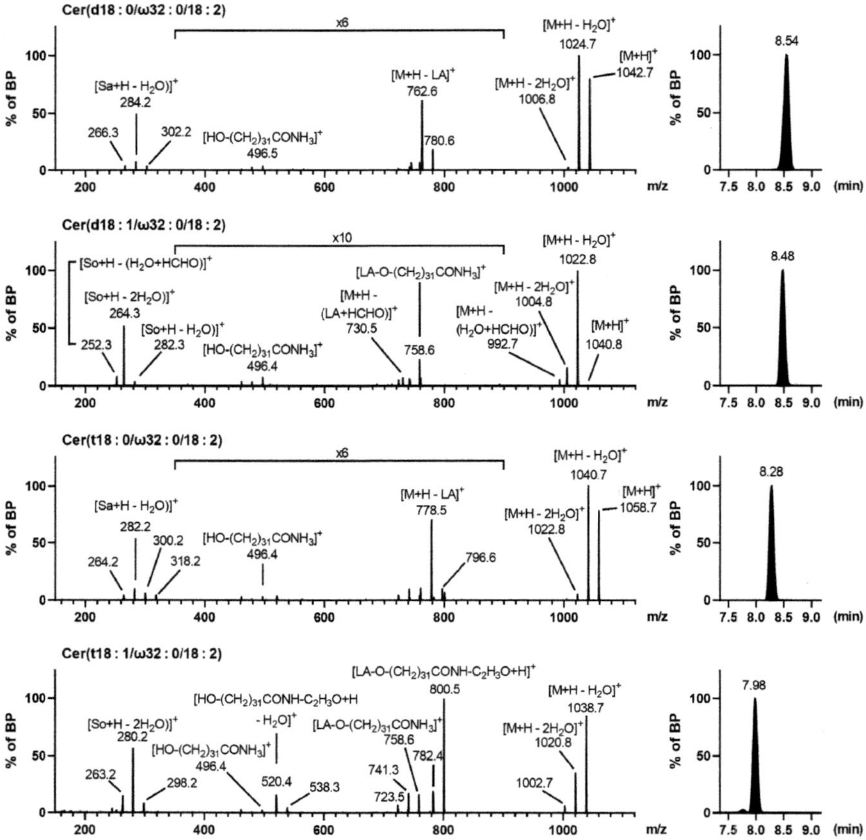

Figure 15.11 Product ion spectra of synthetic protonated ω-esterified ceramides containing linoleic acid (18:2) esterified to ω-hydroxy-dotriacontanic acid (ωh32:0), which is N-linked to either sphinganine (d18:0), sphingosine (d18:1), phytosphingosine (t18:0), or 6-hydroxy sphingosine (t18:1). Product ion spectra were generated from reversed-phase chromatographic peaks (right side) after electrospray ionization and collision induced dissociation at 25 V on a QqQ-instrument with unit resolution.

CID of protonated 1-OAC ions containing a sphingosine base leads to loss of water ($[M+H-H_2O]^+$), of 1-*O*-acylchain (Z'_0,$[M+H-(FA)]^+$), or of both (Z''_0, $[M+H-(FA+H_2O)]^+$). The Z''_0, ion may further lose NH_3 resulting in a signal at *m/z* 247 in case of C18-base. Furthermore, a release of the protonated sphingosine fragment (N″, $[So+H-2H_2O]^+$), and of protonated acyl amide ions (U, $[R\text{-}CONH_2+H]^+$ and T, $[R\text{-}CONH\text{-}CCH+H]^+$) is observed (Figure 15.12). With this information the amount of carbon atoms and double bonds can be attributed to each building block of 1-OACs. The transition of $[M+H]^+$ to $[M+H-(FA)]^+$ has also been used to identify and quantify sphingosine-based 1-OACs in various organs [77, 104]. Using linear ion-trap MS^n with high-resolution MS, the authors estimated more than 700 structurally different 1-OACs to be present in mouse epidermis [105].

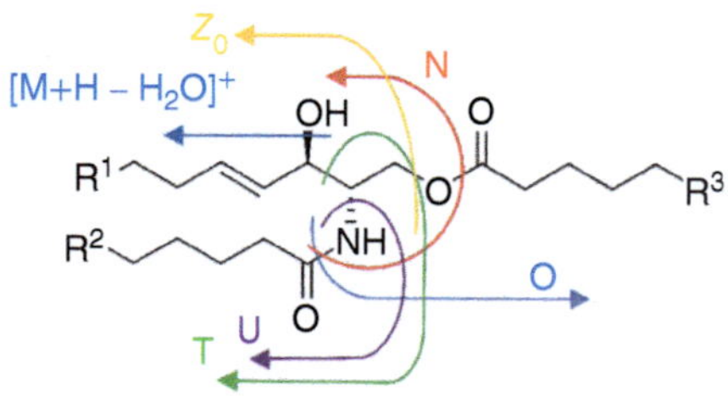

Figure 15.12 Structure of sphingosine-based 1-*O*-acylceramides (1-OACs) including cleavage sites corresponding to observed product ions derived by CID.

15.4.4 Ceramide 1-Phosphates

Cer and DHCer are transformed into Cer1P and DHCer1P by ceramide kinases (CERK) by a direct phosphorylation, which is the only known pathway for their biosynthesis. However, in mice with CERK genetic ablation, approximately 50% of normal levels of Cer1P were found [106], suggesting either the presence of multiple isoforms of this enzyme or an alternative biosynthetic pathway. In theory, Cer1P may be synthesized by a fatty acyl-CoA transfer to S1P or by a degradation of sphingomyelin by a phospholipase D; nevertheless, none of these enzymes were identified in mammalian cells. CERK has been found in multiple cell compartments like cytosol, plasma membrane, or nucleus, but its main localization is in the Golgi apparatus. From its site of biosynthesis, Cer1P is transported by ceramide 1-phosphate transfer protein to a plasma membrane [107]. The balance between Cer and Cer1P is important because both lipids strongly affect cellular processes [14], and it is maintained by CERK phosphorylating Cer and ceramide-1-phosphate phosphatase which is cleaving the phosphate group from Cer1P [108].

HPLC-coupled ESI-MS^2 allows the recognition of multiple species of Cer1P present in cells, which differ in their chain lengths. Previously, only the quantification of total Cer1P was possible using radiolabeling followed by thin layer chromatography (TLC) separation. In most of the published methods for Cer1P quantification using HPLC-coupled ESI-MS^2, the reversed-phase separation with MRM monitoring in either positive or negative mode was used. In positive mode, the transition of protonated molecular ion to the fragment with *m/z* 264.4 $[C18\text{-}So+H-(H_2O)_2]^+$ could be followed [109–112] (Figure 15.13). Such fragmentation provides more structural information about the ceramide part of Cer1P, but suffers from a low signal intensity of the MRM transition. On the contrary, in negative ionization mode, the deprotonated Cer1P ion $[M-H]^-$ provides a characteristic fragment of *m/z* 78.9 belonging to the residue of the phosphate group after a loss of a single water molecule (Figure 15.13). This transition provides limited structural information, but the signal intensity is much stronger than for the MRM transitions in positive ionization mode [113]. The highly amphiphilic nature and the terminal phosphate group of Cer1P brings along multiple problems during separation on normal or reversed phase: broadening of peaks and carryover, which can reach up

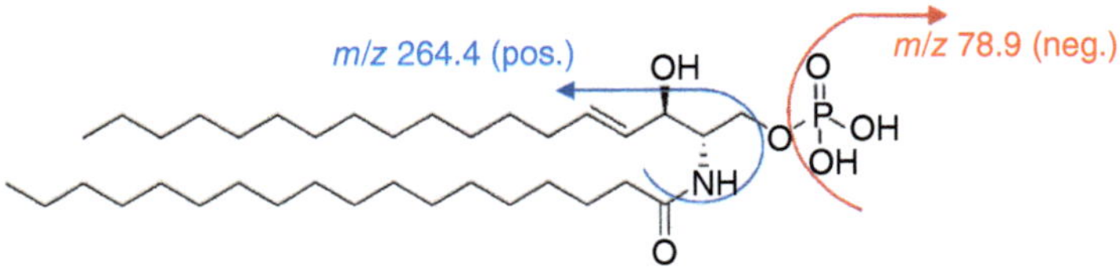

Figure 15.13 Structure of Cer1P(d18:1/16:0) showing characteristic fragmentations observed by CID.

to 10%. These problems were partly suppressed by a selection of C8 reversed-phase column instead of C18, heating the column to 60 °C and by the addition of 2% tetrahydrofuran into the mobile phase [109, 112, 114]. Another option is methylation, which has been used to enhance sensitivity along with peak shape for detection of Cer1P [41, 115, 116]. In some extraction protocols, complications with overestimation of Cer1P compared to the radiolabeling methods may occur. It is caused by an insufficient neutralization of extracts before drying and the excess of Cer1P originates from the decomposition of SM. When the neutralized lipid extract was compared to the extract without neutralization, the Cer1P content was decreased by more than 75% and the corresponding changes in SM concentrations were reported [112].

15.4.5 Sphingomyelins

SM are the most abundant complex SLs in human cells. Unlike glucosyl and galactosylceramides, which are critical for cell functions but not required for cell survival, SMs are indispensable for cell survival [4]. SMs are produced by sphingomyelin synthases SMS1 and SMS2 [117]. SMS1 and SMS2 are localized in trans-Golgi, SMS2 also in plasma membrane, and they transfer a phosphocholine head group from phosphatidylcholine to Cer, providing a molecule of SM and diacylglycerol. Cer, SM, and diacylglycerol are all bioactive lipids and affect cellular fate; thus, their intracellular regulation is essential [118, 119]. Sphingomyelin synthase-related protein SMSr transfers the phosphoethanolamine head group from phosphorylethanolamine to Cer and contributes in maintaining the Cer homeostasis in ER [120]. SM is then decomposed by a family of sphingomyelinases (SMases) leading to a cleavage of a phosphocholine head group and providing Cer and free phosphocholine. Mammalian SMases differ according to their pH optimum and tissue localization and could be divided into acidic (mostly localized in lysosomes) [121–123], neutral (three subtypes are known with a prevalent localization in Golgi and plasma membrane) [124], and alkaline (localized in liver and intestine on a microvillar membrane in endosome-like structures and in the Golgi complexes of human enterocytes) [125] SMases.

Historically, multiple ionization techniques were used for the SM structure determination like chemical ionization with a moving-belt interface [126], thermospray [127], discharge-assisted thermospray [128], APCI [129, 130], fast-atom bombardment [131, 132], or MALDI [133]. This chapter will be focused on an ESI ionization approach.

Compared to other lipids discussed in this chapter, sphingomyelins already contain a charged nitrogen atom in their molecule turning it together with the phosphodiester at neutral pH into a zwitterion. This zwitterion is defined here as the molecule M. Upon positive ESI ionization, signals of protonated SM $[M+H]^+$ easily appear. CID of the protonated molecular ion $[M+H]^+$ provides a characteristic fragment with *m/z* 184 belonging to phosphorylcholine head group (Figure 15.14). Precursor ion scanning with this fragment will also reveal phosphatidylcholines as the head group of both lipids is identical. After fragmentation, the charge is retained on the polar head group, which complicates obtaining the structural information on the ceramide part [134]. In positive ionization mode, further information about an *N*-acyl chain could be obtained from CID of alkaline metal adducts. In the presence of alkaline metals (Li^+, Na^+, K^+), SMs produce with the exception of the abundant $[M+H]^+$ ion also $[M+Cat]^+$ ions. These molecular ions provide identical abundance profiles, which enables the use of such alkaline-adduct ions for semi-quantitation of SM in mixtures. The product ion spectra of $[M+Li]^+$ ion first provide the neutral loss of trimethylamine followed by a neutral loss of cyclophosphane or lithium cyclophosphane to yield a fragment $[M+Li-183]^+$ and $[M+Li-189]^+$, respectively. Further elimination of water or formaldehyde and neutral loss of an acyl moiety or sphingoid base lead to fragments specific for sphingoid bases and fatty acyls and the final structure can be resolved based on these fragments (Figure 15.14). $[M+Na]^+$ and $[M+K]^+$ molecular ions may be used for the structure determination as well, the difference is that fragment ions associated with alkaline metals like potassiated cyclophosphane become more prevalent. Scanning for the constant neutral loss of 59, 183, and 207 is characteristic for SMs [135].

In negative ESI ionization mode, the molecular ion was found as a formiate adduct, i.e. $[M+45]^-$. CID of this molecular ion in negative ESI mode provided a

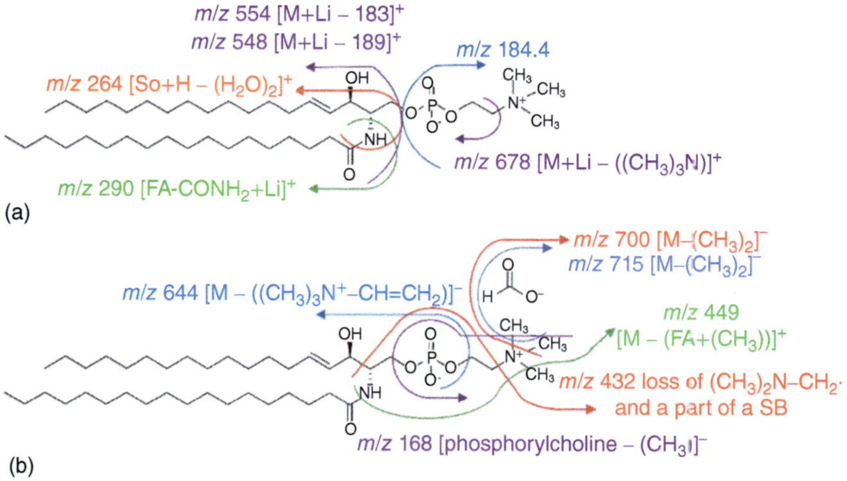

Figure 15.14 Structure of SM(d18:1/16:0) including characteristic fragmentations observed by CID of quasi-molecular cations (a) and anions (b). Please note that this figure is a combination of different fragmentation patterns described by different authors as indicated in the main text.

fragment with the loss of formiate methyl ester (by abstraction of a CH_3^+ group form the choline nitrogen) and a minor fragment corresponding to the additional loss of another methyl group from choline moiety. Further fragmentation also revealed a product ion with *m/z* 168 consisting of phosphorylcholine head group after a loss of a methyl group [134] (Figure 15.14). The demethylated phosphorylcholine fragment was subsequently used to selectively screen for sphingomyelin signals in extracts of biological samples containing abundant phosphatidylcholines. The negative ESI precursor ion scanning mode revealed signals corresponding to [M−15] ions [136]. Since the authors of the latter work performed direct injection of lipids dissolved in methanol/chloroform without additives, it may be assumed that during the ionization and drying processes negatively charged solvents like MeO^- did abstract the CH_3^+-group from the zwitterionic sphingomyelin. Conventional CID methods again cannot provide a detailed information about the composition of Cer backbone of SMs. This problem was overcome by the application of a pseudo MS^3 CID of an acetate or formiate adduct, which revealed the backbone structure specific fragment of *m/z* 449.4 originating from the loss of an acyl chain and a single methyl group from the quaternary nitrogen atom allowing to determine the sphingoid base structure as So(18:1) [109]. Another approach in negative ESI ionization is CID of SM bicarbonate adducts $[M+HCO_3]^-$. Fragments with neutral loss of 62 amu (loss of H_2CO_3), 121 amu (combined loss of $H_2CO_3 + (CH_3)_3N$), and 147 amu (loss of $HCO_3^- \cdot (CH_3)_3N^+ - CH{=}CH_2$) are highly abundant. Contrary to the previous approach, here, the fragment at *m/z* 449 was absent. Instead a fragment at *m/z* 430.7 was found, which results from a subsequent loss of H_2CO_3, *N,N*-dimethylaminomethyl radical and cleavage of the sphingoid base between C_2 and C_3 carbons providing information about the SM *N*-acyl chain (in this case 18:1). The same fragment is found for fully saturated (18:0) *N*-acyl chain with *m/z* value of 432.4. The limit of identification for *N*-acyl composition was better with 25 nM compared to 200 nM for the pseudo MS^3 CID of the formiate adduct. Further information about the positions of C=C double bonds in the *N*-acyl chain and sphingoid base chain could be determined using Paternò–Büchi reaction, and the possible α-hydroxylation of *N*-acyl chain as well as methyl branching could be determined using radical-directed dissociation [137].

15.4.6 Hexosylceramides

In mammals, ceramides can be glycosylated with glucose or galactose by specific transmembrane glycosyltransferases to yield β-glucosylceramide (GlcCer) or β-galactosylceramide (GalCer). Whereas GlcCer is found in most tissues, myelin and kidney are especially rich in GalCer [138]. Lysosomal degradation is performed at intralysosomal vesicles with water-soluble enzymes supported by activator proteins facilitating access to the hydrophobic substrates. Genetic mutations associated with corresponding enzymatic defects or loss of activator protein function lead to lysosomal storage disorders: Gaucher disease with the primary accumulation of GlcCer and GlcSph, and Krabbe disease with the primary accumulation of GalCer and GalSph [139].

GlcCer and GalCer are detected by ESI or MALDI as ion adducts, protonated or deprotonated. With identical ceramide anchors, they are isobaric compounds and CID of protonated GlcCer and GalCer (or corresponding lithium or chloride adducts) give rise to the same set of product ions (mainly loss of the sugar moiety and subsequently fragmentation similar to ceramides) and do not allow unambiguous differentiation between both species [22, 135, 140–144]. Therefore special chromatographic conditions have been set up, with their separation on LC column without [141, 145–147] or with prior derivation [148]. Alternatively, the isobaric compounds can also be separated by ion mobility before MS detection, which allowed the separate quantification of GlcCer and GalCer from plasma and cerebrospinal fluid [149]. Recently, also supercritical fluid chromatography coupled with tandem mass spectrometry revealed separation of isobaric GlcCer and GalCer from human plasma [150].

In contrast to the β-anomeric hexosylceramides, α-GalCer is a potent stimulator of iNKT cells, when presented by CD1-molecules of antigen-presenting cells [151, 152] and has been proposed as an endogenous ligand in iNKT cell maturation [153]. A mechanism of endogenous synthesis remains to be verified [154]. The α- and β-anomeric hexosylceramides may be separately detected by HILIC-MS^2 [141]. However, MS^2-transitions arising from the sodiated molecular ion adducts (Figure 15.15) also help to identify the anomeric structure [155]. Combining the HILIC-MS^2 approach with the MRM tracing of the sodiated molecular ions, bacterial α-galactosylceramides could be quantitatively identified in mouse gut in the presence of more than 70 different endogenous β-hexosylceramides [156].

15.4.7 Neutral Complex Glycosphingolipids

In mammalian cells, it is mainly GlcCer that serves as substrate of glycosyltransferases elongating the neutral sugar chain. GlcCer is substrate for two β4-galactosyltransferases (B4GALT5/6), which yield lactosylceramide (LacCer) at the luminal side of the Golgi apparatus. Six different transferases then use LacCer as substrate guiding GSL synthesis into different GSL series. Two of them, GM3 synthase (ST3GAL5) and cerebroside sulfotransferase (GAL3ST1), add acidic building blocks to the 3-hydroxy group: sialic acids or a sulfate group. However, the other 4 transferases (A4GALT, A3GALT2, B4GALNT1, B3GNT5) add neutral sugars to the 3 or 4 position of the terminal galactosyl-residue, by that elongating the neutral sugar chain into the ganglio-, globo-, isoglobo, or (neo)lacto-series of GSLs. Subsequently, further glycosyltransferases may add step wise additional sugar moieties in a combinatorial manner [6, 157, 158]. Globoseries GSLs for example are abundant in human visceral organs and on erythrocytes, where the most common GSL is globoside (Gb_4Cer, also known as the p antigen). On erythrocytes globo- and neolactoseries GSLs represent several important blood group antigens. The blood group p phenotype is characterized by mutations in A4GALT and the absence of all globoseries GSLs. This may cause hemolytic transfusion reactions and may result in hemolytic disease of the affected mother's fetus or newborn and even in recurrent spontaneous abortions [159, 160]. Otherwise, and in contrast to other known

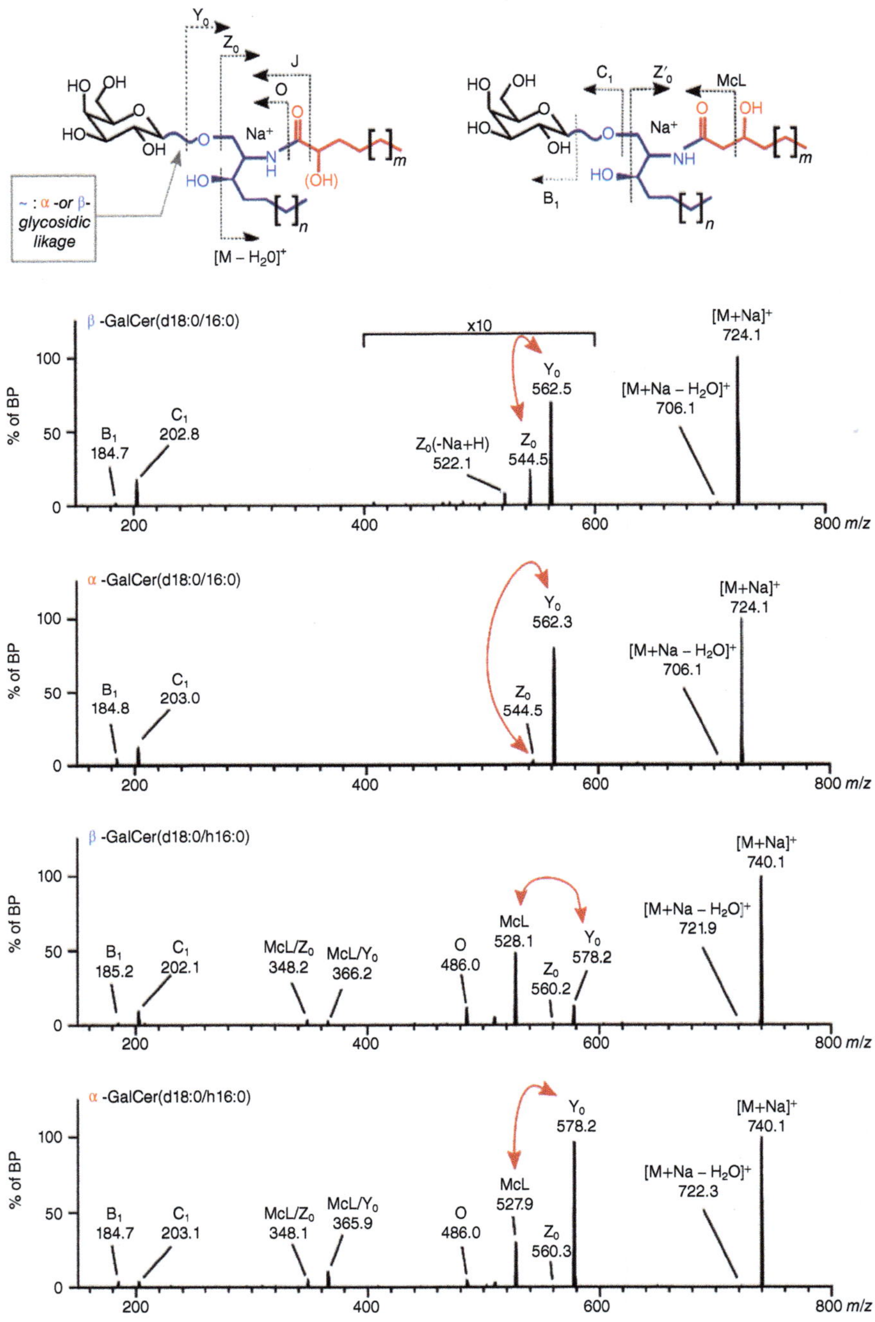

Figure 15.15 Structure of GalCer with indication of CID-fragmentation sites observed from sodiated molecular ions. Product ion spectra of β- and α-GalCer(d18:0/16:0) and β- and α-GalCer(d18:0/βh16:0) were obtained by CID at 45 eV. As described the signal ratio of Y_0 and Z_0 significantly changes between the alpha- and beta-anomeric GalCer with N-linked non-hydroxy acyl chains. In case of GalCer with N-linked β-hydroxy acyl chains the ratio of the McL over the Y_0 product ion is indicative for the anomeric linkage of the galactosyl residue. Data were recorded from synthetic compounds by LC-MS^2 on a unit resolution QqQ-like tandem mass spectrometer.

deficiencies of enzymes in the metabolic pathways of SLs, A4GALT mutations seem to be compatible with the individual's normal life.

The neutral glycan moieties of mammalian GSLs may contain the monosaccharides glucose (Glc), galactose (Gal), *N*-acetylglucosamine (GlcNAc), *N*-acetylgalactosamine (GalNAc), and fucose (Fuc), and hundreds of different neutral GSLs structures have been identified and are collected at LIPID MAPS structure database (LMSD) (http://www.lipidmaps.org/databases/lmsd). The only neutral GSL series arising from GalCer is the Gala-series by the addition of an α4-galactosyl residue to yield galabiaosylceramide (Ga_2Cer, Galα4GalCer) by Gb3 synthase (A4GALT). Similar to N-linked glycans [161], neutral GSLs are detected as molecular ion adducts in positive and negative ESI-MS mode [162, 163]. Likewise, MALDI produces sodium adducts as major molecular ion signals and deprotonated species in negative mode [133, 164, 165].

As Glc and Gal as well as GlcNAc and GalNAc are stereoisomers and therefore isobaric, exact composition of GSLs cannot be attributed by MS^1, and instead the amount of hexoses (Hex) and *N*-acetylhexosamines (HexNAc) has to be annotated. Hence, both LacCer and Ga_2Cer are Hex_2Cer. Likewise, Gb_3Cer and iGb_3Cer transform into Hex_3Cer, whereas Gg_3Cer and Lc_3Cer transform into HexNAc-Hex_2Cer. At the level of tetraosylceramides, all five isomers contain one HexNAc and three Hex residues: Gb_4Cer (GalNAcβ3Galα4Galβ4GlcCer, i.e. HexNAc-Hex_3Cer), iGb4Cer (GalNAcβ3Galα3Galβ4GlcCer, i.e. HexNAc-Hex_3Cer), Gg_4Cer (Galβ3GalNAcβ4Galβ4GlcCer, i.e. Hex-HexNAc-Hex_2Cer), Lc_4Cer (Galβ3GlcNAcβ3Galβ4GlcCer, i.e. Hex-HexNAc-Hex_2Cer), and neolactotetraosylceramide (nLc_4Cer, Galβ4GlcNAcβ3Galβ4GlcCer, i.e. Hex-HexNAc-Hex_2Cer).

CID in tandem mass spectrometry can reveal the ceramide composition of GSLs and the order of the Hex and HexNAc building blocks, but not necessarily their identity and positional or anomeric linkage [135, 166, 167]. Ring cleavage fragments (A- and X-fragments), which help to identify linkage positions, are not always observed and are of minor intensity [145]. Using MALDI-TOF MS with high-energy CID, $^{3,5}A_2$- and $^{3,5}A_3$-ions are observed analyzing $[Gb_3Cer+Na]^+$ ions, which identifies the compound as Hex1,4Hex1,4Hex1,1Cer and distinguishes it from iGb_3Cer with a terminal 1,3-linkage [144].

If the order of Hex and HexNAc residues differs between two GSLs (e.g. Gg_4Cer and Gb_4Cer, Figure 15.16), the intensity ratios of the various fragments will also diverge. For example, with ESI-MS/MS of protonated molecular ions, the proton tends to stick to fragments containing an amide nitrogen like the HexNAc-residue. Therefore an intense B_1 fragment ($[HexNAc+H-H_2O]^+$) is observed with globoside (Gb_4Cer), whereas gangliotetraosylceramide (Gg_4Cer) will reveal a dominant B_2 fragment ($[HexHexNAc+H-H_2O]^+$). Although both can be used in MRM mode to screen for these compounds in biological samples, they are not specific for one or the other, since Gb_4Cer also releases a B_2 fragment ($[HexNAcHex+H-H_2O]^+$) identical in m/z to the B_2 fragment of Gg_4Cer and vice versa Gg_4Cer can undergo double break to produce the B_2/Y_3-ion ($[HexNAc+H-H_2O]^+$) identical in m/z to the B_1 fragment of Gb_4Cer (Figure 15.17).

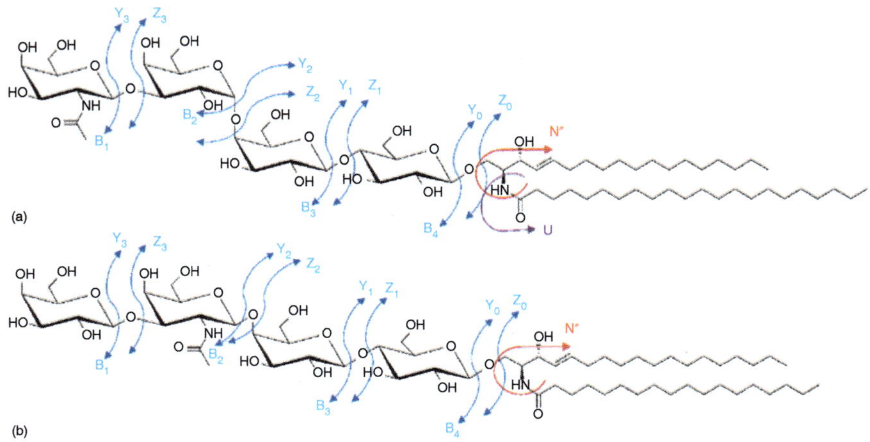

Figure 15.16 Chemical structures of Gb_4Cer (a) and Gg_4Cer/GA1 (b) with indication of major fragmentations upon CID.

Using an ESI-coupled ion trap mass spectrometer, permethylated globoseries and isogloboseries triaosyl- and tetraosylceramides reveal structure-specific fragments at the MS^4 and MS^5 levels, respectively [168, 169]. With this approach, iGb_3Cer could be quantified in the presence of the abundant stereoisomeric Gb_3Cer [169]. Similar to monohexosylceramide, normal-phase liquid chromatography/HILIC-coupled tandem mass spectrometry was successfully applied to separately quantify dihexosylceramides, galabiaosylceramides (Ga_2Cer), and LacCer, in urine of Fabry disease patients [170]. Additional information on double-bond positions within the ceramide moiety of GSLs may be obtained via 193 nm ultraviolet photodissociation [24]. Alternatively, the glycan moieties of neutral GSLs have been released enzymatically. The free (and eventually reduced) glycans then separate on graphite columns and produce also diagnostic cross-ring A-type fragments for identification of structural isomers upon CID in negative ESI mode [171].

15.4.8 Gangliosides

Gangliosides (GGs) derive from GSLs by transfer of one or more sialic acid residues to the glycan chain. Initially, GGs are isolated post mortem from an infantile patient's brain and are enriched in eponymous ganglion cells [6]. Mammalian brain expresses GGs with the sialic acid *N*-acetyl neuraminic acid (Neu5Ac). However, other sialic acids, like *N*-glycolyl neuraminic acid (Neu5Gc), may be found on GGs in other tissues, and composition varies from species to species [172]. Due to a mutation in CMP-sialic acid hydroxylase, humans, however, cannot synthesize Neu5Gc [173, 174]. Traces of GGs with Neu5Gc are nevertheless found in normal human tissue and human tumors, likely due to metabolic incorporation of Neu5Gc derived from dietary uptake [175, 176]. Further modifications of sialic acid, e.g. O-acetylation, increase the heterogeneity of gangliosides [177]. GGs may belong to

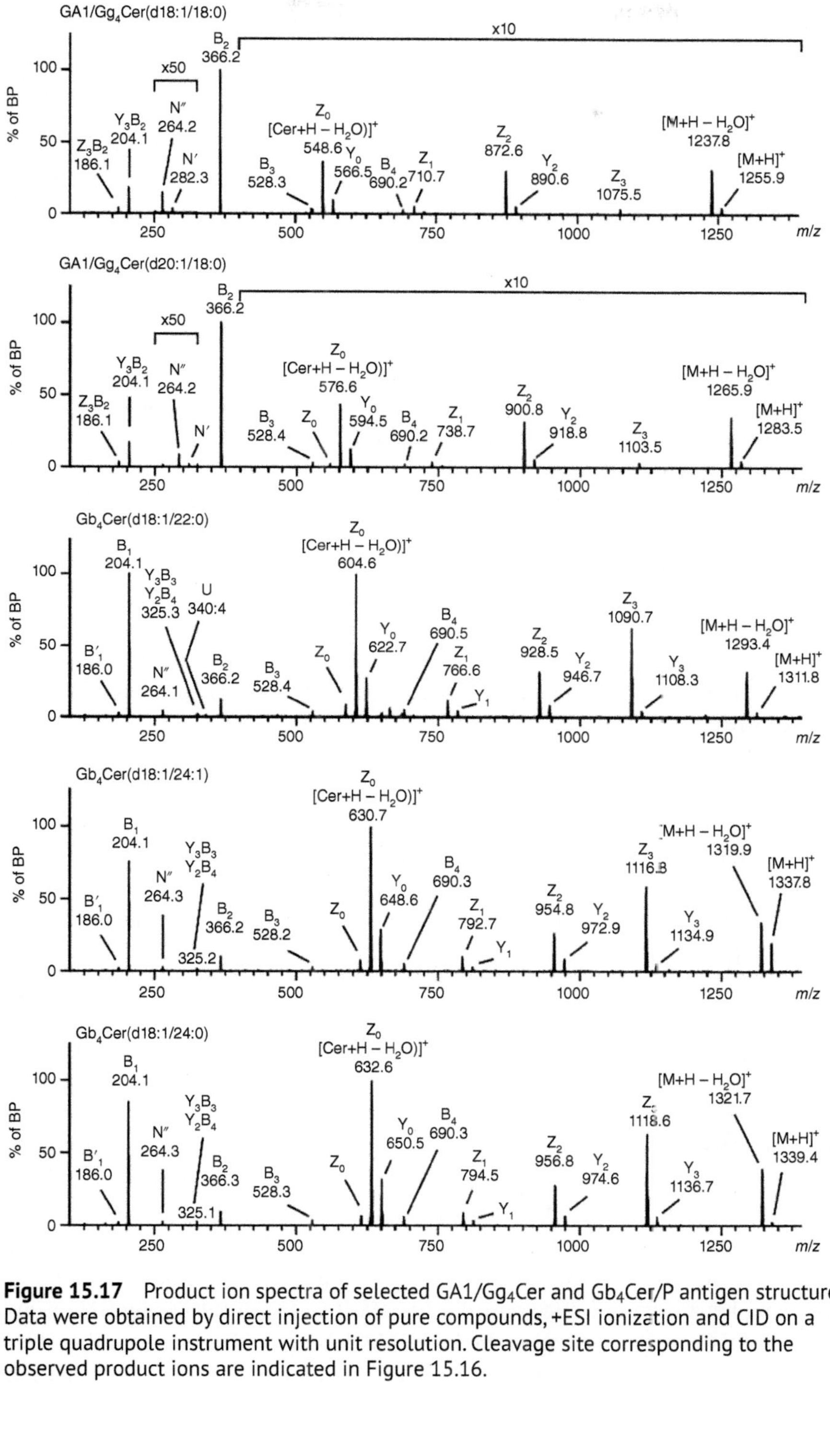

Figure 15.17 Product ion spectra of selected GA1/Gg_4Cer and Gb_4Cer/P antigen structures. Data were obtained by direct injection of pure compounds, +ESI ionization and CID on a triple quadrupole instrument with unit resolution. Cleavage site corresponding to the observed product ions are indicated in Figure 15.16.

different series of GSLs (for example GM1 belongs to the ganglioseries and LM1 belongs to the lactoseries). However, major brain GGs have a ganglioseries backbone. They are derived by stepwise combinatorial synthesis leading mainly to a- and b-series GGs [158]. GM3 synthase catalyzes the first transfer of sialic acid on the growing glycan chain and is necessary for synthesis of both a- and b-series GGs. Loss-of-function mutations in GM3 synthase cause infantile-onset symptomatic epilepsy syndrome [178]. This is not observed with corresponding knockout mice [179] except for hearing loss [180]. Lysosomal degradation of GGs and nGSLs also is a stepwise process, and defects in individual catabolic steps may lead to lysosomal storage diseases like Tay–Sachs disease [139].

Although prone to be analyzed as deprotonated molecular ions in negative mode, the sensitivity of the corresponding GG signals is rather weak compared to corresponding neutral GSLs, and ionization may lead to ISD with loss of sialic acid residues. Furthermore, no sharp peaks are obtained by conventional LC methods. In comparison to fast atom bombardment [166], ionization of per-methylated GGs was improved by positive electrospray with no ion source fragmentation [181]. Subsequent CID results in glycosidic bond cleavage and the terminal sugars are identified by an additional methylation [181].

Using MALDI, GGs are detected in positive mode as adduct ions with sodium and potassium and in negative mode as deprotonated molecular compounds. Upon ionization, underivatized GGs show less adduct formation and fragmentation in negative-ion mode and per-methylation improved stability of GGs in positive mode significantly [165, 182].

Whereas CID of unmodified deprotonated GGs provide mainly B/Y and C/Z fragments [183], only few fragments relate to the hydrophobic ceramide base, which is overcome by ultraviolet photodissociation. In addition a broader range of cross ring cleavages (A/X) is observed facilitating differentiation of isobaric GGs [184]. Another possibility to separate isomeric GGs is ultra-high-resolution ion mobility mass spectrometry, which was successfully applied to base line separate GD1a and GD1b [185].

Release of deprotonated sialic acid fragments or sialic acid containing glycan fragments has been used in targeted (MRM) LC-ESI-MS^2 analysis [186]. Neu5Ac-containing GGs are then detected with the transition to *m/z* 290 ($[Neu5Ac{-}H_3O^+]^-$) and are separated from Neu5Gc carrying GGs, which would release a signal at *m/z* 306 ($[Neu5Gc{-}H_3O^+]^-$). At higher CID energy, a breakdown product common to both Neu5Ac and Neu5Gc is observed at *m/z* 87, which may be used to selectively screen for both types of GGs in biological samples. With these three precursor scans, it was shown that GM2 from mouse liver almost exclusively contains Neu5Gc, whereas GM3 from the same lipid extract was rather a mixture dominated by Neu5Ac-containing GGs. Interestingly, deletion of glycolipid synthesis in albumin expressing hepatocytes of mice only did not affect GM3 but selectively GM2. This indicated that hepatocytes express mainly GM2, but not GM3, whereas other liver resident cells express mainly GM3 with a different sialic acid pattern (Figure 15.18) [187]. In combination with LC, the precursor ion scanning may be translated into multi-channel multiple reaction monitoring to cover the ganglioside

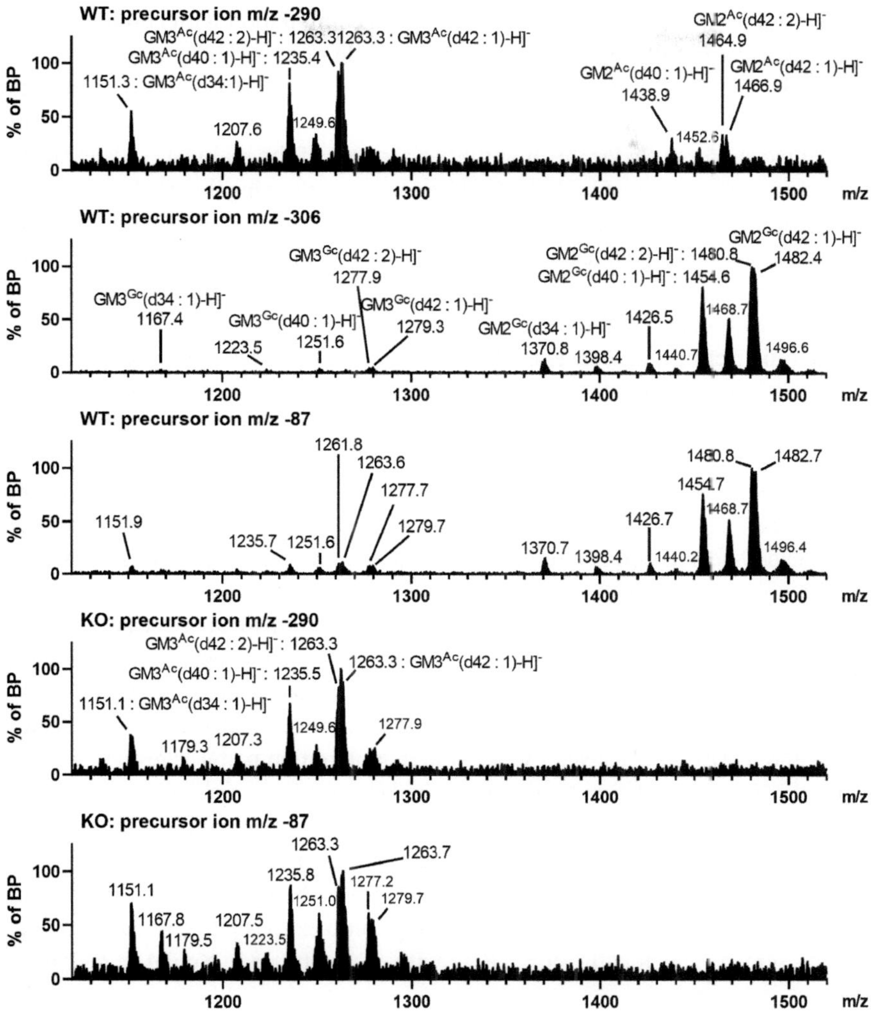

Figure 15.18 Precursor ion scans identifying Neu5Ac and Neu5Gc containing GGs from mouse liver lipid extracts. Wild type (WT) and knockout (KO) correspond to liver from wild type and hepatocyte-specific glucosylceramide synthase knockout mice, respectively. *M/z* 290 and *m/z* 306 relate to $[\text{Neu5Ac}-H_3O^+]^-$ and $[\text{Neu5Gc}-H_3O^+]^-$, respectively; and *m/z* 87 is released at higher energy from both types of sialic acids, possibly $[H_2C{=}C(OH)CO_2]^-$, including the first three carbon atoms of the sialic acid. Note, GM2 contains almost exclusively Neu5Gc (annotated as $GM2^{Gc}$), whereas the most intense signals for GM3 obtained by the precursor *m/z* 87 scan relate to Neu5Ac (annotated as $GM3^{Ac}$). In mice with a hepatocyte-specific deficiency of GSL synthesis, liver GM3 remains, whereas liver GM2 is lost [187]. Data had been recorded by direct injection into a nano-ESI-(QqQ) tandem mass spectrometer. It may be noted that the simple shift of Δ16 in molecular weight not necessarily relates to Neu5Gc instead of Neu5Ac as also the ceramide portion of GGs could be hydroxylated.

profile of tissues in a semi-untargeted manner. In combination with chiral column LC, this was successfully applied to brain ganglioside samples [188].

Using high-resolution mass spectrometers, GGs may be quantified by scanning the whole range of expected GGs *m/z* signals using LC-MS^1. This approach had been applied successfully for the quantitative analysis of the main GG species from 2-day-old and 80-day-old rat brain [189]. The post analysis using the resolving power of the Orbitrap mass to filter for known masses of GG species turned the initially untargeted raw data acquisition of this method into a targeted approach. Nevertheless, the raw data can be analyzed anytime with a different approach to reveal eventually other lipid species.

In order to discriminate isomeric α2–3- and α2–6-sialylated ganglioside species of human granulocytes for relative quantification, S. Kirsch et al. developed a nano-HILIC method based on a TSK-Gel Amide 80 column. Base line-separated α2–6-sialylated gangliosides were identified according to the release of $^{0,4}A_2$–CO_2 ring cleavage ions at nominal *m/z* 306, which occur exclusively from α2–6-sialylated GSLs (Figure 15.19) [190].

Sialic acid *O*-acetylation is one of the most common structural modifications of GGs. The nominal mass shift for acetylation is 42 amu and is the same as with acyl chain length prolongation by C_3H_6. High mass resolution instruments easily distinguish the mass difference of 0.036 amu, but this overlap has to be considered especially with unit resolution QqQ-like instruments. Using a MALDI-QIT-TOF instrument GGs with *O*-acetylated-Neu5Ac residues were detected by their molecular mass and in MS^2 by *O*-AcNeu5Ac-specific fragmentation patterns revealing neutral loss of 315 or 333 amu for inner and terminal *O*-AcNeu5Ac residues as compared to 273 and 291 amu for corresponding Neu5Ac residues. Corresponding ions at *m/z* 314 or *m/z* 332 were not reported [191]. The latter product ion had been detected with a secondary ion mass spectrometer from GD3 with O-acetylation detected in brain of a mouse with a deficiency in expressing complex GGs [192], and it could be used to screen specifically for GGs with O-acetylated Neu5Ac residues.

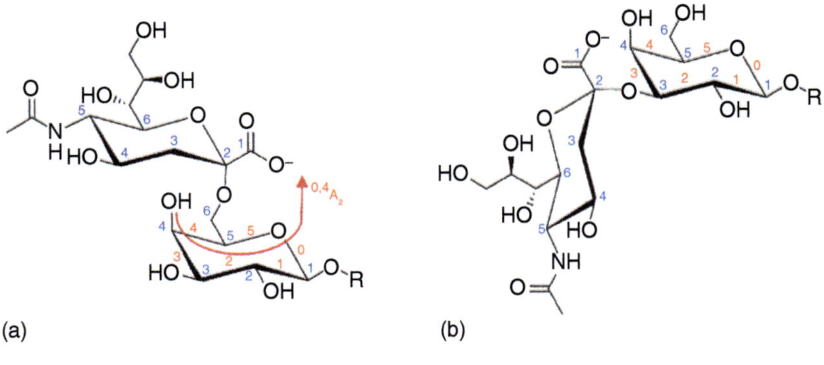

Figure 15.19 Terminal Neu5Ac α2–3- and α2–6-galactosyl residues found on neolacto-GSLs of human granulocytes. Source: Adapted from Kirsch et al. [190]. The $^{0,4}A_2$-CO_2 ring cleavage is indicated, which is specific for α2–6 sialylated GSLs. Numbering of the first six carbon atoms of each residue is indicated in blue and corresponding ring bonds are numbered in red.

These fragment ions however do not indicate the position of acetylation within the sialic acid moiety and different techniques are required to solve this issue [193]. Changes in the O-acetylation pattern of brain GGs due to enzymatic deficiencies and therefore changes in the metabolic pathways of GGs had also been detected subsequently by tandem mass spectrometry [194]. Since O-acetylated ganglioside species are mainly expressed during embryonic development and in the central nervous system in healthy adults, but are re-expressed during cancer development, they are considered as markers of cancers of neuroectodermal origin and as targets for cancer immunotherapy [195].

15.4.9 Sulfatides (Incl. Complex Sulfatides)

Sulfatides are sulfated glycolipids and display a large class of lipids on their own [196]. In mammals, the enzyme cerebroside sulfotransferase (galactose-3-*O*-sulfotransferase 1, GAL3ST1) transfers sulfate groups from 3′-phosphoadenosin-5′-phosphosulfat (PAPS) onto 3-hydroxy groups of terminal galactosyl-residues of GSLs and of galactosylalkylacylglycerol (GalEAG). The most common reactions are those with GalCer to produce sulfatide (SM4s) in myelin sheath and in renal tubular cells, and with GalEAG to produce seminolipid (SM4g) in testicular germ cells [197]. However, more complex GSLs such as renal LacCer and II^3-Sulfo-Gg$_4$Cer (SM1a) are also targets of cerebroside sulfotransferase [198]. Among other functions, expression of sulfatide SM4s within myelin sheath is necessary for proper organization of Ranvier nodes and within kidney tubular epithelial cells, sulfatide expression enables renal adaptation to chronic metabolic acidosis [198, 199]. Seminolipid, however, is required for proper development of sperm cells [199]. Lysosomal degradation of sulfatides requires arylsulfatase A (ARSA) and activator protein saposin B and corresponding protein deficiencies lead to lysosomal sulfatide accumulation associated with metachromatic leukodystrophy [200–202].

Due to the low pK_a value of sulfatides (and lyso-sulfatides), they are easily detected as deprotonated ions with current soft ionization techniques, i.e. MALDI and ESI [167, 203–205]. Product ion spectra obtained by CID of deprotonated sulfatide SM4s are dominated by sulfate-containing fragments, especially HSO_4^- at *m/z* 97 and at higher energies SO_3^- at *m/z* 80 (Figure 15.20). The latter distinguishes at the nominal mass-level sulfatides from phosphate-containing lipids. At lower intensity,

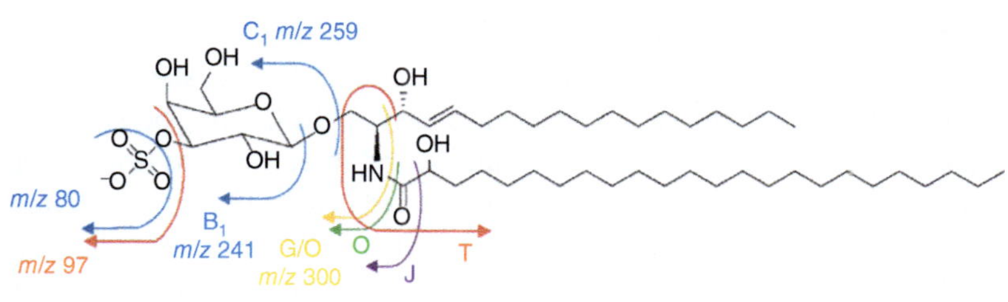

Figure 15.20 Structure of deprotonated Sulfatide SM4s(d18:1/αh24:0) showing characteristic fragmentations observed by CID.

the B- and C-fragments as well as a G/O-fragment will appear. The diagnostic O- and T-fragments, allowing determination of fatty acid and sphingoid base combination, are also observed at quite low relative intensity. The presence of α-hydroxyacyl chains facilitates cleavage between the amide bond and the α-hydroxyl group, i.e. J-cleavage ion at *m/z* 568 (with C18-sphingosine), which will also enhance fragment with subsequent loss of CO and H_2O, i.e. signals *m/z* 540 and *m/z* 522, respectively [206]. Precursor ion scanning with *m/z* 568 (or *m/z* 540) will therefore specifically reveal signals of sulfatides with N-linked α-hydroxyacyl chains [204]. This J-fragment is dependent on *m/z* on the chain length of the sphingoid base, but also on the amount of double bonds. Thereby, it revealed the presence of sulfatides with sphingadienine in renal kidneys [207]. Fragmentation of complex sulfatides, like SB1a (gangliotetraosylceramide-II^3, IV^3 bis-sulfate) with MALDI-TOF tandem mass spectrometry, will reveal mainly ions due to cleavage at B-, Y-, and Z-positions [207] similar to liquid secondary ion mass spectrometry at low and high CID [208].

The sulfate ion with *m/z* 97 is the most abundant ion released by CID of sulfatides and subsequently was used to quantitatively identify simple and complex sulfatides by semi-targeted precursor ion scanning using direct injection nanoESI-(QqQ) tandem mass spectrometry [202]. This precursor ions scanning revealed *de novo* appearing sulfated lactosylceramide (SM3) in brain of genetically modified mice, which could not produce gangliosides from lactosylceramide. The obtained MS pattern revealed a ceramide composition similar to GGs of wild-type brain, but different from simple sulfatide SM4s, which is expressed by myelin-producing oligodendrocytes. These results gave a hint to production of SM3 by neuronal cells [209]. Coupling to LC allowed then sensitive targeted analysis of sulfatides by MRM with transition to *m/z* 97 from various biological samples [109, 188, 210–216].

15.5 Stable Isotope Labeling

Sphingolipid biosynthesis and metabolism could be investigated using stable isotopes, most often using deuterium (D), ^{13}C carbon, or ^{15}N nitrogen by the shotgun or LC-MS/MS approaches. Stable isotope labeling of SLs may be introduced into polar precursors (e.g. serine, choline, or glucose), into fatty acids (e.g. palmitic acid), into sphingoid bases, or into final molecules of SLs. Since stable isotope-labeled SLs are also often used as internal standards for quantification, they have to differ from sphingolipids endogenously labeled to track cellular SL metabolism.

From the polar precursors, serine is the most intuitive precursor suitable for labeling since this molecule enters sphingolipid biosynthesis in the first step and only once per SL molecule. Due to the small molecule of serine, the labeling possibilities are limited: only a single ^{15}N atom, three ^{13}C atoms, and three D atoms (excluding hydrogens on amino and alcohol groups) may be inserted in different combinations. It is worth noting that one ^{13}C and one D atom are cleaved in the first step by SPT (Figure 15.21a) [217]. For sphingolipid labeling, it is beneficial to introduce more than two labeled atoms, because the difference between sphingosine and dihydrosphingosine-based SLs is 2 amu, which may complicate distinguishing

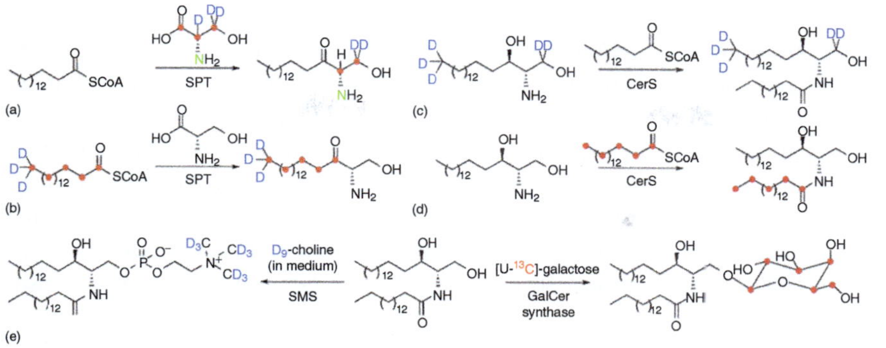

Figure 15.21 Possible stable isotope labeling within sphingolipid biosynthesis. D in blue indicates deuterium, N in green indicates ^{15}N nitrogen and red dots indicate ^{13}C carbons. Note that different combinations are available, e.g. more D atoms or less ^{13}C atoms in lipid chains, no ^{15}N etc. Reaction of unlabeled palmitoyl-CoA with labeled L-serine (a), labeled palmitoyl-CoA with unlabeled L-serine (b), labeled dihydrosphingosine with unlabeled palmitoyl-CoA (c), unlabeled dihydrosphingosine with labeled palmitoyl-CoA (d) and a formation of polar head group-labeled SM and GalCer (e).

between unlabeled dihydrosphingosine and labeled sphingosine. Such complications may then be surpassed using high-resolution instruments [218]. [U-^{13}C,^{15}N]-serine was used to study a *de novo* biosynthesis of dihydrosphingosine-1-phosphate and inhibition of ceramide synthases using FTY720 in human pulmonary artery endothelial cells. The individual precursor and product ions of SBs, Cers, and SB-1-P were followed via MRM transitions of [M+3] precursor and product ions [219, 220]. However, if monitoring of SL synthesis with [U-^{13}C,^{15}N]-serine is performed with MRM transitions, which include loss of C_1 as formaldehyde, as is the case especially for KDS, transition of [M+3] to [P+2] ions has to be used [20]. Wigger et al. used D_3-L-serine in combination with D_3-palmitic acid to follow the sphingolipid biosynthesis in cell-free conditions using rat liver microsomes, which contained all the necessary enzymes for Cer biosynthesis. The products of the D_3-L-serine condensation with D_3-palmitic acid by SPT then contained five D atoms (one D atom is lost during the catalytic cycle by a transfer to a lysine residue of SPT) [217]. Trimethyl-D_9-choline chloride was used to determine the cellular efflux of SM (and phosphatidylcholine [PC] and other lipids) to high-density lipoprotein (HDL_3) in monocyte-derived macrophages. D_9-choline labeled lipids were analyzed by precursor ion scan of *m/z* 193 instead of *m/z* 184 which is typical for choline moiety in molecules of SM and PC [221]. [U-^{13}C]-glucose was used to determine the turnover of GSL species in HEp-2 cells. Individual SLs were monitored by MRM transitions where the *m/z* values for labeled GlcCer were increased by 6 amu compared to unlabeled molecules, by 6 and 12 amu for LacCer, and by 6, 12, and 18 amu for Gb_3 species [222].

Free fatty acids (FFAs), most often palmitate, can be labeled with multiple atoms of both D and ^{13}C to probe the sphingolipid metabolism. [1,2,3,4-$^{13}C_4$]-palmitate was used as a tracer to study the metabolic kinetics of diacylglycerol, Cer, PC, and SM in HL-60 cells by GC-MS. Upon electron impact-induced ionization of the bis-trimethylsilyl derivative of Cer(d18:1/16:0), three major fragments were found with *m/z* 311 (containing So carbons), *m/z* 370 (containing *N*-palmitate carbons), and *m/z* 666 as an [M–(CH_3)] fragment. When [$^{13}C_4$]-palmitate was applied, the [M+4] and [M+8] fragments were found, i.e. *m/z* 315 for labeling in sphingoid part, *m/z* 374 for labeling in *N*-acyl chain, and *m/z* 670 and 674 for labeling Cer in a single or both chains, respectively [223]. [U-^{13}C]-palmitate was used to analyze the sphingolipid biosynthesis using LC-MS^2 in HEK293 cells. Such labeled palmitate was found to be incorporated into the sphingoid part and *N*-acyl chain of the Cer moiety or into both, providing ions of [M+16] and [M+32]. Subsequent mass spectrometric fragmentation enables to distinguish which Cer part contained stable isotopically labeled chain after biosynthesis [224]. [U-^{13}C]-palmitate was also used *in vivo* by an intramuscular infusion to volunteers to study the sphingolipid metabolism. The authors followed mass transitions of a loss of water from [U-^{13}C]-Cer(d18:1/16:0), 554 > 536 and a combined loss of fatty acyl and two water molecules, 554 > 264. Due to a limited enrichment of labeled compounds, the double-labeled (i.e. both in *N*-acyl chain and sphingoid base) compounds were not expected [225]. FFAs, namely D_9-palmitic, D_9-stearic, and D_4-lignoceric acids in combination with [D_3, ^{15}N]-L-serine, were used to study differences in sphingolipid metabolic flux based on the FFA chain length in Hek293T cells, and it was found that very long chain Cer turns over more rapidly than long chain Cer [226]. Palmitate containing three D atoms in combination with labeled serine was discussed before [217].

Sphingolipid metabolism and transport may be monitored using exogenously added labeled Cer. After an incubation of HL60 cells with Cer(d18:1/[$^{13}C_4$]-16:0), the labeled palmitate moiety was not only found as an *N*-acyl chain of Cer but also as a part of a sphingoid base. Such findings indicate that exogenous Cer in cells does not mix with the endogenous pool; the exogenous molecules are rather degraded to palmitate and reused for a *de novo* Cer synthesis [227]. ^{15}N-labeled sphingosine and dihydrosphingosine in combination with ^{13}C labeled FFAs after their metabolization in Clone 15 fibroblast cells were used to study sphingolipid domains in plasma membranes of fibroblasts by secondary ion mass spectrometry and revealed micrometer-scale sphingolipid-enriched patches composed of numerous microdomains in the plasma membrane [228].

15.6 Imaging Mass Spectrometry (IMS) of Sphingolipids

There are several imaging techniques to detect lipids directly on tissue by mass spectrometry preserving the information on their tissue location with more or less good spatial resolution. For details of these techniques including spatial resolution, we refer to Chapter 2 and recent reviews [229–236]. MALDI is perhaps the most popular imaging mass spectrometry (IMS) technique, which enables analyzing the distributions of intact lipid within tissues. Loss of LC-specific retention times as well as interference with matrix signals may be overcome by introducing ion mobility into IMS [237, 238]. However, ISD arising from "larger" GSLs and giving eventually rise to false-positive signals of "smaller" sphingolipids remains a potential danger nevertheless. Another obstacle in IMS is ion suppression of readily ionizable lipids to others, as described, e.g. for renal sulfatides and phosphatidylinositols [211], and mixtures of matrices have been used with MALDI to reduce such effects [239]. With increasing resolution, it became possible to show the complex spatial heterogeneity of lipids in anatomical structures such as the human retina and optic nerve underlining cell and cell-structure-specific organization and function of lipids including SLs [240].

Subsequently IMS provided spatial insight into injury or disease-related sphingolipid changes in (models of) ischemia, Alzheimer's disease, Parkinson's disease, Huntington's disease, lysosomal storage diseases, ichthyosis, mucopolysaccharidosis, Alport syndrome, leprosy, or various cancers [207, 241–253]. The best studied appear to be the easily chargeable gangliosides, sulfatides, and sphingomyelins in tissues with their high concentrations, i.e. in brain and kidney, respectively [237].

Furthermore, IMS provided visual assessment of the contribution of enzymes to the acyl-chain- and region-specific sphingolipid expression: while C18-sphingoid base species of gangliosides are widely distributed throughout the frontal murine brain, the C20-sphingoid base species selectively localize along the entorhinal-hippocampus projections, especially in the molecular layer (ML) of the dentate gyrus (DG). A development- and aging-related accumulation of the C^{-20} species in the ML-DG was observed. This indicates a brain-region-specific regulation of long chain base (LCB) chain length, which depends on the composition of the SPT complex and which may be particularly important for the distinct function in cells of central nervous system (CNS) [254]. Likewise, a differential pattern of storage was observed for GM2 with C18- or C20-sphingoid base content in brain of Hexa-/-&Neu3-/-double-knockout mice [255]. Similarly, the differential localization of sphingomyelin molecular species appears to be regulated by particular ceramide synthase in mouse brains [256]. Investigating murine SMS2- and CerS2-deficient kidneys demonstrated that the localization of sphingomyelin and sulfatide is specific to each acyl chain and sphingoid base type [212, 257]. With higher spatial resolution, not only gross spatial differences like kidney cortex, medulla, and papillae can be distinguished, but, as shown here in Figure 15.22, anatomical structures within the cortex such as the glomerulus can also be subdivided into inner and outer layers demonstrating cell-specific sphingolipid expression, here of sphingomyelins with different size of the ceramide anchors, within these structures. These different sizes could result from differential expression of CerS1 and CerS2, which prefer C18-AcylCoA and C22/24-AcylCoA, respectively.

Likewise, MALDI-IMS visualized the differentiation specific expression of sphingomyelins with ultra-long and polyunsaturated acyl chains during spermatogenesis in murine testis, which is caused by differentiation-dependent expression of ceramide synthase 3 and corresponding fatty acid substrates [79]. Developmental changes were also revealed for oligodendrocyte-derived sulfatides by IMS [258]. Their chain length pattern shifted from short-chain fatty acids in pro-oligodendroblasts to longer fatty acids during oligodendrocyte maturation. Therefore, individual sulfatide species may have unique roles in oligodendrocyte maturation and myelination.

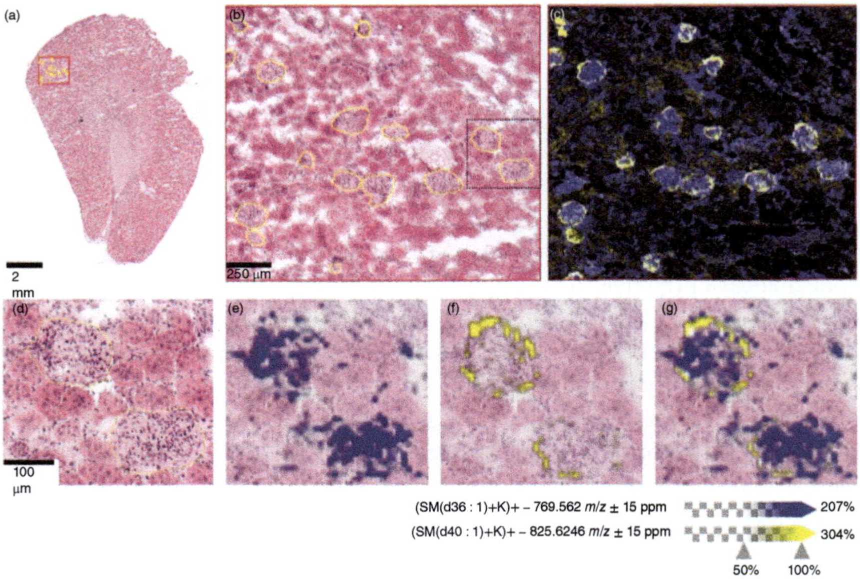

Figure 15.22 Spatial distribution of sphingomyelin SM(d36:1) and SM(d40:1) in renal glomeruli. MALDI Imaging data were acquired on a (tissue imaging mass spectrometer) timsTOF fleX instrument (Bruker Daltonics GmbH & Co. KG, Bremen, Germany) from a fresh frozen mouse kidney section which was sliced at 10 µm thickness. DHB matrix was applied by sublimation and mass spectra were acquired at 10 µm pixel size. Hematoxylin and eosin (H&E) staining of the mouse kidney section after MALDI imaging analysis with glomeruli regions indicated in yellow (a, b, d). The measurement area is indicated by a red rectangle in (a) and enlarged in (b). Ion distribution map of the potassium adducts of sphingomyelin SM(d36:1) in blue and SM(d40:1) in yellow (see color bar at the bottom) with linear interpolation and hot spot removal applied in SCILS Lab 2022a. Ion images are plotted after root mean square normalization (c). Enlarged view (e, f, g) as indicated by a dotted rectangle in (b) on two glomeruli visualizing the distribution of SM(d36:1) (e) and SM(d40:1) (f) or both species (g) overlayed with the histology (low intensity peaks were made transparent). While SM(d36:1) was mainly present in the renal corpuscles' interior, i.e. Mesangium, Glomerulus capillaries, and visceral layer of the Bowman's capsule (Podocytes), SM(d40:1) was localized in the parietal layer of the Bowman's capsule. Lipid annotations are supported by on-tissue MALDI-TIMS MS/MS (data not shown). Source: Images were kindly supplied by Janina Oetjens, from Bruker Daltonics, Bremens.

15.7 Plants, Yeast, Fungi, Bacteria, Marine Organisms, and Sponges

Distinct SLs were found in all the mentioned non-mammalian species. Sphingolipid biosynthesis pathways in these organisms share key features with mammalian biosynthesis; however, multiple variations leading to sphingolipid molecules differing from mammals could be recognized (Figure 15.23). In this part, we are going to comment only on the most pronounced differences characteristic for each species and refer to review articles dealing with this topic in detail.

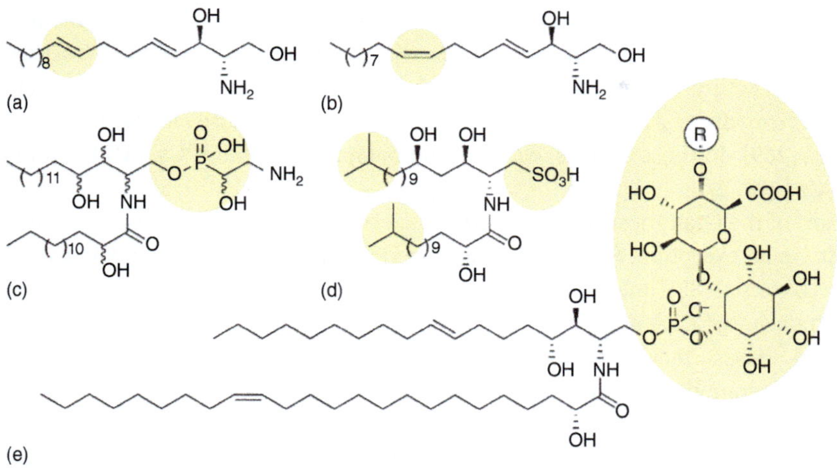

Figure 15.23 Examples of SLs in non-mammalian species. Differences from mammalian SLs are indicated in yellow. Desaturated sphingoid bases in plants (a and b). Sphingophosphonolipid found in *Bdellovibrio stolpii* (c). Sulfonolipid found in *Algoriphagus machipongonensis* (d). Glycosyl inositolphosphoceramide found in plants (e); R substitutes a glycan moiety consisting of mannose, glucosamine, or *N*-acetylglucosamine combinations.

In plants, sphingolipid biosynthesis begins in the same manner as in mammals with condensation of L-serine with palmitoyl-CoA via SPT followed by a reduction of KDS to sphinganine by KDSR. This long chain base may directly undergo a hydroxylation using sphingoid base hydroxylases in position 4 leading to phytosphingosine [259]. An acyl chain (very often α-hydroxylated) is then connected to the free amino group of sphingoid bases to form Cer using ceramide synthases, and these Cers may further be desaturated by long chain base desaturases in position 4 (always trans) or position 8 (both cis and trans are possible) of the sphingoid base. In the Golgi, Cers are modified by glucosylation to form glucosylceramides (by glucosylceramide synthase) [260] or by an attachment of an inositolphosphate group (by inositolphosphorylceramide [IPC] synthases) resulting in IPCs [261]. Subsequently, additional sugar moieties may be attached starting with glucuronic acid and followed by an attachment of mannose, glucosamine, or GlcNAc units to form glycosyl inositolphosphorylceramides (GIPCs) [262–265].

Sphingolipid biosynthesis in *Saccharomyces cerevisiae* remains the same as previous up to the stage of dihydrosphingosine, which may then be hydroxylated in position 4 to phytosphingosine. Both sphingoid bases are N-acylated by a fatty acyl with predominantly 26 carbons and up to 2 hydroxyl groups (on C2 and C3). Dihydroceramide may be hydroxylated to the form of phytoceramide in this stage as well [266]. Complex SLs are formed first by an attachment of inositol phosphate (IPCs), followed by an attachment of mannose to C2 of inositol (MIPC), and the last step is a connection of an additional inositol moiety (M(IP)$_2$C) [264, 267, 268]. In other fungal species like *Cryptococcus neoformans*, *Aspergillus nidulans*, *Fusarium graminearum*, *Candida albicans*, and others, we can find additional modifications like a desaturation of dihydroceramide in position 4 (sphingosine-based ceramide formation), in position 8 or methylation in position 9. To the resulting structure of 9-methyl-4,8-sphingadienine, bearing predominantly an N-linked α-hydroxylated acyl chain, glucose and galactose units may be attached [269]. In these species, we also find complex SLs like IPCs, MIPCs, and $M(IP)_2Cs$ [270, 271].

In most of the Gram-positive or -negative bacterial species, SLs are not commonly present. As an exception, SLs were found in certain *Bacteroides* or *Sphingomonads* strains. Bacterial SPT may utilize different chain lengths of acyl-CoA, ranging between decanoyl-CoA and oleoyl-CoA, including branched acyl chains. Unusual sphingoid bases may be found in certain species like an isobranched α-galactosylceramide in *Bacteroides fragilis* (similar lipids were also found in a maritime sponge *Agelas Mauritianus*) [272], a phosphonate-containing base in *Bdellovibrio stolpii* [273] (which seems to affect the life cycle of this bacteria) or a sulfonate-containing base in a maritime bacterium *Algoriphagus machipongonensis* [271, 274, 275].

References

1 Thudichum, J.L.W. (1884). *A Treatise on the Chemical Constitution of the Brain*. London: Bailliere, Tindall and Cox.

2 Carter, H.E., Glick, F.J., Norris, W.P., and Phillips, G.E. (1947). Biochemistry of the sphingolipides 3. Structure of sphingosine. *J. Biol. Chem.* 170 (1): 285–294.

3 Pruett, S.T., Bushnev, A., Hagedorn, K. et al. (2008). Biodiversity of sphingoid bases ("sphingosines") and related amino alcohols. *J. Lipid Res.* 49 (8): 1621–1639.

4 Gault, C.R., Obeid, L.M., and Hannun, Y.A. (2010). An overview of sphingolipid metabolism: from synthesis to breakdown. *Adv. Exp. Med. Biol.* 688: 1–23.

5 Rabionet, M., Gorgas, K., and Sandhoff, R. (2014). Ceramide synthesis in the epidermis. *Biochim. Biophys. Acta* 1841 (3): 422–434.

6 Sandhoff, R. and Sandhoff, K. (2018). Emerging concepts of ganglioside metabolism. *FEBS Lett.* 592 (23): 3835–3864.

7 Fahy, E., Subramaniam, S., Brown, H.A. et al. (2005). A comprehensive classification system for lipids. *J. Lipid Res.* 46 (5): 839–861.

8 Merrill, A.H. Jr., Wang, M.D., Park, M., and Sullards, M.C. (2007). (Glyco) sphingolipidology: an amazing challenge and opportunity for systems biology. *Trends Biochem. Sci* 32 (10): 457–468.

9 Liebisch, G., Fahy, E., Aoki, J. et al. (2020). Update on LIPID MAPS classification, nomenclature, and shorthand notation for MS-derived lipid structures. *J. Lipid Res.* 61 (12): 1539–1555.

10 Mashima, R., Okuyama, T., and Ohira, M. (2019). Biosynthesis of long chain base in sphingolipids in animals, plants and fungi. *Future Sci. OA* 6 (1): FSO434.

11 Jojima, K., Edagawa, M., Sawai, M. et al. (2020). Biosynthesis of the anti-lipid-microdomain sphingoid base 4,14-sphingadiene by the ceramide desaturase FADS3. *FASEB J.* 34 (2): 3318–3335.

12 Karsai, G., Lone, M., Kutalik, Z. et al. (2020). FADS3 is a Delta14Z sphingoid base desaturase that contributes to gender differences in the human plasma sphingolipidome. *J. Biol. Chem.* 295 (7): 1889–1897.

13 Pilz, R., Opálka, L., Majcher, A. et al. (2022). Formation of keto-type ceramides in palmoplantar keratoderma based on biallelic KDSR mutations in patients. *Hum. Mol. Genet.* 31: 1105–1114.

14 Hannun, Y.A. and Obeid, L.M. (2018). Sphingolipids and their metabolism in physiology and disease. *Nat. Rev. Mol. Cell Biol.* 19 (3): 175–191.

15 Penno, A., Reilly, M.M., Houlden, H. et al. (2010). Hereditary sensory neuropathy type 1 is caused by the accumulation of two neurotoxic sphingolipids. *J. Biol. Chem.* 285 (15): 11178–11187.

16 Astudillo, L., Sabourdy, F., Therville, N. et al. (2015). Human genetic disorders of sphingolipid biosynthesis. *J. Inherit. Metab. Dis.* 38 (1): 65–76.

17 Lone, M.A., Santos, T., Alecu, I. et al. (2019). 1-Deoxysphingolipids. *Biochim. Biophys. Acta, Mol. Cell. Biol. Lipids* 1864 (4): 512–521.

18 Zitomer, N.C., Mitchell, T., Voss, K.A. et al. (2009). Ceramide synthase inhibition by fumonisin B1 causes accumulation of 1-deoxysphinganine: a novel category of bioactive 1-deoxysphingoid bases and 1-deoxydihydroceramides biosynthesized by mammalian cell lines and animals. *J. Biol. Chem.* 284 (8): 4786–4795.

19 Ren, J., Snider, J., Airola, M.V. et al. (2018). Quantification of 3-ketodihydrosphingosine using HPLC-ESI-MS/MS to study SPT activity in yeast *Saccharomyces cerevisiae*. *J. Lipid Res.* 59 (1): 162–170.

20 Wu, J., Ma, S., Sandhoff, R. et al. (2019). Loss of neurological disease HSAN-I-associated gene SPTLC2 impairs $CD8^+$ T cell responses to infection by inhibiting T cell metabolic fitness. *Immunity* 50 (5): 1218–31 e5.

21 Rubino, M.F., Zecca, L., and Sonnino, S. (1992). Characterization of sphingosine long-chain bases by fast atom bombardment and high-energy collision-induced decomposition tandem mass spectrometry. *Org. Mass Spectrom.* 27: 1357–1364.

22 Sullards, M.C. (2000). Analysis of sphingomyelin, glucosylceramide, ceramide, sphingosine, and sphingosine 1-phosphate by tandem mass spectrometry. *Methods Enzymol.* 312: 32–45.

23 Sullards, M.C., Liu, Y., Chen, Y., and Merrill, A.H. Jr. (2011). Analysis of mammalian sphingolipids by liquid chromatography tandem mass spectrometry (LC-MS/MS) and tissue imaging mass spectrometry (TIMS). *Biochim. Biophys. Acta, Mol. Cell. Biol. Lipids* 1811 (11): 838–853.

24 Ryan, E., Nguyen, C.Q.N., Shiea, C., and Reid, G.E. (2017). Detailed structural characterization of sphingolipids via 193 nm ultraviolet photodissociation and ultra high resolution tandem mass spectrometry. *J. Am. Soc. Mass Spectrom.* 28 (7): 1406–1419.

25 Alecu, I., Othman, A., Penno, A. et al. (2017). Cytotoxic 1-deoxysphingolipids are metabolized by a cytochrome P450-dependent pathway. *J. Lipid Res.* 58 (1): 60–71.

26 Kobayashi, T., Mitsuo, K., and Goto, I. (1988). Free sphingoid bases in normal murine tissues. *Eur. J. Biochem.* 172 (3): 747–752.

27 Green, C.D., Maceyka, M., Cowart, L.A., and Spiegel, S. (2021). Sphingolipids in metabolic disease: the good, the bad, and the unknown. *Cell Metab.* 33 (7): 1293–1306.

28 Proia, R.L. and Hla, T. (2015). Emerging biology of sphingosine-1-phosphate: its role in pathogenesis and therapy. *J. Clin. Invest.* 125 (4): 1379–1387.

29 Park, S.W., Kim, M., Chen, S.W. et al. (2010). Sphinganine-1-phosphate attenuates both hepatic and renal injury induced by hepatic ischemia and reperfusion in mice. *Shock* 33 (1): 31–42.

30 Van Brocklyn, J.R., Lee, M.J., Menzeleev, R. et al. (1998). Dual actions of sphingosine-1-phosphate: extracellular through the Gi-coupled receptor Edg-1 and intracellular to regulate proliferation and survival. *J. Cell Biol.* 142 (1): 229–240.

31 Cutignano, A., Chiuminatto, U., Petruzziello, F. et al. (2010). UPLC-MS/MS method for analysis of sphingosine 1-phosphate in biological samples. *Prostaglandins Other Lipid Mediat.* 93 (1, 2): 25–29.

32 Scherer, M., Schmitz, G., and Liebisch, G. (2009). High-throughput analysis of sphingosine 1-phosphate, sphinganine 1-phosphate, and lysophosphatidic acid in plasma samples by liquid chromatography-tandem mass spectrometry. *Clin. Chem.* 55 (6): 1218–1222.

33 Ceglarek, U., Dittrich, J., Helmschrodt, C. et al. (2014). Preanalytical standardization of sphingosine-1-phosphate, sphinganine-1-phosphate and sphingosine analysis in human plasma by liquid chromatography-tandem mass spectrometry. *Clin. Chim. Acta* 435: 1–6.

34 Tang, X., Chen, H., Chen, G. et al. (2020). Validated LC-MS/MS method of sphingosine 1-phosphate quantification in human serum for evaluation of response to radiotherapy in lung cancer. *Thorac. Cancer* 11 (6): 1443–1452.

35 Schmidt, H., Schmidt, R., and Geisslinger, G. (2006). LC-MS/MS-analysis of sphingosine-1-phosphate and related compounds in plasma samples. *Prostaglandins Other Lipid Mediat.* 81 (3, 4): 162–170.

36 Sullards, M.C., Wang, E., Peng, Q., and Merrill, A.H. Jr. (2003). Metabolomic profiling of sphingolipids in human glioma cell lines by liquid chromatography tandem mass spectrometry. *Cell Mol. Biol.* 49 (5): 789–797.

37 Sakai, E., Kurano, M., Morita, Y. et al. (2020). Establishment of a measurement system for sphingolipids in the cerebrospinal fluid based on liquid chromatography-tandem mass spectrometry, and its application in the diagnosis of carcinomatous meningitis. *J. Appl. Lab. Med.* 5 (4): 656–670.

38 Egom, E.E., Fitzgerald, R., Canning, R. et al. (2017). Determination of sphingosine-1-phosphate in human plasma using liquid chromatography coupled with Q-TOF mass spectrometry. *Int. J. Mol. Sci.* 18 (8).

39 Gowda, S.G.B., Nakahashi, A., Yamane, K. et al. (2018). Facile chemoselective strategy toward capturing Sphingoid bases by a unique glutaraldehyde-functionalized resin. *ACS Omega* 3 (1): 753–759.

40 Berdyshev, E.V., Gorshkova, I.A., Garcia, J.G. et al. (2005). Quantitative analysis of sphingoid base-1-phosphates as bisacetylated derivatives by liquid chromatography-tandem mass spectrometry. *Anal. Biochem.* 339 (1): 129–136.

41 Lee, J.W., Nishiumi, S., Yoshida, M. et al. (2013). Simultaneous profiling of polar lipids by supercritical fluid chromatography/tandem mass spectrometry with methylation. *J. Chromatogr. A* 1279: 98–107.

42 Levy, M. and Futerman, A.H. (2010). Mammalian ceramide synthases. *IUBMB Life* 62 (5): 347–356.

43 Kihara, A. (2016). Synthesis and degradation pathways, functions, and pathology of ceramides and epidermal acylceramides. *Prog. Lipid Res.* 63: 50–69.

44 Law, B.A., Hancock, W.D., and Cowart, L.A. (2018). Getting to the heart of the sphingolipid riddle. *Curr. Opin. Physiol.* 1: 111–122.

45 Turpin, S.M., Nicholls, H.T., Willmes, D.M. et al. (2014). Obesity-induced CerS6-dependent C16:0 ceramide production promotes weight gain and glucose intolerance. *Cell Metab.* 20 (4): 678–686.

46 Raichur, S., Wang, S.T., Chan, P.W. et al. (2014). CerS2 haploinsufficiency inhibits beta-oxidation and confers susceptibility to diet-induced steatohepatitis and insulin resistance. *Cell Metab.* 20 (4): 687–695.

47 Michel, C., van Echten-Deckert, G., Rother, J. et al. (1997). Characterization of ceramide synthesis. A dihydroceramide desaturase introduces the 4,5-trans-double bond of sphingosine at the level of dihydroceramide. *J. Biol. Chem.* 272 (36): 22432–22437.

48 Ternes, P., Franke, S., Zahringer, U. et al. (2002). Identification and characterization of a sphingolipid delta 4-desaturase family. *J. Biol. Chem.* 277 (28): 25512–25518.

49 Omae, F., Miyazaki, M., Enomoto, A. et al. (2004). DES2 protein is responsible for phytoceramide biosynthesis in the mouse small intestine. *Biochem. J.* 379 (Pt 3): 687–695.

50 Robson, K.J., Stewart, M.E., Michelsen, S. et al. (1994). 6-Hydroxy-4-sphingenine in human epidermal ceramides. *J. Lipid Res.* 35 (11): 2060–2068.

51 Hinder, A., Schmelzer, C.E., Rawlings, A.V., and Neubert, R.H. (2011). Investigation of the molecular structure of the human stratum corneum ceramides [NP] and [EOS] by mass spectrometry. *Skin Pharmacol. Physiol.* 24 (3): 127–135.

52 Kramer, R., Bielawski, J., Kistner-Griffin, E. et al. (2015). Neurotoxic 1-deoxysphingolipids and paclitaxel-induced peripheral neuropathy. *FASEB J.* 29 (11): 4461–4472.

53 Hornemann, T. (2021). Mini review: lipids in peripheral nerve disorders. *Neurosci. Lett.* 740: 135455.

54 Alecu, I., Tedeschi, A., Behler, N. et al. (2017). Localization of 1-deoxysphingolipids to mitochondria induces mitochondrial dysfunction. *J. Lipid Res.* 58 (1): 42–59.

55 Hannich, J.T., Loizides-Mangold, U., Sinturel, F. et al. (2021). Ether lipids, sphingolipids and toxic 1-deoxyceramides as hallmarks for lean and obese type 2 diabetic patients. *Acta Physiol. (Oxf).* 232 (1): e13610.

56 Ando, A., Oka, M., and Satomi, Y. (2019). Deoxysphingolipids and ether-linked diacylglycerols accumulate in the tissues of aged mice. *Cell Biosci.* 9: 61.

57 Morell, P. and Radin, N.S. (1970). Specificity in ceramide biosynthesis from long chain bases and various fatty acyl coenzyme A's by brain microsomes. *J. Biol. Chem.* 245 (2): 342–350.

58 Shoyama, Y. and Kishimoto, Y. (1978). in vivo metabolism of 3-ketoceramide in rat brain. *J. Neurochem.* 30 (2): 377–382.

59 Couch, L.H., Churchwell, M.I., Doerge, D.R. et al. (1997). Identification of ceramides in human cells using liquid chromatography with detection by atmospheric pressure chemical ionization-mass spectrometry. *Rapid Commun. Mass Spectrom.* 11 (5): 504–512.

60 Gu, M., Kerwin, J.L., Watts, J.D., and Aebersold, R. (1997). Ceramide profiling of complex lipid mixtures by electrospray ionization mass spectrometry. *Anal. Biochem.* 244 (2): 347–356.

61 Sullards, M.C., Allegood, J.C., Kelly, S. et al. (2007). Structure-specific, quantitative methods for analysis of sphingolipids by liquid chromatography-tandem mass spectrometry: "inside-out" sphingolipidomics. *Methods Enzymol.* 432: 83–115.

62 Huang, S., Cheng, T.Y., Young, D.C. et al. (2011). Discovery of deoxyceramides and diacylglycerols as CD1b scaffold lipids among diverse groove-blocking lipids of the human CD1 system. *Proc. Natl. Acad. Sci. U.S.A.* 108 (48): 19335–19340.

63 Hsu, F.F. and Turk, J. (2002). Characterization of ceramides by low energy collisional-activated dissociation tandem mass spectrometry with negative-ion electrospray ionization. *J. Am. Soc. Mass Spectrom.* 13 (5): 558–570.

64 Hsu, F.F. (2016). Complete structural characterization of ceramides as [M−H](−) ions by multiple-stage linear ion trap mass spectrometry. *Biochimie* 130: 63–75.

65 Hsu, F.F. (2021). Electrospray ionization with higher-energy collision dissociation tandem mass spectrometry toward characterization of ceramides as [M+Li](+) ions: mechanisms of fragmentation and structural identification. *Anal. Chim. Acta* 1142: 221–234.

66 Mano, N., Oda, Y., Yamada, K. et al. (1997). Simultaneous quantitative determination method for sphingolipid metabolites by liquid chromatography/ionspray ionization tandem mass spectrometry. *Anal. Biochem.* 244 (2): 291–300.

67 Liebisch, G., Drobnik, W., Reil, M. et al. (1999). Quantitative measurement of different ceramide species from crude cellular extracts by electrospray ionization tandem mass spectrometry (ESI-MS/MS). *J. Lipid Res.* 40 (8): 1539–1546.

68 Jennemann, R., Kaden, S., Sandhoff, R. et al. (2012). Glycosphingolipids are essential for intestinal endocytic function. *J. Biol. Chem.* 287: 32598–32616.

69 de Groot, T., Ebert, L.K., Christensen, B.M. et al. (2019). Identification of Acer2 as a first susceptibility gene for lithium-induced nephrogenic diabetes insipidus in mice. *J. Am. Soc. Nephrol.* 30: 2322–2336.

70 Simons, B., Kauhanen, D., Sylvanne, T. et al. (2012). Shotgun lipidomics by sequential precursor ion fragmentation on a hybrid quadrupole time-of-flight mass spectrometer. *Metabolites* 2 (1): 195–213.

71 van Smeden, J., Hoppel, L., van der Heijden, R. et al. (2011). LC/MS analysis of stratum corneum lipids: ceramide profiling and discovery. *J. Lipid Res.* 52 (6): 1211–1221.

72 Boiten, W., Absalah, S., Vreeken, R. et al. (2016). Quantitative analysis of ceramides using a novel lipidomics approach with three dimensional response modelling. *Biochim. Biophys. Acta* 1861 (11): 1652–1661.

73 t'Kindt, R., Jorge, L., Dumont, E. et al. (2012). Profiling and characterizing skin ceramides using reversed-phase liquid chromatography-quadrupole time-of-flight mass spectrometry. *Anal. Chem.* 84 (1): 403–411.

74 Tsugawa, H., Ikeda, K., Tanaka, W. et al. (2017). Comprehensive identification of sphingolipid species by in silico retention time and tandem mass spectral library. *Aust. J. Chem.* 9: 19.

75 Alvarez-Jarreta, J., Rodrigues, P.R.S., Fahy, E. et al. (2021). LipidFinder 2.0: advanced informatics pipeline for lipidomics discovery applications. *Bioinformatics* 37 (10): 1478–1479.

76 O'Donnell, V.B., Ekroos, K., Liebisch, G., and Wakelam, M. (2020). Lipidomics: current state of the art in a fast moving field. *Wiley Interdiscip. Rev. Syst. Biol. Med.* 12 (1): e1466.

77 Rabionet, M., Bayerle, A., Marsching, C. et al. (2013). 1-*O*-acylceramides are natural components of human and mouse epidermis. *J. Lipid Res.* 54 (12): 3312–3321.

78 Imgrund, S., Hartmann, D., Farwanah, H. et al. (2009). Adult ceramide synthase 2 (CERS2)-deficient mice exhibit myelin sheath defects, cerebellar degeneration, and hepatocarcinomas. *J. Biol. Chem.* 284 (48): 33549–33560.

79 Rabionet, M., Bayerle, A., Jennemann, R. et al. (2015). Male meiotic cytokinesis requires ceramide synthase 3-dependent sphingolipids with unique membrane anchors. *Hum. Mol. Genet.* 24 (17): 4792–4808.

80 Jennemann, R., Rabionet, M., Gorgas, K. et al. (2012). Loss of ceramide synthase 3 causes lethal skin barrier disruption. *Hum. Mol. Genet.* 21 (3): 586–608.

81 Vasireddy, V., Uchida, Y., Salem, N. Jr. et al. (2007). Loss of functional ELOVL4 depletes very long-chain fatty acids (> or =C28) and the unique omega-*O*-acylceramides in skin leading to neonatal death. *Hum. Mol. Genet.* 16 (5): 471–482.

82 Li, W., Sandhoff, R., Kono, M. et al. (2007). Depletion of ceramides with very long chain fatty acids causes defective skin permeability barrier function, and neonatal lethality in ELOVL4 deficient mice. *Int. J. Biol. Sci.* 3 (2): 120–128.

83 Farwanah, H., Wohlrab, J., Neubert, R.H., and Raith, K. (2005). Profiling of human stratum corneum ceramides by means of normal phase LC/APCI-MS. *Anal. Bioanal. Chem.* 383 (4): 632–637.

84 Ohno, Y., Nakamichi, S., Ohkuni, A. et al. (2015). Essential role of the cytochrome P450 CYP4F22 in the production of acylceramide, the key lipid for skin permeability barrier formation. *Proc. Natl. Acad. Sci. U.S.A.* 112 (25): 7707–7712.

85 Lin, M.H., Hsu, F.F., Crumrine, D. et al. (2019). Fatty acid transport protein 4 is required for incorporation of saturated ultralong-chain fatty acids into epidermal ceramides and monoacylglycerols. *Sci. Rep.* 9 (1): 13254.

86 Yamamoto, H., Hattori, M., Chamulitrat, W. et al. (2020). Skin permeability barrier formation by the ichthyosis-causative gene FATP4 through formation of the barrier lipid omega-*O*-acylceramide. *Proc. Natl. Acad. Sci. U.S.A.* 117 (6): 2914–2922.

87 Mizutani, Y., Kihara, A., and Igarashi, Y. (2006). LASS3 (longevity assurance homologue 3) is a mainly testis-specific (dihydro)ceramide synthase with relatively broad substrate specificity. *Biochem. J.* 398 (3): 531–538.

88 Pichery, M., Huchenq, A., Sandhoff, R. et al. (2017). PNPLA1 defects in patients with autosomal recessive congenital ichthyosis and KO mice sustain PNPLA1 irreplaceable function in epidermal omega-*O*-acylceramide synthesis and skin permeability barrier. *Hum. Mol. Genet.* 26 (10): 1787–1800.

89 Ohno, Y., Kamiyama, N., Nakamichi, S., and Kihara, A. (2017). PNPLA1 is a transacylase essential for the generation of the skin barrier lipid omega-*O*-acylceramide. *Nat. Commun.* 8: 14610.

90 Hirabayashi, T., Anjo, T., Kaneko, A. et al. (2017). PNPLA1 has a crucial role in skin barrier function by directing acylceramide biosynthesis. *Nat. Commun.* 8: 14609.

91 Grond, S., Eichmann, T.O., Dubrac, S. et al. (2017). PNPLA1 deficiency in mice and humans leads to a defect in the synthesis of omega-*O*-acylceramides. *J. Invest. Dermatol.* 137 (2): 394–402.

92 Hirabayashi, T., Murakami, M., and Kihara, A. (2019). The role of PNPLA1 in omega-*O*-acylceramide synthesis and skin barrier function. *Biochim. Biophys. Acta, Mol. Cell. Biol. Lipids* 1864 (6): 869–879.

93 Kien, B., Grond, S., Haemmerle, G. et al. (2018). ABHD5 stimulates PNPLA1-mediated omega-*O*-acylceramide biosynthesis essential for a functional skin permeability barrier. *J. Lipid Res.* 59 (12): 2360–2367.

94 Radner, F.P. and Fischer, J. (2014). The important role of epidermal triacylglycerol metabolism for maintenance of the skin permeability barrier function. *Biochim. Biophys. Acta* 1841 (3): 409–415.

95 Krieg, P. and Furstenberger, G. (2014). The role of lipoxygenases in epidermis. *Biochim. Biophys. Acta* 1841 (3): 390–400.

96 Takeichi, T., Hirabayashi, T., Miyasaka, Y. et al. (2020). SDR9C7 catalyzes critical dehydrogenation of acylceramides for skin barrier formation. *J. Clin. Invest.* 130 (2): 890–903.

97 Hsu, F.F., Turk, J., Stewart, M.E., and Downing, D.T. (2002). Structural studies on ceramides as lithiated adducts by low energy collisional-activated dissociation tandem mass spectrometry with electrospray ionization. *J. Am. Soc. Mass Spectrom.* 13 (6): 680–695.

98 Vietzke, J.P., Brandt, O., Abeck, D. et al. (2001). Comparative investigation of human stratum corneum ceramides. *Lipids* 36 (3): 299–304.

99 Zheng, Y.X., Yin, H.Y., Boeglin, W.E. et al. (2011). Lipoxygenases mediate the effect of essential fatty acid in skin barrier formation a proposed role in releasing omega-hydroxyceramide for construction of the corneocyte lipid envelope. *J. Biol. Chem.* 286 (27): 24046–24056.

100 Farwanah, H., Pierstorff, B., Schmelzer, C.E. et al. (2007). Separation and mass spectrometric characterization of covalently bound skin ceramides using LC/APCI-MS and Nano-ESI-MS/MS. *J. Chromatogr. B Analyt. Technol. Biomed. Life Sci.* 852 (1, 2): 562–570.

101 Okabe, H. and Kishimoto, Y. (1977). Invivo metabolism of ceramides in rat-brain – fatty-acid replacement and esterification of ceramide. *J. Biol. Chem.* 252 (20): 7068–7073.

102 Voynova, N.S., Vionnet, C., Ejsing, C.S., and Conzelmann, A. (2012). A novel pathway of ceramide metabolism in *Saccharomyces cerevisiae*. *Biochem. J.* 447 (1): 103–114.

103 Senkal, C.E., Salama, M.F., Snider, A.J. et al. (2017). Ceramide is metabolized to acylceramide and stored in lipid droplets. *Cell Metab.* 25 (3): 686–697.

104 Bayerle, A., Marsching, C., Rabionet, M. et al. (2020). Endogenous levels of 1-*O*-acylceramides increase upon acidic ceramidase deficiency and decrease due to loss of Dgat1 in a tissue-dependent manner. *Biochim. Biophys. Acta, Mol. Cell. Biol. Lipids* 1865 (9): 158741.

105 Lin, M.H., Miner, J.H., Turk, J., and Hsu, F.F. (2017). Linear ion-trap MSn with high-resolution MS reveals structural diversity of 1-*O*-acylceramide family in mouse epidermis. *J. Lipid Res.* 58 (4): 772–782.

106 Mietla, J.A., Wijesinghe, D.S., Hoeferlin, L.A. et al. (2013). Characterization of eicosanoid synthesis in a genetic ablation model of ceramide kinase. *J. Lipid Res.* 54 (7): 1834–1847.

107 Presa, N., Gomez-Larrauri, A., Dominguez-Herrera, A. et al. (2020). Novel signaling aspects of ceramide 1-phosphate. *Biochim. Biophys. Acta, Mol. Cell. Biol. Lipids* 1865 (4): 158630.

108 Boudker, O. and Futerman, A.H. (1993). Detection and characterization of ceramide-1-phosphate phosphatase activity in rat liver plasma membrane. *J. Biol. Chem.* 268 (29): 22150–22155.

109 Shaner, R.L., Allegood, J.C., Park, H. et al. (2009). Quantitative analysis of sphingolipids for lipidomics using triple quadrupole and quadrupole linear ion trap mass spectrometers. *J. Lipid Res.* 50 (8): 1692–1707.

110 Graf, C., Zemann, B., Rovina, P. et al. (2008). Neutropenia with impaired immune response to *Streptococcus pneumoniae* in ceramide kinase-deficient mice. *J. Immunol.* 180 (5): 3457–3466.

111 Faedo, R.R., da Silva, G., da Silva, R.M. et al. (2022). Sphingolipids signature in plasma and tissue as diagnostic and prognostic tools in oral squamous cell carcinoma. *Biochim. Biophys. Acta, Mol. Cell. Biol. Lipids* 1867 (1): 159057.

112 Wijesinghe, D.S., Allegood, J.C., Gentile, L.B. et al. (2010). Use of high performance liquid chromatography-electrospray ionization-tandem mass spectrometry for the analysis of ceramide-1-phosphate levels. *J. Lipid Res.* 51 (3): 641–651.

113 Yoo, H.H., Son, J., and Kim, D.H. (2006). Liquid chromatography-tandem mass spectrometric determination of ceramides and related lipid species in cellular extracts. *J. Chromatogr. B Analyt. Technol. Biomed. Life Sci.* 843 (2): 327–333.

114 Boath, A., Graf, C., Lidome, E. et al. (2008). Regulation and traffic of ceramide 1-phosphate produced by ceramide kinase: comparative analysis to glucosylceramide and sphingomyelin. *J. Biol. Chem.* 283 (13): 8517–8526.

115 Wasslen, K.V., Canez, C.R., Lee, H. et al. (2014). Trimethylation enhancement using diazomethane (TrEnDi) II: rapid in-solution concomitant quaternization of glycerophospholipid amino groups and methylation of phosphate groups via reaction with diazomethane significantly enhances sensitivity in mass spectrometry analyses via a fixed, permanent positive charge. *Anal. Chem.* 86 (19): 9523–9532.

116 Lee, J.C., Byeon, S.K., and Moon, M.H. (2017). Relative quantification of phospholipids based on isotope-labeled methylation by nanoflow ultrahigh performance liquid chromatography-tandem mass spectrometry: enhancement in cardiolipin profiling. *Anal. Chem.* 89 (9): 4969–4977.

117 Huitema, K., van den Dikkenberg, J., Brouwers, J.F., and Holthuis, J.C. (2004). Identification of a family of animal sphingomyelin synthases. *EMBO J.* 23 (1): 33–44.

118 Taniguchi, M. and Okazaki, T. (2014). The role of sphingomyelin and sphingomyelin synthases in cell death, proliferation and migration-from cell and animal models to human disorders. *Biochim. Biophys. Acta* 1841 (5): 692–703.

119 Bienias, K., Fiedorowicz, A., Sadowska, A. et al. (2016). Regulation of sphingomyelin metabolism. *Pharmacol. Rep.* 68 (3): 570–581.

120 Vacaru, A.M., Tafesse, F.G., Ternes, P. et al. (2009). Sphingomyelin synthase-related protein SMSr controls ceramide homeostasis in the ER. *J. Cell Biol.* 185 (6): 1013–1027.

121 Quintern, L.E., Weitz, G., Nehrkorn, H. et al. (1987). Acid sphingomyelinase from human urine: purification and characterization. *Biochim. Biophys. Acta* 922 (3): 323–336.

122 Schissel, S.L., Schuchman, E.H., Williams, K.J., and Tabas, I. (1996). Zn^{2+}–stimulated sphingomyelinase is secreted by many cell types and is a product of the acid sphingomyelinase gene. *J. Biol. Chem.* 271 (31): 18431–18436.

123 Breiden, B. and Sandhoff, K. (2021). Acid sphingomyelinase, a lysosomal and secretory phospholipase C, is key for cellular phospholipid catabolism. *Int. J. Mol. Sci.* 22 (16): 9001.

124 Tani, M. and Hannun, Y.A. (2007). Neutral sphingomyelinase 2 is palmitoylated on multiple cysteine residues. Role of palmitoylation in subcellular localization. *J. Biol. Chem.* 282 (13): 10047–10056.

125 Duan, R.D., Cheng, Y., Hansen, G. et al. (2003). Purification, localization, and expression of human intestinal alkaline sphingomyelinase. *J. Lipid Res.* 44 (6): 1241–1250.

126 Jungalwala, F.B., Evans, J.E., and McCluer, R.H. (1984). Compositional and molecular species analysis of phospholipids by high performance liquid chromatography coupled with chemical ionization mass spectrometry. *J. Lipid Res.* 25 (7): 738–749.

127 Kim, H.Y. and Salem, N. Jr. (1987). Application of thermospray high-performance liquid chromatography/mass spectrometry for the determination of phospholipids and related compounds. *Anal. Chem.* 59 (5): 722–726.

128 Valeur, A., Olsson, N.U., Kaufmann, P. et al. (1994). Quantification and comparison of some natural sphingomyelins by on-line high-performance liquid chromatography/discharge-assisted thermospray mass spectrometry. *Biol. Mass Spectrom.* 23 (6): 313–319.

129 Byrdwell, W.C. and Borchman, D. (1997). Liquid chromatography/mass-spectrometric characterization of sphingomyelin and dihydrosphingomyelin of human lens membranes. *Ophthalmic Res.* 29 (4): 191–206.

130 Karlsson, A.A., Michelsen, P., and Odham, G. (1998). Molecular species of sphingomyelin: determination by high-performance liquid chromatography/mass spectrometry with electrospray and high-performance liquid chromatography/tandem mass spectrometry with atmospheric pressure chemical ionization. *J. Mass Spectrom.* 33 (12): 1192–1198.

131 Murphy, R.C. and Harrison, K.A. (1994). Fast-atom-bombardment mass-spectrometry of phospholipids. *Mass Spectrom. Rev.* 13 (1): 57–75.

132 Ann, Q.H. and Adams, J. (1993). Collision-induced decomposition of sphingomyelins for structural elucidation. *Biol. Mass Spectrom.* 22 (5): 285–294.

133 Harvey, D.J. (1995). Matrix-assisted laser desorption/ionization mass spectrometry of sphingo- and glycosphingo-lipids. *J. Mass Spectrom.* 30: 1311–1324.

134 Kerwin, J.L., Tuininga, A.R., and Ericsson, L.H. (1994). Identification of molecular species of glycerophospholipids and sphingomyelin using electrospray mass spectrometry. *J. Lipid Res.* 35 (6): 1102–1114.

135 Hsu, F.F. and Turk, J. (2001). Structural determination of glycosphingolipids as lithiated adducts by electrospray ionization mass spectrometry using low-energy collisional-activated dissociation on a triple stage quadrupole instrument. *J. Am. Soc. Mass Spectrom.* 12 (1): 61–79.

136 Brugger, B., Erben, G., Sandhoff, R. et al. (1997). Quantitative analysis of biological membrane lipids at the low picomole level by nano-electrospray ionization tandem mass spectrometry. *Proc. Natl. Acad. Sci. U.S.A.* 94 (6): 2339–2344.

137 Zhao, X., Wu, G., Zhang, W. et al. (2020). Resolving modifications on sphingoid base and *N*-acyl chain of sphingomyelin lipids in complex lipid extracts. *Anal. Chem.* 92 (21): 14775–14782.

138 Sandhoff, R., Schulze, H., and Sandhoff, K. (2018). Ganglioside metabolism in health and disease. *Prog. Mol. Biol. Transl. Sci.* 156: 1–62.

139 Breiden, B. and Sandhoff, K. (2019). Lysosomal glycosphingolipid storage diseases. *Annu. Rev. Biochem.* 88: 461–485.

140 Han, X. and Cheng, H. (2005). Characterization and direct quantitation of cerebroside molecular species from lipid extracts by shotgun lipidomics. *J. Lipid Res.* 46 (1): 163–175.

141 von Gerichten, J., Schlosser, K., Lamprecht, D. et al. (2017). Diastereomer-specific quantification of bioactive hexosylceramides from bacteria and mammals. *J. Lipid Res.* 58 (6): 1247–1258.

142 Taketomi, T., Hara, A., Uemura, K., and Sugiyama, E. (1998). Matrix-assisted laser desorption ionization time-of-flight mass spectrometric analysis of glycosphingolipids including gangliosides. *Acta Biochim. Pol.* 45 (4): 987–999.

143 Doering, T., Holleran, W.M., Potratz, A. et al. (1999). Sphingolipid activator proteins are required for epidermal permeability barrier formation. *J. Biol. Chem.* 274 (16): 11038–11045.

144 Tanaka, K., Yamada, M., Tamiya-Koizumi, K. et al. (2011). Systematic analyses of free ceramide species and ceramide species comprising neutral glycosphingolipids by MALDI-TOF MS with high-energy CID. *Glycoconjugate J.* 28 (2): 67–87.

145 Merrill, A.H. Jr., Sullards, M.C., Allegood, J.C. et al. (2005). Sphingolipidomics: high-throughput, structure-specific, and quantitative analysis of sphingolipids by liquid chromatography tandem mass spectrometry. *Methods* 36 (2): 207–224.

146 Boutin, M., Sun, Y., Shacka, J.J., and Auray-Blais, C. (2016). Tandem mass spectrometry multiplex analysis of glucosylceramide and galactosylceramide isoforms in brain tissues at different stages of Parkinson disease. *Anal. Chem.* 88 (3): 1856–1863.

147 Nakajima, K., Akiyama, H., Tanaka, K. et al. (2016). Separation and analysis of mono-glucosylated lipids in brain and skin by hydrophilic interaction chromatography based on carbohydrate and lipid moiety. *J. Chromatogr. B Analyt. Technol. Biomed. Life Sci.* 1031: 146–153.

148 Kaye, E.M. and Ullman, M.D. (1984). Separation and quantitation of perbenzoylated glucocerebroside and galactocerebroside by high-performance liquid chromatography. *Anal. Biochem.* 138 (2): 380–385.

149 Xu, H., Boucher, F.R., Nguyen, T.T. et al. (2019). DMS as an orthogonal separation to LC/ESI/MS/MS for quantifying isomeric cerebrosides in plasma and cerebrospinal fluid. *J. Lipid Res.* 60 (1): 200–211.

150 Yang, L., Nie, H., Zhao, F. et al. (2020). A novel online two-dimensional supercritical fluid chromatography/reversed phase liquid chromatography-mass spectrometry method for lipid profiling. *Anal. Bioanal. Chem.* 412 (10): 2225–2235.

151 Morita, M., Motoki, K., Akimoto, K. et al. (1995). Structure-activity relationship of alpha-galactosylceramides against B16-bearing mice. *J. Med. Chem.* 38 (12): 2176–2187.

152 Kawano, T., Cui, J., Koezuka, Y. et al. (1997). CD1d-restricted and TCR-mediated activation of valpha14 NKT cells by glycosylceramides. *Science* 278 (5343): 1626–1629.

153 Kain, L., Webb, B., Anderson, B.L. et al. (2014). The identification of the endogenous ligands of natural killer T cells reveals the presence of mammalian alpha-linked glycosylceramides. *Immunity* 41 (4): 543–554.

154 Kim, J., Kim, J.H., and Winau, F. (2014). Thinking inside the box: endogenous alpha-anomeric lipid antigens. *Immunity* 41 (4): 505–506.

155 Brennan, P.J., Cheng, T.Y., Pellicci, D.G. et al. (2017). Structural determination of lipid antigens captured at the CD1d-T-cell receptor interface. *Proc. Natl. Acad. Sci. U.S.A.* 114 (31): 8348–8353.

156 von Gerichten, J., Lamprecht, D., Opalka, L. et al. (2019). Bacterial immunogenic alpha-galactosylceramide identified in the murine large intestine: dependency on diet and inflammation. *J. Lipid Res.* 60 (11): 1892–1904.

157 Farwanah, H. and Kolter, T. (2012). Lipidomics of glycosphingolipids. *Metabolites* 2 (1): 134–164.

158 Kolter, T., Proia, R.L., and Sandhoff, K. (2002). Combinatorial ganglioside biosynthesis. *J. Biol. Chem.* 277 (29): 25859–25862.

159 Westman, J.S., Benktander, J., Storry, J.R. et al. (2015). Identification of the molecular and genetic basis of PX2, a glycosphingolipid blood group antigen lacking on Globoside-deficient erythrocytes. *J. Biol. Chem.* 290 (30): 18505–18518.

160 Iwamura, K., Furukawa, K., Uchikawa, M. et al. (2003). The blood group P1 synthase gene is identical to the Gb3/CD77 synthase gene. A clue to the solution of the P1/P2/p puzzle. *J. Biol. Chem.* 278 (45): 44429–44438.

161 Duffin, K.L., Welply, J.K., Huang, E., and Henion, J.D. (1992). Characterization of N-linked oligosaccharides by electrospray and tandem mass spectrometry. *Anal. Chem.* 64 (13): 1440–1448.

162 Ohashi, Y., Ii, T., Matsuzaki, Y. et al. (1993). Negative-ion electrospray mass-spectrometry (ESIMS) of glycolipids - comparison with (−) FABMS. *Glycoconjugate J.* 10 (4): 341–342.

163 Ii, T., Ohashi, Y., and Nagai, K. (1993). Observation of a novel adduct-ion in negative-ion electrospray mass spectrometry of neutral glycolipids. *Org. Mass Spectrom.* 28: 927–928.

164 Egge, H., Peterkatalinic, J., Karas, M., and Stahl, B. (1991). The use of fast-atom-bombardment and laser desorption mass-spectrometry in the analysis of complex carbohydrates. *Pure Appl. Chem.* 63 (4): 491–498.

165 Juhasz, P. and Costello, C.E. (1992). Matrix-assisted laser desorption ionization time-of-flight mass spectrometry of underivatized and permethylated gangliosides. *J. Am. Soc. Mass Spectrom.* 3 (8): 785–796.

166 Domon, B. and Costello, C.E. (1988). Structure elucidation of glycosphingolipids and gangliosides using high-performance tandem mass spectrometry. *Biochemistry* 27 (5): 1534–1543.

167 Levery, S.B. (2005). Glycosphingolipid structural analysis and glycosphingolipidomics. *Methods Enzymol.* 405: 300–369.

168 Li, Y., Teneberg, S., Thapa, P. et al. (2008). Sensitive detection of isoglobo and globo series tetraglycosylceramides in human thymus by ion trap mass spectrometry. *Glycobiology* 18 (2): 158–165.

169 Li, Y., Zhou, D., Xia, C. et al. (2008). Sensitive quantitation of isoglobotriaosylceramide in the presence of isobaric components using electrospray ionization-ion trap mass spectrometry. *Glycobiology* 18 (2): 166–176.

170 Boutin, M., Menkovic, I., Martineau, T. et al. (2017). Separation and analysis of lactosylceramide, galabiosylceramide and globotriaosylceramide by LC-MS/MS in urine of Fabry disease patients. *Anal. Chem.* 89: 13382–13390.

171 Karlsson, H., Halim, A., and Teneberg, S. (2010). Differentiation of glycosphingolipid-derived glycan structural isomers by liquid chromatography/mass spectrometry. *Glycobiology* 20 (9): 1103–1116.

172 Varki, A., Schnaar, R.L., and Schauer, R. (2015). Sialic acids and other nonulosonic acids. In: *Essentials of Glycobiology* (ed. A. Varki, R.D. Cummings, J.D. Esko, et al.), 179–195. Cold Spring Harbor.

173 Irie, A., Koyama, S., Kozutsumi, Y. et al. (1998). The molecular basis for the absence of *N*-glycolylneuraminic acid in humans. *J. Biol. Chem.* 273 (25): 15866–15871.

174 Chou, H.H., Takematsu, H., Diaz, S. et al. (1998). A mutation in human CMP-sialic acid hydroxylase occurred after the Homo-Pan divergence. *Proc. Natl. Acad. Sci. U.S.A.* 95 (20): 11751–11756.

175 Bergfeld, A.K., Pearce, O.M., Diaz, S.L. et al. (2012). Metabolism of vertebrate amino sugars with *N*-glycolyl groups: incorporation of *N*-glycolylhexosamines into mammalian glycans by feeding *N*-glycolylgalactosamine. *J. Biol. Chem.* 287 (34): 28898–28916.

176 Banda, K., Gregg, C.J., Chow, R. et al. (2012). Metabolism of vertebrate amino sugars with *N*-glycolyl groups: mechanisms underlying gastrointestinal incorporation of the non-human sialic acid xeno-autoantigen *N*-glycolylneuraminic acid. *J. Biol. Chem.* 287 (34): 28852–28864.

177 Schauer, R. (2004). Sialic acids: fascinating sugars in higher animals and man. *Zoology* 107 (1): 49–64.

178 Simpson, M.A., Cross, H., Proukakis, C. et al. (2004). Infantile-onset symptomatic epilepsy syndrome caused by a homozygous loss-of-function mutation of GM3 synthase. *Nat. Genet.* 36: 1225–1229.

179 Yamashita, T., Hashiramoto, A., Haluzik, M. et al. (2003). Enhanced insulin sensitivity in mice lacking ganglioside GM3. *Proc. Natl. Acad. Sci. U.S.A.* 100 (6): 3445–3449.

180 Yoshikawa, M., Go, S., Takasaki, K. et al. (2009). Mice lacking ganglioside GM3 synthase exhibit complete hearing loss due to selective degeneration of the organ of Corti. *Proc. Natl. Acad. Sci. U.S.A.* 106 (23): 9483–9488.

181 Reinhold, B.B., Chan, S.Y., Chan, S., and Reinhold, V.N. (1994). Profiling glycosphingolipid structural detail - periodate-oxidation, electrospray, collision-induced dissociation and tandem mass-spectrometry. *Org. Mass Spectrom.* 29 (12): 736–746.

182 Costello, C.E., Juhasz, P., and Perreault, H. (1994). New mass-spectral approaches to ganglioside structure determinations. Biological Function of Gangliosides. *Prog Brain Res.* 101: 45–61.

183 Costello, C.E. and Vath, J.E. (1990). Tandem mass spectrometry of glycolipids. *Methods Enzymol.* 193: 738–768.

184 O'Brien, J.P. and Brodbelt, J.S. (2013). Structural characterization of gangliosides and glycolipids via ultraviolet photodissociation mass spectrometry. *Anal. Chem.* 85 (21): 10399–10407.

185 Wojcik, R., Webb, I.K., Deng, L. et al. (2017). Lipid and glycolipid isomer analyses using ultra-high resolution ion mobility spectrometry separations. *Int. J. Mol. Sci.* 18 (1).

186 Ikeda, K., Shimizu, T., and Taguchi, R. (2008). Targeted analysis of ganglioside and sulfatide molecular species by LC/ESI-MS/MS with theoretically expanded multiple reaction monitoring. *J. Lipid Res.* 49 (12): 2678–2689.

187 Jennemann, R., Rothermel, U., Wang, S. et al. (2010). Hepatic glycosphingolipid deficiency and liver function in mice. *Hepatology* 51 (5): 1799–1809.

188 Fujiwara, Y., Hama, K., and Yokoyama, K. (2020). Mass spectrometry in combination with a chiral column and multichannel-MRM allows comprehensive analysis of glycosphingolipid molecular species from mouse brain. *Carbohydr. Res.* 490: 107959.

189 Fong, B., Norris, C., Lowe, E., and McJarrow, P. (2009). Liquid chromatography-high-resolution mass spectrometry for quantitative analysis of gangliosides. *Lipids* 44 (9): 867–874.

190 Kirsch, S., Muthing, J., Peter-Katalinic, J., and Bindila, L. (2009). On-line nano-HPLC/ESI QTOF MS monitoring of alpha 2-3 and alpha 2-6 sialylation in granulocyte glycosphingolipidome. *Biol. Chem.* 390 (7): 657–672.

191 Cavdarli, S., Yamakawa, N., Clarisse, C. et al. (2020). Profiling of *O*-acetylated gangliosides expressed in neuroectoderm derived cells. *Int. J. Mol. Sci.* 21 (1): 370.

192 Furukawa, K., Aixinjueluo, W., Kasama, T. et al. (2008). Disruption of GM2/GD2 synthase gene resulted in overt expression of 9-*O*-acetyl GD3 irrespective of Tis21. *J. Neurochem.* 105 (3): 1057–1066.

193 Zhang, Q., Wang, Y., Zheng, Q., and Li, J. (2019). Analysis of *O*-acetylated sialic acids in dried blood spots. *Anal. Chem.* 91 (4): 2744–2751.

194 Mlinac, K., Fabris, D., Vukelic, Z. et al. (2013). Structural analysis of brain ganglioside acetylation patterns in mice with altered ganglioside biosynthesis. *Carbohydr. Res.* 382: 1–8.

195 Cavdarli, S., Delannoy, P., and Groux-Degroote, S. (2020). *O*-acetylated gangliosides as targets for cancer immunotherapy. *Cells* 9 (3).

196 Ishizuka, I. (1997). Chemistry and functional distribution of sulfoglycolipids. *Prog. Lipid Res.* 36 (4): 245–319.

197 Honke, K., Hirahara, Y., Dupree, J. et al. (2002). Paranodal junction formation and spermatogenesis require sulfoglycolipids. *Proc. Natl. Acad. Sci. U.S.A.* 99 (7): 4227–4232.

198 Stettner, P., Bourgeois, S., Marsching, C. et al. (2013). Sulfatides are required for renal adaptation to chronic metabolic acidosis. *Proc. Natl. Acad. Sci. U.S.A.* 110 (24): 9998–10003.

199 Honke, K. (2013). Biosynthesis and biological function of sulfoglycolipids. *Proc. Jpn. Acad. Ser. B Phys. Biol. Sci.* 89 (4): 129–138.

200 Gieselmann, V. and Ingeborg, K.-M. (2019). Metachromatic leukodystrophy. In: *The Online Metabolic and Molecular Bases of Inherited Disease* (ed. D.L. Valle, S. Antonarakis, A. Ballabio, et al.). New York, NY: McGraw-Hill Education.

201 Lullmann-Rauch, R., Matzner, U., Franken, S. et al. (2001). Lysosomal sulfoglycolipid storage in the kidneys of mice deficient for arylsulfatase A (ASA) and of double-knockout mice deficient for ASA and galactosylceramide synthase. *Histochem. Cell Biol.* 116 (2): 161–169.

202 Sandhoff, R., Hepbildikler, S.T., Jennemann, R. et al. (2002). Kidney sulfatides in mouse models of inherited glycosphingolipid disorders: determination by nano-electrospray ionization tandem mass spectrometry. *J. Biol. Chem.* 277 (23): 20386–20398.

203 Taketomi, T., Hara, A., Uemura, K. et al. (1996). A microwave-mediated saponification of galactosylceramide and galactosylceramide I3-sulfate and identification of their lyso-compounds by delayed extraction matrix-assisted laser desorption ionization time-of-flight mass spectrometry. *Biochem. Biophys. Res. Commun.* 224 (2): 462–467.

204 Hsu, F.F., Bohrer, A., and Turk, J. (1998). Electrospray ionization tandem mass spectrometric analysis of sulfatide. Determination of fragmentation patterns and characterization of molecular species expressed in brain and in pancreatic islets. *Biochim. Biophys. Acta* 1392 (2, 3): 202–216.

205 Kyogashima, M., Tamiya-Koizumi, K., Ehara, T. et al. (2006). Rapid demonstration of diversity of sulfatide molecular species from biological materials by MALDI-TOF MS. *Glycobiology* 16 (8): 719–728.

206 Hsu, F.F. and Turk, J. (2004). Studies on sulfatides by quadrupole ion-trap mass spectrometry with electrospray ionization: structural characterization and the fragmentation processes that include an unusual internal galactose residue loss and the classical charge-remote fragmentation. *J. Am. Soc. Mass Spectrom.* 15 (4): 536–546.

207 Marsching, C., Eckhardt, M., Grone, H.J. et al. (2011). Imaging of complex sulfatides SM3 and SB1a in mouse kidney using MALDI-TOF/TOF mass spectrometry. *Anal. Bioanal. Chem.* 401 (1): 53–64.

208 Tadano-Aritomi, K., Kubo, H., Ireland, P. et al. (1995). Structural analysis of mono- and bis-sulfated glycosphingolipids by negative liquid secondary ion mass spectrometry with high- and low- energy collision-induced dissociation. *Carbohydr. Res.* 273 (1): 41–52.

209 Yamashita, T., Wu, Y.P., Sandhoff, R. et al. (2005). Interruption of ganglioside synthesis produces central nervous system degeneration and altered axon-glial interactions. *Proc. Natl. Acad. Sci. U.S.A.* 102 (8): 2725–2730.

210 Ikeda, K. and Taguchi, R. (2010). Highly sensitive localization analysis of gangliosides and sulfatides including structural isomers in mouse cerebellum sections by combination of laser microdissection and hydrophilic interaction liquid chromatography/electrospray ionization mass spectrometry with theoretically expanded multiple reaction monitoring. *Rapid Commun. Mass Spectrom.* 24 (20): 2957–2965.

211 Marsching, C., Jennemann, R., Heilig, R. et al. (2014). Quantitative imaging mass spectrometry of renal sulfatides: validation by classical mass spectrometric methods. *J. Lipid Res.* 55 (11): 2343–2353.

212 Marsching, C., Rabionet, M., Mathow, D. et al. (2014). Renal sulfatides: sphingoid base-dependent localization and region-specific compensation of CerS2-dysfunction. *J. Lipid Res.* 55 (11): 2354–2369.

213 Barcenas, M., Suhr, T.R., Scott, C.R. et al. (2014). Quantification of sulfatides in dried blood and urine spots from metachromatic leukodystrophy patients by liquid chromatography/electrospray tandem mass spectrometry. *Clin. Chim. Acta* 433: 39–43.

214 Saville, J.T., Smith, N.J.C., Fletcher, J.M., and Fuller, M. (2017). Quantification of plasma sulfatides by mass spectrometry: utility for metachromatic leukodystrophy. *Anal. Chim. Acta* 955: 79–85.

215 Han, M., Jun, S.H., Song, S.H. et al. (2014). Ultra-performance liquid chromatography/tandem mass spectrometry for determination of sulfatides in dried blood spots from patients with metachromatic leukodystrophy. *Rapid Commun. Mass Spectrom.* 28 (6): 587–594.

216 Mirzaian, M., Kramer, G., and Poorthuis, B.J. (2015). Quantification of sulfatides and lysosulfatides in tissues and body fluids by liquid chromatography-tandem mass spectrometry. *J. Lipid Res.* 56 (4): 936–943.

217 Wigger, D., Gulbins, E., Kleuser, B., and Schumacher, F. (2019). Monitoring the sphingolipid de novo synthesis by stable-isotope labeling and liquid chromatography-mass spectrometry. *Front. Cell Dev. Biol.* 7: 210

218 Ecker, J. and Liebisch, G. (2014). Application of stable isotopes to investigate the metabolism of fatty acids, glycerophospholipid and sphingolipid species. *Prog. Lipid Res.* 54: 14–31.

219 Berdyshev, E.V., Gorshkova, I.A., Usatyuk, P. et al. (2006). De novo biosynthesis of dihydrosphingosine-1-phosphate by sphingosine kinase 1 in mammalian cells. *Cell. Signalling* 18 (10): 1779–1792.

220 Berdyshev, E.V., Gorshkova, I., Skobeleva, A. et al. (2009). FTY720 inhibits ceramide synthases and up-regulates dihydrosphingosine 1-phosphate formation in human lung endothelial cells. *J. Biol. Chem.* 284 (9): 5467–5477.

221 Schifferer, R., Liebisch, G., Bandulik, S. et al. (2007). ApoA-I induces a preferential efflux of monounsaturated phosphatidylcholine and medium chain sphingomyelin species from a cellular pool distinct from HDL(3) mediated phospholipid efflux. *Biochim. Biophys. Acta* 1771 (7): 853–863.

222 Skotland, T., Ekroos, K., Kavaliauskiene, S. et al. (2016). Determining the turnover of glycosphingolipid species by stable-isotope tracer lipidomics. *J. Mol. Biol.* 428: 4856–4866.

223 Tserng, K.Y. and Griffin, R. (2004). Studies of lipid turnover in cells with stable isotope and gas chromatograph-mass spectrometry. *Anal. Biochem.* 325 (2): 344–353.

224 Haynes, C.A., Allegood, J.C., Wang, E.W. et al. (2011). Factors to consider in using [U-C]palmitate for analysis of sphingolipid biosynthesis by tandem mass spectrometry. *J. Lipid Res.* 52 (8): 1583–1594.

225 Blachnio-Zabielska, A.U., Persson, X.M.T., Koutsari, C. et al. (2012). A liquid chromatography/tandem mass spectrometry method for measuring the in vivo incorporation of plasma free fatty acids into intramyocellular ceramides in humans. *Rapid Commun. Mass Spectrom.* 26 (9): 1134–1140.

226 Zelnik, I.D., Volpert, G., Viiri, L.E. et al. (2020). Different rates of flux through the biosynthetic pathway for long-chain versus very-long-chain sphingolipids. *J. Lipid Res.* 61 (10): 1341–1346.

227 Tserng, K.Y. and Griffin, R.L. (2004). Ceramide metabolite, not intact ceramide molecule, may be responsible for cellular toxicity. *Biochem. J.* 380: 715–722.

228 Frisz, J.F., Lou, K., Klitzing, H.A. et al. (2013). Direct chemical evidence for sphingolipid domains in the plasma membranes of fibroblasts. *Proc. Natl. Acad. Sci. U.S.A.* 110 (8): E613–E622.

229 Ellis, S.R., Brown, S.H., In Het Panhuis, M. et al. (2013). Surface analysis of lipids by mass spectrometry: more than just imaging. *Prog. Lipid Res.* 52 (4): 329–353.

230 Prentice, B.M., Caprioli, R.M., and Vuiblet, V. (2017). Label-free molecular imaging of the kidney. *Kidney Int.* 92: 580–598.

231 Samfors, S. and Fletcher, J.S. (2020). Lipid diversity in cells and tissue using imaging SIMS. *Annu. Rev. Anal. Chem. (Palo Alto Calif.)* 13 (1): 249–271.

232 Yalcin, E.B. and de la Monte, S.M. (2015). Review of matrix-assisted laser desorption ionization-imaging mass spectrometry for lipid biochemical histopathology. *J. Histochem. Cytochem.* 63 (10): 762–771.

233 Fernandez, J.A., Ochoa, B., Fresnedo, O. et al. (2011). Matrix-assisted laser desorption ionization imaging mass spectrometry in lipidomics. *Anal. Bioanal. Chem.* 401 (1): 29–51.

234 Neumann, E.K., Djambazova, K.V., Caprioli, R.M., and Spraggins, J.M. (2020). Multimodal imaging mass spectrometry: next generation molecular mapping in biology and medicine. *J. Am. Soc. Mass Spectrom.* 31 (12): 2401–2415.

235 Bodzon-Kulakowska, A. and Suder, P. (2016). Imaging mass spectrometry: instrumentation, applications, and combination with other visualization techniques. *Mass Spectrom. Rev.* 35 (1): 147–169.

236 Engel, K.M., Prabutzki, P., Leopold, J. et al. (2022). A new update of MALDI-TOF mass spectrometry in lipid research. *Prog. Lipid Res.* 86: 101145.

237 Jackson, S.N., Ugarov, M., Egan, T. et al. (2007). MALDI-ion mobility-TOFMS imaging of lipids in rat brain tissue. *J. Mass Spectrom.* 42 (8): 1093–1098.

238 Skraskova, K., Claude, E., Jones, E.A. et al. (2016). Enhanced capabilities for imaging gangliosides in murine brain with matrix-assisted laser desorption/ionization and desorption electrospray ionization mass spectrometry coupled to ion mobility separation. *Methods* 104: 69–78.

239 Wang, J., Wang, C., and Han, X. (2018). Enhanced coverage of lipid analysis and imaging by matrix-assisted laser desorption/ionization mass spectrometry via a strategy with an optimized mixture of matrices. *Anal. Chim. Acta* 1000: 155–162.

240 Zemski Berry, K.A., Gordon, W.C., Murphy, R.C., and Bazan, N.G. (2014). Spatial organization of lipids in the human retina and optic nerve by MALDI imaging mass spectrometry. *J. Lipid Res.* 55 (3): 504–515.

241 Wang, W.X. and Whitehead, S.N. (2020). Imaging mass spectrometry allows for neuroanatomic-specific detection of gangliosides in the healthy and diseased brain. *Analyst* 145 (7): 2473–2481.

242 Kaya, I., Jennische, E., Lange, S. et al. (2020). Brain region-specific amyloid plaque-associated myelin lipid loss, APOE deposition and disruption of the myelin sheath in familial Alzheimer's disease mice. *J. Neurochem.* 154 (1): 84–98.

243 Gessel, M.M., Spraggins, J.M., Voziyan, P.A. et al. (2019). Two specific sulfatide species are dysregulated during renal development in a mouse model of Alport syndrome. *Lipids* 54 (6, 7): 411–418.

244 Jones, E.E., Dworski, S., Canals, D. et al. (2014). On-tissue localization of ceramides and other sphingolipids by MALDI mass spectrometry imaging. *Anal. Chem.* 86 (16): 8303–8311.

245 Kaya, I., Brinet, D., Michno, W. et al. (2017). Delineating amyloid plaque associated neuronal sphingolipids in transgenic Alzheimer's disease mice (tgArcSwe) using MALDI imaging mass spectrometry. *ACS Chem. Neurosci.* 8 (2): 347–355.

246 Blank, M. and Hopf, C. (2021). Spatially resolved mass spectrometry analysis of amyloid plaque-associated lipids. *J. Neurochem.* 159 (2): 330–342.

247 Roy, S., Touboul, D., Brunelle, A. et al. (2006). Imaging mass spectrometry: a new tool for the analysis of skin biopsy. Application in Fabry's disease. *Ann. Pharm. Fr.* 64 (5): 328–334.

248 Whitehead, S.N., Chan, K.H., Gangaraju, S. et al. (2011). Imaging mass spectrometry detection of gangliosides species in the mouse brain following transient focal cerebral ischemia and long-term recovery. *PLoS One* 6 (6): e20808.

249 Grove, K.J., Voziyan, P.A., Spraggins, J.M. et al. (2014). Diabetic nephropathy induces alterations in the glomerular and tubule lipid profiles. *J. Lipid Res.* 55 (7): 1375–1385.

250 Dufresne, M., Guneysu, D., Patterson, N.H. et al. (2017). Multimodal detection of GM2 and GM3 lipid species in the brain of mucopolysaccharidosis type II mouse by serial imaging mass spectrometry and immunohistochemistry. *Anal. Bioanal. Chem.* 409 (5): 1425–1433.

251 Jones, E.E., Zhang, W., Zhao, X. et al. (2017). Tissue localization of glycosphingolipid accumulation in a Gaucher disease mouse brain by LC-ESI-MS/MS and high-resolution MALDI imaging mass spectrometry. *SLAS Discov.* 22 (10): 1218–1228.

252 Goto-Inoue, N., Hayasaka, T., Zaima, N. et al. (2012). Imaging mass spectrometry visualizes ceramides and the pathogenesis of dorfman-chanarin syndrome due to ceramide metabolic abnormality in the skin. *PLoS One* 7 (11): e49519.

253 de Macedo, C.S., Anderson, D.M., Pascarelli, B.M. et al. (2015). MALDI imaging reveals lipid changes in the skin of leprosy patients before and after multidrug therapy (MDT). *J. Mass Spectrom.* 50 (12): 1374–1385.

254 Sugiura, Y., Shimma, S., Konishi, Y. et al. (2008). Imaging Mass Spectrometry Technology and Application on Ganglioside Study; Visualization of age-dependent accumulation of C20-ganglioside molecular species in the mouse hippocampus. *PLoS One* 3 (9): e3232.

255 Seyrantepe, V., Demir, S.A., Timur, Z.K. et al. (2018). Murine Sialidase Neu3 facilitates GM2 degradation and bypass in mouse model of Tay-Sachs disease. *Exp Neurol.* 299 (Pt A): 26–41.

256 Sugimoto, M., Shimizu, Y., Yoshioka, T. et al. (2015). Histological analyses by matrix-assisted laser desorption/ionization-imaging mass spectrometry reveal differential localization of sphingomyelin molecular species regulated by particular ceramide synthase in mouse brains. *Biochim. Biophys. Acta* 1851 (12): 1554–1565.

257 Sugimoto, M., Wakabayashi, M., Shimizu, Y. et al. (2016). Imaging mass spectrometry reveals acyl-chain- and region-specific sphingolipid metabolism in the kidneys of sphingomyelin synthase 2-deficient mice. *PLoS One* 11 (3): e0152191.

258 Hirahara, Y., Wakabayashi, T., Mori, T. et al. (2017). Sulfatide species with various fatty acid chains in oligodendrocytes at different developmental stages determined by imaging mass spectrometry. *J. Neurochem.* 140 (3): 435–450.

259 Chen, M., Markham, J.E., Dietrich, C.R. et al. (2008). Sphingolipid long-chain base hydroxylation is important for growth and regulation of sphingolipid content and composition in Arabidopsis. *Plant Cell* 20 (7): 1862–1878.

260 Adem, A.A., Belete, A., Soboleva, A. et al. (2021). Structural characterization of plant glucosylceramides and the corresponding ceramides by UHPLC-LTQ-orbitrap mass spectrometry. *J. Pharm. Biomed. Anal.* 192: 113677.

261 Mina, J.G., Okada, Y., Wansadhipathi-Kannangara, N.K. et al. (2010). Functional analyses of differentially expressed isoforms of the Arabidopsis inositol phosphorylceramide synthase. *Plant Mol. Biol.* 73 (4, 5): 399–407.

262 Liu, N.J., Hou, L.P., Bao, J.J. et al. (2021). Sphingolipid metabolism, transport, and functions in plants: recent progress and future perspectives. *Plant Commun.* 2 (5): 100214.

263 Luttgeharm, K.D., Kimberlin, A.N., and Cahoon, E.B. (2016). Plant sphingolipid metabolism and function. *Subcell Biochem.* 86: 249–286.

264 Marques, J.T., Marinho, H.S., and de Almeida, R.F.M. (2018). Sphingolipid hydroxylation in mammals, yeast and plants - an integrated view. *Prog. Lipid Res.* 71: 18–42.

265 Pata, M.O., Hannun, Y.A., and Ng, C.K. (2010). Plant sphingolipids: decoding the enigma of the sphinx. *New Phytol.* 185 (3): 611–630.

266 Grilley, M.M., Stock, S.D., Dickson, R.C. et al. (1998). Syringomycin action gene SYR2 is essential for sphingolipid 4-hydroxylation in Saccharomyces cerevisiae. *J. Biol. Chem.* 273 (18): 11062–11068.

267 Dickson, R.C. (2008). Thematic review series: sphingolipids. New insights into sphingolipid metabolism and function in budding yeast. *J. Lipid Res.* 49 (5): 909–921.

268 Megyeri, M., Riezman, H., Schuldiner, M., and Futerman, A.H. (2016). Making sense of the yeast sphingolipid pathway. *J. Mol. Biol.* 428 (24 Pt A): 4765–4775.

269 Barreto-Bergter, E., Sassaki, G.L., and de Souza, L.M. (2011). Structural analysis of fungal cerebrosides. *Front. Microbiol.* 2: 239.

270 Singh, A. and Del Poeta, M. (2016). Sphingolipidomics: an important mechanistic tool for studying fungal pathogens. *Front. Microbiol.* 7: 501.

271 Olsen, I. and Jantzen, E. (2001). Sphingolipids in bacteria and fungi. *Anaerobe* 7 (2): 103–112.

272 Natori, T., Morita, M., Akimoto, K., and Koezuka, Y. (1994). Agelasphins, novel antitumor and immunostimulatory cerebrosides from the marine sponge *Agelas-Mauritianus. Tetrahedron* 50 (9): 2771–2784.

273 Watanabe, Y., Nakajima, M., Hoshino, T. et al. (2001). A novel sphingophosphonolipid head group 1-hydroxy-2-aminoethyl phosphonate in *Bdellovibrio stolpii. Lipids* 36 (5): 513–519.

274 Alegado, R.A., Brown, L.W., Cao, S. et al. (2012). A bacterial sulfonolipid triggers multicellular development in the closest living relatives of animals. *Elife* 1: e00013.

275 Harrison, P.J., Dunn, T.M., and Campopiano, D.J. (2018). Sphingolipid biosynthesis in man and microbes. *Nat. Prod. Rep.* 35 (9): 921–954.

16

Sterol Lipids

William J. Griffiths, Eylan Yutuc, and Yuqin Wang

Swansea University, Swansea University Medical School, Singleton Park, Swansea, SA2 8PP Wales, UK

16.1 Introduction

Sterol lipids were one of the first categories of biomolecules to be investigated using mass spectrometry (MS) [1]. They represent a broad range of molecules starting with C_{30} lanosterol [2], or cycloartenol in organisms of photosynthetic linage [3], proceeding through cholesterol in animals, ergosterol in yeast, and sitosterol, campesterol, and brassicasterol in many plants, to bile acids and steroid hormones [4–6]. Bile acids are discussed in a separate chapter, and this chapter will concentrate on mass spectrometric methods for the analysis of cholesterol, its precursors, and its oxidized derivatives, i.e. oxysterols, particularly in the context of human and mammalian biology. Methodologies described for these molecules are equally applicable to those present in yeast and plants.

According to the LIPID MAPS nomenclature system, sterols lipids (ST) are the category of lipids embracing all molecules based on the cyclopentanoperhydrophenanthrene structure [2], and this chapter focuses on the sterol class (ST01), which includes the subclass cholesterol and derivatives (ST0101). This subclass embraces cholesterol and its precursors and oxidized derivatives (Figure 16.1). Steryl esters fall into a separate subclass (ST0102). Bile acids fall into a separate class all together, bile acids and derivatives (ST04), although a number of C_{27} members of this class (subclass ST0403) can also be classified in ST0101. For simplicity in this chapter, we will use the term sterol to refer to all members of the ST0101 subclass excluding oxysterols, which will be referred to specifically.

Sterols/oxysterols can be found in the non-esterified form or esterified to fatty acids (ST0102) or sulfuric acid and can also form acetal links with sugars (ST05). This results in huge structural and physicochemical complexity and extremes of hydrophobicity and hydrophilicity, ranging from hydrophobic cholesteryl esters to

Mass Spectrometry for Lipidomics: Methods and Applications, First Edition.
Edited by Michal Holčapek and Kim Ekroos.

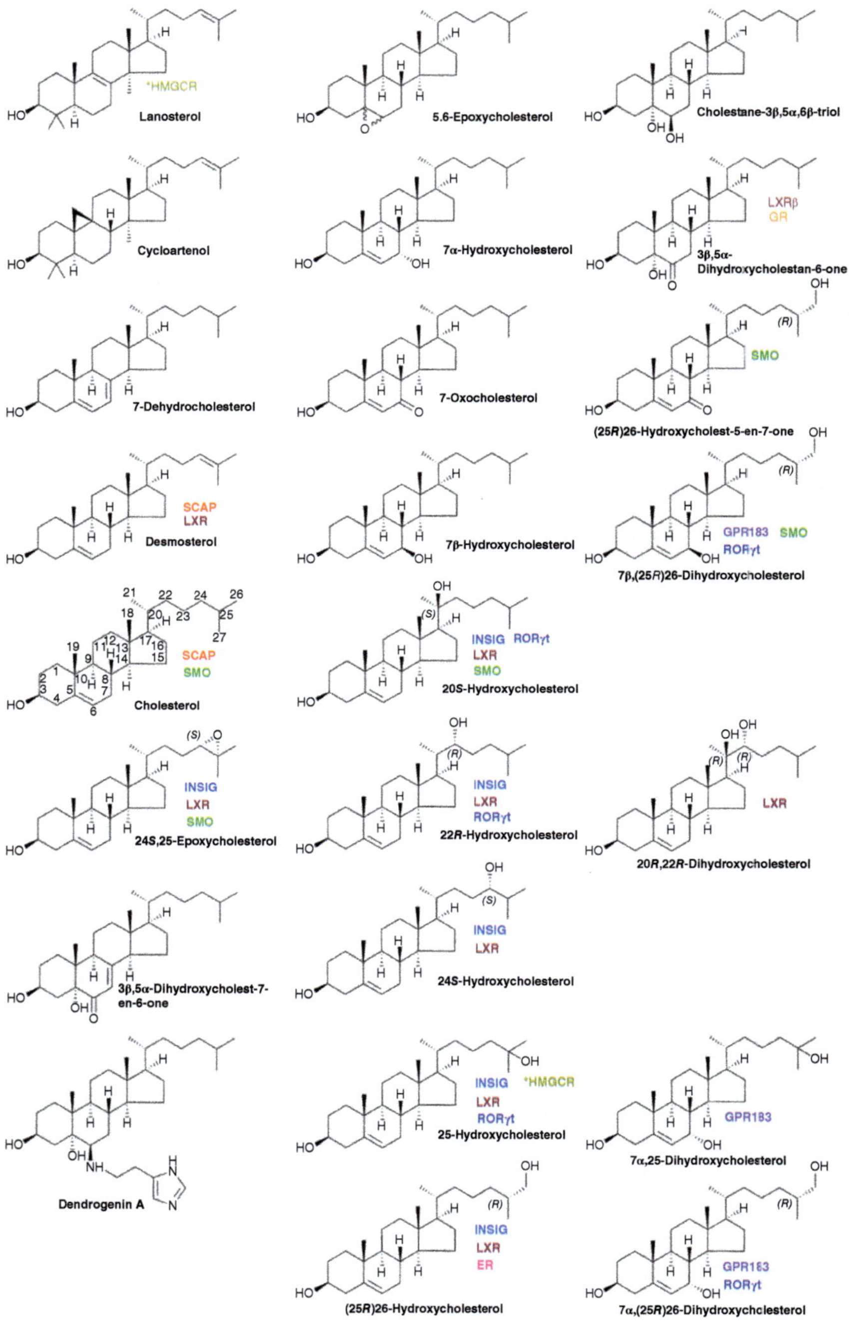

Figure 16.1 Structures of biologically active and related sterols and oxysterols. Protein receptors are indicated in colored text. The sterol numbering system is shown for cholesterol. *HMGCR indicates that the protein is activated for degradation. 24*S*,25-Epoxycholesterol and 3β,5α-dihydroxycholest-7-en-6-one are oxysterols formed from cholesterol precursors. Dendrogenin A is a histamine conjugate of 5α,6α-epoxycholesterol.

water-soluble oxysterol glucuronides or sulfates [5, 7]. In reality, this means that few if any lipidomic studies cover the entire sterolome ST, but rather attention is focused on particular sterol classes, e.g. ST01 or subclasses ST0101.

16.1.1 Sterol in Cells

Why are sterols of interest to the lipidomic scientist? As most lipidomic studies are framed in a mammalian context, this question can be answered in relation to human biochemistry and health. Cholesterol itself is an essential lipid, being present at a level of about 10 fmol/cell in animals [3], where it provides a structural component of membranes. It is particularly abundant in the plasma membrane where its level may be as high as 40–50 mol% of total lipids [8, 9]. In fact, 60–90% of the cells' cholesterol may be found in the plasma membrane [8, 10]. Excess cholesterol can be stored esterified to fatty acids within lipid droplets. Cholesterol is metabolized to bioactive steroid hormones and bile acids, while one of its immediate precursors, 7-dehydrocholesterol (7-DHC), is converted to vitamin D_3, which is ultimately processed to the active component 1,25-dihydroxyvitamin D_3. It is interesting to note that while the majority of cholesterol is found in the plasma membrane [8], it is synthesized in the endoplasmic reticulum [3], where much of its metabolism is also initiated [6]. Clearly, an important component of cellular cholesterol homeostasis is cholesterol transport, and that of its metabolites, between membranes and different organelles [11–13]. Insufficient cellular cholesterol will lead to membrane defects while excess cellular cholesterol is cytotoxic.

Cholesterol regulates its own synthesis and uptake by a mechanism uncovered by Brown, Goldstein, and coworkers and termed convergent inhibition [14–16]. In brief, the transcription factor sterol regulatory element-binding protein-2 (SREBP-2) is synthesized in the endoplasmic reticulum where it binds to the transport protein SREBP-cleavage activating protein (SCAP). The function of SCAP is to transport SREBP-2 to the Golgi for processing to its active form (n-SREBP-2), which translocates to the nucleus and acts as the master regulator of transcription of genes of the cholesterol biosynthesis pathway and of the LDL receptor (Figure 16.2). This process occurs when the endoplasmic reticulum is deficient in cholesterol, i.e. <5 mol% endoplasmic reticulum lipids [17], and the result is increased cholesterol synthesis and uptake via the LDL receptor and restoration of homeostasis. When the cholesterol levels in the endoplasmic reticulum exceed 5 mol%, cholesterol binds to SCAP, which then binds to the endoplasmic reticulum resident protein INSIG (insulin-induced gene) tethering the SCAP-SREBP-2 complex in the endoplasmic reticulum and preventing transport to the Golgi for processing of SREBP-2 to its active form. The result is cholesterol synthesis and LDL receptor expression are turned down, and cholesterol homeostasis is restored. It is surprising that although the majority of cholesterol in a cell is found in the plasma membrane, it is the cholesterol content of the endoplasmic reticulum that is the key determinant of cholesterol homeostasis. This conundrum can be explained by the concept of "active" or "accessible" cholesterol [18–20].

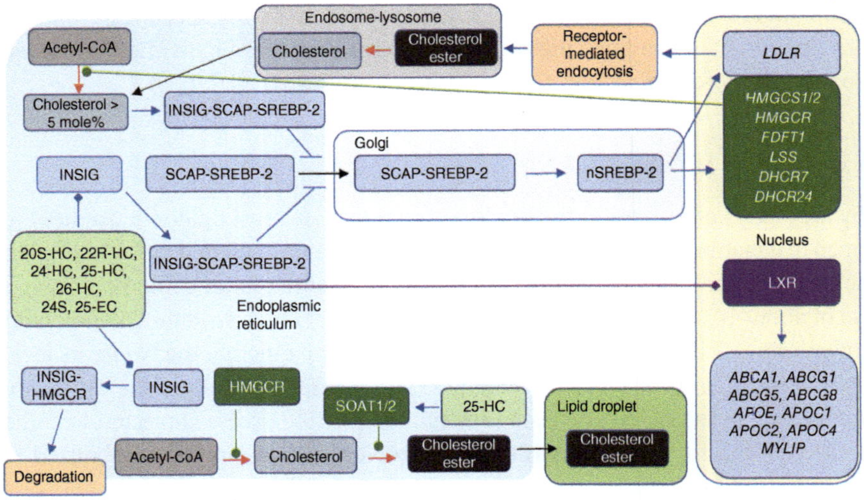

Figure 16.2 Schematic summary of the mechanism of cholesterol synthesis, uptake and storage, and the involvement of cholesterol and side-chain oxysterols in cholesterol homeostasis. The cell can synthesize cholesterol via the mevalonate pathway under regulation of SREBP-2 or it can be taken up as cholesteryl esters via lipoprotein receptors. Excess cholesterol can be exported by ABC transporters onto extracellular lipoprotein particles or stored as cholesteryl esters in lipid droplets. Cell compartments are color coded; endoplasmic reticulum, light blue; Golgi, light gray; endosomal–lysosomal compartment, gray; nucleus, pale yellow; and lipid droplet, green. Some of the cholesterol biosynthesis genes are shown as are LXR target genes. Side-chain oxysterols are on a light green background, and enzymes are on a dark green background. Blue arrows indicate a process, red arrows indicate a chemical reaction, black arrows represent transport, green oval arrowheads indicate catalysis, and diamond arrowheads connect ligands to receptors. Source: Griffiths & Wang/Elsevier/CC BY 4.0.

Accessible or active cholesterol is a descriptive term for cholesterol in membranes, which, by virtue of its environment, has high thermodynamic chemical activity [19]. Unlike the majority of cholesterol in membranes, which is complexed with phospholipids (sphingolipids and glycerophospholipids) or proteins and has a low chemical activity, accessible cholesterol is available to be transported between membranes [20]. When the level of accessible cholesterol in the plasma membrane becomes elevated as a consequence of, e.g., excessive LDL uptake, accessible cholesterol becomes transported to the endoplasmic reticulum and switches off the SREBP-2 pathway [20].

One of the consequences of cholesterol over accumulation in cells is inflammasome activation and induction of inflammation [21]; this is a defense mechanism against microbial infection but has also been linked to autoimmune disease. Remarkably, cellular cholesterol homeostasis can be modulated as part of a defense mechanism by immune cells and epithelial cells in their protection against microbial infection, where a reduction in the plasma membrane-accessible cholesterol provides protection against bacterial toxins, bacterial spread, and also viral entry [22–25].

16.1.2 Oxysterols

While cholesterol regulates its own synthesis, its metabolites, i.e. oxysterols, also play a role in the fine-tuning of cellular cholesterol levels [26] and in modulating the levels of accessible cholesterol in epithelial cells [24, 27]. Side-chain oxysterols will inhibit the processing of SREBP-2 to its active form as a transcription factor by binding to INSIG causing a conformational change resulting in INSIG binding to SCAP and blocking the transport of SREBP-2 to the Golgi for processing to its active form (Figure 16.2) [28]. Side-chain oxysterols will also activate the esterification of cholesterol with fatty acids in a reaction catalyzed by the endoplasmic reticulum enzymes sterol O-acyltransferase (SOAT) [29] and enhance the degradation of HMG-CoA reductase (HMGCR), the enzyme catalyzing the rate-determining step of cholesterol synthesis, probably by binding to INSIG [15, 30]. In addition, side-chain oxysterols activate the liver X receptors (LXRα and LXRβ) whose target genes include ATP-binding cassette transporters ABCA1 and ABCG5/8, which mediate cholesterol efflux from cells, and the cholesterol carrier protein including APOE [31]. In combination, these activities lead to reduced cellular non-esterified cholesterol.

The Hedgehog (Hh) signaling pathway is important in determining cell fate during development [19] and accessible cholesterol, and a number of oxysterols have been shown to be activators of this pathway through binding to the G protein-coupled receptor (GPCR) smoothened (SMO) [32–34]. These ligands include oxysterols oxidized at carbon-7 and in the side chain [34, 35]. Oxysterols with a hydroxy group at carbon-7 and in the side chain have also been shown to be ligands toward a second GPCR known as the Epstein Barr virus-induced gene 2 (EBI2) or GPR183 [36, 37]. This receptor is expressed in immune cells and which move up a concentration gradient toward their chemoattractant oxysterol ligands [36–38].

Ring oxysterols *per se* are also biologically active. One group of oxysterols derived from 5,6-epoxycholesterol appears to be particularly important in relation to breast cancer. Both 5α,6α-epoxycholesterol and 5β,6β-epoxycholesterol can be converted enzymatically to 3β,5α-dihydroxycholestan-6-one (oncosterone, Figure 16.1) [39]. Oncosterone is a tumor promoter acting through the glucocorticoid (GR) and LXRβ receptors. Another product, exclusively of 5α,6α-epoxycholesterol, is dendrogenin A (DDA), a tumor suppressor. DDA is formed by the nucleophilic addition of histamine to the 5α,6α-epoxy function [40]. Whether 5α,6α-epoxycholesterol is formed enzymatically is still unknown, there is strong evidence for its formation *in vivo* via non-enzymatic oxidation by its presence in plasma from patients with lysosomal storage disorders [41]. Another related ring oxysterol, this time derived from the cholesterol precursor 7-DHC, is 3β,5α-dihydroxycholest-7-en-6-one, which inhibits Hh signaling by binding to SMO [42].

16.1.3 Precursors of Cholesterol

In a lipidomic context, cholesterol precursors are of interest as markers of cholesterol biosynthesis [43], while plant sterols represent markers of cholesterol absorption [44]. Profiling for cholesterol precursors is also important clinically in the identification of inborn errors of cholesterol biosynthesis [4]. Like oxysterols, cholesterol precursors are biologically active. Desmosterol will activate LXRs and also bind to SCAP and inhibit SREBP-2 processing [28, 45], while 4,4-dimethylated sterols, including lanosterol, augment HMG-CoA reductase degradation [30].

16.1.4 Sterols and Oxysterol in Blood Plasma and Serum

Despite the cellular origin of cholesterol, its precursors and oxysterols, for practical reasons most lipidomic studies are performed on body fluids. In blood, cholesterol and oxysterols are transported in association with lipoprotein particles [46], and the presence of the enzyme lecithin–cholesterol acyltransferase (LCAT) in LDL particles dictates that they are mostly found in the esterified form [46–48]. Steryl esters can be analyzed directly by mass spectrometry [7], or following ester hydrolysis with alkaline [46, 49]. Care must always be taken with interpretation of data following either alkaline or acid hydrolysis as some functions on the oxysterol skeleton are labile, e.g. 7-hydroxy-3-oxo-4-ene which will dehydrate [5, 50].

Clinically, sterol/oxysterol profiling of plasma/serum is important in the identification of inborn errors of metabolism [4, 51–53]. The deficiency of an enzyme in either the cholesterol biosynthesis or metabolic pathways will result in elevated levels of its substrates. These can be monitored in the plasma; however, their metabolites formed in reactions independent of the defective enzyme may provide even better biomarkers [54–56].

16.1.5 Analytical Challenges

There are a number of challenges in the lipidomic analysis of sterols/oxysterols, one of which is dynamic range. Other than cholesterol and its fatty acid esters, sterols are minor lipids. While in plasma the concentration of total cholesterol, i.e. the combination of free and esterified cholesterol, is about 2 mg/ml, the levels of its precursors and metabolites are about 1000 times lower [47, 48], this ratio is similarly reflected in tissue. This provides a challenge to mass spectrometry (MS) analysis whether linked online to chromatography or not. The solution is usually to perform a separate analysis of cholesterol from that of its metabolites. The challenge is accentuated by the propensity of cholesterol to become oxidized by air during sample handling [57]. To minimize the formation of oxysterols during sample preparation, cholesterol is often separated from oxysterols in an early step of the sample preparation protocol [46, 48, 58, 59]. A third challenge is the lack of volatility of sterols/oxysterols, important in gas chromatography (GC) separations, and a lack of a basic or acidic group for protonation or deprotonation, important for atmospheric pressure ionization (API)-MS methods. These physicochemical deficiencies can be overcome by derivatization [46, 48, 53–55, 58–62], addition of additives to the mobile phase, and exploitation of multiple reaction monitoring (MRM) [47, 49, 63].

16.2 Analytical Methods

MS-based methods for the analysis of sterols/oxysterols fall into four categories: [1] GC-MS methods, [2] liquid chromatography (LC) – tandem MS (MS/MS) methods, [3] LC-MS/MS methods with derivatization, and [4] MS/MS methods for steryl esters.

16.2.1 Classical GC-MS Methods for Sterol and Oxysterol Analysis

Many of the GC-MS methods used today for sterol and oxysterol analysis are based on the methods of Björkhem et al. and Dzeletovic et al., respectively, developed for plasma/serum analysis [46, 64]. The methods usually incorporate a saponification step and so are appropriate for the analysis of total sterols and total oxysterols [59, 65]. The sterol analysis method has been modified by Acimovic et al. specifically for analysis of sterols extracted from cells [66], while the oxysterol method can be modified for analysis of tissues by inclusion of an extraction step into chloroform/methanol (2 : 1) [67, 68]. The methods are all essentially similar, and here, we will highlight the method of Dzeletovic et al. (Figure 16.3) [46].

In brief, blood is collected into EDTA-containing evacuated tubes, plasma is prepared by centrifugation, and the antioxidant butylated hydroxytoluene (BHT) is added. Isotope-labeled internal standards are added at this point, and saponification is performed in 0.35 M KOH in ethanol for two hours at room temperature. The solution is neutralized by the addition of H_3PO_4, and cholesterol and oxysterols are extracted into chloroform. The extract is then dried and reconstituted in toluene. Oxysterols are then separated from cholesterol on a normal-phase silica solid-phase extraction (SPE)

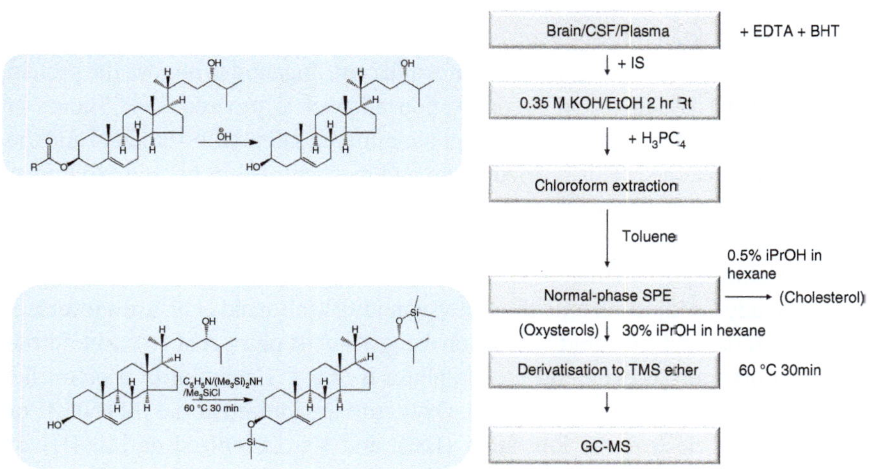

Figure 16.3 Extraction and GC-MS analysis of oxysterols based on the method of Dzeletovic et al. [46].

cartridge, cholesterol, and sterols to a similar polarity elute with 0.5% propan-2-ol in hexane and oxysterols in 30% propan-2-ol in hexane. In Dzeletovic et al.'s original method, cholesterol is discarded at this point [46], but assuming isotope-labeled cholesterol is added as one of the internal standards, cholesterol can be derivatized in parallel to oxysterols. Oxysterols are derivatized to their trimethylsilyl (TMS) ethers using a mixture of pyridine/hexamethyldisilazane/trimethylchlorosilane (3:2:1, v:v:v) at 60 °C for 30 minutes. Solvent is removed and the residue is dissolved in hexane ready for GC-MS analysis. To analyze unesterified oxysterols, the hydrolysis step is omitted. Most oxysterols are chromatographically separated on an appropriate GC column and those which are not can be monitored by different fragment ions generated by the electron impact ionization source. Quantification is performed by exploiting selected ion monitoring (SIM) with the use of appropriate isotope-labeled standards. Today, most of the common oxysterols and cholesterol precursors and their isotope labeled analogues can be purchased from Avanti Polar Lipids or Cayman Chemical Company, allowing accurate sterol and oxysterol quantification by isotope dilution mass spectrometry. It is noteworthy that sterol and oxysterol fractions are derivatized separately, and this avoids the possibility of oxysterols being generated from cholesterol in the derivatization process and contaminating the oxysterol analysis.

16.2.2 LC-MS/MS Analysis of Sterols and Oxysterols

Perhaps, the most impressive study of sterols and oxysterols performed to date using LC-MS/MS is that of Stiles et al. who measured the abundance of 60 sterols and oxysterols in a sample set of 3230 serum samples from the Dallas Heart study [63]. They adopted the methodology developed by McDonald et al., which is briefly described below (Figure 16.4) [49].

Serum (or plasma) is added to a 1:1 solution of dichloromethane and methanol containing a cocktail of isotope-labeled internal standards and the antioxidant BHT. After incubation at 30 °C for 10 minutes and centrifugation to remove the protein precipitate, KOH is added and saponification allowed to proceed for 1.5 hours at 35 °C. After hydrolysis is completed, phosphate-buffered saline (PBS) is added and the organic layer is retained. This is dried down and reconstituted in hexane. Non-polar lipids are then separated from sterols and oxysterols on a normal-phase aminopropyl SPE cartridge. After elution of non-polar lipids with hexane, sterols and oxysterols are eluted with chloroform:methanol, 23:1. This fraction is dried and reconstituted in 90% methanol ready for injection onto the LC system [49]. McDonald et al. use a reversed-phase C_{18} column and with gradient elution using a mobile phase A of 70% acetonitrile in 5 mM ammonium acetate and mobile phase B of a 1:1 solution of acetonitrile: 5 mM ammonium acetate in propan-2-ol. Oxysterols are ionized as the $[M+NH_4]^+$ or $[M+H]^+$ ion by electrospray ionization (ESI), and sterols ionized as $[M+H]^+$ or $[M+H-H_2O]^+$ by atmospheric pressure chemical ionization (APCI). MS/MS analysis exploits timed-MRM with the dominant transitions being for the loss of water. McDonald et al. provide a list of transitions in their publication [49]. Quantification is performed by isotope dilution against added isotope-labeled standards.

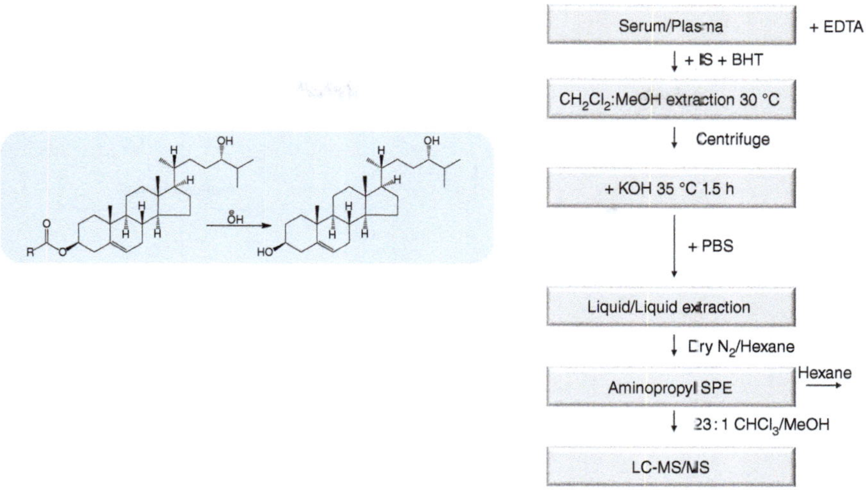

Figure 16.4 Schematic illustrating the sample preparation method for sterol/oxysterol analysis by LC-MS/MS. *Source:* Adapted from McDonald et al. [49].

The advantages of the method of McDonald et al. are that it is sensitive with an on-column detection of 50 pg and a lower limit of detection of about 1 ng/ml, and it is applicable for a wide range of sterols and oxysterols [49]. Impressively, Stiles et al. were able to profile for 60 analytes of which about 20 are routinely detected in plasma/serum, and the method, although somewhat laborious, is applicable to automation [49, 63]. All methods also have their limitation; in this case, they include the lack of formation of diagnostic fragment ions, leading to the use of non-specific MRM transitions to $[M+H-(H_2O)_n]^+$ fragment ions for oxysterols, and the reliance on retention time and the availability of authentic standards for oxysterol identification. Fortunately, a large number of authentic oxysterols are currently commercially available.

LC-MS/MS methods based on the protocol of McDonald et al. have been used in many interesting biological studies in recent years, including the investigation of the role of (25R)26-hydroxycholesterol (26-HC, better known by the non-systematic name 27-HC) in breast cancer [69], uncovering the role of 25-HC in immune response [21, 70], and identifying oxysterols that bind to SMO and modulate Hh signaling [34, 71].

16.2.3 LC-MS/MS Analysis of Sterols and Oxysterols Incorporating Derivatization

A major challenge to LC-MS/MS analysis of sterols in general is their comparatively poor ionization characteristics. This is true for cholesterol and its precursors and for oxysterols, but not for bile acids, which give $[M-H]^-$ ions, or for sterol/oxysterol sulfate or glucuronide conjugates. The problem can be overcome to some extent by the addition of additives to the mobile phase and exploiting MRM. An alternative approach is to incorporate a derivatization step to improve the ionization of the target analyte. There are many different derivatization protocols applicable for sterols and oxysterols and only a few will be discussed here.

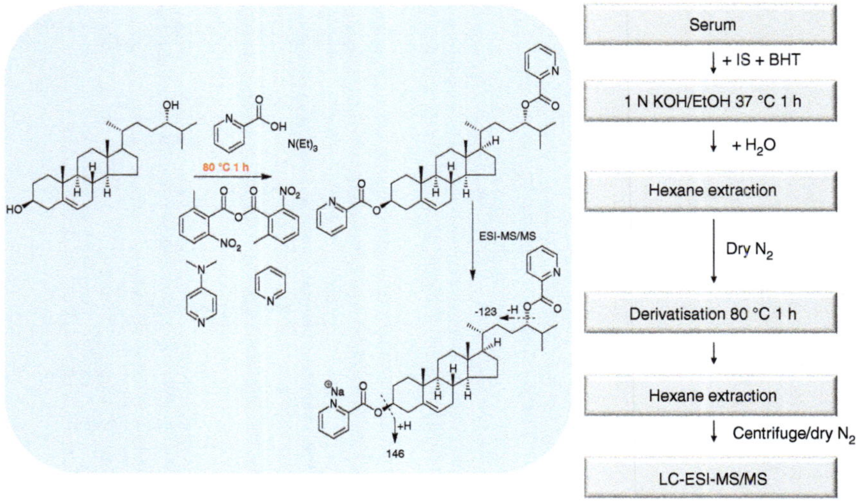

Figure 16.5 Extraction and derivatization of oxysterols using the method of Honda et al. [55].

16.2.3.1 Derivatization to Picolinyl and Nicotinyl Esters

Picolinic acid (pyridine-2-carboxylic acid) and nicotinic acid (pyridine-3-carboxylic acid) are isomers. They both form esters with sterols and oxysterols which are better Lewis bases (electron donors) than the parent molecule. Honda et al. developed methodologies for the analysis of cholesterol and its precursors and also related phytosterols and for oxysterols by derivatization to picolinyl esters [54, 55]. The basic method for oxysterol analysis in serum is described below (Figure 16.5).

16.2.3.1.1 Picolinyl Esters

An internal standard mix and the antioxidant, BHT, are first added to the serum, and saponification is carried out in 1N ethanolic KOH at 37 °C for one hour. Water is added and oxysterols are extracted into hexane. The extract is dried. This is important as the derivatization reaction requires anhydrous conditions. The derivatization reagent consists of a freshly prepared mixture of 2-methyl-6-nitrobenzoic anhydride, 4-(*N*,*N*-dimethylamino)pyridine, picolinic acid, pyridine, and triethylamine, this is added to the oxysterol extract and the mixture incubated at 80 °C for 60 minutes. The incubation temperature represents a potential pitfall to the method as cholesterol will be present in the oxysterol extract and is susceptible to oxidation in air, which can potentially generate oxysterol artifacts. Derivatized oxysterols are then extracted into hexane. Excess derivatization reagent precipitates out. This represents a major advantage of the method alleviating further clean-up steps and minimizing LC column contamination. The hexane extract is then dried down and derivatized oxysterols redissolved in acetonitrile [55]. Chromatographic separation is performed on a reversed-phase C_{18} column using a mobile phase of acetonitrile/methanol/water containing 0.1% formic acid. Oxysterols with two hydroxy groups become doubly derivatized and upon ESI give $[M+Na]^+$ ions. These fragment upon MS/MS to give ions corresponding to the Na^+ adduct of picolinic acid at *m/z* 146 or $[M+Na\text{-picolinic acid}]^+$ ($[M+Na-123]^+$), which are used for MRM. Quantification is performed by isotope-dilution utilizing the standard mix added to serum [55].

The method of Honda et al. is very sensitive with on-column detection of less than 10 fg, which is sufficient for oxysterol quantification from 5 μl of serum with a limit of detection less than 1 ng/ml [55]. While the MRM method offers some selectivity, it is only at the level of lipids derivatized with picolinic acid and not specific for sterols or oxysterols. As with most LC–MS/MS methods for oxysterols analysis, exact identification relies on matching chromatographic retention time with that of internal standards.

16.2.3.1.2 Nicotinyl Esters

A related derivatization method to that of Honda et al. [55] is that developed by Sidhu et al. who replaced picolinic acid with nicotinic acid [61]. Sidhu et al.'s method was developed in the context of measuring non-esterified 24S-HC in cerebrospinal fluid (CSF) as a biomarker in support of a clinical trial for the treatment of Niemann Pick type C1 (NPC1) disease [61]. NPC1 disease is a neurodegenerative disease in the young, which is a result of a deficiency in NPC1 protein, important for cholesterol transport in the cell. An interesting aspect of the method developed by Sidhu et al. is the addition to CSF during its collection of 2.5% 2-hydroxypropyl-β-cyclodextrin (HPβCD) as a chelating reagent to reduce analyte losses to plastic ware. The method is also applicable to analysis of 24S-HC in plasma. In brief, an internal standard mix is added in methanol to the biofluid and oxysterols extracted into methyl *tert*-butyl ether (MTBE, Figure 16.6). After centrifugation, any water in the organic phase is removed by cooling the solution in a dry-ice ethanol bath (−78 °C)

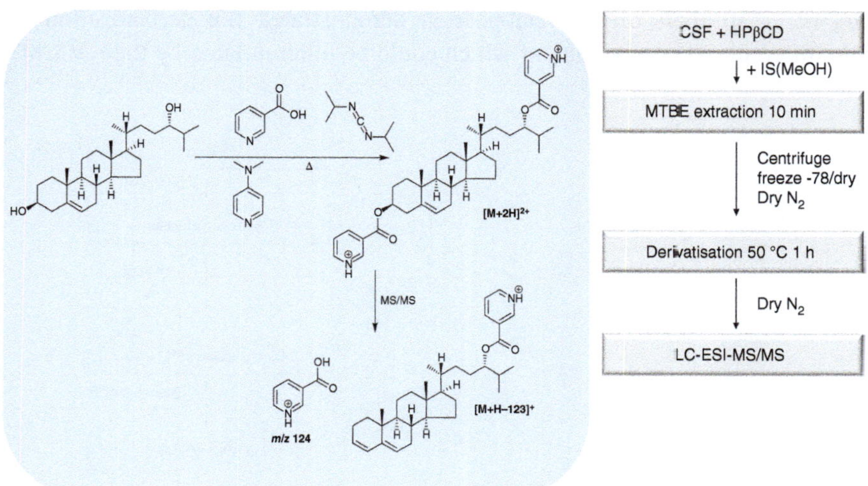

Figure 16.6 Extraction of 24S-HC from CSF and derivatization with nicotinic acid according to the method of Sidhu et al. [61]. *Source:* Adapted from Sidhu et al. [61].

where it freezes, before collecting the upper phase. The organic phase is dried down and oxysterols derivatized with a reagent mixture of *N,N*-diisopropylcarbodiimide, nicotinic acid, and 4-(*N,N*-dimethylamino)pyridine in chloroform. The mixture is incubated at 50 °C for one hour. The solvent is removed and the sample is dissolved in methanol for LC-MS/MS analysis [61]. LC separation is performed on a reversed-phase C_{18} column using mobile phases of (A) 0.1% formic acid and (B) 0.1% formic acid in propan-2-ol: acetonitrile (1 : 2) with gradient elution. Upon ESI, oxysterol nicotinyl esters give $[M+H]^+$ and $[M+2H]^{2+}$ ions, while MS/MS of both of these ions give protonated nicotinic acid (*m/z* 124) and [M+H-nicotinic acid ($[M+H-123]^+$) as major fragment ions. Sidhu et al. found the MRM transition $[M+2H]^{2+} \rightarrow 124$ provided the greatest sensitivity. Using the isotope-dilution method for quantification, Sidhu et al.'s method gives a lower limit of quantification in plasma of 1 ng/ml and in CSF of 0.025 ng/ml. As with the other LC-MS/MS methods discussed, oxysterol identification is based on chromatographic retention time and the availability of authentic standards, this is not a problem for 24S-HC. The advantage of the method is its high sensitivity and simplicity. The chromatography system developed did not separate 25-HC and 26-HC, but this was not a problem as the method is targeted to 24S-HC. Obviously, chromatography would need to be modified for oxysterol profiling.

16.2.3.2 Derivatization to Dimethylglycyl Esters

Derivatization of oxysterols to their dimethylglycyl esters represents one of the more popular derivatization methods for subsequent analysis by LC-MS/MS. Jiang, Ory, and Han developed a derivatization method using a similar catalytic system to that exploited by Sidhu et al. [61] but used dimethylglycine as the carboxylic acid rather than nicotinic acid and 1-ethyl-3-(3-dimethylaminopropyl)carbodiimide rather than *N,N*-diisopropylcarbodiimide as the coupling reagent (Figure 16.7) [60]. Jiang et al. in their original publication demonstrated the derivatization of seven different oxysterols, each of which could be differentiated by their MS/MS

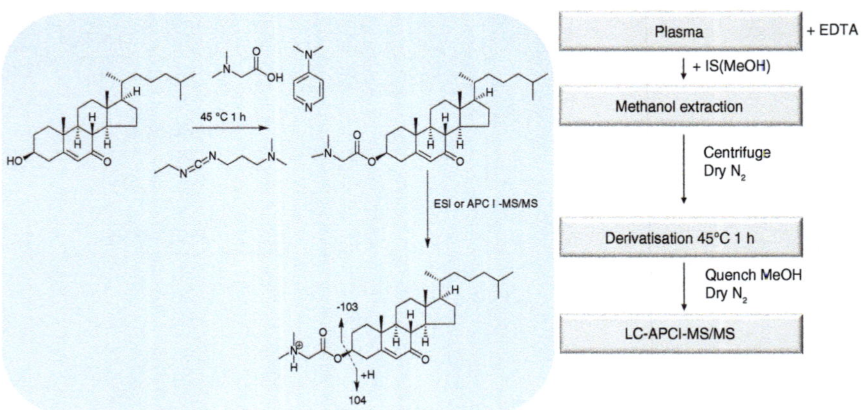

Figure 16.7 Extraction of oxysterols from plasma and derivatization to dimethylglycyl esters.

spectrum. The method has mostly been used in the context of identifying Niemann Pick disease and other lysosomal disorders [62, 72–74].

Niemann Pick disease results in an increased cholesterol content in lysosomes and a lack of its transport to the plasma membrane and endoplasmic reticulum. The consequence is enhanced formation of cholestane-3β,5α,6β-triol and 7-oxocholesterol (7-OC, also called 7-ketocholesterol), probably via non-enzymatic oxidation, and increased concentrations of these two oxysterols in blood. Jiang et al. exploited their derivatization method in combination with LC-MS/MS to quantify these two oxysterols in plasma [62]. In brief, EDTA-plasma is added to methanol containing isotope-labeled internal standards; after centrifugation, the supernatant is taken to dryness and derivatized with a mixture of dimethylglycine, 4-(*N,N*-dimethylamino) pyridine, and 1-ethyl-3-(3-dimethylaminopropyl)carbodiimide in chloroform. After incubation for one hour at 45 °C, the reaction is quenched with methanol, and the mixture is dried down and reconstituted in methanol for LC-MS/MS analysis. Cholestane-3β,5α,6β-triol becomes doubly derivatized while 7-OC, just having one hydroxy group, is singly derivatized. Jiang et al. used a C_{18} column and a gradient with the mobile phases (A) 0.015% trichloroacetic acid and 0.5% acetic acid in water and (B) 0.015% trichloroacetic acid and 0.5% acetic acid in acetonitrile. Ionization is by APCI and the MRM transitions $[M+H]^+ \rightarrow 104$ monitored for quantification, where *m/z* 104 corresponds to protonated dimethylglycine. To add some specificity, the transition $[M+H]^+ \rightarrow [M+H-103]^+$ is also monitored as a qualifier where the product ion corresponds to the loss of dimethylglycine from the precursor. This offers specificity to molecules derivatized with dimethylglycine [62]. The attraction of Jiang et al.'s method is its simplicity and sensitivity [60, 62].

The study of Jiang et al. is important as it unequivocally established cholestane-3β,5α,6β-triol and 7-OC as plasma markers of NPC1 disease [62]. Before this publication, there had always been concern over the biological relevance of the identification and quantification of these two molecules in plasma [57], as they can be formed by non-enzymatic oxidation of cholesterol in air [75–77], and at the time of Jiang et al.'s study, enzymatic pathways for their formation had not been defined, although soon after Goyal et al. demonstrated that 7-OC could be formed from 7-DHC in a reaction catalyzed by CYP7A1 [78]. Elevated levels of cholestane-3β,5α,6β-triol and 7-OC are not unique to NPC1 disease and are also seen with other lysosomal storage disorders [72–74] and also the inborn error of cholesterol biosynthesis Smith-Lemli-Opitz syndrome (SLOS) [35]. More recently, a bile acid metabolite of cholestane-3β,5α,6β-triol has proved to be a more attractive biomarker of NPC1 disease as, unlike cholestane-3β,5α,6β-triol and 7-OC, it cannot be formed from cholesterol in air but requires *in vivo* enzymatic reactions for its formation from cholestane-3β,5α,6β-triol [53, 79].

16.2.3.3 Derivatization with Girard Hydrazine Reagents

The Girard reagents, (2-hydrazinyl-2-oxoethyl)-trimethylazanium chloride (Girard T, GT) and 1-(2-hydrazinyl-2-oxoethyl)pyridin-1-ium bromide (Girard P, GP) [80], have been used in steroid analysis for decades (Figure 16.8) [81, 82]. The hydrazine group of the Girard reagent will react with a carbonyl group on a steroid skeleton and the

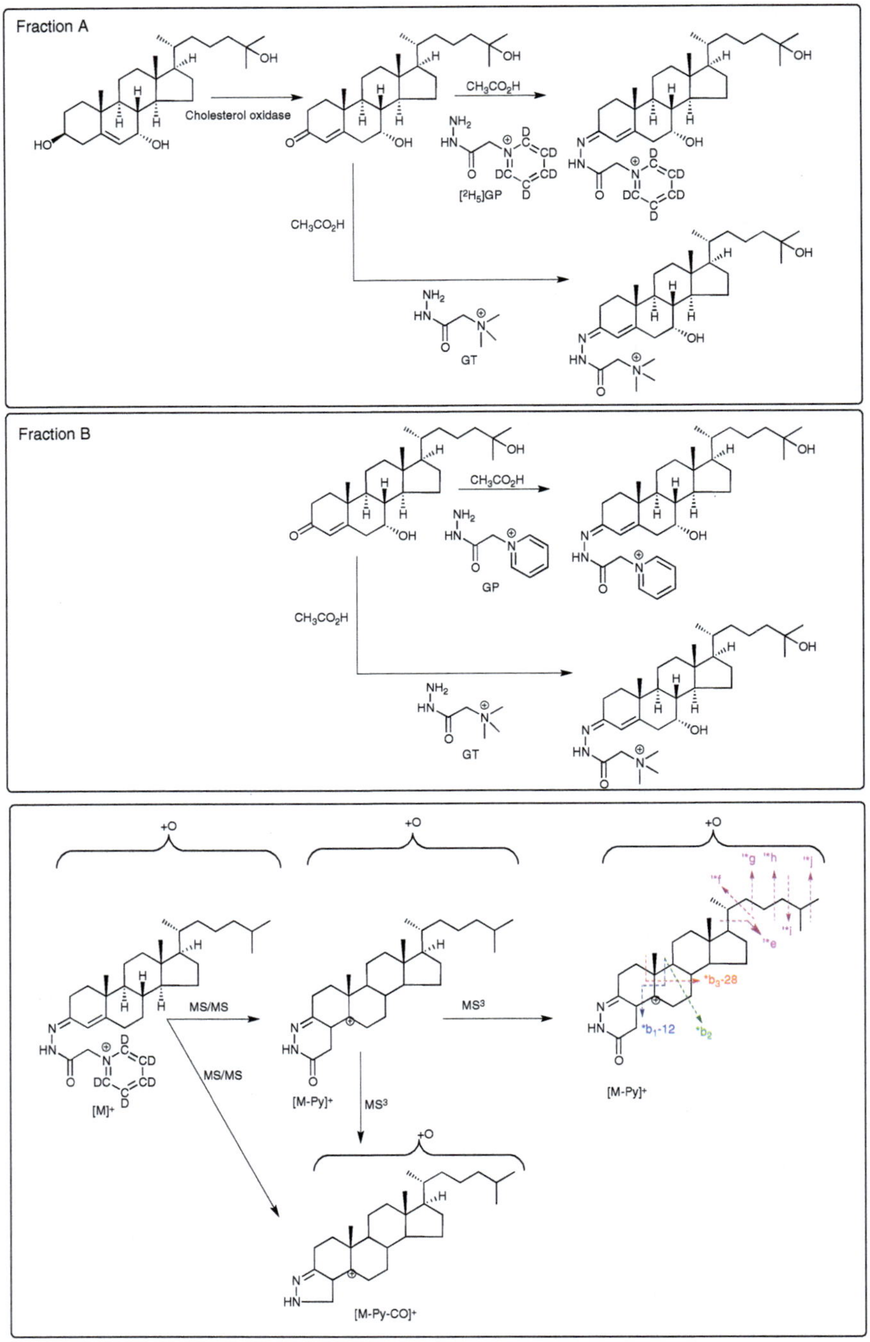

Figure 16.8 Derivatization of oxysterols using Girard reagents. Top panel, oxysterols without a carbonyl group are treated with cholesterol oxidase and then derivatized with the GP- or GT-hydrazine reagent. Center panel, oxysterols with a carbonyl group are derivatized directly with Girard hydrazine. Bottom panel, fragmentation of Girard derivatized oxysterols.

presence of the charged quaternary nitrogen will improve solubility in aqueous solvents [81]. With respect to ESI-MS, the presence of the permeant charge on the derivatized molecule will add sensitivity to MS analysis. While C_{21} and C_{18} steroids and their pharmaceutical relatives often possess carbonyl groups, these functional groups are less common in sterols and oxysterols of biological interest. However, DeBarber et al. have exploited the Girard P reagent for the LC-MS/MS analysis of oxysterols with a 3-oxo-4-ene function in the diagnostic testing for the inborn error of cholesterol metabolism cerebrotendinous xanthomatosis (CTX) [83, 84], while Lavrynenko et al. have explored the Girard T, P and a third reagent Girard C (potassium hydrazincarbonyl acetate) in the study of ectysteroids in *Drosophila* [85].

In the past, biochemists have exploited bacterial cholesterol oxidase enzymes for the *ex vivo* conversion of sterols with a 3β-hydroxy-5-ene function to their 3-oxo-4-ene equivalents [86, 87]. 3-Oxo-4-ene is not only a better chromophore for sterol detection by LC-UV [87] but also provides a reactive handle for coupling with the Girard reagent [58]. The strategy of utilizing Girard derivatization following a cholesterol oxidase step is gaining popularity for LC-MS/MS analysis of oxysterols [88–92]. The method is also applicable to matrix-assisted laser desorption/ionization (MALDI)-MS/MS [93] and to mass spectrometry imaging [94, 95]. We named the strategy "enzyme-assisted derivatization for sterol analysis" (EADSA) and have exploited it widely for the analysis of not only oxysterols but also cholesterol and its precursors [96], in plasma [35, 41, 48, 97], CSF [48], cells [98], and tissue samples [58, 99, 100]. The method is described in detail in reference [48] and described in brief below (Figure 16.9).

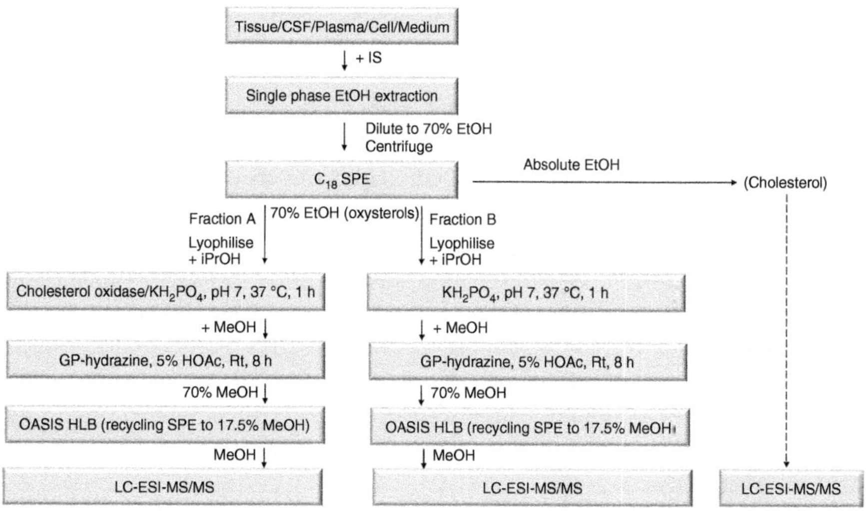

Figure 16.9 Extraction of sterols/oxysterols and derivatization with Girard reagents. The cholesterol fraction is derivatized in parallel to the oxysterol fraction. If different isotopic versions of the Girard reagent are used to derivatize fractions-A and -B, they may be combined after the second SPE step and analyzed together by LC-MS. *Source:* Adapted from Yutuc et al. [48].

Sterols and oxysterols are extracted (single phase) from tissue, cells, cell medium, plasma, or CSF into absolute ethanol containing a standard mix. The solution is diluted to 70% ethanol and centrifuged to remove the precipitated matter. The resultant solution is passed through a C_{18} Sep-Pak cartridge, and oxysterols elute in 70% ethanol; cholesterol and similarly lipophilic sterols are eluted with absolute ethanol. This is an important step as it minimizes the possibility of oxysterols being formed from cholesterol *ex vivo* and contaminating the oxysterol fraction. The oxysterol and cholesterol fractions are then treated identically but separately. For simplicity, only further treatment of the oxysterol fraction is described below.

If the investigator is interested in oxysterols that contain a natural 3β-hydroxy-5-ene group and also those that contain a 3-oxo-4-ene function, each fraction is divided into two sub-fractions A and B. Alternatively, if only oxysterols that contain a 3β-hydrox-5-ene function are of interest and equivalent 3-oxo-4-ene oxysterols are not in the sample, this division can be omitted. All fractions are dried under vacuum and reconstituted into propan-2-ol. To the A-fractions, 10 volumes (cf. that of propan-2-ol) of 50 mM KH_2PO_4, pH 7, containing 0.26 units of cholesterol oxidase (1 unit will oxidize 1 μmole of cholesterol to its 4-en-3-one in 1 minute) from *Streptomyces sp* are added, and the mixture is incubated for one hour at 37 °C. Methanol is added to give a 2:1 methanol-to-buffer ratio, followed by glacial acetic acid to make the solution 5% acetic acid. A molar excess of Girard reagent is then added, and the mixture incubated overnight in the dark. B-fractions are prepared, if required, in an identical manner but in the absence of cholesterol oxidase. Our preference is to derivatize fraction-A with $[^2H_5]$GP and fraction-B with $[^2H_0]$GP (Figure 16.8). By doing this, A- and B-fractions can be analyzed together by LC-MS in a single LC-MS run. If the same isotopic version of the Girard reagent is used for both A- and B-fractions, they must be analyzed in separate LC-MS runs. To minimize deterioration of the LC column, maximize reproducibility of retention time, and if different isotopic versions of the Girard reagent are utilized to avoid scrambling of the derivatization tags, we include a clean-up step following derivatization and before LC-MS analysis to remove excess Girard reagent. This step is performed on an Oasis HLB cartridge where the individual fractions are added to separate cartridges in about 70% methanol, and the eluent collected, diluted to 35% methanol, and recycled through the column; this is repeated to give a 17.5% methanol solution, and after this has passed through the column, all oxysterol or sterols are extracted by the column. The column is then washed with 10% methanol and oxysterols/sterols eluted with methanol. If C_{18}–C_{21} steroids are of interest, it is necessary to perform one extra step of recycling. An aliquot of the eluate is diluted to 60% methanol ready for LC-MS analysis. If A- and B-fractions were derivatized with different isotopically labeled Girard reagents, the A- and B- fractions can be combined ready for LC-MS analysis in a single run.

The sample preparation protocol described above is for non-esterified sterols and oxysterols. If total sterols and oxysterols are of interest, single-phase extraction is into ethanolic KOH 0.35 M, rather than absolute ethanol, followed by a two hour incubation, and then neutralization with glacial acetic acid is included. The solution is then diluted to 70% ethanol and processed as described above.

We perform LC separation on a reversed-phase Hypersil Gold C_{18} column from Thermo Scientific [48]. A gradient is employed with mobile phase A as 33.3% methanol and 16.7% acetonitrile containing 0.1% formic acid, and mobile phase B as 63.3% methanol and 31.7% acetonitrile containing 0.1% formic acid. The cholesterol and oxysterol fractions are analyzed separately; this is advisable as cholesterol is often a 1000 times more abundant than a given oxysterol. The high abundance of cholesterol compared to its precursors also challenges the analytical system as the precursors are often of similar abundance to oxysterols. GP- and GT-derivatized sterols and oxysterols give $[M]^+$ ions upon ESI. MS/MS of $[M]^+$ ions of GP and GT derivatives lead to major $[M\text{-}Py]^+$ and $[M\text{-}Py\text{-}CO]^+$ or $[M\text{-}(CH_3)_3N]^+$ and $[M\text{-}(CH_3)_3N\text{-}CO]^+$ fragment ions, respectively (Figure 16.8) [101, 102]. Structurally informative fragment ions generated by cleavage across the ring system and side chain are also evident but of low abundance (Figure 16.8) [102]. These ions are more readily visible when MS^n fragmentation is performed on ion-trap instruments (Figure 16.10) [48, 58] and can be used for the identification of oxysterols with unknown structures [103]. A disadvantage of Girard derivatization is that the resulting hydrazones can have *syn* or *anti* conformations, resulting in twin chromatographic peaks for most oxysterols complicating the chromatography. Quantification is performed against isotope-labeled standards, and this can be at the level of the $[M]^+$ ions using high-resolution instruments or at the MS/MS level monitoring one of the major fragment ions on lower resolution MS/MS instruments. Importantly, the quantities provided by analysis B-fractions are for endogenous molecules with a 3-oxo-4-ene function, and in the case of 7-OC a 7-oxo-5-ene function, while A-fractions give quantities for the sum of endogenous 3-oxo-4-ene compounds and their 3β-hydroxy-5-ene endogenous equivalents oxidized *ex vivo* by cholesterol oxidase to 3-oxo-4-enes. The concentrations of endogenous 3β-hydroxy-5-ene compounds can simply be determined by subtracting concentrations measured for fraction-B from those measured from fraction-A, i.e. A-B.

In summary, the major advantages of the Girard derivatization methods are increased sensitivity and structurally informative fragmentation spectra, particularly those derived by MS^3. The disadvantages are the laborious nature of the procedure, the reliance on an enzyme-catalyzed reaction, and as with other derivatization methods the potential for unexpected side reactions.

16.2.3.4 Derivatization with 4-Phenyl-1,2,4-triazoline-3,5-dione (PTAD)

The PTAD (4-phenyl-1,2,4-triazoline-3,5-dione) derivatization method was originally designed to target the conjugated diene structure of vitamin D through a Diels–Alder reaction [104]. However, it is also applicable to derivatize sterols such as 7-DHC with a diene structure in the B-ring (Figure 16.11) [105]. By careful choice of derivatization conditions, PTAD can react via an “ene”-type addition reaction to target double bonds as in desmosterol [105]. The PTAD derivatives give $[M+H]^+$ ions upon APCI and fragment in MS/MS by loss of the PTAD derivatizing group, which can be exploited in MRM assays.

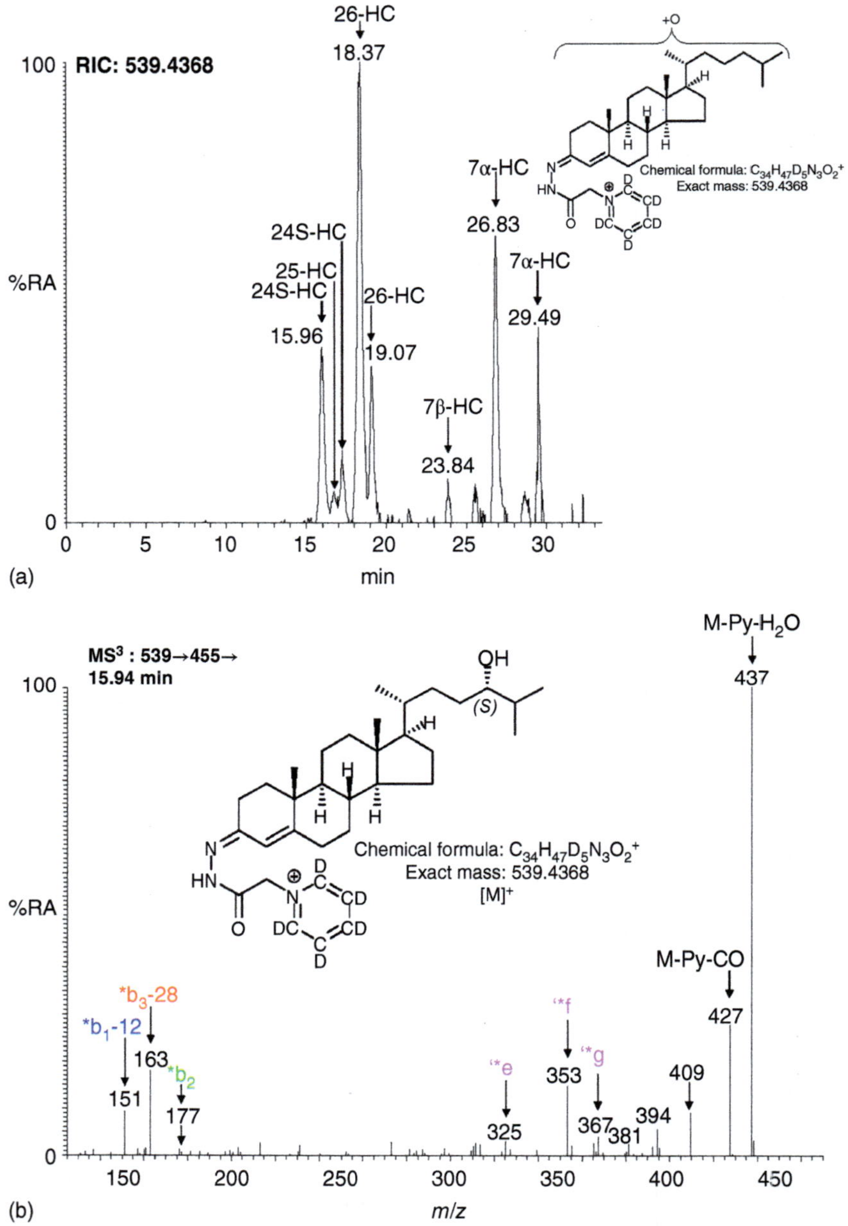

Figure 16.10 LC-MS(MS^3) analysis of oxysterols in plasma. (a) Reconstructed ion chromatogram of the $[M]^+$ ion of $[^2H_5]$GP-derivatized monhydroxycholesterols following cholesterol oxidase treatment. (b) MS^3 ($[M]^+ \rightarrow [M\text{-}Py]^+ \rightarrow$) fragmentation of 24*S*-hydroxycholesterol.

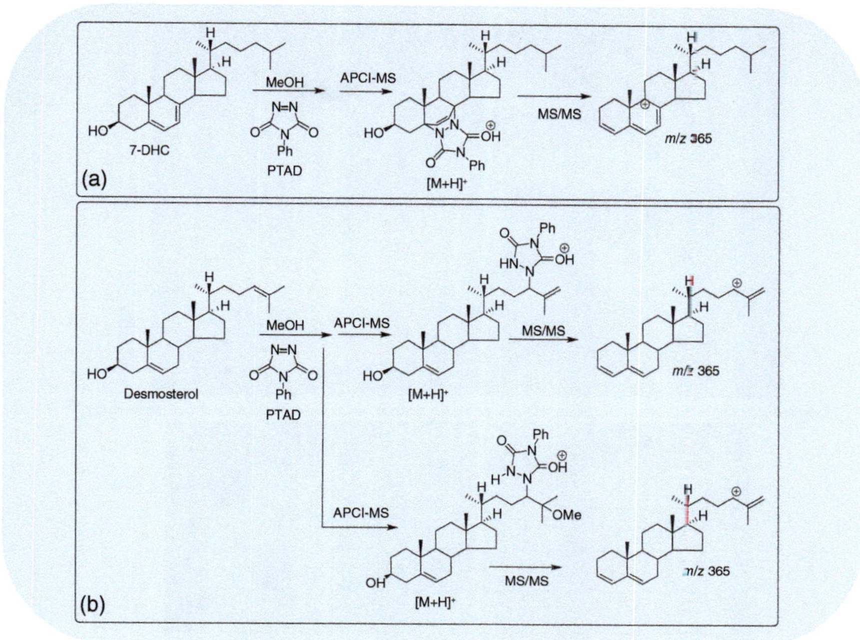

Figure 16.11 Scheme illustrating PTAD derivatization and MS/MS of 7-DHC (a) and desmosterol (b).

16.2.4 Mass Spectrometry Imaging of Cholesterol and Oxysterols in Tissue

Mass spectrometry imaging (MSI) provides spatial information for molecules in tissues. The method works particularly well for lipids that possess a permanent charge [106]. Cholesterol and oxysterols are neutral molecules, as are steroid hormones, so these molecules do not ionize well by MALDI, the most common ionization method for MSI. However, many C_{21}–C_{18} steroid and steroidal drugs possess a carbonyl group amenable to on-tissue Girard derivatization and MSI. This has been cleverly exploited by Cobice et al., Shimma et al., and Barr et al. for the MSI of steroids [107–109]. Cholesterol does not possess a carbonyl group but can be oxidized with cholesterol oxidase to contain one. We have been able to perform MSI imaging on both cholesterol and oxysterols by incorporating an on-tissue cholesterol oxidation step before on-tissue GP derivatization. We have utilized this method to image cholesterol in rodent brain exploiting MALDI (Figure 16.12) [95]. Imaging oxysterols in tissues provides an extra challenge of isomer separation. Side-chain hydroxycholesterol show only subtle differences in geometry between isomers and to date cannot be separated by ion-mobility devices, which are sometimes used in combination with MSI. Instead, we have performed low-resolution (800 μm pixel size) MSI using liquid extraction for surface analysis (LESA) following on-tissue EADSA and incorporating LC separation of the extract before MS(MS^3) analysis. We are able to perform quantification by spraying on a layer of isotope-labeled standard before on-tissue EADSA is carried out [94].

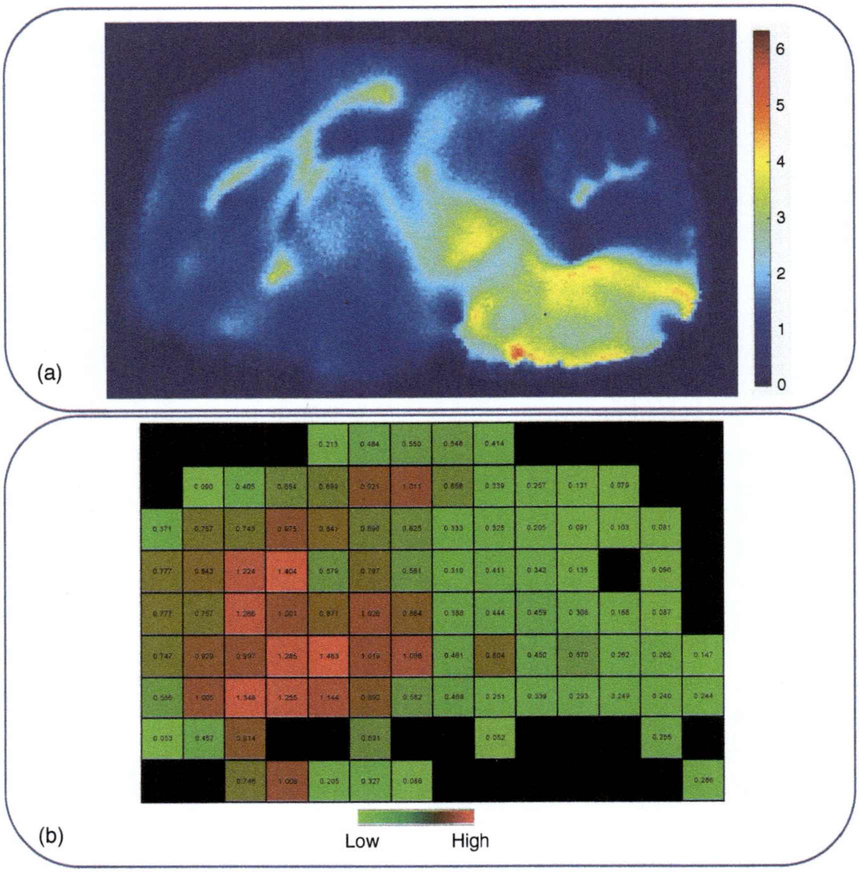

Figure 16.12 MSI of cholesterol and oxysterols from a 10 μm sagittal section of mouse brain. GP derivatization is performed by on-tissue EADSA. (a) MALDI-MSI of cholesterol at 50 μm pixel size performed on an Orbitrap IDX instrument at 100 000 mass resolution with 5 ppm mass accuracy. Cholesterol areal density varies from about 100 ng/mm^2 in the cortex to 300 ng/mm^2 in the midbrain, pons, and medulla. (b) LESA-LC-ESI-MSI of 24S-HC. Pixel size 800 μm, 100 000 mass resolution, and 5 ppm mass accuracy. 24S-HC areal density varies from about 0.1 ng/mm^2 in the cerebellum to 1 ng/mm^2 in the striatum. Quantification was performed against sprayed-on isotope-labeled standards.

16.2.5 Analysis of Steryl Esters

Cholesterol esters, where cholesterol is esterified to a fatty acid, are an abundant class of lipid. They are present in many cells and in blood [47]. They can be analyzed by shotgun MS/MS methods, LC-MS/MS, and by MALDI-MSI. Cholesterol esters are typically ionized as the NH_4^+ adduct and monitored in MS/MS by the fragment ion corresponding to dehydrated protonated cholesterol at *m/z* 369 [7, 110]. When LC is performed, it is often normal phase with a flow of ammonium acetate in aqueous acetonitrile introduced following the column but before the ionization source. Oxysterols linked to fatty acids can be similarly analyzed but by monitoring fragment ions at *m/z* 385 [110].

In MALDI-MSI, cholesterol esters tend to be ionized as Na^+ adducts; however, Na^+ adducts tend to give only a minor fragment ion at m/z 369; instead, the sodiated acyl fragment dominates [110].

16.3 Conclusions

The sterols are an important class of lipid (ST01). In this chapter, we have focused on the analysis of the sub-class cholesterol and derivatives (ST0101). We have emphasized methods targeted at sterols and oxysterols of importance in human biology, but the methods are applicable to these molecules in other species and kingdoms. As the reader will have realized, we have a bias toward methods based on derivatization chemistry, which have been used extensively in our own laboratory; however, methods avoiding derivatization are also very widely used and are excellent when accompanied by sound chromatography and the use of authentic standards.

Acknowledgments

This work was supported by the Biotechnology and Biological Sciences Research Council, the European Union through European Structural Funds (ESF), as part of the Welsh Government funded Academic Expertise for Business project (to WJG and YW). Members of the European Network for Oxysterol Research (ENOR, https://www.oxysterols.net) are thanked for informative discussions.

Conflict of Interest Statement

WJG and YW are listed as inventors on the patent "Kit and method for quantitative detection of steroids" US9851368B2. WJG, EY, and YW are shareholders in CholesteniX Ltd.

References

1 Ryhage, R. and Stenhagen, E. (1960). Mass spectrometry in lipid research. *J. Lipid Res.* 1: 361–390.

2 Fahy, E., Subramaniam, S., Brown, H.A. et al. (2005). A comprehensive classification system for lipids. *J. Lipid Res.* 46 (5): 839–861.

3 Nes, W.D. (2011). Biosynthesis of cholesterol and other sterols. *Chem. Rev.* 111 (10): 6423–6451.

4 Shackleton, C.H. (2012). Role of a disordered steroid metabolome in the elucidation of sterol and steroid biosynthesis. *Lipids* 47 (1): 1–12.

5 Griffiths, W.J. and Sjovall, J. (2010). Bile acids: analysis in biological fluids and tissues. *J. Lipid Res.* 51 (1): 23–41.
6 Russell, D.W. (2003). The enzymes, regulation, and genetics of bile acid synthesis. *Annu. Rev. Biochem.* 72: 137–174.
7 Murphy, R.C., Leiker, T.J., and Barkley, R.M. (2011). Glycerolipid and cholesterol ester analyses in biological samples by mass spectrometry. *Biochim. Biophys. Acta* 1811 (11): 776–783.
8 Lange, Y., Swaisgood, M.H., Ramos, B.V., and Steck, T.L. (1989). Plasma membranes contain half the phospholipid and 90% of the cholesterol and sphingomyelin in cultured human fibroblasts. *J. Biol. Chem.* 264 (7): 3786–3793.
9 van Meer, G., Voelker, D.R., and Feigenson, G.W. (2008). Membrane lipids: where they are and how they behave. *Nat. Rev. Mol. Cell Biol.* 9 (2): 112–124.
10 de Duve, C. (1971). Tissue fraction-past and present. *J. Cell Biol.* 50 (1): 20d–55d.
11 Olkkonen, V.M. and Hynynen, R. (2009). Interactions of oxysterols with membranes and proteins. *Mol. Aspects Med.* 30 (3): 123–133.
12 Trinh, M.N., Brown, M.S., Goldstein, J.L. et al. (2020). Last step in the path of LDL cholesterol from lysosome to plasma membrane to ER is governed by phosphatidylserine. *Proc. Natl. Acad. Sci. U.S.A.* 117 (31): 18521–18529.
13 Sandhu, J., Li, S., Fairall, L. et al. (2018). Aster proteins facilitate nonvesicular plasma membrane to ER cholesterol transport in mammalian cells. *Cell* 175 (2): 514–29.e20.
14 Goldstein, J.L. and Brown, M.S. (2009). The LDL receptor. *Arterioscler. Thromb. Vasc. Biol.* 29 (4): 431–438.
15 Goldstein, J.L., DeBose-Boyd, R.A., and Brown, M.S. (2006). Protein sensors for membrane sterols. *Cell* 124 (1): 35–46.
16 Horton, J.D., Goldstein, J.L., and Brown, M.S. (2002). SREBPs: activators of the complete program of cholesterol and fatty acid synthesis in the liver. *J. Clin. Invest.* 109 (9): 1125–1131.
17 Radhakrishnan, A., Goldstein, J.L., McDonald, J.G., and Brown, M.S. (2008). Switch-like control of SREBP-2 transport triggered by small changes in ER cholesterol: a delicate balance. *Cell Metab.* 8 (6): 512–521.
18 Lange, Y. and Steck, T.L. (2020). Active cholesterol 20 years on. *Traffic* 21 (11): 662–674.
19 Radhakrishnan, A., Rohatgi, R., and Siebold, C. (2020). Cholesterol access in cellular membranes controls hedgehog signaling. *Nat. Chem. Biol.* 16 (12): 1303–1313.
20 Das, A., Brown, M.S., Anderson, D.D. et al. (2014). Three pools of plasma membrane cholesterol and their relation to cholesterol homeostasis. *Elife* 3.
21 Dang, E.V., McDonald, J.G., Russell, D.W., and Cyster, J.G. (2017). Oxysterol restraint of cholesterol synthesis prevents AIM2 inflammasome activation. *Cell* 171 (5): 1057–71 e11.
22 Zhou, Q.D., Chi, X., Lee, M.S. et al. (2020). Interferon-mediated reprogramming of membrane cholesterol to evade bacterial toxins. *Nat. Immunol.* 21 (7): 746–755.

23 Abrams, M.E., Johnson, K.A., Perelman, S.S. et al. (2020). Oxysterols provide innate immunity to bacterial infection by mobilizing cell surface accessible cholesterol. *Nat. Microbiol.* 5 (7): 929–942.

24 Ormsby, T.J.R., Owens, S.E., Horlock, A.D. et al. (2021). Oxysterols protect bovine endometrial cells against pore-forming toxins from pathogenic bacteria. *FASEB J.* 35 (10): e21889.

25 Wang, S., Li, W., Hui, H. et al. (2020). Cholesterol 25-hydroxylase inhibits SARS-CoV-2 and other coronaviruses by depleting membrane cholesterol. *EMBO J.* e106057.

26 Gill, S., Chow, R., and Brown, A.J. (2008). Sterol regulators of cholesterol homeostasis and beyond: the oxysterol hypothesis revisited and revised. *Prog. Lipid Res.* 47 (6): 391–404.

27 Griffiths, W.J. and Wang, Y. (2021). Cholesterol metabolism: from lipidomics to immunology. *J. Lipid Res.* 100165.

28 Radhakrishnan, A., Ikeda, Y., Kwon, H.J. et al. (2007). Sterol-regulated transport of SREBPs from endoplasmic reticulum to Golgi: oxysterols block transport by binding to Insig. *Proc. Natl. Acad. Sci. U.S.A.* 104 (16): 6511–6518.

29 Cheng, D., Chang, C.C.Y., Qu, X.-m., and Chang, T.-Y. (1995). Activation of acyl-coenzyme A: cholesterol acyltransferase by cholesterol or by oxysterol in a cell-free system. *J. Biol. Chem.* 270 (2): 685–695.

30 Chen, L., Ma, M.Y., Sun, M. et al. (2019). Endogenous sterol intermediates of the mevalonate pathway regulate HMGCR degradation and SREBF-2 processing. *J. Lipid Res.* 60 (10): 1765–1775.

31 Wang, B. and Tontonoz, P. (2018). Liver X receptors in lipid signalling and membrane homeostasis. *Nat. Rev. Endocrinol.* 14 (8): 452–463.

32 Nachtergaele, S., Mydock, L.K., Krishnan, K. et al. (2012). Oxysterols are allosteric activators of the oncoprotein smoothened. *Nat. Chem. Biol.* 8 (2): 211–220.

33 Kinnebrew, M., Iverson, E.J., Patel, B.B. et al. (2019). Cholesterol accessibility at the ciliary membrane controls hedgehog signaling. *Elife* 8.

34 Raleigh, D.R., Sever, N., Choksi, P.K. et al. (2018). Cilia-associated oxysterols activate smoothened. *Mol. Cell* 72 (2): 316–27 e5.

35 Abdel-Khalik, J., Hearn, T., Dickson, A.L. et al. (2020). Bile acid biosynthesis in Smith-Lemli-Opitz syndrome bypassing cholesterol: potential importance of pathway intermediates. *J. Steroid Biochem. Mol. Biol.* 206: 105794.

36 Hannedouche, S., Zhang, J., Yi, T. et al. (2011). Oxysterols direct immune cell migration via EBI2. *Nature* 475 (7357): 524–527.

37 Liu, C., Yang, X.V., Wu, J. et al. (2011). Oxysterols direct B-cell migration through EBI2. *Nature* 475 (7357): 519–523.

38 Cyster, J.G., Dang, E.V., Reboldi, A., and Yi, T. (2014). 25-Hydroxycholesterols in innate and adaptive immunity. *Nat. Rev. Immunol.* 14 (11): 731–743.

39 de Medina, P., Diallo, K., Huc-Claustre, E. et al. (2020). The 5,6-epoxycholesterol metabolic pathway in breast cancer: emergence of new pharmacological targets. *Br. J. Pharmacol.* 178 (16): 3248–3260.

40 de Medina, P., Paillasse, M.R., Segala, G. et al. (2013). Dendrogenin A arises from cholesterol and histamine metabolism and shows cell differentiation and anti-tumour properties. *Nat. Commun.* 4: 1840.

41 Griffiths, W.J., Yutuc, E., Abdel-Khalik, J. et al. (2019). Metabolism of non-enzymatically derived oxysterols: clues from sterol metabolic disorders. *Free Radic. Biol. Med.* 144: 124–133.

42 Sever, N., Mann, R.K., Xu, L. et al. (2016). Endogenous B-ring oxysterols inhibit the Hedgehog component smoothened in a manner distinct from cyclopamine or side-chain oxysterols. *Proc. Natl. Acad. Sci. U.S.A.* 113 (21): 5904–5909.

43 Brown, A.J., Ikonen, E., and Olkkonen, V.M. (2014). Cholesterol precursors: more than mere markers of biosynthesis. *Curr. Opin Lipidol.* 25 (2): 133–139.

44 Weingärtner, O., Pinsdorf, T., Rogacev, K.S. et al. (2010). The relationships of markers of cholesterol homeostasis with carotid intima-media thickness. *PLoS One* 5 (10): e13467.

45 Yang, C., McDonald, J.G., Patel, A. et al. (2006). Sterol intermediates from cholesterol biosynthetic pathway as liver X receptor ligands. *J. Biol. Chem.* 281 (38): 27816–27826.

46 Dzeletovic, S., Breuer, O., Lund, E., and Diczfalusy, U. (1995). Determination of cholesterol oxidation products in human plasma by isotope dilution-mass spectrometry. *Anal. Biochem.* 225 (1): 73–80.

47 Quehenberger, O., Armando, A.M., Brown, A.H. et al. (2010). Lipidomics reveals a remarkable diversity of lipids in human plasma. *J. Lipid Res.* 51 (11): 3299–3305.

48 Yutuc, E., Dickson, A.L., Pacciarini, M. et al. (2021). Deep mining of oxysterols and cholestenoic acids in human plasma and cerebrospinal fluid: quantification using isotope dilution mass spectrometry. *Anal. Chim. Acta* 1154: 338259.

49 McDonald, J.G., Smith, D.D., Stiles, A.R., and Russell, D.W. (2012). A comprehensive method for extraction and quantitative analysis of sterols and secosteroids from human plasma. *J. Lipid Res.* 53 (7): 1399–1409.

50 Saeed, A., Floris, F., Andersson, U. et al. (2014). 7alpha-hydroxy-3-oxo-4-cholestenoic acid in cerebrospinal fluid reflects the integrity of the blood-brain barrier. *J. Lipid Res.* 55 (2): 313–318.

51 Clayton, P.T. (2011). Disorders of bile acid synthesis. *J. Inherit. Metab. Dis.* 34 (3): 593–604.

52 Heubi, J.E., Setchell, K.D., and Bove, K.E. (2007). Inborn errors of bile acid metabolism. *Semin. Liver Dis.* 27 (3): 282–294.

53 Jiang, X., Sidhu, R., Mydock-McGrane, L. et al. (2016). Development of a bile acid-based newborn screen for Niemann–Pick disease type C. *Sci. Transl. Med.* 8 (337): 337ra63.

54 Honda, A., Yamashita, K., Miyazaki, H. et al. (2008). Highly sensitive analysis of sterol profiles in human serum by LC-ESI-MS/MS. *J. Lipid Res.* 49 (9): 2063–2073.

55 Honda, A., Yamashita, K., Hara, T. et al. (2009). Highly sensitive quantification of key regulatory oxysterols in biological samples by LC-ESI-MS/MS. *J. Lipid Res.* 50 (2): 350–357.

56 Schule, R., Siddique, T., Deng, H.X. et al. (2010). Marked accumulation of 27-hydroxycholesterol in SPG5 patients with hereditary spastic paresis. *J. Lipid Res.* 51 (4): 819–823.

57 Kudo, K., Emmons, G.T., Casserly, E.W. et al. (1989). Inhibitors of sterol synthesis. Chromatography of acetate derivatives of oxygenated sterols. *J. Lipid Res.* 30 (7): 1097–1111.

58 Karu, K., Hornshaw, M., Woffendin, G. et al. (2007). Liquid chromatography-mass spectrometry utilizing multi-stage fragmentation for the identification of oxysterols. *J. Lipid Res.* 48 (4): 976–987.

59 Schott, H.F. and Lutjohann, D. (2015). Validation of an isotope dilution gas chromatography-mass spectrometry method for combined analysis of oxysterols and oxyphytosterols in serum samples. *Steroids* 99 (Pt B): 139–150.

60 Jiang, X., Ory, D.S., and Han, X. (2007). Characterization of oxysterols by electrospray ionization tandem mass spectrometry after one-step derivatization with dimethylglycine. *Rapid Commun. Mass Spectrom.* 21 (2): 141–152.

61 Sidhu, R., Jiang, H., Farhat, N.Y. et al. (2015). A validated LC-MS/MS assay for quantification of 24(*S*)-hydroxycholesterol in plasma and cerebrospinal fluid. *J. Lipid Res.* 56 (6): 1222–1233.

62 Jiang, X., Sidhu, R., Porter, F.D. et al. (2011). A sensitive and specific LC-MS/MS method for rapid diagnosis of Niemann-Pick C1 disease from human plasma. *J. Lipid Res.* 52 (7): 1435–1445.

63 Stiles, A.R., Kozlitina, J., Thompson, B.M. et al. (2014). Genetic, anatomic, and clinical determinants of human serum sterol and vitamin D levels. *Proc. Natl. Acad. Sci. U.S.A.* 111 (38): E4006–E4014.

64 Björkhem, I., Blomstrand, R., and Svensson, L. (1974). Serum cholesterol determination by mass fragmentography. *Clin. Chim. Acta* 54 (2): 185–193.

65 Lutjohann, D., Bjorkhem, I., Friedrichs, S. et al. (2019). First international descriptive and interventional survey for cholesterol and non-cholesterol sterol determination by gas- and liquid-chromatography-urgent need for harmonisation of analytical methods. *J. Steroid Biochem. Mol. Biol.* 190: 115–125.

66 Acimovic, J., Lovgren-Sandblom, A., Monostory, K. et al. (2009). Combined gas chromatographic/mass spectrometric analysis of cholesterol precursors and plant sterols in cultured cells. *J. Chromatogr. B Analyt. Technol. Biomed. Life Sci.* 877 (22): 2081–2086.

67 Lutjohann, D., Breuer, O., Ahlborg, G. et al. (1996). Cholesterol homeostasis in human brain: evidence for an age-dependent flux of 24*S* -hydroxycholesterol from the brain into the circulation. *Proc. Natl. Acad. Sci. U.S.A.* 93 (18): 9799–9804.

68 Heverin, M., Bogdanovic, N., Lutjohann, D. et al. (2004). Changes in the levels of cerebral and extracerebral sterols in the brain of patients with Alzheimer's disease. *J. Lipid Res.* 45 (1): 186–193.

69 Wu, Q., Ishikawa, T., Sirianni, R. et al. (2013). 27-hydroxycholesterol promotes cell-autonomous, ER-positive breast cancer growth. *Cell Rep.* 5 (3): 637–645.

70 Reboldi, A., Dang, E.V., McDonald, J.G. et al. (2014). Inflammation. 25-Hydroxycholesterol suppresses interleukin-1-driven inflammation downstream of type I interferon. *Science* 345 (6197): 679–684.

71 Qi, X., Liu, H., Thompson, B. et al. (2019). Cryo-EM structure of oxysterol-bound human smoothened coupled to a heterotrimeric Gi. *Nature* 571 (7764): 279–283.

72 Klinke, G., Rohrbach, M., Giugliani, R. et al. (2015). LC-MS/MS based assay and reference intervals in children and adolescents for oxysterols elevated in Niemann-Pick diseases. *Clin. Biochem.* 48 (9): 596–602.

73 Pajares, S., Arias, A., Garcia-Villoria, J. et al. (2015). Cholestane-3beta,5alpha,6beta-triol: high levels in Niemann-Pick type C, cerebrotendinous xanthomatosis, and lysosomal acid lipase deficiency. *J. Lipid Res.* 56 (10): 1926–1935.

74 Potter, J.E., Petts, G., Ghosh, A. et al. (2021). Enzyme replacement therapy and hematopoietic stem cell transplant: a new paradigm of treatment in Wolman disease. *Orphanet. J. Rare Dis.* 16 (1): 235.

75 Schroepfer, G.J. Jr. (2000). Oxysterols: modulators of cholesterol metabolism and other processes. *Physiol. Rev.* 80 (1): 361–554.

76 Murphy, R.C. and Johnson, K.M. (2008). Cholesterol, reactive oxygen species, and the formation of biologically active mediators. *J. Biol. Chem.* 283 (23): 15521–15525.

77 Iuliano, L. (2011). Pathways of cholesterol oxidation via non-enzymatic mechanisms. *Chem. Phys. Lipids* 164 (6): 457–468.

78 Goyal, S., Xiao, Y., Porter, N.A. et al. (2014). Oxidation of 7-dehydrocholesterol and desmosterol by human cytochrome P450 46A1. *J. Lipid Res.* 55 (9): 1933–1943.

79 Mazzacuva, F., Mills, P., Mills, K. et al. (2016). Identification of novel bile acids as biomarkers for the early diagnosis of Niemann-Pick C disease. *FEBS Lett.* 590 (11): 1651–1662.

80 Girard, A. and Sandulesco, G. (1936). Sur une nouvelle série de réactifs du groupe carbonyle, leur utilisation à l'extraction des substances cétoniques et à la caractérisation microchimique des aldéhydes et cétones. *Helv. Chim. Acta* 19 (1): 1095–1107.

81 Wheeler, O.H. (1968). The Girard reagents. *Chem. Educ.* 45: 6.

82 Shackleton, C.H., Chuang, H., Kim, J. et al. (1997). Electrospray mass spectrometry of testosterone esters: potential for use in doping control. *Steroids* 62 (7): 523–529.

83 DeBarber, A.E., Connor, W.E., Pappu, A.S. et al. (2010). ESI-MS/MS quantification of 7α-hydroxy-4-cholesten-3-one facilitates rapid, convenient diagnostic testing for cerebrotendinous xanthomatosis. *Clin. Chim. Acta* 411 (1): 43–48.

84 DeBarber, A.E., Sandlers, Y., Pappu, A.S. et al. (2011). Profiling sterols in *Cerebrotendinous xanthomatosis*: utility of Girard derivatization and high resolution exact mass LC-ESI-MS(n) analysis. *J. Chromatogr. B Analyt. Technol. Biomed. Life Sci.* 879 (17, 18): 1384–1392.

85 Lavrynenko, O., Nedielkov, R., Möller, H.M., and Shevchenko, A. (2013). Girard derivatization for LC-MS/MS profiling of endogenous ecdysteroids in Drosophila. *J. Lipid Res.* 54 (8): 2265–2272.

86 MacLachlan, J., Wotherspoon, A.T., Ansell, R.O., and Brooks, C.J. (2000). Cholesterol oxidase: sources, physical properties and analytical applications. *J. Steroid Biochem. Mol. Biol.* 72 (5): 169–195.

87 Zhang, Z., Li, D., Blanchard, D.E. et al. (2001). Key regulatory oxysterols in liver: analysis as delta4-3-ketone derivatives by HPLC and response to physiological perturbations. *J. Lipid Res.* 42 (4): 649–658.

88 Roberg-Larsen, H., Lund, K., Vehus, T. et al. (2014). Highly automated nano-LC/MS-based approach for thousand cell-scale quantification of side chain-hydroxylated oxysterols. *J. Lipid Res.* 55 (7): 1531–1536.

89 Solheim, S., Hutchinson, S.A., Lundanes, E. et al. (2019). Fast liquid chromatography-mass spectrometry reveals side chain oxysterol heterogeneity in breast cancer tumour samples. *J. Steroid Biochem. Mol. Biol.*

90 Hutchinson, S.A., Websdale, A., Cioccoloni, G. et al. (2021). Liver x receptor alpha drives chemoresistance in response to side-chain hydroxycholesterols in triple negative breast cancer. *Oncogene* 40 (16): 2872–2883.

91 Soroosh, P., Wu, J., Xue, X. et al. (2014). Oxysterols are agonist ligands of RORgammat and drive Th17 cell differentiation. *Proc. Natl. Acad. Sci. U.S.A.* 111 (33): 12163–12168.

92 Russo, L., Muir, L., Geletka, L. et al. (2020). Cholesterol 25-hydroxylase (CH25H) as a promoter of adipose tissue inflammation in obesity and diabetes. *Mol. Metab.* 39: 100983.

93 Wang, Y., Hornshaw, M., Alvelius, G. et al. (2006). Matrix-assisted laser desorption/ionization high-energy collision-induced dissociation of steroids: analysis of oxysterols in rat brain. *Anal. Chem.* 78 (1): 164–173.

94 Yutuc, E., Angelini, R., Baumert, M. et al. (2020). Localization of sterols and oxysterols in mouse brain reveals distinct spatial cholesterol metabolism. *Proc. Natl. Acad. Sci. U.S.A.* 117 (11): 5749–5760.

95 Angelini, R., Yutuc, E., Wyatt, M.F. et al. (2021). Visualizing cholesterol in the brain by on-tissue derivatization and quantitative mass spectrometry imaging. *Anal. Chem.* 93 (11): 4932–4943.

96 Griffiths, W.J., Abdel-Khalik, J., Crick, P.J. et al. (2017). Sterols and oxysterols in plasma from Smith-Lemli-Opitz syndrome patients. *J. Steroid Biochem. Mol. Biol.* 169: 77–87.

97 Crick, P.J., William Bentley, T., Abdel-Khalik, J. et al. (2015). Quantitative charge-tags for sterol and oxysterol analysis. *Clin. Chem.* 61 (2): 400–411.

98 Blanc, M., Hsieh, W.Y., Robertson, K.A. et al. (2013). The transcription factor STAT-1 couples macrophage synthesis of 25-hydroxycholesterol to the interferon antiviral response. *Immunity* 38 (1): 106–118.

99 Höflinger, P., Hauser, S., Yutuc, E. et al. (2021). Metabolic profiling in serum, cerebrospinal fluid and brain of patients with *Cerebrotendinous xanthomatosis*. *J. Lipid Res.* 62: 100078.

100 Raselli, T., Hearn, T., Wyss, A. et al. (2019). Elevated oxysterol levels in human and mouse livers reflect nonalcoholic steatohepatitis. *J. Lipid Res.* 60 (7): 1270–1283.

101 Griffiths, W.J., Liu, S., Alvelius, G., and Sjovall, J. (2003). Derivatisation for the characterisation of neutral oxosteroids by electrospray and matrix-assisted laser desorption/ionisation tandem mass spectrometry: the Girard P derivative. *Rapid Commun. Mass Spectrom.* 17 (9): 924–935.

102 Griffiths, W.J., Wang, Y., Alvelius, G. et al. (2006). Analysis of oxysterols by electrospray tandem mass spectrometry. *J. Am. Soc. Mass Spectrom.* 17 (3): 341–362.

103 Abdel-Khalik, J., Crick, P.J., Yutuc, E. et al. (2018). Identification of 7alpha, 24-dihydroxy-3-oxocholest-4-*en*-26-oic and 7alpha,25-dihydroxy-3-oxocholest-4-*en*-26-oic acids in human cerebrospinal fluid and plasma. *Biochimie* 153: 86–98.

104 Higashi, T., Shimada, K., and Toyo'oka, T. (2010). Advances in determination of vitamin D related compounds in biological samples using liquid chromatography-mass spectrometry: a review. *J. Chromatogr. B Analyt. Technol. Biomed. Life Sci.* 878 (20): 1654–1661.

105 Liu, W., Xu, L., Lamberson, C. et al. (2014). A highly sensitive method for analysis of 7-dehydrocholesterol for the study of Smith-Lemli-Opitz syndrome. *J. Lipid Res.* 55 (2): 329–337.

106 Berry, K.A., Hankin, J.A., Barkley, R.M. et al. (2011). MALDI imaging of lipid biochemistry in tissues by mass spectrometry. *Chem. Rev.* 111 (10): 6491–6512.

107 Cobice, D.F., Livingstone, D.E., Mackay, C.L. et al. (2016). Spatial localization and quantitation of androgens in mouse testis by mass spectrometry imaging. *Anal. Chem.* 88 (21): 10362–10367.

108 Shimma, S., Kumada, H.O., Taniguchi, H. et al. (2016). Microscopic visualization of testosterone in mouse testis by use of imaging mass spectrometry. *Anal. Bioanal. Chem.* 408 (27): 7607–7615.

109 Barre, F.P., Flinders, B., Garcia, J.P. et al. (2016). Derivatization strategies for the detection of triamcinolone acetonide in cartilage by using matrix-assisted laser desorption/ionization mass spectrometry imaging. *Anal. Chem.* 88 (24): 12051–12059.

110 Hutchins, P.M., Moore, E.E., and Murphy, R.C. (2011). Electrospray MS/MS reveals extensive and nonspecific oxidation of cholesterol esters in human peripheral vascular lesions. *J. Lipid Res.* 52 (11): 2070–2083.

17

Bile Acids

Sebastian Simstich[1] and Günter Fauler[2]

[1] *Medical University of Graz, Clinical Institute of Medical and Chemical Laboratory Diagnostics, Auenbruggerplatz 34, 8036, Graz, Austria*
[2] *University Hospital Graz, Clinical Institute of Medical and Chemical Laboratory Diagnostics, Auenbruggerplatz 34, 8036, Graz, Austria*

17.1 Introduction

Bile acids (BAs) are a group of cholesterol metabolites with different combinations of hydroxyl groups at the cholestane-backbone and a carboxyl group at the end of an aliphatic side chain, on carbon atom 24 in human, or up to carbon atom 27 in lower vertebrates. This terminal carboxylic acid group can be further conjugated with the amino acids glycine (G) or taurine (T) (Figure 17.1). This special structure leads to a characteristic amphiphilic nature, allowing the promotion of micelle formation and thus digestion of dietary triacylglycerols and further complex lipids, as well as their delivery to the liver [1, 2]. In humans, the entire BA pool of about 2–5 g is composed of individual inputs of biosynthesis, dietary intake, and reabsorption.

In human hepatocytes BAs normally are synthesized via two pathways. Starting from cholesterol, synthesis takes place in multiple intracellular compartments of hepatocytes in a multi-step process (at least 17 individual enzymes are involved), resulting in the formation of the two "primary" BAs cholic acid (CA) and chenodeoxycholic acid (CDCA), as well as their conjugation.

The "classical," or "neutral," pathway is initiated by cytochrome CYP7A1 in the endoplasmic reticulum, forming in a first step 7α-hydroxy-cholesterol. After a series of different enzymatic reactions, via the important precursor C4 (7α-hydroxy-4-cholesten-3-one), mainly CA is generated and subsequently conjugated to a high degree (up to 98%). A second, "alternative" or "acidic," pathway starts with hydroxylation at position 24 and mainly produces CDCA, but, in comparison to CA, normally to a minor extent. However, this process can be upregulated, when the classical pathway is obstructed [3].

Mass Spectrometry for Lipidomics: Methods and Applications, First Edition.
Edited by Michal Holčapek and Kim Ekroos.

After transportation of BAs from hepatocytes into the gallbladder, where they are intermediately stored, a postprandial contraction allows their release into the duodenum, where they can exert their main physiological function. In the intestinal tract, the two primary BAs can be bacterially modified (through various deconjugation, dehydroxylation, and epimerization processes) to unconjugated "secondary" BAs, mainly deoxycholic acid (DCA) and ursodeoxycholic acid (UDCA), originating from CA, as well as lithocholic acid (LCA), from CDCA. In mouse intestine, however, BAs can be also converted to taurohyocholic acid (THCA), o-muricholic acid, taurohyodeoxycholic acid (THDCA), and tauroursodeoxycholic acid (TUDCA) [4].

To a major part (about 95%) BAs are then reabsorbed, together with alimentary lipids and fat-soluble vitamins, in order to enter an enterohepatic circulation back to hepatocytes where they can be recycled, restored, and reconjugated. Only a minor degree of the human BA pool is excreted with stool.

As BAs circulate back to the liver, they are also target of the so-called hepatic first-pass effect. This can result in further sulfation and/or glucuronidation (mainly, but not exclusively, on position 3, ["R," in Figure 17.1]), to increase their water solubility for a better urinary excretion, which, however, normally remains a minor part of human BA elimination. These conjugated products sometimes are called "tertiary" BAs [2].

The synthesis of BAs, starting from cholesterol and subsequent partial excretion, is the major route for the elimination of cholesterol from the human body, but by far not the only function of BAs as metabolic regulators.

Due to their amphiphilic nature, BAs are able to interact with apolar lipids, like triacylglycerols, cholesterol esters, and fat-soluble vitamins (–A, –D, –E, –K), by the formation of lipid micelles in the surrounding aqueous medium. This process facilitates the digestion of dietary triacylglycerols due to a better access to pancreatic lipases.

While intermediately stored in the gallbladder, BAs, together with phospholipids, also execute their function to solubilize cholesterol, thus preventing its precipitation to gall stones.

In the last decades, more and more insights into the biological activities of BAs have been elucidated. They are not only involved in the control of their own metabolism and transport via the enterohepatic circulation, but they also act as metabolically active signaling molecules, thus providing signals to coordinate triacylglycerol, glucose, and energy homeostasis [5, 6]. Furthermore, they are also part of numerous different receptor activities.

Due to the rapid growing knowledge in signaling capacities, some "axes" have been defined in the literature, like the gut–brain axis, where alterations in the gut microbiome can have an impact to the cognitive status of patients. Also in this context, data from a metabolic network analysis have been published from post mortem brain samples of Alzheimer's disease patients, showing that BA synthesis and cholesterol metabolism, as well as taurine transport, develop differences in comparison to cognitively normal individuals [7, 8]. Further examples of BA signaling resulted also in a gut–liver axis, or a gut–kidney axis, where, among others, also cardiac effects can be observed. At the end, a so-called "gut–organ axis" (or "BA-mediated

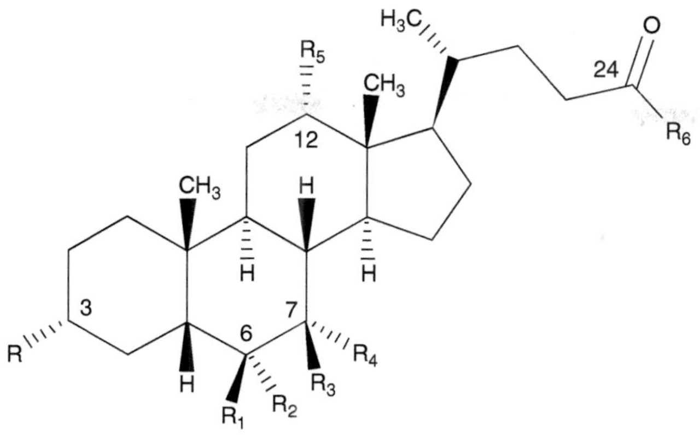

	R	R1	R2	R3	R4	R5	R6
CA	OH	H	H	H	OH	OH	OH
CDCA	OH	H	H	H	OH	H	OH
DCA	OH	H	H	H	H	OH	OH
LCA	OH	H	H	H	H	H	OH
UDCA	OH	H	H	OH	H	H	OH
HDCA	OH	H	OH	H	H	H	OH
α-MCA	OH	OH	H	H	OH	H	OH
β-MCA	OH	OH	H	OH	H	H	OH
γ-MCA (HCA)	OH	H	OH	H	OH	H	OH
o-MCA	OH	H	OH	OH	H	H	OH
Glyco-Conj							$-NHCH_2COOH$
Tauro-Conj							$-NH(CH_2)_2SO_3H$
Sulfated	$HOSO_2O-$						
Glucuronidated	$C_6H_2O_6O-$						

Figure 17.1 Molecular structure and composition of usual human and rodent BA isoforms. Abbreviations: cholic acid (CA), chenodeoxycholic acid (CDCA), deoxycholic acid (DCA), lithocholic acid (LCA), ursodeoxycholic acid (UDCA), hyodeoxycholic acid (HDCA), hyocholic acid (HCA), muricholic acids (MCA, alpha [α], beta [β], gamma [γ], omega [o]).

host–gut microbiota crosstalk" [9]) can be defined as a summary of all of these effects [10, 11].

Moreover, in a recent review [12] the authors raised the question, whether some intermediates in the BA synthesis are even more interesting and important in signaling than the BAs itself, which opens a further dimension in the area of BA research and analysis. Using BAs as biomarker for the determination of abnormalities in BA metabolism, or for dysregulation of BA transport, as well as for impaired receptor signaling, can lead to an association to different diseases, like intrahepatic cholestasis, non-diabetic metabolic syndrome, diabetes mellitus type II, or non-alcoholic fatty liver disease (NAFLD), to mention only a few of them.

A further approach in the clinical interest of BA determination is the use of UDCA as a medication in NAFLD patients, or in pregnant women, developing intrahepatic cholestasis [13]. Data demonstrate that UDCA treatment strongly affects cholesterol and BA synthesis [14]. This also indicates that BA determination can be a valuable part in the diagnosis, as well as in the follow-up, of many different diseases.

Another, quite different, but widely used application of BA determination by mass spectrometry (MS) is traditional Chinese medicine (TCM), both in detection of diverse components in bile from different animals, used for TCM therapy [15, 16], and control of human BAs under a certain TCM therapy [17].

Finally, since the recent progress in microbiome research, which initiated the definition of the gut–organ–, or host–gut–axis, together with the development of even more sensitive and accurate MS instrumentation, the interest in determination of all kinds of BAs again has led to an increasing importance of BA analysis, not only in healthcare and but also in fundamental metabolic and analytical research. This resulted in the number of about 3000 publications, dealing with bile acids and MS since the beginning of mass spectrometric BA-analytics in the 1960s, and, out of these, in more than 1400 contributions in the last decade.

Considering the biological and clinical importance of BAs, a reliable and efficient technology for a robust and precise analysis is important to reveal all their physiological roles. However, the development of a sensitive and accurate analytical methodology still is a challenge due to huge differences in chemical properties of BAs, the broad spectrum of biological concentrations (differences of more than six decades), as well as the molecular diversity and complexity of the biological materials [2, 18].

17.2 Analytical Methods and Applications

Since the beginning of BA analysis using MS in the 1960s, numerous publications have been released, dealing with all kinds of determination of usual human and animal BAs, as well as of unusual and rare species, plus BA precursors and metabolites, all extracted from a huge number of different materials. In order to reach an organization in the cavalcade of contributions, here, in conjunction with the development of analytical MS instrumentation, a chronological release is reflected in the course of scientific papers and reviews.

Moreover, to get a general regime of the interesting methodology, we decided to split remarkable articles into three main subgroups, namely gas chromatography mass spectrometry (GC-MS), liquid chromatography mass spectrometry (LC-MS), and supercritical fluid chromatography (SFC). As LC-MS is the more complex methodology, there will be a further partition in triple-quad technology and high-resolution techniques.

Table 17.1 gives a very brief overview of a few of different selected methods and applications, out of the unmanageable long list of contributions, together with a

Table 17.1 Overview of selected methods and applications for BA-analysis, together with indications to sample preparation from different matrices and a subjective comment of the authors, to highlight outstanding features of these papers.

Instruments	Sample preparation	Method	Material	Comments	References
GC-MS					
	Preparative TLC	Silica-packed column	Human feces	Derivatization of BAs, comparison with standards	[19]
			Human biol. Fluids	Critical **summary** of methods	[20]
	Lse, liquid–gel chrom.	**Review**	Serum	Patients with liver disease	[21]
		Review	Biological fluids	Quantitative determination	[22]
		Review of MS methods	Biol. fluids and tissues	GC-MS LC-MS	[23]
	Spe	Chemical ionization Tandem MS	Animal tissue	Camphor −10-sulfonic acid as catalyst	[24]
	Lle, Two step derivatization	EI, Tandem MS	Urine	High stability and abundance of fragments	[25]
LC-MS					
		FAB		Magnetic field, electric field	[26]
	Spe, anion exchange	FAB, ESI	Urine	Capillary column-LC-MS	[27]
		Review of methods	Biological samples		[28]
	Ethanol extract	LC-ESI-MS	Human feces	Inflammatory bowel disease	[29]
		Drift tube ion mobility		Subclasses of BAs, nitrogen drift gas	[30]
		MALDI-Imaging	Rat liver tissue	Ischemia	[31]
		Review, MALDI-MS	Mammalian systems	Inborn errors, Girard reagent	[32]

(Continued)

Table 17.1 (Continued)

Instruments	Sample preparation	Method	Material	Comments	References
Triple-Quad					
		Review of methods	Biological samples	Fatty acids, BAs, steroids	[33]
		LC-ESI-MS/MS	Human plasma	Human adult normal values	[34]
	Spe, protein precipitation	LC-MS/MS	Mouse tissues and fluids	Mouse BAs,	[35]
		LC-ESI-MS/MS, SRM	Animal bile, gallstone	TCM	[16]
	Dbs extraction	LC-ESI-MS/MS	Dried blood spots	Newborn,	[36]
		LC-ESI-MS/MS	Rat plasma, liver, bile, intestine	BA homeostasis in enterohepatic circulation	[37]
		UHPLC-QqQ-MS/MS, linear ion trap, MRM	Snake bile	TCM	[38]
	Enzyme digestion	LC-ESI-MS/MS QTOF-MS	Serum, urine	70 Unconjugated Bas	[9]
		Review	Human biol. Samples	BAs analysis from 2000 to 2020	[39]
	Trap column	LC-ESI-MS/MS	Human urine	Nieman-Pick disease.	[40]
	Derivatization	LC-ESI-MS/MS	Rat serum	Over 200 BAs	[41]
HR					
		HRMS		Chem. synthesis of BAs, IS	[42]
		LC-HR-MS/MS	Rat urine, feces	491 BAs 4D data mining	[43]
FT/ORBITRAP					
	Protein precipitation	LC-ESI-HRMS, full scan	Human plasma, urine, feces, bile	Formic acid adducts of unconjugated BAs	[44]

(Continued)

Table 17.1 (Continued)

Instruments	Sample preparation	Method	Material	Comments	References
		ESI-LTQ-FTMS in full scan	Biol. fluids and tissues	17 BAs plus conjugates	[45]
		LC-ESI-HRMS	Biol. fluids mult. species	50 BAs in high-throughput-mode	[46]
		LC-LTQ-ORBI-MS	Rat bile	Age-dependent changes	[47]
		LC-HR-MS	Human plasma	Liver diseases	[48]
		LC-QExactive-ORBI-MS, statistical analysis	Rat	T2DM, Ginseng berry	[49]
		LC-LTQ-ORBI-MS, statistical analysis	Rat-liver	hepatotoxicity	[50]
	Protein precipitation	LC-QExactive-ORBI-MS	Human serum	Reference values, children	[51]
TOF					
		LC-TOF-MS	Rat plasma	Liver injuries	[52]
	Methanol precipitation	LC-TOF-MS LC-MS/MS	Biofluids	145 BAs	[18]
	Hollow fiber micro extract.	LC-ion trap-TOF	Human plasma	Hepatotoxic compounds	[53]
		LC-ESI-QTOF-MS/MS	Snake bile	TCM	[15]
		Multiplex HR-triple TOF-MS	Plasma	sulfate/ glucuronide	[54]
		LC-QTOF-MS	Rat	Tibet Kefir Milk	[55]
SFC					
		Cyanopropyl-bonded silica column	Bile, hepatobiliary diseases	Methodological study	[56]
		SFC-ESI-MS/MS	Rat serum	24 BAs	[57]
		UHR-SFC-MS		Method description	[58]
		Review	Clinical applications	Lipid biomarker	[59]

subjective comment of the authors, to highlight outstanding features of these papers. For detailed information the reader is kindly referred to relevant review articles of the particular fields.

Before starting to implement new BA-analytics, a final selection of a distinctive methodology for BA-determination depends on different criteria. The first question is whether there is an interest in all over qualitative data of an untargeted approach, or one concerns in quantitative results of defined targeted compounds. In the first case, the interest is directed in more complex spectral data of the unknown compounds, to eventually get deeper information about their structure. Therefore, an instrumentation with relative strong ionization technology that delivers spectral data, will be selected. Electron impact (EI) ionization is the choice when using GC-MS, or high-resolution full-scan measurement, eventually combined with different fragmentation experiments, will be considered in LC-MS detection. If quantitative results are of importance, the choice will go toward the use of softer ionization processes, like chemical ionization (CI) in GC-MS, where less, but more, intense fragments are generated. In LC-MS, ionization will be generated mainly by the use of electro-spray-ionization (ESI), or sometimes atmospheric pressure chemical ionization (APCI) can be used.

For quantitative MS, by the use of triple-quadrupole technology, multiple reaction monitoring (MRM) transitions will be the method of choice, whereas, when using HR instruments, the exact mass of ions from full-scan data can be used also for quantitative results, eventually again in combination with defined fragmentation experiments.

One big advantage of quantitative MS-methods over other detection principles is the possibility of using internal standards (ISs), by introduction of stable isotopes (i.e. deuterium, D, or ^{2}H, ^{13}C, ^{15}N, ^{17}O) into the molecular structure of the interesting molecule.

One disadvantage in the use of stable isotope labeled ISs may be the relative less availability and, sometimes, high costs of commercial products. A knowledge of organic synthesis of distinctive compounds eventually can circumvent these problems [42, 44].

Another big point that has to be discussed is sample preparation. As BAs occur in many different matrices, like blood, bile, urine, stool, cell suspensions, food, and animal tissues, to list only a few of the possibilities, and, as the analytical methods also are manifold, sample preparation is a complex theme and has to be selected carefully, according to the defined analytical problem. A comprehensive discussion of different early techniques for extraction experiments has been published [60], for example, in one of a series of interesting books, which furthermore offer many aspects of BA background and applications of that period.

Among the preferred methods used for sample preparation are simple protein precipitation (using methanol or acetonitrile), liquid–liquid extraction (different organic solvents), ion-exchange extraction, as well as solid-phase extraction, and solid-phase microextraction. For more detailed information, a comprehensive review has been presented by Liu et al. [4].

Finally, regarding substantial and fundamental information of all kinds of chromatography and MS, the interested reader is kindly referred to the appropriate Chapters 2–7 of this book.

17.2.1 Gas Chromatography–Mass Spectrometry (GC-MS)

Fundamental aspects in GC-MS are undecomposed evaporation of target molecules in the gas phase and ionization of these molecules for MS detection. The mainly used ionization processes in GC-MS are electron impact (EI) ionization, or, as an alternative, chemical ionization (CI), under the assistance of distinct reagent gases, such as methane.

As native BAs are, at least partially, present in a conjugated form, a cleavage of these conjugates often is necessary before GC-MS-analysis. For example, G- and T-conjugates usually are hydrolyzed with cholylglycine hydrolase [61, 62], sometimes also applied after pretreatment of ion-exchange purification. BA-glucuronides and -sulfates, however, can be cleaved by β-glucuronidase from *Helix pomatia* [63].

In GC-MS-analysis of BAs there is a strong necessity to derivatize the polar components of the molecules to make them more volatile and/or accessible to better ionization. This can be seen as a disadvantage, because of time-consuming sample preparation, but could also be a challenge for introduction of favorable properties, e.g. for an improved ionization, or for a better separation of similar compounds, like structural isomers. Usual reagents that have been applied for derivatization of hydroxyl-groups are alkyl-silyl compounds, mainly trimethylsilyl-delivering reagents (*N*-Methyl-*N*-(trimethylsilyl)trifluoroacetamide, [MSTFA], or *N,O*-bis-(trimethylsilyl)trifluoroacetamide, [BSTFA]), or higher alkyl-silyl-derivatization reagents. In some cases acetylation also has been used. For derivatization of the carboxyl group, mainly methylation, or butylation, was preferred. More basic information about sample preparation in GC-MS can be found in a comprehensive review of Griffith and Sjövall [23].

In the 1960s, among many others, the same group [19, 64] published GC-MS methods for determination of BAs in human feces, by the use of different methods for derivatization and/or ionization.

Summaries and critical reviews of the quantitation methods from plasma, urine, bile, and stool, which have been used until the end of the century, are given by the groups of Street et al. [20], Setchell and Matsui [21], and Batta and Salen [22], as well as again by Griffith and Sjövall [23].

In the last decade, two remarkable methods [24, 65] have been published, where the authors performed chemical ionization. They now applied tandem MS, hyphenated to GC, instead of single quadrupole MS. The latter work described methods for electron impact ionization, as well as positive and negative chemical ionization in GC-triple quadrupole determination of oxysterols and BAs from plasma samples. In a different approach, Kumar et al. [25] developed an EI-method for analysis of BAs, together with oxysterols, in a two-step-derivatization process, reporting highly improved stability and high abundances of MS-fragment ions.

To sum up, GC-MS at the beginning of the introduction of MS in BA analytics was the method of choice, which then successively was bypassed by LC-MS methods (comprehensive and detailed review of GC-MS as well as LC-MS are also given by Griffith and Sjövall [23, 66]).

But the recent introduction of high-resolution and/or tandem MS instrumentation, hyphenated to modern gas chromatographs, could lead to new challenges in GC-MS analysis of complex mixtures of unconjugated bile acids, also under the aspect of the use of specific derivatization methods. The application of extensive structural information, given by electron impact ionization and fragmentation, and the large number of published reference spectra can lead to a certain renaissance of this technology in the future.

17.2.2 Liquid Chromatography–Mass Spectrometry (LC-MS)

17.2.2.1 Early Technologies and ESI-Quadrupole MS

In LC target-molecules are dissolved in water and/or organic solvents and simultaneously have to be vaporized into the gas phase and ionized to undergo further MS analysis.

Among the first technologies for ionization of molecules by direct infusion (a kind of precursor-technology for "shot-gun"-MS) have been thermospray-ionization and fast atom bombardment (FAB). As an example, a method for FAB-ionization of conjugated BAs and their simultaneous determination via magnetic field plus electric sector field instrumentation was published in 1991 [26], whereas a comparative study, using FAB- or ES-ionization and collision induced dissociation (CID) was presented by Yang et al. [27]. Summaries and applications of these methods are listed by Roda et al. [28, 67].

A very important step in direction of sensitivity and specificity was done by the introduction of soft ionization methods, particularly electrospray ionization (ESI), permitting the use of mass spectrometers as detectors in high-performance liquid chromatography (HPLC). A comprehensive review of these early methods, together with attempts to different fragmentation reactions, including discussion of methods for CID and tandem MS has been published by Griffith and Sjövall [23].

In the following years a huge number of publications, all including LC-MS-methods for BA-analysis, but covering many different aspects of applications and goals have been published. All details of these methods can be found in the comprehensive reviews from Dutta et al. [2] and Griffiths and colleagues [32, 66].

A quite different approach to BA analysis is TCM. As an example, two groups investigated animal bile [16], or snake bile [38], for therapeutic reasons.

A more recent review, covering a broad range of different methods of BA analysis from divers human biological samples of the last 20 years, has been shown by Dosedělová et al. [39].

The detection of biomarkers of different kinds of diseases, at any time, was a driving force for an increasing interest in analysis of BAs, thus pushing the development of state-of-the-art-methods during the decades.

At the beginning, in 1972 Eriksson et al. [68] reported details of neutral steroids in urine and feces of women with intrahepatic cholestasis, whereas Peter

Clayton [69] reviewed the urinary analysis of some inborn errors of BA-synthesis, using ESI-tandem-MS for BAs and bile alcohols.

Also in the search for biomarkers of BA-related diseases, among others, Maekawa et al. [40] discussed new urinary BA-biomarkers for detection of Nieman–Pick type C disease, using a trapping column for concentration and purification of samples, and a second, analytical column, for detection of BAs in ESI-tandem MS.

17.2.2.2 High-Resolution Mass Spectrometry (HR-MS)

HR-MS instruments utilize a complete different system for detection of ions and are either time-of-flight-(TOF)-MS, Fourier transform ion cyclotron resonance (FT-ICR) instruments, or working in Orbitrap-technology. These instruments have the advantage of measuring the exact mass of target ions with or without preceding fragmentation. They further can be combined with one, (or more) quadrupole(s), which offers the opportunity to a more sensitive determination of target ions. However, in case of isobaric molecules, these instruments also require separation by LC before isomers can enter the HR-MS.

As an example, one contribution [45] measured the profile of BAs in biofluids and tissues, using reversed-phase HPLC and ESI-linear ion trap Fourier transform mass spectrometry (LTQ-FTMS) in full scan mode.

In the last few years different works have been published with variable healthcare aspects by the use of Orbitrap-technology. [44, 47–50].

Relatively less data have been shown, regarding reference ranges of serum BAs in healthy children and adolescents [51] as well as in adults [34, 70] (in notes from the Mayo Clinic Laboratories). In the latter contributions, however, researchers used triple-quad technology.

Also the use of variable combinations of LC with TOF instruments is the basis of BA-analysis in a wide field of interest. For example 145 isoforms of BAs could be detected in biofluids [18], or using HR-MS in a high-throughput-mode, 50 different BA-isoforms have been analyzed from different species and matrices [46].

Recent contributions for BA determination, all using TOF-technology, have been described by the groups of Minato et al. [52] (five rodent BAs in rat plasma), de Paiva et al. [53] (hollow fiber-based liquid-phase micro-extraction), Zhang et al. [15] (snake bile in TCM), as well as by Gao et al. [55] (effects of Tibet kefir milk on rat gut microbiota), and by Rago et al. [54], in this case a triple-TOF-instrument was used.

17.2.3 Supercritical Fluid Chromatography (SFC)

SFC is a chromatographic separation technique, complementary to HPLC, which uses a supercritical fluid as mobile phase, such as carbon dioxide, under specially defined conditions. Nowadays SFC is as simple and robust as HPLC, but fractionized collection for MS is more convenient, because the mobile phase evaporates, leaving only the separated molecules and a small volume of polar co-solvent.

In 1992 Scalia and Games [56] published first attempts to analyze BAs, using SFC and a cyanopropyl-bonded silica column in a methodical study.

Later, two groups [57, 58] combined SFC with MS for a comprehensive BA-profiling, whereas a recent review of the method has been published by Gordillo [59].

17.3 Conclusions and Outlook

A common interest in reliable qualitative and quantitative data of a huge, and still increasing, number of BA-isoforms has been clearly demonstrated, but, last not least, due to the improvements in microbiome research during the past years, the inquiry of BA-analysis is a yet rising segment. Therefore, it is still an urgent need to develop proper analytical methods for identification and quantification of individual BAs. Despite the widespread use of ELISAs in clinical laboratories, with the main disadvantage of delivering only data for total BA-concentrations, the need for quantitation of distinct isoforms in research, as well as particular biomarkers of an increasing number of different diseases, is rising. The combination of modern chromatography methods, together with the advantages of MS in nearly every particular analytical specification, provides a good basis for a sophisticated analysis of BA-isoforms in all their structural and physico-chemical diversity. Due to the relative stability of the BA-molecule in several matrices, preanalytics (sample storage and transport) can be managed in the same way, as it is usual for biological samples. Furthermore, as can be seen in Table 17.1, nearly every common chromatography- and MS-instrumentation can be applied, which enables a broad field of analytical possibilities.

Sometimes discussions are emerging in the MS community, whether ion trap (Orbi-) instruments may effect less excellent performance, in comparison to triple quad instruments, especially regarding quantitative experiments. As a contribution to the status in BA-analytics, Figure 17.2 shows a comparison of an identical routine BA-sample, measured in both methods, on an HR-FS-Orbi-instrument, as well as using a MRM-method on a triple quad instrument. Although the total ion chromatogram (TIC) of the HR-FS-experiment on the Orbi instrument seems to indicate a higher background, the chromatograms of the extracted HR-ions show a quite comparable performance to the pseudo-MRMs ($[M-H]^- \rightarrow [M-H]^-$, usually the method for unconjugated BAs) of the CA-molecule. In addition, in both methods there are formic acid (FA) adducts detectable ($[M-H+FA]^-$) in case of unconjugated BAs, as can be seen in the chromatograms of the extracted ions and in the inserted HR-mass spectra of CA-, plus added d_4-CA-ions (IS). Finally, a comparison of quantitative results of more than 100 of human routine samples showed a mean deviation of about 5 % over all measured BA-isoforms. As a consequence, this means that both types of instrumentations are equally feasible for BA-analysis.

One advantage of high-resolution instruments, when operated in full-scan-mode, is the possibility of a retrospective view of already-measured samples, in the case subsequent interests of new, or different, compounds arise. Due to the recorded FS spectra in HR-mode, new isoforms of BAs, for instance, can be detected, if the retention time in the chromatography can be estimated.

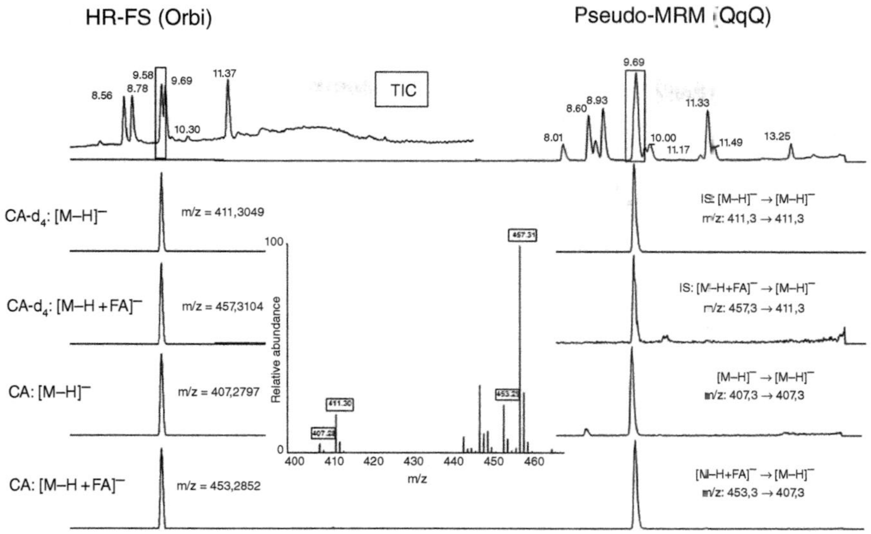

Figure 17.2 Comparison of an identical routine BA-sample, measured in both methods, on a HR-FS-Orbi-instrument, as well as using a MRM-method on a triple quad instrument. The chromatograms of extracted HR-ions are shown in comparison to the chromatograms of the pseudo-MRMs ($[M-H]^- \rightarrow [M-H]^-$) of the unconjugated CA-molecule. In addition, in both methods there are formic acid adducts detectable ($[M-H+FA]^-$), as shown also in the inserted HR-mass spectra of CA-, and d_4-CA-ions (IS).

Due to complex biological matrices in different samples, which may interfere with the performance of the chromatography, easy, reliable and, if possible, automatable sample preparation steps more and more will gain importance, especially when sample numbers are rising.

However, all these circumstances will lead to an expected high number of contributions, including an unmanageable number of results, where a comparability of data will be almost impossible, if it this not even the case already in this moment.

One important step toward a solution can be the installation of ring-trials, where every interested laboratory can compare their own results from identical, central delivered, samples, with any other group. Such a comparison typically is performed routinely in clinical laboratories for a huge number of different parameters.

As a first attempt in this direction, a few years ago an international ring trial was organized, where 12 laboratories tested a commercial BA-kit [71], but it was limited to the use of that kit. In the next future another ring trial will be launched by the group of Markus Wenk from Singapore [72], which should be more comprehensive.

After such a validation-process, to get inter-laboratory comparison of data, universal valid normal values for healthy people could be determined, which would be important for comparison of results between health care institutions and/or studies. There are still some published data, dealing with normal values for adults and children [34, 51, 70], but the comparability with other laboratories is not yet proven.

Regarding future perspectives, it will be interesting how the implementation of present developments and new inventions can bring a further enhancement for BA-analysis. In GC-MS-analytics, for instance, the implementation of GC-HR/MSMS-methods for detection of new isoforms of BAs can lead to new challenges, eventually amplified by the use of specialized derivatization procedures.

Again, a derivatization of compounds for analysis in LC-MSMS eventually can contribute to highlight different, low concentrated isoforms in specialized assays, due to a remarkable increase of the sensitivity of the target compounds, as can be seen, for instance, by a study from Hu et al. [41].

Moreover, the implementation of ion mobility-technology seems to be an interesting approach in the separation of isobaric isomers, to avoid long and complex chromatography [30], also multi-dimensional data mining for profiling of BAs could be a promising approach [43], especially, when huge data sets get nearly unmanageable.

Another question will be, whether, and if, in which extent, new LC-MSMS-technologies, like Imaging MALDI [31], or liquid extraction surface analysis-(LESA-) LC-ESI-MS, here exemplarily shown for determination of triacylglycerol in milk [73], or different ionization methods, like "sweep jet collection laser-induced acoustic desorption atmospheric pressure photoionization" [74] can be introduced in BA-analytics. Also, thus not being a MS-method, but eventually a complementary approach, the determination of BAs by the use of NMR-technology, could be a challenge in the future.

One further question is an eventual enhancement of new materials for chromatography, like micro-, or nano-HPLC, or chip-based-HPLC [75].

Regarding SFC-(HR)-MSMS, there are still questions open on whether future developments and applications will bring this promising technology into routine laboratories.

As a final annotation for the importance of BA-analytics, in a recently published, high-ranked, study [76] researchers described that the microbiome of some groups of very old people (over 100 years old) consists of a different pattern of bacteria, which have been shown to be responsible for the synthesis of special secondary BAs. These isoforms are proposed to be responsible for special immune effects that seem to prevent people from certain inflammatory processes. Such findings could function as basis for future diagnosis and/or medication. Such discoveries can serve as excellent examples that substantiated BA-analytics will remain of great and still raising interest in the next future.

Abbreviations

7-DHC	7-dehydrocholesterole
dbs	dried blood spot
AD	Alzheimer's disease

APCI	atmospheric pressure chemical ionization
BA	bile acid
BSTFA	*N,O*-Bis(trimethylsilyl)trifluoroacetamide
C4	7α-OH-4-cholestene-3-one
CA	cholic acid
CDCA	chenodeoxycholic acid
CID	collision-induced dissociation
DCA	deoxycholic acid
EI	electron impact ionization
ESI	electrospray ionization
FA	formic acid
FAB	fast atom bombardment
FTICR	Fourier-transform ion cyclotron resonance
GC	gas chromatography
GLC	gas liquid chromatography
HDCA	hyodeoxycholic acid
HPLC	high-performance liquid chromatography
HR-MS	high-resolution mass spectrometry
IS	internal standard
LCA	lithocholic acid
lle	liquid–liquid extraction
MCA	muricholic acid (alpha [α], beta [β], gamma [γ], omega [ο]; T, tauro-; G, glyco-)
MRM	multiple reaction monitoring
MS	mass spectrometry
MSTFA	*N*-Methyl-*N*-(trimethylsilyl)trifluoroacetamide
NAFLD	nonalcoholic fatty liver disease
NCI	negative chemical ionization
QqQ	triple quadrupole
SFC	supercritical fluid chromatography
SIM	single-ion monitoring
spe	solid-phase extraction
SRM	single reaction monitoring
T2DM	type 2 diabetes mellitus
TCM	traditional Chinese medicine
TIC	total ion current (total ion chromatogram)
TLC	thin-layer chromatography
TOF	time-of-flight
UDCA	ursodeoxycholic acid
UPLC	ultra-high-performance liquid chromatography

References

1 Chiang, J.Y. (2013). Bile acid metabolism and signaling. *Compr. Physiol.* 3 (3): 1191–1212.

2 Dutta, M., Cai, J., Gui, W. et al. (2019). A review of analytical platforms for accurate bile acid measurement. *Anal. Bioanal. Chem.* 411 (19): 4541–4549.

3 Li, T. and Apte, U. (2015). Bile acid metabolism and signaling in cholestasis, inflammation, and cancer. *Adv. Pharmacol.* 74: 263–302.

4 Liu, Y., Rong, Z., Xiang, D. et al. (2018). Detection technologies and metabolic profiling of bile acids: a comprehensive review. *Lipids Health Dis.* 17 (1): 121.

5 Trauner, M., Claudel, T., Fickert, P. et al. (2010). Bile acids as regulators of hepatic lipid and glucose metabolism. *Dig. Dis.* 28 (1): 220–224.

6 Claudel, T. and Trauner, M. (2020). Bile acids as signaling molecules. *Liver* 299–312.

7 MahmoudianDehkordi, S., Arnold, M., Nho, K. et al. (2019). Altered bile acid profile associates with cognitive impairment in Alzheimer's disease – an emerging role for gut microbiome. *Alzheimers Dement* 15 (1): 76–92.

8 Baloni, P., Funk, C.C., Yan, J. et al. (2020). Metabolic network analysis reveals altered bile acid synthesis and metabolism in Alzheimer's disease. *Cell Rep. Med.* 1 (8): 100138.

9 Zhu, P., Zhang, J., Chen, Y. et al. (2018). Analysis of human C24 bile acids metabolome in serum and urine based on enzyme digestion of conjugated bile acids and LC-MS determination of unconjugated bile acids. *Anal. Bioanal. Chem.* 410 (21): 5287–5300.

10 Ahlawat, S., Asha, and Sharma, K.K. (2021). Gut-organ axis: a microbial outreach and networking. *Lett. Appl. Microbiol.* 72 (6): 636–668.

11 Wahlström, A., Sayin, S.I., Marschall, H.U. et al. (2016). Intestinal crosstalk between bile acids and microbiota and its impact on host metabolism. *Cell Metab.* 24 (1): 41–50.

12 Wang, Y., Yutuc, E., and Griffiths, W.J. (2021). Cholesterol metabolism pathways – are the intermediates more important than the products? *FEBS J.* 288: 3727–3745.

13 Manna, L.B., Ovadia, C., Lövgren-Sandblom, A. et al. (2019). Enzymatic quantification of total serum bile acids as a monitoring strategy for women with intrahepatic cholestasis of pregnancy receiving ursodeoxycholic acid treatment: a cohort study. *BJOG* 126 (13): 1633–1640.

14 Mueller, M., Thorell, A., Claudel, T. et al. (2015). Ursodeoxycholic acid exerts farnesoid X receptor-antagonistic effects on bile acid and lipid metabolism in morbid obesity. *J. Hepatol.* 62 (6): 1398–1404.

15 Zhang, J., Peng, J., Chen, X. et al. (2016). Rapid identification of bile acids in snake bile using ultrahigh-performance liquid chromatography with electrospray ionization quadrupole time-of-flight tandem mass spectrometry. *J. Chromatogr. B Analyt. Technol. Biomed. Life Sci.* 1036, 1037: 157–169.

16 Qiao, X., Ye, M., Pan, D.L. et al. (2011). Differentiation of various traditional Chinese medicines derived from animal bile and gallstone: simultaneous

determination of bile acids by liquid chromatography coupled with triple quadrupole mass spectrometry. *J. Chromatogr. A* 1218 (1): 107–117.

17 Yang, Q., Yang, F., Tang, X. et al. (2015). Chlorpromazine-induced perturbations of bile acids and free fatty acids in cholestatic liver injury prevented by the Chinese herbal compound Yin-Chen-Hao-Tang. *BMC Complement. Altern. Med.* 15: 122.

18 Sarafian, M.H., Lewis, M.R., Pechlivanis, A. et al. (2015). Bile acid profiling and quantification in biofluids using ultra-performance liquid chromatography tandem mass spectrometry. *Anal. Chem.* 87 (19): 9662–9670.

19 Eneroth, P., Gordon, B., and Sjövall, J. (1966). Characterization of trisubstituted cholanoic acids in human feces. *J. Lipid Res.* 7 (4): 524–530.

20 Street, J.M., Trafford, D.J., and Makin, H.L. (1983). The quantitative estimation of bile acids and their conjugates in human biological fluids. *J. Lipid Res.* 24 (5): 491–511.

21 Setchell, K.D. and Matsui, A. (1983). Serum bile acid analysis. *Clin. Chim. Acta* 127 (1): 1–17.

22 Batta, A.K. and Salen, G. (1999). Gas chromatography of bile acids. *J. Chromatogr. B Biomed. Sci. Appl.* 723 (1, 2): 1–16.

23 Griffiths, W.J. and Sjövall, J. (2010). Bile acids: analysis in biological fluids and tissues. *J. Lipid Res.* 51 (1): 23–41.

24 Tsai, S.J., Zhong, Y.S., Weng, J.F. et al. (2011). Determination of bile acids in pig liver, pig kidney and bovine liver by gas chromatography-chemical ionization tandem mass spectrometry with total ion chromatograms and extraction ion chromatograms. *J. Chromatogr. A* 1218 (3): 524–533.

25 Kumar, B.S., Chung, B.C., Lee, Y.J. et al. (2011). Gas chromatography-mass spectrometry-based simultaneous quantitative analytical method for urinary oxysterols and bile acids in rats. *Anal. Biochem.* 408 (2): 242–252.

26 Wood, K.V., Sun, Y., and Elkin, R.G. (1991). Differentiation of isomeric conjugated bile acids using positive-ion B/E linked scans. *Anal. Chem.* 63 (3): 247–250.

27 Yang, Y., Griffiths, W.J., Nazer, H. et al. (1997). Analysis of bile acids and bile alcohols in urine by capillary column liquid chromatography-mass spectrometry using fast atom bombardment or electrospray ionization and collision-induced dissociation. *Biomed. Chromatogr.* 11 (4): 240–255.

28 Roda, A., Piazza, F., and Baraldini, M. (1998). Separation techniques for bile salts analysis. *J. Chromatogr. B Biomed. Sci. Appl.* 717 (1, 2): 263–278.

29 Cai, X., Liu, Y., Zhou, X. et al. (2012). An LC-ESI-MS method for the quantitative analysis of bile acids composition in fecal materials. *Biomed. Chromatogr.* 26 (1): 101–108.

30 Poland, J.C., Leaptrot, K.L., Sherrod, S.D. et al. (2020). Collision cross section conformational analyses of bile acids via ion mobility-mass spectrometry. *J. Am. Soc. Mass Spectrom.* .

31 Shibata, K., Hayasaka, T., Hashimoto, S. et al. (2020). Imaging mass spectrometry reveals the changes in the taurine conjugates of dihydroxycholanoic acid during hepatic warm ischemia and reperfusion in a rat model. *Transplant. Proc.* 52 (6): 1880–1883.

32 Griffiths, W.J. and Wang, Y. (2019). Sterolomics in biology, biochemistry, medicine. *TrAC, Trends Anal. Chem.* 120: 115280.

33 Griffiths, W.J. (2003). Tandem mass spectrometry in the study of fatty acids, bile acids, and steroids. *Mass Spectrom. Rev.* 22 (2): 81–152.

34 Tagliacozzi, D., Mozzi, A.F., Casetta, B. et al. (2003). Quantitative analysis of bile acids in human plasma by liquid chromatography-electrospray tandem mass spectrometry: a simple and rapid one-step method. *Clin. Chem. Lab. Med.* 41 (12): 1633–1641.

35 Alnouti, Y., Csanaky, I.L., and Klaassen, C.D. (2008). Quantitative-profiling of bile acids and their conjugates in mouse liver, bile, plasma, and urine using LC-MS/MS. *J. Chromatogr. B Analyt. Technol. Biomed. Life Sci.* 873 (2): 209–217.

36 Bleyle, L., Huidekoper, H.H., Vaz, F.M. et al. (2016). Update on newborn dried bloodspot testing for *Cerebrotendinous xanthomatosis*: an available high-throughput liquid-chromatography tandem mass spectrometry method. *Mol. Genet. Metab. Rep.* 7: 11–15.

37 Yang, T., Shu, T., Liu, G. et al. (2017). Quantitative profiling of 19 bile acids in rat plasma, liver, bile and different intestinal section contents to investigate bile acid homeostasis and the application of temporal variation of endogenous bile acids. *J. Steroid. Biochem. Mol. Biol.* 172: 69–78.

38 Zhang, J., Fan, Y., Gong, Y. et al. (2017). Simultaneous determination of nine kinds of dominating bile acids in various snake bile by ultrahigh-performance liquid chromatography with triple quadrupole linear iontrap mass spectrometry. *J. Chromatogr. B Analyt. Technol. Biomed. Life Sci.* 1068, 1069: 245–252.

39 Dosedělová, V., Itterheimová, P., and Kubáň, P. (2021). Analysis of bile acids in human biological samples by microcolumn separation techniques: a review. *Electrophoresis* 42 (1, 2): 68–85.

40 Maekawa, M., Shimada, M., Iida, T. et al. (2014). Tandem mass spectrometric characterization of bile acids and steroid conjugates based on low-energy collision-induced dissociation. *Steroids* 80: 80–91.

41 Hu, T., Li, H., Xu, B. et al. (2021). Parallel derivatization strategy for comprehensive profiling of unconjugated and glycine-conjugated bile acids using ultra-high performance liquid chromatography-tandem mass spectrometry. *J. Steroid Biochem. Mol. Biol.* 214: 105986.

42 Donazzolo, E., Gucciardi, A., Mazzier, D. et al. (2017). Improved synthesis of glycine, taurine and sulfate conjugated bile acids as reference compounds and internal standards for ESI-MS/MS urinary profiling of inborn errors of bile acid synthesis. *Chem. Phys. Lipids* 204: 43–56.

43 Lin, M., Chen, X., Wang, Z. et al. (2020). Global profiling and identification of bile acids by multi-dimensional data mining to reveal a way of eliminating abnormal bile acids. *Anal. Chim. Acta* 1132: 74–82.

44 Amplatz, B., Zohrer, E., Haas, C. et al. (2017). Bile acid preparation and comprehensive analysis by high performance liquid chromatography-high-resolution mass spectrometry. *Clin. Chim. Acta* 464: 85–92.

45 Bobeldijk, I., Hekman, M., de Vries-van der Weij, J. et al. (2008). Quantitative profiling of bile acids in biofluids and tissues based on accurate mass high resolution LC-FT-MS: compound class targeting in a metabolomics workflow. *J. Chromatogr. B Analyt. Technol. Biomed. Life Sci.* 871 (2): 306–313.

46 Sangaraju, D., Shi, Y., Van Parys, M. et al. (2021). Robust and comprehensive targeted metabolomics method for quantification of 50 different primary, secondary, and sulfated bile acids in multiple biological species (human, monkey, rabbit, dog, and rat) and matrices (plasma and urine) using liquid chromatography high resolution mass spectrometry (LC-HRMS) analysis. *J. Am. Soc. Mass Spectrom.* 32 (8): 2033–2049.

47 Lee, G., Lee, H., Hong, J. et al. (2016). Quantitative profiling of bile acids in rat bile using ultrahigh-performance liquid chromatography-orbitrap mass spectrometry: alteration of the bile acid composition with aging. *J. Chromatogr. B Analyt. Technol. Biomed. Life Sci.* 1031: 37–49.

48 Shi, Y., Gao, Y., Parys, M.V. et al. (2018). Definitive profiling of plasma bile acids as potential biomarkers for human liver diseases using UPLC-HRMS. *Bioanalysis* 10 (12): 917–932.

49 Wang, H., Huang, R., Li, H. et al. (2021). Serum metabolomic analysis of the anti-diabetic effect of Ginseng berry in type II diabetic rats based on ultra high-performance liquid chromatography-high resolution mass spectrometry. *J. Pharm. Biomed. Anal.* 196: 113897.

50 Jiang, P., Sun, Y., and Cheng, N. (2020). Liver metabolomic characterization of Sophora flavescens alcohol extract-induced hepatotoxicity in rats through UPLC/LTQ-Orbitrap mass spectrometry. *Xenobiotica* 50 (6): 670–676.

51 Jahnel, J., Zöhrer, E., Scharnagl, H. et al. (2015). Reference ranges of serum bile acids in children and adolescents. *Clin. Chem. Lab. Med.* 53 (11): 1807–1813.

52 Minato, K., Suzuki, M., Nagao, H. et al. (2015). Development of analytical method for simultaneous determination of five rodent unique bile acids in rat plasma using ultra-performance liquid chromatography coupled with time-of-flight mass spectrometry. *J. Chromatogr. B Analyt. Technol. Biomed. Life Sci.* 1002: 399–410.

53 de Paiva, M.J., Menezes, H.C., da Silva, J.C. et al. (2015). New method for the determination of bile acids in human plasma by liquid-phase microextraction using liquid chromatography-ion-trap-time-of-flight mass spectrometry. *J. Chromatogr. A* 1388: 102–109.

54 Rago, B., Tierney, B., Rodrigues, A.D. et al. (2018). A multiplex HRMS assay for quantifying selected human plasma bile acids as candidate OATP biomarkers. *Bioanalysis* 10 (9): 645–657.

55 Gao, J., Mao, K., Wang, X. et al. (2021). Tibet Kefir Milk regulated metabolic changes induced by high-fat diet via amino acids, bile acids, and equol metabolism in human-microbiota-associated rats. *J. Agric. Food Chem.* 69 (23): 6720–6732.

56 Scalia, S. and Games, D.E. (1992). Analysis of conjugated bile acids by packed-column supercritical fluid chromatography. *J. Chromatogr.* 574 (2): 197–203.

57 Taguchi, K., Fukusaki, E., and Bamba, T. (2013). Simultaneous and rapid analysis of bile acids including conjugates by supercritical fluid chromatography coupled to tandem mass spectrometry. *J. Chromatogr. A* 1299: 103–109.

58 Jones, M.D., Rainville, P.D., Isaac, G. et al. (2014). Ultra high resolution SFC-MS as a high throughput platform for metabolic phenotyping: application to metabolic

profiling of rat and dog bile. *J. Chromatogr. B Analyt. Technol. Biomed. Life Sci.* 966: 200–207.
59 Gordillo, R. (2021). Supercritical fluid chromatography hyphenated to mass spectrometry for metabolomics applications. *J. Sep. Sci.* 44 (1): 448–463.
60 Sjövall, J. and Setchell, K. (1988). Techniques for extraction and group separation of bile acids. In: *The Bile Acids: Chemistry, Physiology, and Metabolism* (ed. K.D.R. Setchell, D. Kritchevsky and P.P. Nair), 1–42. Springer.
61 Karlaganis, G. and Paumgartner, G. (1979). Determination of bile acids in serum by capillary gas-liquid chromatography. *Clin. Chim. Acta* 92 (1): 19–26.
62 Huijghebaert, S.M. and Hofmann, A.F. (1986). Influence of the amino acid moiety on deconjugation of bile acid amidates by cholylglycine hydrolase or human fecal cultures. *J. Lipid Res.* 27 (7): 742–752.
63 Almé, B. and Sjövall, J. (1980). Analysis of bile acid glucuronides in urine. Identification of 3 alpha, 6 alpha, 12 alpha-trihydroxy-5 beta-cholanoic acid. *J. Steroid Biochem.* 13 (8): 907–916.
64 Eneroth, P., Gordon, B., Ryhage, R. et al. (1966). Identification of mono- and dihydroxy bile acids in human feces by gas-liquid chromatography and mass spectrometry. *J. Lipid Res.* 7 (4): 511–523.
65 Matysik, S. and Schmitz, G. (2013). Application of gas chromatography-triple quadrupole mass spectrometry to the determination of sterol components in biological samples in consideration of the ionization mode. *Biochimie* 95 (3): 489–495.
66 Griffiths, W.J. and Sjövall, J. (2010). Analytical strategies for characterization of bile acid and oxysterol metabolomes. *Biochem. Biophys. Res. Commun.* 396 (1): 80–84.
67 Clayton, P.T., Leonard, J.V., Lawson, A.M. et al. (1987). Familial giant cell hepatitis associated with synthesis of 3 beta, 7 alpha-dihydroxy-and 3 beta,7 alpha, 12 alpha-trihydroxy-5-cholenoic acids. *J. Clin. Invest.* 79 (4): 1031–1038.
68 Eriksson, H., Gustafsson, J.A., Sjövall, J. et al. (1972). Excretion of neutral steroids in urine and faeces of women with intrahepatic cholestasis of pregnancy. *Steroids Lipids Res.* 3 (1): 30–48.
69 Clayton, P.T. (2011). Disorders of bile acid synthesis. *J. Inherit. Metab. Dis.* 34 (3): 593–604.
70 Mayo Clinic Laboratories. (2021). Test Definition: BAFS: Bile Acids, Fractioned and ToT. https://www.mayocliniclabs.com/test-catalog/pod/MayoTestCatalog-Rochester--SortedByTestName-duplex-interpretive.pdf.
71 Pham, H.T., Arnhard, K., Asad, Y.J. et al. (2019). Inter-laboratory robustness of next-generation bile acid study in mice and humans: international ring trial involving 12 laboratories. *J.Appl. Lab. Med.* 1 (2): 129–142.
72 Wenk, M., Bowden, J., Koellensperger, G., et al. 2021. International Lipidomics Society, Reference materials and biological reference ranges. Bile Acids. https://lipidomicssociety.org/interest_groups/reference-materials-and-biological-reference-ranges.
73 Koulman, A., Furse, S., Baumert, M. et al. (2019). Rapid profiling of triglycerides in human breast milk using liquid extraction surface analysis Fourier transform mass

spectrometry reveals new very long chain fatty acids and differences within individuals. *Rapid Commun. Mass Spectrom.* 33 (15): 1267–1276.

74 Benham, K., Fernández, F.M., and Orlando, T.M. (2019). Sweep jet collection laser-induced acoustic desorption atmospheric pressure photoionization for lipid analysis applications. *J. Am. Soc. Mass Spectrom.* 30 (4): 647–658.

75 Thurmann, S., Lotter, C., Heiland, J.J. et al. (2015). Chip-based high-performance liquid chromatography for high-speed enantio separations. *Anal. Chem.* 87 (11): 5568–5576.

76 Sato, Y., Atarashi, K., Plichta, D.R. et al. (2021). Novel bile acid biosynthetic pathways are enriched in the microbiome of centenarians. *Nature* 599 (7885): 458–464.

Part III

Lipidomic Applications

18

Lipidomic Profiling in a Large-Scale Cohort

Daisuke Saigusa[1,2]

Teikyo University, Faculty of Pharma-Science, 2-11-1 Kaga, Itabashi-ku, Tokyo, 173-8605, Japan
Tohoku University, Tohoku Medical Megabank Organization, 2-1 Seiryo-machi, Aoba-ku, Sendai, 980-8573, Japan

18.1 Lipidomic Profiling in a Large-Scale Cohort Project

Lipidomic changes are directly associated with phenotypic variations that can describe and define the characteristics of each individual, which are generated through interactions between genes and the environment, such as lifestyle, food intake, and the gut microbiome [1]. Therefore, lipidomic profiling is a promising approach to identify disease-specific predictive molecules in biological samples and understand disease expression and progression.

The methodology of lipidomic profiling has been optimized by the recent remarkable development of mass spectrometry (MS)-based techniques and has been widely used for biomarker discovery [2]. However, the expression of common diseases is derived from complex environmental factors, and it is impossible to estimate only by tens or hundreds of individual analyses.

Recent national biobank projects have incorporated molecular profiling, including lipid species, in a large-scale clinical and/or prospective cohort, such as the UK biobank [3], European Prospective Investigation into Cancer and Nutrition (EPIC) [4], Cooperative Health Research in the Region of Augsburg (KORA) [5], and Tohoku Medical Megabank (TMM) projects [6]. However, the optimal strategy and methodology for tens of hundreds of sample analyses in a large-scale cohort must be established to overcome several bottlenecks and issues in conducting MS-based lipidomic profiling.

For instance, it is necessary to reduce the significant fluctuations in lipid species due to preanalytical conditions during sample collection and storage and consider establishing an analytical workflow in high-throughput sample preparation and MS analyses and the technical intricacy of data processing and normalization. Moreover, the utilization of the large-scale dataset for data sharing with researchers

Mass Spectrometry for Lipidomics: Methods and Applications, First Edition.
Edited by Michal Holčapek and Kim Ekroos.

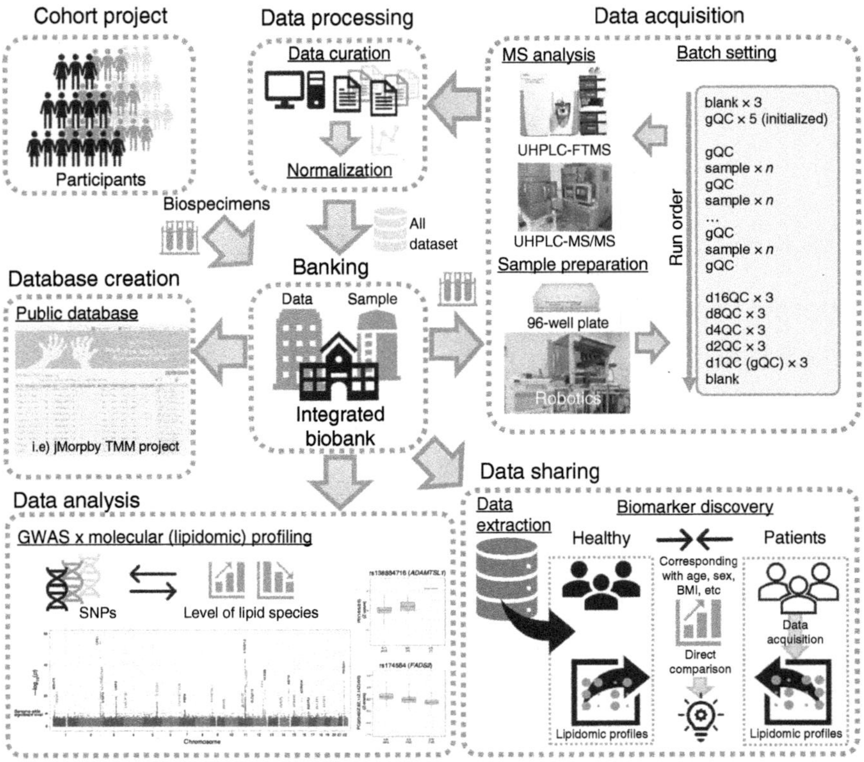

Figure 18.1 The summarized process of a lipidomic profiling project in a large-scale cohort. Picture Sources: jMorp. https://jmorp.megabank.tohoku.ac.jp/, Thermo Fisher Scientific Inc.

and companies should be considered after creating a database including the reference values of lipid species. Therefore, to adopt MS-based lipidomic profiling in a large-scale cohort, original protocols must be developed before starting the project. The summarized process of a lipidomic profiling project in a large-scale cohort is shown in Figure 18.1.

18.2 Sample Collection

Lipidomic profiling in a large-scale cohort begins with the collection and preparation of biofluid samples. In most cases, the method and consistency of collection and storage are of paramount importance. For instance, blood samples should be maintained at 4 °C after collection and immediately stored at −80 °C within four hours to reduce the variations in lipid species [7]. A detailed example of a procedure for sample collection in biobank storage is described below.

In the case of the TMM project, blood samples (34 ml) were collected from participants of the cohort study into four distinct types of tubes (EDTA, ethylenediaminetetraacetic acid disodium salt,2-hydrate-2Na tubes for plasma and buffy coat collection, heparin tubes for peripheral blood mononuclear cell collection, two plain tubes with a separation gel for serum collection and clinical biochemical analysis, and sodium fluoride tubes for blood glucose level analysis), and urine (3–10 ml) was also collected. The plasma fraction was transferred to a liquid handling machine, dispensed into four 1.0-ml 2D barcoded screw tubes (700 μl, each), and stored at −80 °C in a TMM biobank [8]. For molecular profiling, including lipid species, the plasma sample in each dispensed tube (700 μl) was further divided into six tubes (approximately 120 μl per tube) and stored at −80 °C. Their information was managed by a laboratory information management system [6].

18.3 Sample Preparation

The selection of lipid extraction method should consider the type of lipid species to be detected by MS. Since the development of liquid–liquid extraction (LLE) techniques, such as those developed by Bligh and Dyer and Folch [9, 10], LLE has been applied for general lipidomic profiling; hydrophilic lipid species, such as lysophospholipids (LPLs), are extracted with relatively low abundances by LLE methods, whereas hydrophobic lipid species, such as triacylglycerols (TGs) and cholesterylesters (CEs), are extracted with relatively high abundances.

In contrast, the deprotenization method using an organic solution, such as methanol or acetonitrile, can extract LPLs with higher abundances than LLE, whereas TGs and CEs are not extracted from the sample as efficiently. Recently, kit-based metabolomics (Kit-Met) consisting of a consumable 96-well preparation plate that includes several internal standards (ISs) and an optimal ultrahigh-performance liquid chromatography triple quadrupole mass spectrometry (UHPLC-MS/MS) method in strict accordance with a standard operating procedure involving detailed documentation for sample preparation, instrument setup, system suitability testing, and data analysis, was established [11]. Kit-Met enables us to obtain quantified values of several hundred molecules, including wide coverage of lipid species with a simple derivatization process on a 96-well plate, and has been conducted in large-scale studies [7].

An additional essential factor of sample preparation is reproducibility over tens of hundreds of analyses, as systematic differences in sample handling can introduce an apparent abundant signal into one batch or 96-well plate, and the profiling results are then affected by batch-to-batch differences and produce biased outcomes [12]. Although most national biobank projects have been installing robotic systems considering the advantages in a large-scale study, the cost of system installation, time of setting up optimal programs, and maintaining the system will be present for several years. Notably, customizing a robotic system for the LLE method using a 96-well plate is generally difficult because the volume of clots in the sample after centrifugation is different between samples. Therefore, the optimization step to select the extraction methodology manually or with robotic system is essential in a large-scale cohort.

18.3.1 Analytical Platform

The various lipidomic profiling methodologies have been established by LC-MS with separation using an octadecylsilyl column, which requires 20–30 minutes for data acquisition for one sample. Although several rapid methods have been established for large-scale cohort/clinical studies, batch-to-batch differences can be observed. In addition, the quantification values of lipids had large variation among the laboratories based on the different protocols for the analysis, and the standardized guideline of the method was required [13].

In this context, Kit-Met has recently been used in clinical and cohort studies for biomarker discovery, and it was demonstrated to have the potential to reduce batch-to-batch differences [11]. Indeed, Kit-Met consists of two separate methodologies. Most lipid species (such as acylcarnitines, ceramides [Cers], CEs, diacylglycerols [DGs]), dihydroceramides, and glycerophospholipids (including lysophosphatidylcholines [LPCs] and phosphatidylcholines [PCs], glycosylceramides, sphingolipids, and TGs) and sugars can be detected in biological samples in flow injection analysis (FIA)-MS/MS mode, whereas the representative metabolites (such as amino acids, amino acid-related metabolites, bile acids, biogenic amines, cresol, fatty acids, hormones, indole derivatives, nucleobases, and vitamins) are detected in UHPLC-MS/MS using an analytical column and are utilized for data acquisition in the large-scale cohort study [14].

18.3.2 Data Acquisition

A global quality control (gQC) sample must be prepared before beginning the analyses by covering more than estimated batches in the large-scale samples. In fact, batch-to-batch differences could be observed at most acquisitions by MS-based global lipidomic profiling, widely targeted metabolomics (WT-Met) analysis, or Kit-Met. The intensity of molecules in samples often has to be normalized by using their intensity from intermediate gQC analysis. Although a mixture of all samples in the project as the gQC would be optimal, it is impossible to obtain all samples from the participants at the beginning of the project. Recently, human-derived products, such as plasma and urine, have been available as verified matrices from companies to prepare large volumes of a gQC sample. Since the components and profile of lipid species in the product are not completely similar to the cohort-derived samples, these products are utilized to some extent for a large-scale cohort study [12].

In the process of data acquisition, the cohort samples should be injected in randomized run orders in every batch, and higher rates of gQC injections in a batch resulted in better data correction but at the expense of a longer total time, and the best signal correction results are achieved with at least one gQC injection every two hours of cohort injections. For instance, gQC was injected into every eight samples for 15 minutes of data acquisition per sample by LC-MS analysis. Additionally, five consecutive injections of gQC are prepared at the start of the chromatographic run to initialize the analytical column [12].

A large number of the features are often detected as contaminants and are expected from plate, tube, and other consumable products by global lipidomics as well as global metabolomics. The analysis of dilution QC (dQC) samples has the potential to eliminate such contaminants, noise, and artifact spectra. For instance, 2-fold (d2QC), 4-fold (d4QC), 8-fold (d8QC), and 16-fold (d16QC) samples are prepared and analyzed at the end of batch, and approximately 50% of the features are often eliminated from the dataset because of the disproportionality between the dQC dilution and feature intensities (e.g. the higher intensities of d8QC or d16QC injections over others, and vice versa) [12]. The recommended run order of the data acquisition in a batch is shown at "Data acquisition" in Figure 18.1.

Recently, the National Institute of Standards and Technology (NIST) has started providing the reference material (Metabolites in human plasma NIST® SRM® 1950, Merck-Sigma-Aldrich), including the information of several molecules by representative platforms, such as LC-MS, gas chromatography MS, and nuclear magnetic resonance spectroscopy [15]. Therefore, it is possible to evaluate the quality of the quantified values of molecules by using these NIST reference materials. A previous publication indicated that quantified values of lipid species detected in the NIST reference material can be utilized for the normalization of interlaboratory studies [11], and worldwide challenges named ring trials, which analyzed the same samples at several institutes, demonstrated that the NIST reference material works for data normalization between the interlaboratory differences to notably enhance the international project between a large number of sample analyses [16]. Therefore, at least one NIST reference material should be analyzed in each batch or plate, and the results can be utilized for the interlaboratory study and establishment of a database as the reference values of lipid species in a large-scale cohort study. The details of its importance are described in the database creation section.

18.3.3 Data Processing

After collecting the dataset, normalization of the inter/intrabatch effect is essential in a large-scale cohort study. Since the locally weighted scatter plot smoothing (LOWESS) normalization technique was developed, the intensity variation of detected molecules, including lipid species, was different (such as increasing or decreasing along with the acquisition numbers), depending on the molecular polarity and other artifacts derived from ion suppression with coeluted molecules within the batch [12]. Therefore, intensity-based normalization by gQC analysis of each molecule, including lipid species, is recommended instead of LOWESS normalization. The hypothesis equation is shown in Figure 18.2 (reprinted with permission from ref. [12], copyright 2016, Creative Commons license).

Although the median intensities seemed to be parallel after gQC normalization, the variation derived from batch-to-batch differences did not completely disappear

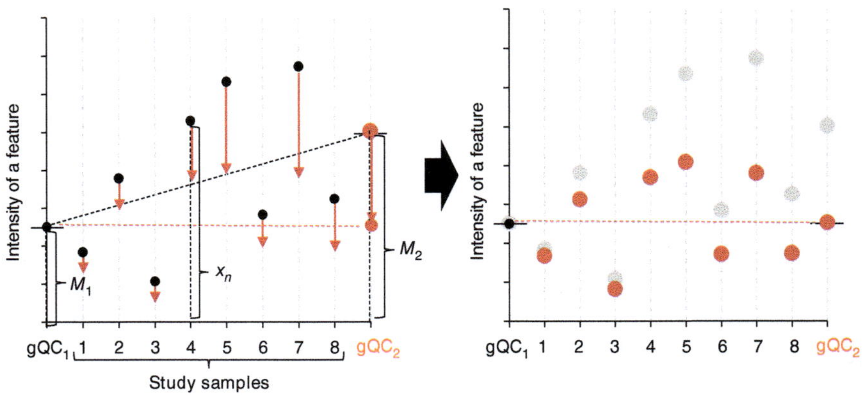

Figure 18.2 Visualized images of gQC normalization: two neighboring gQC (gQC_1 and gQC_2) injections were deemed reliable. The intensities of a feature are described with black and red dots before and after normalization, respectively. x_n: Intensity of a feature (n = Injection number) in a "Study sample: 1~8" between gQCs; M_1 and M_2: Intensity of a feature in gQC_1 and gQC_2 between injections in eight study samples. Equation: $y = x_n M_1 \{n(M_2 - M_1)/9 + M_1\}^{-1}$; x_n = Intensity of a feature in a "Study sample: 1~8" between gQCs; n = Injection number of a "Study sample: 1~8" between gQCs; M_1 = Intensity of a feature in gQC_1: before injection in eight study samples; M_2 = Intensity of a feature in gQC_2: after injection in eight study samples.

in the dataset. Then, the degree of their variation must be manually checked, and cutoff criteria of data must be defined for all detected molecules. For instance, molecules with relatively low abundance were often removed from the range of calibration under the lower limit of detection by decreasing the intensities over several batch analyses because of system contaminants. In the case of the TMM project, the inside of the quadrupoles in the MS instrument is cleaned at least every three months to maintain the quality of the data.

Then, the finalized dataset after processing and original raw files is stored at a supercomputer in a biobank. Additionally, the data processing in a large-scale cohort must be carefully tailored to the volume of the dataset. High-resolution mass spectrometry (HRMS)-based lipidomic profiling generally includes more complex spectra and heavier volumes than multiple reaction monitoring (MRM)-based WT-Met. For instance, a dataset of approximately 5000 samples including gQC sample analysis by LC-HRMS required 15 minutes/run for more than one month and then required transfer of the data for software processing, and automated process of alignment and peak picking to extract features by software was required because of the complexity of the large dataset. Therefore, a computer system should be prepared for large data processing, such as a supercomputer, to avoid data processing issues in a large-scale cohort study. In contrast, a WT-Met dataset includes simple MRM spectra and is approximately 10 times smaller than a HRMS dataset. In fact, the data processing detected over 500 molecules, including 350 lipid species, by MRM in more than 8000 plasma analyses only required half a day for data normalization.

18.3.4 Database Creation

Typical molecular data description commonly includes statistical frequency in a database. These statistical values are verified and utilized to observe age-related and sex-related changes in molecules, including lipid species. Of course, all datasets of molecules derived from each individual without personal information can be verified by researchers and companies based on agreement with the institute or biobank, which organizes the database for the combination project with cohort studies as big data analysis [8].

Indeed, the level of quantification is the most essential point as the reference values of molecules, including lipid species, in the database. The displayed format of values is often shown as intensity level, relative intensity level after normalization to the optimal internal standard/s intensity, and quantified values. However, most users of databases are interested in the quantified value of molecules. In fact, it is impossible to obtain the quantified values of lipid species from healthy humans in hospital and clinical studies in general. These institutions often obtain samples from volunteers that do not correspond with the basic biological information, such as age, gender, and body mass index (BMI). This is the biggest disadvantage of disease biomarker discovery projects in the clinic. Therefore, quantified values as references derived from healthy subjects have been required for a long time.

For instance, in the case of the TMM project, the dataset of plasma cohort lipidomic profiling by LC-HRMS was first obtained and published the original database, the Japanese multi omics reference panel (jMorp) [12, 17]; however, it was impossible to compare the other researcher's dataset and produce a similar level of analysis by LC-HRMS by themselves due to differences in systems. Therefore, the quantified values of molecules, including lipid species, in the large-scale cohort plasma by Kit-Met were obtained and published in jMorp. A representative view of the lipidomic profiling information in jMorp is shown in Figure 18.3.

Additionally, the NIST reference plasma sample was analyzed in one well of each 96-well plate in every batch [14], and the mean quantified values of lipid species in all plates were also displayed on the database (Figure 18.3, dots square). In fact, NIST reference plasma is commercially verified and available as described above, and researchers can analyze these standard references with their samples, such as in a batch of disease plasma samples. Then, the quantified values of lipid species of interest in NIST plasma samples can be compared with the mean quantified values in the database, and all quantified values derived from disease samples can be normalized to the values in the NIST plasma samples. In this process, the disease level of lipid species of interest can be directly compared to the healthy level. Indeed, if there is significance in this comparison, researchers will request extracting the optimal number of data points of lipid species of interest in the database corresponding to basic biological information such as age, sex, and BMI. This will enhance further studies for disease biomarker discovery as shown at "Data sharing" in Figure 18.1 (modified and reprinted with permission from ref. [18], copyright 2021, YODOSHA).

Although batch-to-batch correction was achieved by Kit-Met, it is essential for evaluating the absolute quantified values of lipid species. In fact, lipidomic profiling

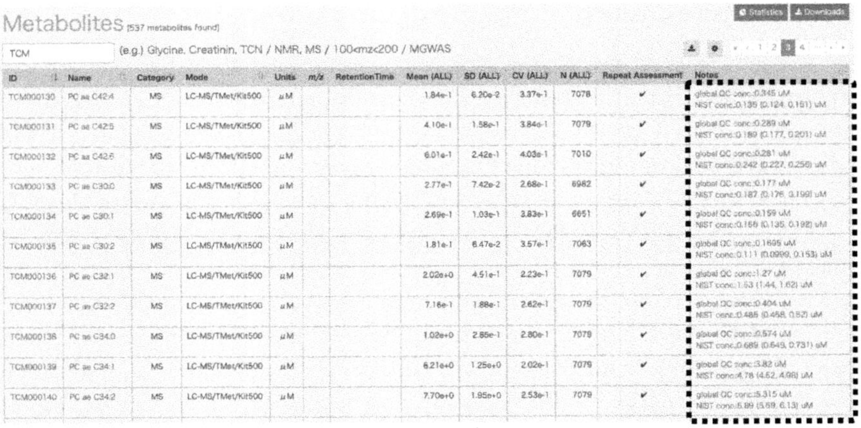

Metabolites [537 metabolites found]

TCM (e.g.) Glycine, Creatinin, TCN / NMR, MS / 100<mz<200 / MGWAS

ID	Name	Category	Mode	Units	m/z	RetentionTime	Mean (ALL)	SD (ALL)	CV (ALL)	N (ALL)	Repeat Assessment	Notes
TCM000130	PC aa C42:4	MS	LC-MS/TMet/Kit500	µM			1.84e-1	6.20e-2	3.37e-1	7078	✔	global QC conc.:0.345 uM NIST conc.:0.135 (0.124, 0.151) uM
TCM000131	PC aa C42:5	MS	LC-MS/TMet/Kit500	µM			4.10e-1	1.58e-1	3.84e-1	7079	✔	global QC conc.:0.289 uM NIST conc.:0.189 (0.177, 0.201) uM
TCM000132	PC aa C42:6	MS	LC-MS/TMet/Kit500	µM			6.01e-1	2.42e-1	4.03e-1	7010	✔	global QC conc.:0.281 uM NIST conc.:0.242 (0.227, 0.256) uM
TCM000133	PC ae C30:0	MS	LC-MS/TMet/Kit500	µM			2.77e-1	7.42e-2	2.68e-1	6982	✔	global QC conc.:0.177 uM NIST conc.:0.187 (0.176, 0.199) uM
TCM000134	PC ae C30:1	MS	LC-MS/TMet/Kit500	µM			2.69e-1	1.03e-1	3.83e-1	6651	✔	global QC conc.:0.159 uM NIST conc.:0.166 (0.135, 0.192) uM
TCM000135	PC ae C30:2	MS	LC-MS/TMet/Kit500	µM			1.81e-1	6.47e-2	3.57e-1	7063	✔	global QC conc.:0.1695 uM NIST conc.:0.111 (0.0999, 0.153) uM
TCM000136	PC ae C32:1	MS	LC-MS/TMet/Kit500	µM			2.02e+0	4.51e-1	2.23e-1	7079	✔	global QC conc.:1.27 uM NIST conc.:1.53 (1.44, 1.62) uM
TCM000137	PC ae C32:2	MS	LC-MS/TMet/Kit500	µM			7.16e-1	1.88e-1	2.62e-1	7079	✔	global QC conc.:0.404 uM NIST conc.:0.485 (0.458, 0.52) uM
TCM000138	PC ae C34:0	MS	LC-MS/TMet/Kit500	µM			1.02e+0	2.85e-1	2.80e-1	7079	✔	global QC conc.:0.574 uM NIST conc.:0.689 (0.649, 0.731) uM
TCM000139	PC ae C34:1	MS	LC-MS/TMet/Kit500	µM			6.21e+0	1.25e+0	2.02e-1	7079	✔	global QC conc.:3.82 uM NIST conc.:4.78 (4.52, 4.98) uM
TCM000140	PC ae C34:2	MS	LC-MS/TMet/Kit500	µM			7.70e+0	1.85e+0	2.53e-1	7079	✔	global QC conc.:5.315 uM NIST conc.:5.89 (5.59, 6.13) uM

Figure 18.3 Overview of lipid species listed in Japanese multi omics reference panel, jMorp (https://jmorp.megabank.tohoku.ac.jp). Quantified values of lipid species in NIST plasma samples were surrounded by the dots square. Source: jMorp. https://jmorp.megabank.tohoku.ac.jp/.

by Kit-Met is conducted by flow FIA-MS/MS, which is considered a shotgun approach, and most annotated lipid species detected by Kit-Met include several isobaric and isomeric compounds that cannot be separated by the applied FIA approach [14]. Indeed, the chromatograms obtained via FIA-MS/MS showed that these isomers overlapped, and low-abundance lipid species were affected by ion suppression. In a previous comparison study of quantified values of lipid species in NIST plasma between Kit-Met and supercritical fluid chromatography (SFC)-MS/MS, the values of LPCs, PCs, Cers, DGs, and TGs quantified by SFC-MS/MS showed better correspondence with the reported consensus values than the values determined by Kit-Met [14]. Notably, most lipid species of CEs might be strongly suppressed by ionization using Kit-Met. Therefore, it should deeply consider how accurate the current quantitative estimates are relative to absolute values by means of different methodologies and improve the accurate values by the gQC normalization procedure.

18.3.5 Combination of Genome-Wide Association Studies

Genome-wide association studies (GWASs) have accelerated investigations of the genetic architecture of complex diseases. However, it cannot be expected only by GWAS and accounts for most of the heritability of disease expression and phenotypic changes in common diseases because rare variants or environmental factors modify genetic risk [19]. Therefore, recent large-scale prospective cohort studies are underway to combine GWAS with molecular profiling, including lipid species, named the metabolome genome-wide association study (MGWAS), to demonstrate the correlation between the expression of single-nucleotide polymorphisms

(SNPs) and the value of lipid species [6, 20]. In fact, a large number of significant lipid species have already been annotated with the MGWAS in several cohort studies.

For instance, an MGWAS was first conducted in a KORA population cohort study based on 363 metabolites in the serum of 284 male participants [21]. Four genetic variants in the genes coding enzymes, including fatty acid desaturase 1 (FADS1), hepatic lipase, short-chain acyl-CoA dehydrogenase, and medium-chain acyl-CoA dehydrogenase, were identified to be associated with blood metabolites, including lipid species levels, in a previous study. This association, particularly that of FADS1, has been replicated many times with related metabolic traits, such as polyunsaturated fatty acids (PUFAs) 20:3 and 20:4 [22]. The FADS1 gene encodes fatty acid delta-5 desaturase, which is a key enzyme in the metabolism of polyunsaturated omega-3 and omega-6 fatty acids [21]. The minor allele variant of the rs174548 SNP resulted in a reduced efficiency of the fatty acid delta-5 desaturase reaction in a previous study [21].

In the case of the TMM project, 42 significant associations of 38 plasma metabolites, including lipid species, with 33 genetic variants (in 26 loci) were identified by an MGWAS of 1008 participants using whole-genome sequence datasets [20]. Well-known associations of the *FADS* locus with LPC20:4 and lysophosphatidylethanolamine 16:0 were included, and the associated genes of *AGBL4* and *ZNF385D* with lysophosphatidic acid 20:3 and phosphatidyl inositol 38:4 were not involved directly in the metabolic pathways. In contrast, synonymous variations in these genes were associated with unrelated plasma PLs and/or lipid-related diseases. In fact, SNPs in *AGBL4* were associated with cardiometabolic risk and dyslipidemia. A variant near the *ZNF385D* gene was associated with an increase in blood levels of Cers and an increased risk of arterial and venous thrombosis. These results imply that *ZNF385D* may contribute to lipid metabolism in a currently unknown manner [20]. Therefore, the combination with GWASs for lipidomic profiling in a large-scale cohort has the potential to estimate missing heritability in common diseases. The visualized figure and image were shown at "Data analysis" in Figure 18.1 (reprinted with permission from ref. [20], copyright 2020, Creative Commons license).

18.4 Conclusion

The technological development of lipidomic profiling by MS enables us to observe the variation in a wide range of lipid species correlated with phenotypic changes. Although lipidomic profiling has the potential to identify novel disease biomarkers by using clinical samples, the quantified values of lipid species in healthy subjects are still low, and reference values are unknown. Thus, the reference values of large numbers of lipid species derived from healthy subjects in a large-scale cohort can play an essential role in evaluating disease-specific changes and enhancing next-generation biomarker discovery in future precision medicine.

References

1 Forouhi, N.G., Koulman, A., Sharp, S.J. et al. (2014). Differences in the prospective association between individual plasma phospholipid saturated fatty acids and incident type 2 diabetes: the EPIC-InterAct case-cohort study. *Lancet Diabetes Endocrinol.* 2 (10): 810–818. https://doi.org/10.1016/S2213-8587(14)70146-9.

2 Han, X. and Gross, R.W. (2021). The foundations and development of lipidomics. *J. Lipid Res.* 63 (2): 100164. Advance online publication. https://doi.org/10.1016/j.jlr.2021.100164.

3 McGurk, K.A., Williams, S.G., Guo, H. et al. (2021). Heritability and family-based GWAS analyses of the *N*-acyl ethanolamine and ceramide plasma lipidome. *Hum. Mol. Genet.* 30 (6): 500–513. https://doi.org/10.1093/hmg/ddab002.

4 Schmidt, J.A., Fensom, G.K., Rinaldi, S. et al. (2020). Patterns in metabolite profile are associated with risk of more aggressive prostate cancer: a prospective study of 3,057 matched case-control sets from EPIC. *Int. J. Cancer* 146 (3): 720–730. https://doi.org/10.1002/ijc.32314.

5 Chak, C.M., Lacruz, M.E., Adam, J. et al. (2019). Ageing investigation using two-time-point metabolomics data from KORA and CARLA studies. *Metabolites* 9 (3): 44. https://doi.org/10.3390/metabo9030044.

6 Koshiba, S., Motoike, I., Saigusa, D. et al. (2018). Omics research project on prospective cohort studies from the Tohoku Medical Megabank Project. *Genes Cells* 23 (6): 406–417. https://doi.org/10.1111/gtc.12588.

7 Saigusa, D., Matsukawa, N., Hishinuma, E. et al. (2021). Identification of biomarkers to diagnose diseases and find adverse drug reactions by metabolomics. *Drug Metab. Pharmacokinet.* 37: 100373. https://doi.org/10.1016/j.dmpk.2020.11.008.

8 Ogishima, S., Nagaie, S., Mizuno, S. et al. (2021). dbTMM: an integrated database of large-scale cohort, genome and clinical data for the Tohoku Medical Megabank Project. *Hum. Genome Var.* 8 (1): 44. https://doi.org/10.1038/s41439-021-00175-5.

9 Bligh, E.G. and Dyer, W.J. (1959). A rapid method of total lipid extraction and purification. *Can. J. Biochem. Physiol.* 37 (8): 911–917. https://doi.org/10.1139/o59-099.

10 Folch, J., Lees, M., and Sloane Stanley, G.H. (1957). A simple method for the isolation and purification of total lipides from animal tissues. *J. Biol. Chem.* 226 (1): 497–509.

11 Siskos, A.P., Jain, P., Römisch-Margl, W. et al. (2017). Interlaboratory reproducibility of a targeted metabolomics platform for analysis of human serum and plasma. *Anal. Chem.* 89 (1): 656–665. https://doi.org/10.1021/acs.analchem.6b02930.

12 Saigusa, D., Okamura, Y., Motoike, I.N. et al. (2016). Establishment of protocols for global metabolomics by LC-MS for biomarker discovery. *PLoS One* 11 (8): e0160555. https://doi.org/10.1371/journal.pone.0160555.

13 Bowden, J.A., Heckert, A., Ulmer, C.Z. et al. (2017). Harmonizing lipidomics: NIST interlaboratory comparison exercise for lipidomics using SRM 1950-metabolites in

frozen human plasma. *J. Lipid Res.* 58 (12): 2275–2288. https://doi.org/10.1194/jlr.M079012.

14 Saigusa, D., Hishinuma, E., Matsukawa, N. et al. (2021). Comparison of kit-based metabolomics with other methodologies in a large cohort, towards establishing reference values. *Metabolites* 11 (10): 652. https://doi.org/10.3390/metabo11100652.

15 Simón-Manso, Y., Lowenthal, M.S., Kilpatrick, L.E. et al. (2013). Metabolite profiling of a NIST Standard Reference Material for human plasma (SRM 1950): GC-MS, LC-MS, NMR, and clinical laboratory analyses, libraries, and web-based resources. *Anal. Chem.* 85 (24): 11725–11731. https://doi.org/10.1021/ac402503m.

16 Thompson, J.W., Adams, K.J., Adamski, J. et al. (2019). International ring trial of a high resolution targeted metabolomics and lipidomics platform for serum and plasma analysis. *Anal. Chem.* 91 (22): 14407–14416. https://doi.org/10.1021/acs.analchem.9b02908.

17 Tadaka, S., Saigusa, D., Motoike, I.N. et al. (2018). jMorp: Japanese multi omics reference panel. *Nucleic Acids Res.* 46 (D1): D551–D557. https://doi.org/10.1093/nar/gkx978.

18 Saigusa, D. (2021). Metabolomics in a large-scale cohort. In: *Metabolomics Jissen Guide* (ed. T. Bamba et al.), 305–307. Japan: YODOSHA.

19 Manolio, T.A., Collins, F.S., Cox, N.J. et al. (2009). Finding the missing heritability of complex diseases. *Nature* 461 (7265): 747–753. https://doi.org/10.1038/nature08494.

20 Koshiba, S., Motoike, I.N., Saigusa, D. et al. (2020). Identification of critical genetic variants associated with metabolic phenotypes of the Japanese population. *Commun. Biol.* 3 (1): 662. https://doi.org/10.1038/s42003-020-01383-5.

21 Gieger, C., Geistlinger, L., Altmaier, E. et al. (2008). Genetics meets metabolomics: a genome-wide association study of metabolite profiles in human serum. *PLos Genet.* 4 (11): e1000282. https://doi.org/10.1371/journal.pgen.1000282.

22 Tanaka, T., Shen, J., Abecasis, G.R. et al. (2009). Genome-wide association study of plasma polyunsaturated fatty acids in the InCHIANTI Study. *PLos Genet.* 5 (1): e1000338. https://doi.org/10.1371/journal.pgen.1000338.

19

Cancer Lipidomics – From the Perspective of Analytical Chemists

Denise Wolrab, Ondřej Peterka, Michaela Chocholoušková, and Michal Holčapek

University of Pardubice, Faculty of Chemical Technology, Department of Analytical Chemistry, Studentská 573, 532 10, Pardubice, Czech Republic

19.1 Introduction

The WHO defines cancer as a group of diseases that can affect any part of the body and involve the uncontrolled growth of abnormal cells, which then invade adjacent parts or spread to other parts of the body, destroying healthy tissues [1]. Cancer is the second leading cause of death worldwide, with about 10 million cancer deaths annually [2]. For 2020, approximately 18.1 million new cancer cases were estimated worldwide, based on breast, lung, colorectal, prostate, and stomach cancers being the most frequently diagnosed cancers considering both genders [2]. The highest mortality rate for pancreas, lung, colorectal, and liver cancers is observed worldwide for both genders.

Lipids represent important biomolecules that are involved in various biological processes. The storage of chemical energy, the formation of cell membranes, and metabolic signaling are examples of lipid functions. The fast proliferation of abnormal cells requires constituents of cell membranes and energy. Therefore, the involvement of lipids is a logical conclusion. Lipidomics can be used to investigate the cancer mechanism through exploratory studies or to compare case and control samples to discover biomarkers used for diagnosis or treatment control, as discussed in recent reviews [3–6].

Here, we describe general considerations for clinical lipidomics and provide personal insight into our experiences with the development of a cancer screening approach based on lipidomics profiling. Figure 19.1 provides an overview of the aspects that must be considered for the development of a clinical method intended to be used for clinical applications, such as preanalytics, sample preparation, analysis method, method validation, results verification, reproducibility, data processing, and statistical analysis.

Mass Spectrometry for Lipidomics: Methods and Applications, First Edition.
Edited by Michal Holčapek and Kim Ekroos.

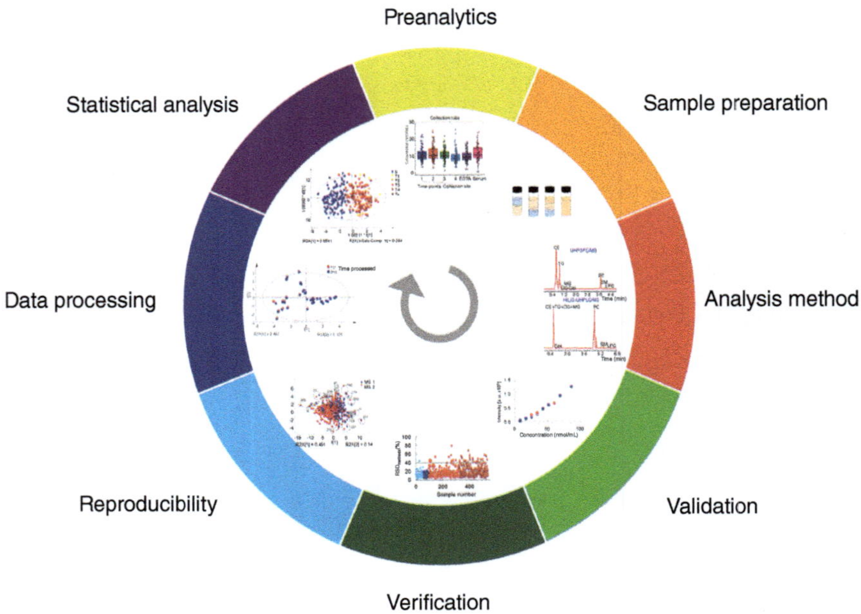

Figure 19.1 Overview of considerations in method development for clinical applications.

19.2 Investigating Dysregulated Lipids in Biological Samples

Changes in lipid concentrations in biological samples may reflect characteristic pathological states. For diagnostic purposes, noninvasive methods with high accuracy are desired, which should reduce patient risk in contrast to invasive methods, such as biopsy, and should increase patient comfort, leading to an increased number of preventive medical health care checks. Regular medical checks can allow the detection of malignant tumors in the early stages, increasing the survival rate. The ability of a method to correctly assign healthy patients as healthy and diseased as diseased is called specificity and sensitivity, respectively. Mostly observational studies are performed, such as prospective cohort studies or case–control studies, sometimes also called retrospective studies. The difference in the design of the clinical study is whether the samples were specifically collected for the clinical study or already present in the house with a known classification determined with the latest methods. Guidelines for the development of cancer biomarkers with minor adoption can be applied to develop a diagnostic test based on lipidomic profiling [7]. The first step in the development of a clinical test at the research level is the discovery of differences between the case and control samples. In the discovery phase, usually only tens of samples are investigated. Subsequently, the results are confirmed with more samples using different methods and, eventually, by different laboratories, which is called the qualification phase. The verification of the assay is then done by applying not only

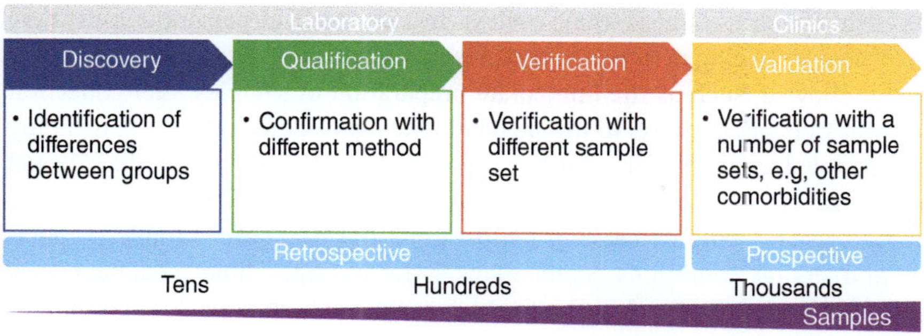

Figure 19.2 Possible workflow to test the applicability of lipidomic profiling to cancer diagnosis.

more samples from different clinical sites, including early stages, but also testing potentially interfering sample groups with a certain risk to develop cancer, such as samples obtained from patients with diabetes, obesity, and chronic inflammation. The final stage is clinical translation (Figure 19.2). In all phases carried out in the laboratory, it is also advantageous to have blinded samples with unknown classification. However, for applicability testing, the sample set can also be split into training and validation sets to determine sensitivity, specificity, accuracy, and receiver operating characteristics (ROC) in an independent cohort.

19.3 Samples

Generally, all types of biological materials can be used for the differentiation of healthy and cancerous samples or to evaluate the efficiency of cancer treatment, such as cells, tissue, urine, plasma, serum, and cerebrospinal fluid. However, noninvasive or minimally invasive methods are preferred due to reduced patient risk. Samples are usually collected at a clinical site, such as a hospital, in the frame of regular medical checks, i.e. high-risk patients, or because of the presence of health problems that require diagnostic confirmation. For the investigation of clinical tests, it is important to collect not only a suitable sample, but also patient characteristics, such as sex, age, body-mass index, disease information (stage and type of cancer), treatment, and diagnosis history (for Kaplan–Meier survival analysis).

19.4 Preanalytical Considerations

The collection, preparation, and storage of the sample belong to the preanalytical treatment of the sample. The whole procedure should always be performed in the same way, and all details are provided to the research team, such as sample collection details (amount, collection tubes, etc.), sample preparation details (time until isolation, centrifugation speed, time, and temperature), and sample storage details (temperature, freeze–thaw cycles). In fact, medical personnel must take care of the well-being of patients, and it is unrealistic that a constant protocol is strictly followed, such as the immediate preparation of samples after collection and flash freezing to preserve the metabolomic state. Consequently, it is highly recommended to evaluate the stability of the investigated analytes in the selected sample material. Parameters that can affect the integrity of the sample include materials used during processing, temperature and time until sample preparation, and storage conditions. Some analytes, for example, lipid mediators, may require the addition of stabilizers to the sample, such as radical scavengers or antioxidants, to maintain the integrity of the sample. In addition, testing of the impact of the collection site on the analyte stability is preferred. In reality, the exact impact of preanalytics is difficult to determine, as the same biological sample cannot be drawn multiple times at different collection sites due to dynamic changes in the metabolome. Evaluation of a statistically significant number of samples from various patients may overcome this limitation. However, a medicinal test should be highly diagnostically conclusive and not altered by preanalytics. Consequently, only highly stable analytes should be investigated for clinical tests.

The impact of the blood collection tube and the collection site on the lipidomic profile was investigated for the most common lipid classes, such as cholesteryl esters (CE), triacylglycerols (TG), diacylglycerols (DG), monoacylglycerols (MG), ceramides (Cer), phosphatidylcholines (PC), sphingomyelines (SM), and lysoglycerophosphatidylcholines (LPC) [8]. Furthermore, the stability of the lipidome was investigated in a period of time of one year for 99 volunteers. Results show that the collection tubes used to isolate plasma, i.e. heparin or EDTA, have only a minor impact on the lipid concentrations. Comparison with serum reveals in general higher lipid concentrations in serum than in plasma. The difference between plasma and serum is that serum is the liquid that remains after coagulation of the blood. In other words, plasma minus clotting factors (i.e. fibrinogen) results in serum. The difference in sample preparation is that plasma can be immediately isolated after blood draw, while for serum, the collection tube must be kept for some time to ensure complete blood coagulation. The biological variance is much greater than the changes in lipidomic concentration caused by the collection tube, collection site, or changes over a year. SM, LPC, and Cer lipid concentrations were observed to be generally stable over a period of a year and therefore are more suitable as potential biomarkers, in contrast to the glycerolipids TG, DG, and MG. A similar trend was observed for the collection site. The highest changes in lipid concentrations were observed depending on the collection site for TG and comparable concentrations for phospholipids and sphingolipids. All results were confirmed by two independent methods (ultrahigh-performance supercritical fluid chromatography [UHPSFC]/MS and hydrophilic interaction liquid chromatography [HILIC]-ultrahigh-performance liquid chromatography [UHPLC]/MS) [8].

19.5 Sample Preparation

Human biological samples are highly complex, containing in addition to lipids, many more biomolecules such as metabolites and proteins. Mass spectrometry (MS) as standalone or hyphenated to chromatography are the most commonly used analytical techniques for lipidomic profiling. However, to reduce the complexity necessary to gain sensitivity and minimize contamination using MS, an adequate sample preparation protocol must be applied. Liquid–liquid extraction is one of the most common protocols used in lipidomics, such as Folch [9], MTBE [10], and BuME [11]. The most common extraction protocols for lipidomics were compared [12], such as Folch, MTBE, BuME, BuHe (BuOH/heptane), MMC (single phase extraction using MeOH/MTBE/CHCl3 [13], and 3-phase [14]. Generally, recovery for the main lipid classes such as CE, TG, DG, MG, Cer, PC, SM, and LPC was comparable. For clinical lipidomics, it is essential to use reproducible and high-throughput sample preparation protocol. Consequently, the protocol has to be straightforward, and every step must always be done in the same way. Preferentially, the sample preparation protocol can be automated, using pipette robots that can ensure sample preparation quality and throughput. Sample preparation protocols based on nonhalogenated extraction solvents are preferred due to easier translation for automatization, as the upper organic layer contains lipids. The use of halogenated solvents, such as carcinogenic chloroform or dichloromethane, is minimalized for safety reasons, although the most common liquid–liquid extraction protocol proved to be beneficial for most lipid classes in terms of recovery. The use of internal standards allows for the evaluation of the sample preparation performance, the determination of recovery, matrix effect, and reproducibility, and the comparison of quantitative results necessary for intra- and interlaboratory studies.

19.6 Method Requirements

A clinical method for cancer screening should be of high accuracy. Furthermore, the method is desired to be relatively inexpensive, robust, and of high throughput. The applied samples should be readily available and applicable. Today, lipidomics is performed mainly using MS as a standalone method or as a hyphenated method for chromatography. By nature, this method is already highly sophisticated and complicated and requires trained operators. Furthermore, MS is prone to fluctuations in sensitivity due to contamination over time. It is necessary to establish strategies that allow monitoring of method performance, which should allow immediate action in the event of inacceptable method performance. For this purpose, the signal response of blanks, quality control, and system suitability samples, as well as the response of internal standards in real samples, are generally controlled. Mostly, cleaning the ion source and ion guides before the mass analyzer is required to return to a sufficient sensitivity. However, these cleaning procedures require experience and should only be performed by trained persons. Consequently, the method should be robust and suitable for at least hundreds of samples to avoid malfunction and maintenance, which would be in contrast to high throughput.

Furthermore, it is important to evaluate the limits of the method by comparing the results with similar methods. The verification should ideally be performed within intra- and interlaboratory studies. As clinical samples are usually too valuable and only a limited amount is available, intra- or interlaboratory studies can be performed on a common sample, i.e. the NIST SRM 1950 reference material [15]. This sample represents a pool of human plasma from 100 individuals reflecting the racial distribution of the US population and an equal number of men and women in the age range between 40 and 50 years. To draw conclusions, the impact of the sample preparation protocol, the internal standards used, and the impact of the analysis method on the lipidomic results should be carefully considered. Ideally, a harmonized sample preparation protocol and an internal standard mixture are used to study the impact of the analysis method and evaluate its integrity. The comparison of lipid class separation methods HILIC-UHPLC/MS and UHPSFC/MS showed that the quantitative results, including method repeatability, stability, and precision, are comparable, with slightly better overall performance of UHPSFC/MS [8, 12, 16]. The comparison of two different mass spectrometers in HILIC-UHPLC/MS or UHPSFC/MS configuration shows [16] that the effect of mass spectrometer is negligible and could be reduced further by the normalization to a reference sample, i.e. NIST SRM 1950.

Evaluation of the limits of the method will help assess its suitability for clinical applications. The reproducibility of the method is of unmet importance, as the clinical test should work independent of the laboratory. Certain differences are expected by using different methods for lipidomic quantitation and can be overcome by normalization. However, the same trend and order of magnitude for lipid concentrations should be achieved even for different lipidomic methods based on MS, provided that the established lipidomic guidelines of International Lipidomics Society (ILS) and Lipidomics Standard Initiative (LSI) are followed.

19.7 Validation and Quality Control

The validation of the method allows the determination of the applicability of the method to the analysis of biological samples. The limitations and strengths of the sample preparation and method can be determined by performing a validation and may illustrate the need to adjust or tune the method. Validation is well known for pharmaceutical and nutrient applications, but less common in lipidomics, which is used to analyze complex biological samples. The FDA and EMA provide guidelines for the validation of bioanalytical methods. For lipidomics analysis, slight adjustments of the guidelines are necessary, since i.e. there is no blank matrix without lipids in biological samples. Consequently, validation is performed for representatives of each lipid class, i.e. exogenous internal standards, which have physicochemical properties similar to those of the endogenous lipids belonging to the same lipid class. For clinical applications, especially, the precision of the method and sample preparation is important and should meet relative standard deviation (RSD) <15%. For a better comparability of results between instruments and laboratories, it is important to determine molar concentrations, since the signal responses of instruments may vary significantly. Quantitation is usually performed using internal standards for each lipid class with known concentration. It is advised to use more than one internal standard (IS) per lipid class, allowing the evaluation of the overall quantitation error of the method by quantifying the IS of the known concentration with the others and vice versa. The quantitation error is of interest to evaluate the performance of the IS and the limitations of the method.

A quality control system allows you to monitor the quality of the acquired data and that evaluation allows you to assess whether the generated results can be used for clinical conclusions or if the analysis should be repeated. There are different types of quality control samples, such as calibration, solvent blank, system standard, quality control (QC), and NIST samples, which should be measured regularly, i.e. after a defined number of samples, depending on the study size and method stability. A proportion of 10–20% belonging to QC samples may present a good compromise with regard to sample throughput. Clinical samples are usually not easy to obtain in research laboratories; the amount is limited, or a fast analysis of results is expected. Consequently, it is important to perform the analysis with reasonable quality. The response of the QC samples and samples should be continuously checked to assess performance. The signal response of IS, other standards, or selected endogenous lipids can be monitored and plotted for QC and study samples. This allows to assess the measurement and sample preparation performance, as the IS should show a constant signal response. Monitoring of the sensitivity drop over time due to contamination and evaluation if the IS can still compensate for the drop by calculating the ratio of endogenous and exogenous signal response, which allows immediate action in case of unsatisfactory performance. After maintenance of the instrument, the measurement and comparison with previously measured QC samples allows one to assess the success of the service and to decide for qualified instrument performance for continuation.

19.8 Data Processing, Statistical Analysis, and Data Reporting

The manual data processing is not practical for more than hundred lipids in hundreds of samples commonly analyzed in lipidomic studies. Consequently, software solutions capable of automatically identifying and quantifying lipids are commonly used for lipidomic studies. However, a manual check of plausibility, validity, and repeatability is recommended, after randomization of the order of sample data processing [17]. The correctness could be checked by comparing the list of identified lipids with the quantitative results for selected lipid species in the QC sample measured at the beginning and at the end of the sequence once processed with the software and once manually. The agreement of the results obtained manually and by the software solution shows the applicability of the software.

The annotation of identified lipids should correspond to the ability of the methodology and experimental settings, as highlighted by the LSI [18, 19].

Statistical analysis of the samples can allow us to assess the performance of the overall method and determine significant differences between the sample groups. Generally, the statistical analysis applied should be transparent and described in detail to allow repetition. To verify the results, more than one statistical software is applied, and the results are compared with each other. Independent statistical evaluation may also help to ensure unbiased statistical results.

19.9 Lipidomic Analysis in Cancer Research

The study of lipid structure, quantity, and function in cells, body fluids, and tissues at a certain time in biological systems is called lipidomics. Lipids are crucial for many biological processes and may have multiple functions. The dysregulation of lipids has been found to be involved in many pathogenic disorders, such as cancer [3]. The metabolic pathways reported to be affected in cancer lipidomics are the sphingolipid, glycerophospholipid, and glycerolipid metabolism pathways, which influence cellular processes such as cell proliferation, cell growth, cell differentiation, and signaling processes for cell function [20]. Up- or down-regulation of lipid concentrations depends on the type of cancer and the biological material studied. Lipidomic analysis of lung, breast, colorectal, kidney, prostate, ovarian, liver, thyroid, bladder, gastric, pancreatic, hepatocellular, and esophageal cancers has been reported in serum, plasma, tissue, and urine samples [21]. The summary of the most regulated lipid species considering all studies included in the review revealed that PC 34:1, SM 34:1, and PC 32:1 were mostly upregulated and LPC 18:0, PE 36:4, PC 38:4, SM 40:1, LPC 16:0, and LPC 18:1 were mostly down regulated in cancer. However, LPC 16:0 and LPC 18:1 were also frequently reported as upregulated, illustrating the complexity of lipidomics according to the type of cancer, the sample matrix, and can also depend on the applied methodology [21]. It is recommended to follow the guidelines established by the Lipidomic Standard Initiative to ensure the quality of the data from the lipidomic analysis and the comparability of various studies. Furthermore, since lipid species and lipid classes are interrelated in both healthy and diseased states, the discovery of individual biomarker candidates for a specific cancer type may be illusionary compared to the comprehensive lipidomic profile. In addition to efforts to apply lipidomics profiling for cancer screening, lipidomics or certain lipid species can also be applied for cancer treatment control or anticancer drugs [5].

We selected UHPSFC/MS as the method of choice for the analysis of clinical samples, the effect of preanalytics [8], the optimization of the sample preparation protocol [12], the validation of the method, the impact and limits of the method and instrument [12, 16], the development of a quality control system and the complement of a software for data processing were investigated [22]. The lipidomic analysis of kidney, prostate, breast, and pancreatic cancer and healthy control samples was performed using UHPSFC/MS as the main methodology [23, 24]. Lipidomic profiling was used to differentiate between cancerous and control samples. Generally, statistical analysis revealed that the performance of gender-separated models is preferable, which may be due to slight gender-specific differences in the lipidome. Furthermore, the sample set should always be divided into a training set and a validation set to allow independent statistical evaluation. However, for simplification reasons, the training and validation set, as well as male and female samples, was combined for the comparison shown in Figure 19.3. ROC, sensitivity, and specificity of lipidomic profiling for the detection of pancreatic ductal adenocarcinom (PDAC) were compared with carbohydrate antigen 19-9 (CA19-9), as the state-of-art diagnostic marker, the combination of lipidomics and CA19-9, and CancerSeek, as one of the most promising cancer diagnosis tests based on liquid biopsy recently published [25] (Figure 19.3). The specificity was slightly higher for CancerSeek and CA19-9 with 97% for both compared to 90% and 92% for lipidomics and lipidomics combined with CA19-9, respectively. However, sensitivity was generally higher for lipidomics or the combination of lipidomics and CA19-9, with 96% for both, compared to 70% and 75% for CA19-9 and CancerSeek. Furthermore, lipid

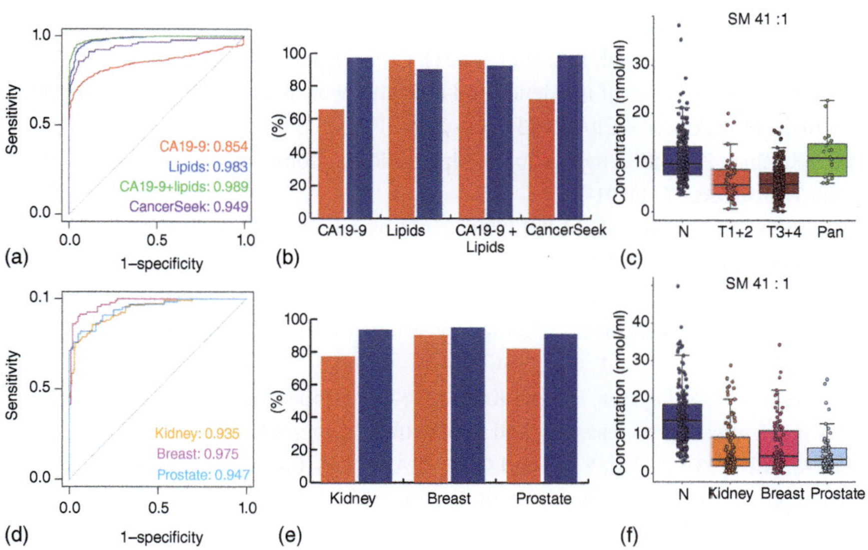

Figure 19.3 Lipidomic profiling for the differentiation of cancerous and control samples using UHPSFC/MS. Comparison of (a) ROC, (b) sensitivity (red), and specificity (blue) obtained for the analysis of CA19-9, lipidomic, or the combination of lipidomic and CA19-9 using serum samples obtained from PDAC and control volunteers, as well as data from the literature for a liquid biopsy test named CancerSeek. (c) Boxplot of SM 41:1, illustrating that the concentration is similar for the cancer stage and similar for pancreatitis and control samples. (d) ROC curves for kidney, breast, and prostate cancer. (e) Comparison of sensitivity (red) and specificity (blue) for kidney, breast, and prostate cancer. (f) Boxplot SM 41:1 comparing the concentrations in healthy and cancerous samples.

concentrations were shown not to be altered by cancer stage, and the concentrations obtained for the pancreatitis samples were similar to those of the control samples, as illustrated for SM 41:1 in Figure 19.3c. Generally, most of the lipids were down-regulated in cancerous samples. Taking into account that nonpolar lipids, such as tri- and diacylglycerols, are also altered by dietary habits, the most important lipid classes to differentiate pancreatic cancerous from healthy samples were SMs, ceramides, and (lyso) PCs. As the results for the detection of pancreatic cancer were promising, the applicability of lipidomic profiling to the detection of breast, prostate, and kidney cancer was also investigated. Generally, a differentiation of kidney, breast, or prostate cancer and healthy samples was achieved depending on the lipid profile with a specificity greater than 90% and a sensitivity ranging from 87% to 93%. The statistically significant lipid species for the differentiation of breast, prostate, or kidney cancer are CE 16:0, Cer 42:1, LPC 18:2, PC 36:2, PC 36:3, SM 32:1, and SM 41:1. However, the results showed that it is preferable to build a multivariate data analysis model for each cancer type separately, since the average accuracy to predict not only the pathological state, but also the cancer type was only 70%.

References

1 World Health Organization (WHO). Cancer. https://www.who.int/health-topics/cancer#tab=tab_1 (accessed 23 June 2022).

2 Sung, H., Ferlay, J., Siegel, R.L. et al. (2021). Global Cancer Statistics 2020: GLOBOCAN estimates of incidence and mortality worldwide for 36 cancers in 185 countries. *CA Cancer J. Clin.* 71 (3): 209–249.

3 Yan, F., Zhao, H., and Zeng, Y. (2018). Lipidomics: a promising cancer biomarker. *Clin. Transl. Med.* 7 (1): 21–23.

4 Fernandis, A.Z. and Wenk, M.R. (2009). Lipid-based biomarkers for cancer. *J. Chromatogr. B* 877 (26): 2830–2835.

5 Butler, L.M., Perone, Y., Dehairs, J. et al. (2020). Lipids and cancer: emerging roles in pathogenesis, diagnosis and therapeutic intervention. *Adv. Drug Delivery Rev.* 159: 245–293.

6 Matsushita, Y., Nakagawa, H., and Koike, K. (2021). Lipid metabolism in oncology: why it matters, how to research, and how to treat. *Cancers* 13 (3): 474.

7 Ali, A., Ali, Z., Yousafzai, Y.M., and Oien, K.A. (2018). Cancer biomarker development from basic science to clinical practice. *J. Postgrad. Med. Inst.* 32 (1): 1–8.

8 Wolrab, D., Chocholoušková, M., Jirásko, R. et al. (2020). Determination of one year stability of lipid plasma profile and comparison of blood collection tubes using UHPSFC/MS and HILIC-UHPLC/MS. *Anal. Chim. Acta* 1137: 74–84.

9 Folch, J., Lees, M., and Stanley, G.H.S. (1957). A simple method for the isolation and purification of total lipides from animal tissues. *J. Biol. Chem.* 226 (1): 497–509.

10 Matyash, V., Liebisch, G., Kurzchalia, T.V. et al. (2008). Lipid extraction by methyl-*tert*-butyl ether for high-throughput lipidomics. *J. Lipid Res.* 49 (5): 1137–1146.

11 Löfgren, L., Forsberg, G.-B., and Ståhlman, M. (2016). The BUME method: a new rapid and simple chloroform-free method for total lipid extraction of animal tissue. *Sci. Rep.* 6: 27688.

12 Wolrab, D., Chocholoušková, M., Jirásko, R. et al. (2020). Validation of lipidomic analysis of human plasma and serum by supercritical fluid chromatography-mass spectrometry and hydrophilic interaction liquid chromatography-mass spectrometry. *Anal. Bioanal. Chem.* 412 (10): 2375–2388.

13 Gil, A., Zhang, W., Wolters, J.C. et al. (2018). One- vs two-phase extraction: re-evaluation of sample preparation procedures for untargeted lipidomics in plasma samples. *Anal. Bioanal. Chem.* 410: 5859–5870. https://doi.org/10.1007/s00216-018-1200-x.

14 Shibusawa, Y., Yamakawa, Y., Noji, R. et al. (2006). Three-phase solvent systems for comprehensive separation of a wide variety of compounds by high-speed counter-current chromatography. *J. Chromatogr. A.* 1133: 119–125.

15 Bowden, J.A., Heckert, A., Ulmer, C.Z. et al. (2017). Harmonizing lipidomics: NIST interlaboratory comparison exercise for lipidomics using SRM 1950-Metabolites in Frozen Human Plasma. *J. Lipid Res.* 58 (12): 2275–2288.

16 Chocholoušková, M., Wolrab, D., Jirásko, R. et al. (2021). Intra-laboratory comparison of four analytical platforms for lipidomic quantitation using hydrophilic interaction liquid chromatography or supercritical fluid chromatography coupled to quadrupole–time-of-flight mass spectrometry. *Talanta* 231 (15): 122367.

17 Köfeler, H.C., Eichmann, T.O., Ahrends, R. et al. (2021). Quality control requirements for the correct annotation of lipidomics data. *Nat. Commun.* 12: 4771.

18 Liebisch, G., Fahy, E., Aoki, J. et al. (2020). Update on LIPID MAPS classification, nomenclature, and shorthand notation for MS-derived lipid structures. *J. Lipid Res.* 61 (12): 1539–1555.

19 Liebisch, G., Vizcaíno, J.A., Köfeler, H. et al. (2013). Shorthand notation for lipid structures derived from mass spectrometry. *J. Lipid Res.* 54 (6): 1523–1530.

20 Pan, M., Qin, C., and Han, X. (2021). Lipid metabolism and lipidomics applications in cancer research. *Adv. Exp. Med. Biol.* 1316: 1–24.

21 Wolrab, D., Jirásko, R., Chocholoušková, M. et al. (2019). Oncolipidomics: mass spectrometric quantitation of lipids in cancer research. *TrAC, Trends Anal. Chem.* 120: 115480.

22 Wolrab, D., Cífková, E., Čáň, P. et al. (2021). LipidQuant 1.0: automated data processing in lipid class separation–mass spectrometry quantitative workflows. *Bioinformatics* 37: 4591–4592.

23 Wolrab, D., Jirásko, R., Peterka, O. et al. (2021). Plasma lipidomic profiles of kidney, breast and prostate cancer patients differ from healthy controls. *Sci. Rep.* 11 (1): 20322.

24 Wolrab, D., Jirásko, R., Cífková, E. et al. (2022). Lipidomic profiling of human serum enables detection of pancreatic cancer. *Nat. Commun.* 13 (124): 1–16.

25 Cohen, J.D., Li, L., Wang, Y. et al. (2018). Detection and localization of surgically resectable cancers with a multi-analyte blood test. *Science* 359 (6378): 926–930.

20

Lipidomics in Clinical Diagnostics

Jayashree Selvalatchmanan[1,2], *Markus R. Wenk*[1,2], *and Anne K. Bendt*[2]

[1] *National University of Singapore, Yong Loo Lin School of Medicine, Department of Biochemistry, 10 Medical Drive, 117597, Singapore*
[2] *Life Sciences Institute, 28 Medical Drive, 117456, Singapore*

20.1 What Do We Mean by "Clinical Diagnostics"?

This would likely be a rather short chapter if we solely defined "lipidomics in clinical diagnostics" as the systems-level analysis of lipids (i.e. "lipidomics") for the real-life diagnosis of a patient's condition by a clinical laboratory. Indeed, although there are (very few) examples where lipidomics (or at least a subset of lipidomes) is used to diagnose patient conditions in clinical laboratories (see below), most lipid analyses currently performed for clinical diagnostics do not qualify as "lipidomics." Instead, current diagnostic applications in hospital settings that are based on lipid analytes measure either individual, singular lipid markers or small multiplexed panels of select lipids, but not the complete lipidome. Prominent examples of such lipid metabolite analyses for patient diagnostics include vitamin D metabolites, sex steroid hormones, and acylcarnitine panels (see Section 20.4).

If we broaden our definition of clinical diagnostics from the "real patient" to include (pre-) clinical research, we find examples aplenty. In clinical research, lipidomic screens are used to (i) advance our understanding of underlying disease pathophysiology, (ii) discover, develop, and validate novel biomarkers, (iii) identify (personalized) therapeutic targets, and (iv) follow treatments responses. Promising recent applications have provided insight into the mechanisms of metabolic disease (cardiovascular disease [1] [see Section 20.2.2], diabetes [2], non-alcoholic fatty liver disease (NAFLD) [3, 4]), neurodegenerative disease (Alzheimer's disease [5], multiple sclerosis [6]), cancer (for review, see [7]), and a host of other rare and specific inflammatory conditions and viral infections, such as asthma [8], atopic dermatitis [9], Fabry disease [10], and COVID-19 infection [11]. Most of these pre-clinical studies are undertaken on retrospective disease cohorts or biobanked samples [12]. More recently, prospective studies based on longitudinal design and even preemptive testing on

Mass Spectrometry for Lipidomics: Methods and Applications, First Edition.
Edited by Michal Holčapek and Kim Ekroos.

leftover clinical samples from routine laboratory medicine analyses are being used. Commercial options include biobank samples from 700 well-characterized individuals provided by the American Association for Clinical Chemistry (AACC) [13].

20.2 Mass Spectrometry as an Enabler for Lipid-Based Clinical Tests

The primary objective of clinical diagnostics is to accurately describe a patient's condition. Accuracy and specificity of clinical tests is hence a major concern for diagnostic laboratories. In this section, we use two examples to illustrate how the superior performance characteristics of liquid chromatography mass spectrometry (LC-MS) enables accuracy and specificity in clinical tests over competing conventional detection methods. A comprehensive overview of currently available tests will be given in Section 20.4.

20.2.1 Vitamin D and Its Metabolites

A growing body of literature describes roles for the secosteriod vitamin D in a variety of diseases, including cancer [14], autoimmune disease [15], neurological disease [16], and metabolic disease [17]. Indeed, the assessment of a patient's vitamin D status to diagnose a potential hypovitaminosis remains a common clinical test, and one that provides a major source of income for clinical testing laboratories. The clinically accepted marker to quantify a patient's vitamin D status is 25-hydroxyvitamin D3 (25(OH)D3). Traditionally, 25(OH)D3 quantification has been done using vitamin D-binding protein or antibodies in protein-binding assays. However, these assays have substantial limitations in terms of accuracy and analytical reliability [18]. Such inaccuracies are especially problematic in differentiating among the various vitamin D metabolites, particularly as a growing number of countries seek to fortify food by the addition of D2 (Vitamin D occurs as vitamin D3 in mammals [cholecalciferol] but as D2 in plants [ergocalciferol]). Furthermore, an additional diagnostic complication occurs as a result of epimers – structurally identical isomers of 25(OH)D3 and 25(OH)D2 at the C3 position – as they cannot be differentiated from the main 25(OH)D3 using mainstream antibody-based tests. This inaccuracy results in overestimating 25(OH)D levels, which is of particular concern in pediatric populations given that children have significant concentrations of C3 epimers [19].

The superior performance characteristics of MS in terms of specificity and sensitivity have been a major driving factor in slowly replacing conventional vitamin D3 assays by LC-MS/MS-based detection over the past few decades. Nowadays, LC-MS/MS is considered the gold standard for analysis of vitamin D metabolites [20]. The enormous commercial interest to capitalize on MS is illustrated by the recent launch of the first fully automated LC-MS/MS based analyzer for vitamin D quantification (ThermoFisher "Cascadion"). For many clinical labs worldwide, the implementation of LC-MS for vitamin D testing was their first step in establishing clinical MS capabilities, leading to subsequent expansion of their MS-based clinical test portfolios toward the testing of other analytes (i.e. sex steroid hormones, immunosuppressants, metanephrines). Over the past 10 years, there have been concerns about the actual clinical relevance of measuring 25(OH)D. Indeed, there have been suggestions that alternative metabolites, including 1,25OH2D and 24,25OH2D, are more clinically useful if we are to truly unravel the pathophysiology of vitamin D-related disease [20]. Again, the versatility of MS-based workflows to modify methods and include newly identified vitamin D metabolites demonstrates the value of these clinical diagnostics over competing immunoassay platforms [21].

20.2.2 The Trailblazing Ceramides

No write-up on clinical lipidomics would be complete without discussing ceramides and their emergent role in routine clinical diagnostics. Ceramides comprise a family of waxy lipids with principal roles in a host of biological functions and disease processes. In particular, these lowly abundant, bioactive molecules are associated with a spectrum of cardiometabolic diseases (for an overview, see the chapter on "Sphingolipids" in this volume). Over the past decade, there has been a surge in evidence to show the additional diagnostic utility of ceramides in predicting cardiovascular disease; for example, in cardiometabolic risk stratification. The initial broad lipidomics screening of cardiovascular disease cohorts pointed to a group of lipids as being strongly associated with cardiovascular disease. A group of scientists from the Finnish company Zora Biosciences – together with their academic and clinical collaborators – then sought to advance the use of ceramides as purely associative biomarkers with prognostic value for routine clinical practice. To this end, targeted screens based on multiple reaction monitoring (MRM) were developed, and these screens helped to narrow the search for specific molecular species with the strongest prognostic power for risk stratification in different coronary artery disease (CAD) cohorts [22]. The resulting panel comprised four ceramide species: d18:1/16:0, d18:1/18:0, d18:1/24:0, and d18:1/24:1. Importantly, to increase ease of use and facilitate clinical adoption of this test, a score was developed, which avoided the need to report and interpret the molecular concentrations of ceramide species within their individual references ranges. This score, since coined CERT1 ("cardiac event risk test"), is based on concentrations of individual species and their ratios. This score is easy to grasp and allows patients with stable coronary heart disease (CHD) and post-acute coronary syndrome (ACS) to be classified according to four risk categories: low, moderate, elevated, and high. Validation studies in various cohorts have since shown that CERT1 serves as a better predictor of patients who have a high risk of cardiovascular disease (CVD) over conventional clinical markers such as low-density lipoprotein cholesterol (LDL-C) [23], hence highlighting the potential of MS-based ceramide testing. This is a powerful example of how MS can enable all major stages of clinical translation: discovery, validation, test development, and deployment for clinical use. Indeed, heralded early on as the "next cholesterol" [24], the commercial potential of ceramide testing has been a major driver in this rather rapid development of a biomarker from discovery to commercialization: soon after the first publication by Zora Biosciences scientists [23], the CERT1 ceramide assay was licensed to the Mayo clinic and is now also widely available in the private healthcare system in Finland [25].

Perhaps not surprisingly, improvements to the ceramide assay are ongoing: two phosphatidylcholine species, selected for their analytical stability and discriminatory power across different multicenter cohorts, have been added to the four-ceramide species panel [26]. The resultant score, CERT2, has even better discriminatory power to stratify CHD patients, especially for their risk of death. Crucially, the utility of this diagnostic has been validated worldwide [27]. Again, the versatility of MS over other traditional diagnostics (e.g. antibody-based testing) enables the rather rapid modification of incorporating additional lipid species to the test. Furthermore, converting lipid molecular concentrations into scores – with intuitive graphical visualization – can provide ease of use for clinicians (and patients), thereby increasing adaptability and clinical utility.

A slightly different approach to biomarker screening was recently employed by Poss et al. [28]. They used a broader, comprehensive sphingolipid panel to screen for biomarkers associated with CAD. In a case–control study of familial CAD, along with control cases from the same (primarily Mormon) population in Utah, USA, they discovered that 30 out of 32 screened sphingolipids were elevated in patients with CAD. Using unbiased machine learning tools such as least absolute shrinkage and selection operator (LASSO) and Random Forrest for variable feature reduction, they narrowed down their panel to sphingolipid species with the most discriminatory ability. The resulting sphingolipid-inclusive CAD risk score, termed SIC, showed better performance than other conventional biomarkers for the demarcation of CAD [28]. This example also beautifully highlights the importance of advanced bioinformatics tools for biomarker development.

As discussed in an accompanying commentary [29], the critical next steps are performance validations in prospective cohorts of more diverse participants in terms of ethnicity, age, gender, and confounding comorbidities. Furthermore, to successfully integrate ceramide measurements into widely adopted, routine clinical diagnostics, it will be important to include updated guidelines for clinical decision making and regulations for financial reimbursement [25].

With a growing number of international studies and clinical trials quantifying ceramides, one wouldn't be surprised to see the next iteration of the panel fine-tuned even further to defined pathophysiology or specific populations. This would pave the way for more individualized patient care – particularly in this era of precision medicine – both for primary prevention (i.e. personalized therapeutic or lifestyle interventions) and secondary prevention (i.e. identification of CHD patients at greatest risk for secondary events) [25].

From a technological point of view, two performance characteristics of MS were the main enabler of this emerging ceramide success story. First, concentrations of circulating ceramides are so low that only MS-based screens are able to initially detect these molecules during the research phase. Second, the versatility of being able to rapidly modify MS assays to include additional molecular species is a major advantage of MS over antibody-based technologies.

20.3 Bringing Lipidomics to the Clinic: Overcoming Current Challenges and Barriers

The examples set out in Section 20.2 describe assays for single lipid metabolites or small lipid panels, but cannot account for wide lipidomic coverage. This begs the question: What would it take to bring comprehensive lipidomics to clinical diagnostics? And what advantages would be brought about by the coverage of a broader lipidomic readout?

The overarching promise of applying comprehensive, broad lipidomics to clinical diagnostics is to use this powerful technology for the benefit of patient clinical diagnostics; i.e. more precise, personalized, or plainly novel diagnostic capabilities. To achieve successful clinical translation, lipidomics tests would not only need to be technologically and clinically reliable, but also cost-effective, and – perhaps, most importantly – actionable and relevant for patient management [30]. In this section, we discuss some of the barriers that need to be overcome to successfully translate lipidomics readouts into the clinic.

20.3.1 Raison D'être for Lipidomics in Patient Care: What Is the Clinical Utility?

First, we need to address why it would be useful to implement lipidomic readouts into clinical diagnostics. The clear value of lipidomics in a research context (i.e. "being interesting") does not necessarily translate into "being useful" in a medical context. Indeed, for any omics-based readout to have clinical utility, the results/readouts need to be actionable; for example, they need to be truly relevant for a differential diagnosis, risk stratification, or therapeutic intervention [30–32].

Lipidomics – with its promise to measure everything there is – offers a unique opportunity to depict in a systems-wide manner of what is happening in the human body and thus paint a picture of fundamental ongoing physiological processes, including those in metabolism and inflammation. Covering the involved pathways broadly (by measuring all involved lipid metabolites) holds the advantages of providing a comprehensive depiction of all metabolic systems in process, and enables an understanding of the underlying pathophysiology and possibly opportunities for therapeutic interventions.

A prime example of this is an understanding of inflammation and its resolution: two processes that are molecularly orchestrated by lipid mediators. These lipid mediators, such as oxylipins, stem primarily from the oxidation of polyunsaturated fatty acids (PUFAs) either enzymatically (i.e. via the oxidative enzymes lipooxygenases [LOX], cyclooxygenases [COX], and cytochrome P450 epoxygenases/hydrolases [33]) or non-enzymatically (i.e. via free radical-induced peroxidation), and leading to the formation of isoprostanoids, such as F2-isoprostanes (F2-isoPs) [34]. Profiling these markers for inflammation and oxidative stress – along with their sheer number of metabolites – promises to enable a deep understanding of the pathophysiology of inflammatory conditions [35]. Potential medically relevant applications include therapeutics (i.e. targeting the COX pathway via non-steroidal anti-inflammatory drugs like Aspirin), lifestyle interventions (i.e. dietary changes [36, 37]), and diagnostics and prognostics (i.e. biomarker development). Indeed, disease-associated oxylipin biomarkers have been described in multiple cohort studies for a variety of diseases; examples include oxylipins in plasma (i.e. blood pressure [38], nonalcoholic steatohepatitis [NASH] [39]) and serum (CAD outcomes [40]), tears (i.e. ocular inflammation [41]), and urine (i.e. asthma [42]).

For F2-isoP levels, numerous cohort studies have shown their strong associations with diseases such as diabetes types 1 [43] and 2 [44], hypertension [45], CHD [46], and neurodegenerative diseases, such as Alzheimer's disease [47], and age-related brain injuries [48]. As such, F2-isoP is regarded as the gold standard marker for (inflammation-associated) oxidative stress. Indeed, leading clinical laboratories, such as Quest and the Mayo clinic, offer LC-MS/MS-based clinical quantification of these biomarkers (see also Section 20.4.7). Yet, despite this commercial success, the clinical utility of measuring F2-isoPs is debated, as elevated levels of these molecules only indicate a generalized systemic oxidative stress, and fail to specify any particular disease [49]. This lack of specificity severely limits the utility of this marker, as there can be no specific actionable result [50].

Similarly, the diagnostic clinical utility for oxylipins is limited. It is tempting to speculate that a broader molecular depiction of an inflammatory condition (i.e. even a personalized fingerprint for an inflammatory condition resulting from a particular disease) might enable a targeted medical decision; i.e. allow for specific, personalized therapeutic interventions. But while promising biomarkers have emerged in disease-specific, case–control studies, these typically do not show the same differentiating power in prospective population studies, where a variety of confounding inflammatory conditions are also prevalent. For example, a promising diagnostic application is the use of eicosanoids (an oxylipin) to stratify severe asthmatic patients for treatment planning (e.g. with Omalizumab, Mepolizumab). Initial cross-sectional studies have suggested that eicosanoids are useful biomarkers of asthma type 2 inflammation [42]; however, a real-life study with Omalizumab showed little ability of urinary eicosanoids to predict individual response to treatment (Craig Wheelock, Karolinska Institute; personal communications).

In addition to a lack of disease-specificity, another major challenge of using these lipid mediators for clinical diagnostics is their bioreactivity. This is particularly observed for the enzymatically derived lipid mediators of the eicosanoid family, which are prone generating *ex vivo* artifacts during sample collection [51] and storage [52]. Hence, developing preanalytical workflows (including sampling, storage, use of antioxidants, and anticoagulants) that preserve the metabolite compositional integrity of the sample while still being robust and simple enough to be compatible with routine clinical workflows poses a major unsolved challenge [53–56]. Thus, despite having undisputed potency in mediating physiological responses and their fundamental role in many disease processes, the clinical utility of markers for inflammation and oxidative stress remains unclear.

Independent of the exact lipid metabolites to be measured, to work toward establishing clinical utility, a dialogue between analytical laboratory professionals and physicians is crucial to assess the (hitherto unmet) clinical needs of future tests. The following questions remain: Where in the clinical pathway would having a lipidomic readout provide an advantage to clinical care and patient outcome? What will be the key performance characteristics of future tests (necessary sensitivities, turnaround time, throughput, robustness, etc.)? The framework established by the European Federation for Laboratory Medicine (EFLM) for an "unmet needs assessment" will be a particularly useful guideline and online tool for answering these questions [57, 58].

20.3.2 The Reproducibility Issue: Is It Time to Harmonize?

Regardless of the measured analytes themselves, be they a complex lipidomic pattern or simpler individual analytes, the obtained measurements need to be accurate, precise, reproducible, and preferably independent of the analytical platform. However, the plethora of available lipidomic workflows and analytical approaches has led to the reporting of conflicting results for lipid values, as shown in efforts coordinated by the LIPID MAPS Consortium [59, 60] and the Interlaboratory Lipidomics Comparison Exercise [59, 60]. This variability in results might explain why preclinical studies to date have generated lots of promising biomarkers but few translatable ones: results are often not reproducible across time and space [61]. This issue is also particularly challenging for untargeted workflows [62].

Consequently, there is a growing awareness that the field would benefit from harmonized analytical workflows [63, 64]. To address this need, the international community has started to collaborate toward generating reproducible and repeatable measurements for lipid values (preferably in molar concentrations). Another approach to derive normalized data across different analytical platforms could be to use shared reference materials, as described by Triebl et al. [65] in their comparison of platforms used in lipidomic labs in Australia and Singapore. Similarly, normalization to reference material or quality control samples would help to harmonize lipid molar concentrations across different analytical platforms [66, 67]. This has already been explored by approaches to harmonize workflows using ring trials and interlaboratory studies for analytes with either demonstrated or potential clinical use such as corticosteroids [68], total oxylipins [69], vitamin D metabolites [70, 71], and cholesterol and non-cholesterol sterols [72]. Typically, these studies use the same shared reference material and more or less defined workflows.

There are other notable ongoing harmonization efforts coordinated by the International Lipidomics Society (ILS) in partnership with National Institute of Standards and Technology (NIST) and lipid standard manufacturers. These efforts include inter-laboratory ring trials for a small panel (four lipid molecular species) of ceramides, for a slightly larger panel (two dozen) of bile acids, and an upcoming ring trial for the "lipidome panel" comprising about 160 molecular species, representing major lipid metabolic pathways. Importantly, all participating laboratories use the same shared reference materials; in this case, NIST's standard reference material (SRM) 1950 and additional materials in development [73]. They also all follow well-described, "harmonized" workflows. These community-driven efforts, with over 30 laboratories globally participating, can give us an understanding of the similarity of molar concentrations among many different laboratories. Ideally, harmonized workflows would lead to less variability in reported lipid concentrations and, hence, tighter consensus values.

In addition to the above-described ring trials and inter-laboratory studies, which offer a relative freedom in the choice of analytical platform, another way to obtain tighter consensus values could be to use commercial platforms with validated and "locked-in" methods. This has recently been highlighted using products from different companies such as the Lipidyzer platform [74] and commercially available kits [75–77]. The use of corporate service laboratories, primarily based on shotgun lipidomics, might offer another alternative to deliver high-throughput plasma lipidomics for future (pre-)clinical applications [78].

Finally, harmonization is also challenged by difficulties associated with integrating large-scale, interlaboratory data sets; one possible solution is LipidLynxX by Fedorova's group [79]. In other efforts, the Interlaboratory Lipidomics Comparison Exercise, Ulmer et al. developed LipidQC, a method validation tool for visual comparison [80], and Izumi et al. published the challenging data integration of an interlaboratory study for cellular metabolite metabolites [81]. The Lipidomics Standards Initiative (LSI) is spearheading the drafting of common standards for minimum acceptable data quality and reporting [63]: this remains a major challenge on the path to the clinical translation of lipidomic workflows, and will need joint efforts by all stakeholders to establish internationally accepted guidelines for the harmonized reporting of lipidomics data from various laboratories and workflows.

20.3.3 From Consensus Values to Reference Intervals and True Values

To make lipidomic readouts useful for clinical applications, the biological boundaries or ranges of natural variation in the human population need to be known [35]. Hence, as a next step, reference ranges – otherwise known as reference intervals – for lipids of clinical significance must be established [82]. A general overview of the suitable approaches on how to obtain reference intervals has been assembled by the Clinical and Laboratory Standards Institute (CLSI) [83].

Of course, for analytes already measured routinely in clinical diagnostics, the reference values and biological variations (i.e. natural variability) have already been established [84, 85]. Notably, with advances in the analytical performance of MS-based methods, reference intervals need to be updated, as demonstrated by Ceglarek's group for sex steroid hormones [86]. A non-exhaustive list of lipid metabolites for which reference values exist includes fatty acids [87, 88], acylcarnitines [89], oxysterols [90], and sphingosine-1-phosphate [91].

The LIPID MAPS Consortium [59, 60] and the Interlaboratory Lipidomics Comparison Exercise [59, 60] set the groundwork for establishing consensus values for hundreds of plasma lipids. These attempts, by means of ongoing community efforts such as the aforementioned ring trials and interlaboratory studies, will hopefully result in tighter consensus values, if not even one day accepted reference intervals. As a next important step, molar concentrations need to be determined in plasma samples that represent a more clinically relevant "real-life" situation as opposed to the processed reference materials from pooled donors such as SRMs. A notable recent effort in this regard is described by Ghorasaini et al., bringing us one a step closer to internationally trusted reference values for hundreds of plasma lipids [74].

Concordantly, the above-described efforts by the International Lipidomics Society to harmonize ceramide measurements will require the work of multiple laboratories using harmonized workflows to quantify ceramide concentrations in various diverse populations; these findings will pave the way to establishing ceramide reference intervals.

For the vast majority of plasma lipids quantified in previous and in ongoing studies, only (community-derived) consensus values are known, and not their true, accurate values. Hence, one can only judge how *precise* these analytical workflows are (based on CoV of repeated measurements of the same material) and not how accurately they provide true concentrations. To obtain (commonly accepted) true values, a standardized "reference measurement procedure" (RMP) would need to be established, as is the case for traditional clinical markers such as cholesterol and vitamin D [92]. For an overview of this concept and available RMPs, please refer to the work of the Joint Committee for Traceability in Laboratory Medicine (JCTLM). The development of such RMPs and accompanying certified reference materials with assigned concentration values for each lipid will facilitate the future development of external quality assurance (EQA) programs. For select analytes of clinical interest, such as total triacylglycerols, vitamin D, sex steroid hormones, cholesterols, and others, this kind of program already exists; for instance, as organized by the Centers for Disease Control and Prevention (CDC) [93]. An internationally available EQA program for novel plasma lipid assays might therefore be the next logical step toward clinical implementation.

20.4 Examples of Currently Existing Diagnostic Applications

In this section, we describe select lipid metabolite assays that are currently available for routine clinical diagnostics. These are assays for single lipid metabolites or for small lipid panels, but not for the lipidome in its entirety. What connects all these analytes is that their diagnostic value (and hence clinical utility) has already been ascertained: test results have a direct impact on the physician's decision for patient treatment. (This demonstrated diagnostic value justifies the considerable efforts of establishing and maintaining a routine MS-based test for these analytes.) In some cases, such as for sex steroid hormones, analytes were initially measured with radio-immunoassays and methods other than MS. However, once MS-based assays were shown to deliver superior patient care (i.e. in terms of better sensitivity and specificity), the assays were replaced by MS; this knowledge provides a valuable signpost for the research community to understand where MS can add value.

20.4.1 Mitochondrial Fatty Acid β-Oxidation and Organic Acid Metabolism

The newborn screening (NBS) program is a public health initiative first established in the USA in the 1960s to detect phenylketonuria, an inborn error in phenylalanine amino acid metabolism that is characterized by a marked increase in phenylalanine and its metabolites in the blood. This screening program was launched with a preventative approach in mind; i.e. the ability to detect congenital issues before signs of clinical manifestations in order to intervene early and achieve a better clinical outcome [94, 95].

Over the years, the NBS program has evolved to mirror technological advances in the field and the progressive changes in our understanding of various genetic conditions. Notably, the advent of LC-MS methods in the 1990s enabled the simultaneous diagnostic monitoring of multiple metabolites [96]. Acylcarnitines, for example, are lipid biomarkers routinely measured today in the USA (and in many other countries) within the first 24–72 hours of birth to detect for signs of two groups of disorders: (i) fatty acid oxidation disorders and (ii) organic acid disorders (organic acidemia). Both disorders are autosomal recessive conditions, which, when left unchecked, can result in seizures, liver damage, intellectual disability, and in the worst case, death [95]. Due to the multiplexing approach used with LC-MS, one analytical run typically incorporates a panel of acylcarnitines and amino acids, and can screen between 30 and 50 inherited disorders [97, 98]. Abnormal acylcarnitine levels, especially those following a certain pattern, while not diagnostic, may be suggestive of a particular disorder [99]. For instance, elevated levels of multiple medium-chain acylcarnitines, such as octanoyl (C8), hexanoyl (C6), and decanoyl (C10) carnitine, may be indicative of medium-chain acyl-CoA dehydrogenase (MCAD) deficiency, which is the most common of the fatty acid oxidation disorders [100, 101]. On the other hand, abnormal values of hexadecanoyl (C16) or octadecenoyl (C18:1) carnitines may point toward either a carnitine/acylcarnitine translocase deficiency (CACT) or carnitine palmitoyltransferase deficiency, type 2 (CPT2) [102]. If initial MS tests show abnormal signs in certain acylcarnitine levels, further testing is required to confirm a diagnosis [103]. LC-MS analysis of dried blood spots is commonly used for NBS, whereas urine acylcarnitine analysis via gas-chromatography MS (GC-MS) is preferred for confirming organic acidemia [104, 105].

In the past, many of these inherited metabolic disorders were not well understood and since affected infants have a high risk of fatality during the first few episodes of hypoglycemia, many of these infant deaths were wrongly attributed to sudden infant death syndrome (SIDS). These days, such misdiagnoses are preventable, and early diagnostic testing of lipid biomarkers offers doctors an opportunistic intervention and treatment window [96].

It is important to note that, although we discuss acylcarnitines as part of the NBS program, this program is not restricted to the disorders listed above. There are many others, and different countries or regions within the same country will often screen for different conditions. For instance, Fabry disease, Gaucher disease, and congenital adrenal hyperplasia (CAH) are other disorders also part of the NBS program, whether through routine testing or as part of a pilot project [106, 107].

20.4.2 Fabry Disease

Fabry disease is an X-linked lysosomal storage disorder characterized by impaired enzymatic activity of alpha-galactosidase A (α-Gal A, also known as GLA and α-GAL). This impaired enzyme activity, which occurs in response to mutations in the gene encoding for the enzyme, causes a systemic accumulation of the glycosphingolipid globotriaosylceramide (Gb3) that primarily affects the heart, kidney, and nervous system. As bearers of a single X chromosome, males are typically more severely impacted by the condition than females; females have two X chromosomes, and thus a healthy copy is able to compensate (to some extent) for the enzyme deficiency caused by the mutated copy [108, 109]. There are over 1000 reported mutations of the *GLA* gene responsible for Fabry disease; the severity and range of symptoms for both males and females depend on the effect of each specific mutation on the levels of α-Gal A activity in the body [110].

There are two major phenotypes of Fabry disease: classic and late-onset. Patients with little to no α-Gal A activity will show clinical symptoms in early childhood, which may include acroparesthesia, angiokeratoma, and corneal opacity. Patients with some residual enzyme activity will develop late-onset Fabry disease, which often causes milder symptoms and only manifests at a much later stage in life (between 30 and 70 years of age). For both phenotypes, there may be progressive organ dysfunction leading to eventual renal failure and various cardiac and cerebral complications [111, 112].

Males with the classic phenotype can be reliably diagnosed by measuring α-Gal A levels in different matrices, such as in plasma, dried blood spots, and leukocytes. Even those who have reduced enzyme activity – and thus present an attenuated phenotype – will still show values that are well below reference values. This diagnosis is not as straightforward for females; heterozygous women may have normal levels of α-Gal A activity and will thus require complementary diagnostic testing [113]. This is where MS comes in. Such complementary diagnostic tests involve MS-based measurements of glycosphingolipids, such as Gb3 in urine [114] and deacylated Gb3 or lyso-globotriaosylceramide (LysoGb3) in plasma/serum [115]. Gb3 and LysoGb3 are diagnostic markers that enable differentiation of phenotypes caused by the various mutations, especially if symptoms are overlapping with other X-linked lysosomal storage disorders, or to assess individuals with novel *GLA* variants [113]. Further, these tests may help to monitor disease progression by taking metabolite measurements before and after treatment [116].

20.4.3 Gaucher Disease

Gaucher disease (GD), like Fabry disease, is an X-linked lysosomal storage disorder. However, it is an autosomal recessive condition, caused by inheriting two copies of the mutated beta-glucocerebrosidase (*GBA*) gene, one from each parent. The absence of a sufficient level of GBA activity results in an accumulation of its natural substrate, glucosylceramide (Gb1), most predominantly in macrophages [117]. Three types of GD exist, based on the absence (GD1) or presence (GD2 or GD3) of neurological involvement. However, the clinical manifestations of GD are dramatically heterogenous: they range from being asymptomatic to having fatal neonatal complications [118, 119]. This heterogeneity may, at least in part, be attributed to the different adaptions in metabolism that occur during GBA deficiency, one of which involves active deacylation of Gb1 to lyso-Gb1 [120]. Lyso-Gb1 was first isolated from the spleen of a GD patient in 1974 [121]. Since then, lyso Gb1 accumulation continues to be discovered in various tissues, including the brain, liver, and spleen [120]. Lyso-Gb1 is therefore highly valued as a diagnostic biomarker for GD; indeed, the development of efficient and accurate MS methods has anchored its place in GD testing [122, 123]. In 2018, Centogene, a company that specializes in rare diseases, published a paper in collaboration with researchers from the Gaucher unit at Hebrew University that confirms lyso-Gb1 as the best response biomarker with which to monitor disease progression after commencing enzyme replacement therapy (ERT) [124].

20.4.4 Minimally Invasive Diagnostic Testing for NAFLD/NASH

Up until recently, diagnosing NAFLD from NASH posed a huge challenge to health care practitioners, as both conditions remain largely asymptomatic until an advanced stage of disease progression. NASH is a more progressive condition of NAFLD, and may eventually lead to fibrosis, cirrhosis, and hepatocellular carcinoma. It is estimated that, in Western countries, the prevalence of NAFLD ranges between 10% and 40%, with approximately 10% to 30% of those with NAFLD progressing to develop NASH [125].

Prior to the inception of MS-based tests, the gold standard for discriminating between the two phenotypes was a liver biopsy. Yet, the benefit of performing such an invasive test was often questioned, particularly given that there are no approved NAFLD-specific treatments, except certain proposed lifestyle modifications (regardless of the stage of progression) [126].

Presently, the OWLiver Test offers a non-invasive, blood-based MS test to differentiate between NAFLD and NASH, with the goal to aid in early detection for personalized follow-up and lifestyle modifications. The OWLiver Test is based on a Spanish study with discovery ($n = 467$) and validation ($n = 192$) cohorts. The study measured over 500 lipids and amino acids in serum samples matched with proven biopsies from subjects with normal liver (NL), NAFLD, and NASH. The group identified two panels with a total of 28 triacylglycerols (TGs): 11 TGs formed the first panel and could significantly distinguish between NL and NAFLD, but not NASH and NAFLD, whereas the second panel, comprising 20 TGs, distinguished with high accuracy NASH and NAFLD, but not NL from NASH [125].

Although this simple serum lipidomic test can establish the degree of severity of steatosis/steatohepatitis, its results need to be interpreted with caution. More validation is required when applying this test to other cohorts. Bril et al. found that the OWLiver test worked sub-optimally for a multi-ethnic cohort with type 2 diabetes mellitus (T2DM) but performed well when a selective subset of subjects was chosen to reflect the demographic profile of the original cohort [127].

20.4.5 Intrahepatic Cholestasis of Pregnancy

Intrahepatic cholestasis of pregnancy (ICP) or obstetric cholestasis (OC) is a reversible but pregnancy-specific hepatic condition that spontaneously resolves after delivery [128]. It occurs, on average, in about 0.1% to 2% of all pregnancies, but its prevalence is geographically varied, with countries like Chile and Bolivia reporting higher incidence rates [129]. The most common symptom is severe itching on the hands and feet but may sometimes include symptoms of dark urine, loss of appetite, nausea, or jaundice. ICP raises the risk of perinatal complications, such as meconium staining of amniotic fluid, a slowed fetal heart rate, and stillbirth [128, 130].

Cholestasis refers to a condition in which the flow of bile from the liver to the intestinal tract is greatly reduced or blocked; this, in turn, leads to an accumulation of bile acids in the liver and systemic circulation [131]. As such, determining fractionated and total serum bile acid levels is a common diagnostic measure for ICP. Five bile acids make up 99% of the total bile composition, including ursodeoxylcholic acid, deoxycholic acid, chenodeoxycholic acid, cholic acid, lithocholic, and their conjugates. Even though enzymatic methods are still widely used to measure bile acid levels, LC-MS-based assays can be ordered from medical centers such as the Mayo clinic and Labcorp [132, 133], usually with four of bile acids (and their taurine and glycine conjugates) able to be measured. Of note, although measuring bile acids aids in the evaluation of liver function, bile acid levels may also be elevated in patients suffering from other hepatobiliary illnesses and other causes of liver dysfunction, which may need to be excluded in the diagnosis of ICP [128].

20.4.6 Steroid Hormone Measurements for CAH and Vitamin D Deficiency

20.4.6.1 Congenital Adrenal Hyperplasia

The accurate analysis of adrenal and sex steroid hormones is crucial for the correct diagnosis of several endocrine disorders, such as CAH, hyperandrogenism, Cushing's syndrome, and polycystic ovarian syndrome (PCOS) [134]. MS methods and immunoassays (IAs) for steroid analyses were developed in parallel; the implementation of both methods dates to more than five decades ago [135, 136]. Between the two, it was IA that was more widely adopted initially: users cited its simplicity, cost, and sensitivity as predominant reasons for its preferential use. A major drawback, however, was its cross-reactivity, caused by structurally related endogenous or exogenous (e.g. medicative drugs) molecules. Such cross-reactivity led to false-positive results [107].

CAH is an inherited disorder, commonly (95%) caused by a deficiency in 21-hydroxylase [137]. CAH is primarily diagnosed by elevated levels of 17-hydroxyprogesterone (17-OHP) in the blood or serum. NBS for CAH commenced worldwide upon the development of a radioimmunoassay by Pang et al. in 1977 [138]. Since then, other forms of immunoassays – enzyme-linked immunosorbent assays (ELISA) and automated time-resolved dissociation-enhanced lanthanide fluoroimmunoassay (DELIFA) – have been adopted to counteract the high false-positive rates (1–2%) associated with this initial screening technique; albeit, without much success [139, 140]. This prompted scientists to introduce a two-tiered approach for diagnosis: all dried blood samples were required to go through an initial IA to measure 17-OHP, with positive cases then followed up with a second tier of steroid profiling. First described by Lacey et al. in 2004, second-tier steroid profiling using LC-MS/MS remains the gold standard in this field [141]. The Endocrine Society issued its updated guidelines in 2018 recommending the use of LC-MS/MS over all other methods to improve the positive predictive value of CAH screening [140]. Although the initial method [141] measures three analytes (17-OHP, androstenedione [4-AD], and cortisol), subsequent reports show that measuring additional analytes, such as 21-deoxycortisol and 11-deoxycortisol, may improve screening results [107, 140, 142].

20.4.6.2 Vitamin D Deficiency

As mentioned earlier (see Section 20.2.1 Vitamin D and its metabolites), Vitamin D deficiency is generally accepted as a worldwide health concern associated with myriad conditions, such as type 1 diabetes, cardiovascular disease, some cancers, osteoporosis, bone disorders, frailty, and rickets (in children) [143, 144]. We elaborate a little more here on the MS measurements of Vitamin D.

The two major forms of Vitamin D – Vitamin D2 (ergocalciferol) and Vitamin D3 (cholecalciferol) – can be gained from the diet or from supplements, whereas Vitamin D3 is the only form that is derived from the conversion of 7-dehydrocholesterol in the skin upon exposure to UV light. In the liver, 25-hydroxylase catalyzes the formation of 25-hydroxyvitamin D2/D3 collectively known as 25-OHD. Further metabolism in the kidney generates the metabolites 1-alpha,25-dihydroxyvitamin-D3 ($1\alpha,25(OH)_2D3$), 1-alpha,25-dihydroxyvitamin-D2 ($1\alpha,25(OH)_2D2$), and a minor metabolite, $24,25(OH)_2D3$. Although more than 50 Vitamin D metabolites have been discovered – enabling us to dwell into the whole Vitamin D metabolome – only the ones mentioned above have been recognized as being clinically relevant [144, 145]. These major vitamin D forms can undergo epimerization at the C3 position, with some studies indicating a role for Vitamin D C3-epimers in inflammatory diseases, such as rheumatoid arthritis [146].

Vitamin D analysis is rife with problems. Although LC-MS/MS is currently considered the gold standard for Vitamin D metabolite analysis, financial constraints, an increasing Vitamin D workload, and the higher level of expertise required to run LC-MS workflows keep many clinical laboratories from swapping out high-throughput, easy-to-use IAs for LC-MS. However, LC-MS methods are also not without their issues; 25(OH)D3 needs to be chromatographically separated from its C3 epimer for accurate measurements (especially in children under one year) and not all LC-MS workflows provide that kind of resolution [147]. Albeit, the major labs, such as the Mayo Clinic, Quest Diagnostics Inc., Arup Laboratories, and LabCorp, all offer 25(OH)D2 and D3 testing via LC-MS/MS [148–151].

20.4.7 F2-Isoprostanes as Markers of Oxidative Stress

Free radicals are typically generated during normal cellular metabolism, but are also produced when the body is exposed to UV radiation, cigarette smoke, pollution, an improper diet, among other factors. They are highly reactive atoms containing one or more unpaired electrons, and mainly belong to one of the three main families of reactive species: reactive nitrogen species (RNS), reactive oxygen species (ROS), and reactive chloride species (RCS) [152, 153]. Oxidative stress in the body is said to occur when there is an imbalance of reactive species (most commonly ROS) over antioxidants. Free radicals attack biomolecules such as proteins, lipids, and DNA, creating a cascade of additional free radicals along the way, causing damage to cells, and, consequently, a multitude of human ailments [154].

How can we assess oxidative stress in the body? F2-isoprostanes (F2-IsoPs) are a series of prostaglandin-like compounds that arise upon free-radical-mediated oxidation of arachidonic acid (AA) in plasma membranes. These molecules, first discovered in 1990 [155], have been studied extensively and shown over time to be reliable and stable biomarkers for measuring oxidative stress [156, 157]. They are ubiquitously present within the body and can be detected in all human biological fluids, such as plasma, urine, cerebrospinal fluid, and bile. Although plasma and urine are most commonly chosen for F2-isoprostane measurements for their non-invasive nature, the latter is still predominantly used to quantify F2-isoporostanes and characterize the levels of systemic oxidative stress [154].

Clinically, oxidative stress is assessed by measuring the ratio of 15-F2t-isoprostane to creatinine via either GC-MS or LC-MS in urine, and elevated levels are associated with elevated risks of atherosclerosis, hypercholesterolemia, diabetes, obesity, and rheumatoid arthritis, among others [49, 154, 158]. Since F2-IsoPS are unaffected by dietary lipids, their measurements in biological fluids provide an estimate of overall oxidative status whereas esterified F2-IsoPs in tissues offer insight into localized stress [159]. This test demonstrates superior sensitivity in measuring the oxidative status of the body but is not specific to any particular condition, thereby limiting its clinical utility.

20.5 Final Comments

Lipid metabolites are supreme candidates for depicting what is going on in the human body since they are intimately involved in (if not responsible for) all major metabolic pathways and processes that govern health and disease. The growing awareness of a diagnostic potential of comprehensive metabolite analyses ("metabolomics") is evident in the growing number of scientific publications describing its use for clinical applications. Demonstrating this point, the leading international body representing clinical diagnostics professionals, the International Federation for Clinical Chemistry, IFCC, recently created a new working group under their "Emerging Technologies" division which is chartered with exploring the potential of metabolomics for clinical diagnostics [160] (disclaimer: one of us, AKB, is an elected member/serves on this WG). To bring these exciting emerging initiatives to fruition, what is needed next is communication between all stakeholders, including professionals from analytical laboratories, clinicians, data scientists, manufacturers, and regulatory bodies, to jointly tackle the multiple hurdles facing the clinical implementation of lipidomics [161] in order to ultimately provide a positive impact to patients and society at large.

References

1 McGranaghan, P., Kirwan, J.A., Garcia-Rivera, M.A. et al. (2021). Lipid metabolite biomarkers in cardiovascular disease: discovery and biomechanism translation from human studies. *Metabolites* 11: 621.

2 Alesi, S., Ghelani, D., Rassie, K. et al. (2021). Metabolomic biomarkers in gestational diabetes mellitus: a review of the evidence. *Int. J. Mol. Sci.* 22: https://doi.org/10.3390/ijms22115512.

3 Vvedenskaya, O., Rose, T.D., Knittelfelder, O. et al. (2021). Nonalcoholic fatty liver disease stratification by liver lipidomics. *J. Lipid Res.* 62: 100104. https://doi.org/10.1016/j.jlr.2021.100104.

4 He, D., Su, Y., Meng, D. et al. (2021). A pilot study optimizing metabolomic and lipidomic acquisition in serum for biomarker discovery in nonalcoholic fatty liver disease. *J. Mass Spectrom. Adv. Clin. Lab* 22: 17–25. https://doi.org/10.1016/j.jmsacl.2021.10.001.

5 Kim, J. and Jeong, Y. (2013). Augmentation of sensory-evoked hemodynamic response in an early Alzheimer's disease mouse model. *J. Alzheimers Dis.* 37: 857–868. https://doi.org/10.3233/jad-121900.

6 Ferreira, H.B., Melo, T., Monteiro, A. et al. (2021). Serum phospholipidomics reveals altered lipid profile and promising biomarkers in multiple sclerosis. *Arch. Biochem. Biophys.* 697: 108672. https://doi.org/10.1016/j.abb.2020.108672.

7 Agarwala, P.K., Aneja, R., Kapoor, S. et al. (2022). Lipidomic landscape in cancer: actionable insights for membrane-based therapy and diagnoses. *Med. Res. Rev.* 42: 983–1018. https://doi.org/10.1002/med.21868.

8 Wang, S., Tang, K., Lu, Y. et al. (2021). Revealing the role of glycerophospholipid metabolism in asthma through plasma lipidomics. *Clin. Chim. Acta* 513: 34–42. https://doi.org/10.1016/j.cca.2020.11.026.

9 Wang, H., Cui, L., Jia, Y. et al. (2020). Application of lipidomics to reveal differences of facial skin surface lipids between atopic dermatitis and healthy infants. *J. Cosmet. Dermatol.* 19: 1528–1534. https://doi.org/10.1111/jocd.13188.

10 Yazd, H.S., Bazargani, S.F., Vanbeek, C.A. et al. (2021). LC-MS lipidomics of renal biopsies for the diagnosis of Fabry disease. *J. Mass Spectrom. Adv Clin. Lab* 22: 71–78. https://doi.org/10.1016/j.jmsacl.2021.11.004.

11 Song, J.W., Lam, S.M., Fan, X. et al. (2020). Omics-driven systems interrogation of metabolic dysregulation in covid-19 pathogenesis. *Cell Metab.* 32: 188–202.e185. https://doi.org/10.1016/j.cmet.2020.06.016.

12 Kirwan, J.A., Kaddurah-Daouk, R., Mitchell, T. et al. (2019). Biobanking for metabolomics and lipidomics in precision medicine. *Clin. Chem.* 65: 827–832. https://doi.org/10.1373/clinchem.2018.298620.

13 Chemistry, A.A.f.C (2022). Universal sample bank. https://www.aacc.org/science-and-research/universal-sample-bank (accessed 18 February 2022).

14 Cook, N.R. (2013). Vitamin d and cancer: can we believe the evidence from observational studies? *Clin. Chem.* 59: 726–728. https://doi.org/10.1373/clinchem.2013.203158.

15 Wiśniewska, A. and Szypowska, A. (2021). The role of vitamin D in selected autoimmune diseases. *Rocz. Panstw. Zakl. Hig.* 72: 111–121. https://doi.org/10.32394/rpzh.2021.0156.

16 Janjusevic, M., Gagno, G., Fluca, A.L. et al. (2022). The peculiar role of vitamin D in the pathophysiology of cardiovascular and neurodegenerative diseases. *Life Sci.* 289: 120193. https://doi.org/10.1016/j.lfs.2021.120193.

17 Park, J.E., Pichiah, P.B.T., Cha, Y.S. et al. (2018). Vitamin D and metabolic diseases: growing roles of vitamin D. *J. Obes. Metab. Syndr.* 27: 223–232. https://doi.org/10.7570/jomes.2018.27.4.223.

18 Vogeser, M. (2010). Quantification of circulating 25-hydroxyvitamin D by liquid chromatography-tandem mass spectrometry. *J. Steroid Biochem. Mol. Biol.* 121: 565–573. https://doi.org/10.1016/j.jsbmb.2010.02.025.

19 Strathmann, F.G., Sadilkova, K., Laha, T.J. et al. (2012). 3-*epi*-25 hydroxyvitamin D concentrations are not correlated with age in a cohort of infants and adults. *Clin. Chim. Acta* 413: 203–206. https://doi.org/10.1016/j.cca.2011.09.028.

20 Zelzer, S., Goessler, W., Herrmann, M. et al. (2018). Measurement of vitamin D metabolites by mass spectrometry, an analytical challenge. *J. Lab. Precis. Med.* 3.

21 Ketha, H. and Singh, R.J. (2017). Vitamin D metabolite quantitation by LC-MS/MS, Chapter 9. In: *Mass Spectrometry for the Clinical Laboratory* (ed. H. Nair and W. Clarke), 181–204. San Diego: Academic Press.

22 Laaksonen, R., Ekroos, K., Sysi-Aho, M. et al. (2016). Plasma ceramides predict cardiovascular death in patients with stable coronary artery disease and acute coronary syndromes beyond LDL-cholesterol. *Eur. Heart J.* 37: 1967–1976. https://doi.org/10.1093/eurheartj/ehw148.

23 Havulinna, A.S., Sysi-Aho, M. et al. (2016). Circulating ceramides predict cardiovascular outcomes in the population-based Finrisk 2002 Cohort. *Arterioscler. Thromb. Vasc. Biol.* 36: 2424–2430. https://doi.org/10.1161/atvbaha.116.307497.

24 Summers, S.A. (2018). Could ceramides become the new cholesterol? *Cell Metab.* 27: 276–280. https://doi.org/10.1016/j.cmet.2017.12.003.

25 Hilvo, M., Vasile, V.C., Donato, L.J. et al. (2020). Ceramides and ceramide scores: clinical applications for cardiometabolic risk stratification. *Front. Endocrinol.* 11: https://doi.org/10.3389/fendo.2020.570628.

26 Hilvo, M., Meikle, P.J., Pedersen, E.R. et al. (2020). Development and validation of a ceramide- and phospholipid-based cardiovascular risk estimation score for coronary artery disease patients. *Eur. Heart J.* 41: 371–380. https://doi.org/10.1093/eurheartj/ehz387.

27 Hilvo, M., Wallentin, L., Lakic, T.G. et al. (2020). Prediction of residual risk by ceramide-phospholipid score in patients with stable coronary heart disease on optimal medical therapy. *J. Am. Heart Assoc.* 9: e015258. https://doi.org/10.1161/JAHA.119.015258.

28 Poss, A.M., Maschek, J.A., Cox, J.E. et al. (2020). Machine learning reveals serum sphingolipids as cholesterol-independent biomarkers of coronary artery disease. *J. Clin. Invest.* 130: 1363–1376. https://doi.org/10.1172/JCI131838.

29 Echouffo-Tcheugui, J.B., Jain, M., Cheng, S. et al. (2020). Breaking through the surface: more to learn about lipids and cardiovascular disease. *J. Clin. Invest.* 130: 1084–1086. https://doi.org/10.1172/JCI134696.

30 Parker, L.A., Chilet-Rosell, E., Hernández-Aguado, I. et al. (2018). Diagnostic biomarkers: are we moving from discovery to clinical application? *Clin. Chem.* 64: 1657–1667. https://doi.org/10.1373/clinchem.2018.292854.

31 Heaney, L.M., Jones, D.J., Suzuki, T. et al. (2017). Mass spectrometry in medicine: a technology for the future? *Future Sci. OA* 3: FSO213. https://doi.org/10.4155/fsoa-2017-0053.

32 Banerjee, S. (2020). Empowering clinical diagnostics with mass spectrometry. *ACS Omega* 5: 2041–2048. https://doi.org/10.1021/acsomega.9b03764.

33 Hajeyah, A.A., Griffiths, W.J., Wang, Y. et al. (2020). The biosynthesis of enzymatically oxidized lipids. *Front. Endocrinol.* 11: https://doi.org/10.3389/fendo.2020.591819.

34 Galano, J.M., Lee, Y.Y., Oger, C. et al. (2017). Isoprostanes, neuroprostanes and phytoprostanes: an overview of 25 years of research in chemistry and biology. *Prog. Lipid Res.* 68: 83–108. https://doi.org/10.1016/j.plipres.2017.09.004.

35 Gladine, C. and Fedorova, M. (2021). The clinical translation of eicosanoids and other oxylipins, although challenging, should be actively pursued. *J. Mass Spectrom. Adv. Clin. Lab* 21: 27–30. https://doi.org/10.1016/j.jmsacl.2021.08.003.

36 Simopoulos, A.P., Serhan, C.N., Bazinet, R.P. et al. (2021). The need for precision nutrition, genetic variation and resolution in COVID-19 patients. *Mol. Aspects Med.* 77: 100943. https://doi.org/10.1016/j.mam.2021.100943.

37 Furse, S., Virtue, S., Snowden, S.G. et al. (2022). Dietary pufa drives diverse systems-level changes in lipid metabolism. *Mol. Metab.* 101457. https://doi.org/10.1016/j.molmet.2022.101457.

38 Palmu, J., Watrous, J.D., Mercader, K. et al. (2020). Eicosanoid inflammatory mediators are robustly associated with blood pressure in the general population. *J. Am. Heart Assoc.* 9: e017598. https://doi.org/10.1161/JAHA.120.017598.

39 Loomba, R., Quehenberger, O., Armando, A. et al. (2015). Polyunsaturated fatty acid metabolites as novel lipidomic biomarkers for noninvasive diagnosis of nonalcoholic steatohepatitis. *J. Lipid Res.* 56: 185–192. https://doi.org/10.1194/jlr.P055640.

40 Huang, C.-C., Chang, M.-T., Leu, H.-B. et al. (2020). Association of arachidonic acid-derived lipid mediators with subsequent onset of acute myocardial infarction in patients with coronary artery disease. *Sci. Rep.* 10: 8105. https://doi.org/10.1038/s41598-020-65014-z.

41 Ambaw, Y.A., Timbadia, D.P., Raida, M. et al. (2020). Profile of tear lipid mediator as a biomarker of inflammation for meibomian gland dysfunction and ocular surface diseases: standard operating procedures. *Ocul. Surf.* https://doi.org/10.1016/j.jtos.2020.09.008.

42 Kolmert, J., Gómez, C., Balgoma, D. et al. (2021). Urinary leukotriene E_4 and prostaglandin D_2 metabolites increase in adult and childhood severe asthma characterized by type 2 inflammation. A Clinical Observational Study. *Am. J. Respir. Crit. Care Med.* 203: 37–53. https://doi.org/10.1164/rccm.201909-1869OC.

43 Kauffman, L.D., Sokol, R.J., Jones, R.H. et al. (2003). Urinary F_2-isoprostanes in young healthy children at risk for type 1 diabetes mellitus. *Free Radic. Biol. Med.* 35: 551–557. https://doi.org/10.1016/s0891-5849(03)00333-2.

44 Il'yasova, D., Wong, B.J., Waterstone, A. et al. (2017). Systemic F_2-isoprostane levels in predisposition to obesity and type 2 diabetes: emphasis on racial differences. *Divers. Equal. Health Care* 14: 91–101. https://doi.org/10.21767/2049-5471.100098.

45 Melton, C.D., Luo, R., Wong, B.J. et al. (2017). Urinary F_2-isoprostanes and the risk of hypertension. *Ann. Epidemiol.* 27: 391–396. https://doi.org/10.1016/j.annepidem.2017.05.005.

46 Davies, S.S. and Roberts, L.J. 2nd (2011). F2-isoprostanes as an indicator and risk factor for coronary heart disease. *Free Radic. Biol. Med.* 50: 559–566. https://doi.org/10.1016/j.freeradbiomed.2010.11.023.

47 Montine, T.J., Peskind, E.R., Quinn, J.F. et al. (2011). Increased cerebrospinal fluid F_2-isoprostanes are associated with aging and latent Alzheimer's disease as identified by biomarkers. *Neuromol. Med.* 13: 37–43. https://doi.org/10.1007/s12017-010-8126-6.

48 Peskind, E.R., Li, G., Shofer, J.B. et al. (2014). Influence of lifestyle modifications on age-related free radical injury to brain. *JAMA Neurol.* 71: 1150–1154. https://doi.org/10.1001/jamaneurol.2014.1428.

49 Laboratories, M.C. (2013). F_2-isoprostanes, random, urine. https://www.mayocliniclabs.com/test-catalog/overview/607698.

50 Menzel, A., Samouda, H., Dohet, F. et al. (2021). Common and novel markers for measuring inflammation and oxidative stress ex vivo in research and clinical practice-which to use regarding disease outcomes? *Antioxidants (Basel)* 10: https://doi.org/10.3390/antiox10030414.

51 Polinski, K.J., Armstrong, M., Manke, J. et al. (2021). Collection and storage of human plasma for measurement of oxylipins. *Metabolites* 11: 137.

52 Koch, E., Mainka, M., Dalle, C. et al. (2020). Stability of oxylipins during plasma generation and long-term storage. *Talanta* 217: 121074. https://doi.org/10.1016/j.talanta.2020.121074.

53 Hyötyläinen, T. and Orešič, M. (2015). Optimizing the lipidomics workflow for clinical studies – practical considerations. *Anal. Bioanal. Chem.* 407: 4973–4993. https://doi.org/10.1007/s00216-015-8633-2.

54 Hahnefeld, L., Gurke, R., Thomas, D. et al. (2020). Implementation of lipidomics in clinical routine: can fluoride/citrate blood sampling tubes improve preanalytical stability? *Talanta* 209: 120593. https://doi.org/10.1016/j.talanta.2019.120593.

55 Lehmann, R. (2021). From bedside to bench – practical considerations to avoid pre-analytical pitfalls and assess sample quality for high-resolution metabolomics and lipidomics analyses of body fluids. *Anal. Bioanal. Chem.* 413: 5567–5585. https://doi.org/10.1007/s00216-021-03450-0.

56 Salvagno, G.L., Danese, E., Lippi, G. et al. (2017). Preanalytical variables for liquid chromatography-mass spectrometry (LC-MS) analysis of human blood specimens. *Clin. Biochem.* 50: 582–586. https://doi.org/10.1016/j.clinbiochem.2017.04.012.

57 Monaghan, P.J., Lord, S.J., St John, A. et al. (2016). Biomarker development targeting unmet clinical needs. *Clin. Chim. Acta* 460: 211–219. https://doi.org/10.1016/j.cca.2016.06.037.

58 Monaghan, P.J., Robinson, S., Rajdl, D. et al. (2018). Practical guide for identifying unmet clinical needs for biomarkers. *EJIFCC* 29: 129–137.

59 Bowden, J.A., Heckert, A., Ulmer, C.Z. et al. (2017). Harmonizing lipidomics: NIST interlaboratory comparison exercise for lipidomics using SRM 1950-metabolites in frozen human plasma. *J. Lipid Res.* 58: 2275–2288. https://doi.org/10.1194/jlr.M079012.

60 Quehenberger, O., Armando, A.M., Brown, A.H. et al. (2010). Lipidomics reveals a remarkable diversity of lipids in human plasma. *J. Lipid Res.* 51: 3299–3305. https://doi.org/10.1194/jlr.M009449.

61 Dias, D.A. and Koal, T. (2016). Progress in metabolomics standardisation and its significance in future clinical laboratory medicine. *EJIFCC* 27: 331–343.

62 Ulmer, C.Z., Maus, A., Hines, J. et al. (2021). Challenges in translating clinical metabolomics data sets from the bench to the bedside. *Clin. Chem.* 67: 1581–1583. https://doi.org/10.1093/clinchem/hvab210.

63 Liebisch, G., Ahrends, R., Arita, M. et al. (2019). Lipidomics needs more standardization. *Nat. Metabol.* 1: 745–747. https://doi.org/10.1038/s42255-019-0094-z.

64 Meikle, T.G., Huynh, K., Giles, C. et al. (2021). Clinical lipidomics: realizing the potential of lipid profiling. *J. Lipid Res.* 62: 100127. https://doi.org/10.1016/j.jlr.2021.100127.

65 Triebl, A., Burla, B., Selvalatchmanan, J. et al. (2020). Shared reference materials harmonize lipidomics across ms-based detection platforms and laboratories. *J. Lipid Res.* 61: 105–115. https://doi.org/10.1194/jlr.D119000393.

66 Chocholoušková, M., Wolrab, D., Jirásko, R. et al. (2021). Intra-laboratory comparison of four analytical platforms for lipidomic quantitation using hydrophilic interaction liquid chromatography or supercritical fluid chromatography coupled to quadrupole – time-of-flight mass spectrometry. *Talanta* 231: 122367. https://doi.org/10.1016/j.talanta.2021.122367.

67 Cajka, T., Smilowitz, J.T., Fiehn, O. et al. (2017). Validating quantitative untargeted lipidomics across nine liquid chromatography–high-resolution mass spectrometry platforms. *Anal. Chem.* 89: 12360–12368. https://doi.org/10.1021/acs.analchem.7b03404.

68 Fanelli, F., Cantù, M., Temchenko, A. et al. (2022). Report from the harmoster study: impact of calibration on comparability of LC-MS/MS measurement of circulating cortisol, 17OH-progesterone and aldosterone. *Clin. Chem. Lab. Med.* https://doi.org/10.1515/cclm-2021-1028.

69 Mainka, M., Dalle, C., Pétéra, M. et al. (2020). Harmonized procedures lead to comparable quantification of total oxylipins across laboratories. *J. Lipid Res.* 61: 1424–1436. https://doi.org/10.1194/jlr.RA120000991.

70 Wise, S.A., Tai, S.S.-C., Nelson, M.A. et al. (2019). Interlaboratory comparison for the determination of 24,25-dihydroxyvitamin D3 in human serum using liquid chromatography with tandem mass spectrometry. *J. AOAC Int.* 100: 1308–1317. https://doi.org/10.5740/jaoacint.17-0183.

71 Wise, S.A., Camara, J.E., Burdette, C.Q. et al. (2022). Interlaboratory comparison of 25-hydroxyvitamin d assays: vitamin d standardization program (VDSP) intercomparison study 2 - part 1 liquid chromatography – tandem mass spectrometry (LC-MS/MS) assays - impact of 3-*epi*-25-hydroxyvitamin D_3 on assay performance. *Anal. Bioanal.Chem.* 414: 333–349. https://doi.org/10.1007/s00216-021-03576-1.

72 Lütjohann, D., Björkhem, I., Friedrichs, S. et al. (2019). First international descriptive and interventional survey for cholesterol and non-cholesterol sterol determination by gas- and liquid-chromatography-urgent need for harmonisation of analytical methods. *J. Steroid Biochem. Mol. Biol.* 190: 115–125. https://doi.org/10.1016/j.jsbmb.2019.03.025.

73 Aristizabal-Henao, J.J., Jones, C.M., Lippa, K.A. et al. (2020). Nontargeted lipidomics of novel human plasma reference materials: hypertriglyceridemic, diabetic, and African-American. *Anal. Bioanal. Chem.* 412: 7373–7380. https://doi.org/10.1007/s00216-020-02910-3.

74 Ghorasaini, M., Mohammed, Y., Adamski, J. et al. (2021). Cross-laboratory standardization of preclinical lipidomics using differential mobility spectrometry and multiple reaction monitoring. *Anal. Chem.* 93: 16369–16378. https://doi.org/10.1021/acs.analchem.1c02826.

75 Thompson, J.W., Adams, K.J., Adamski, J. et al. (2019). International ring trial of a high resolution targeted metabolomics and lipidomics platform for serum and plasma analysis. *Anal. Chem.* 91: 14407–14416. https://doi.org/10.1021/acs.analchem.9b02908.

76 Siskos, A.P., Jain, P., Römisch-Margl, W. et al. (2017). Interlaboratory reproducibility of a targeted metabolomics platform for analysis of human serum and plasma. *Anal. Chem.* 89: 656–665. https://doi.org/10.1021/acs.analchem.6b02930.

77 Saigusa, D., Hishinuma, E., Matsukawa, N. et al. (2021). Comparison of kit-based metabolomics with other methodologies in a large cohort, towards establishing reference values. *Metabolites* 11: 652.

78 Surma, M.A., Herzog, R., Vasilj, A. et al. (2015). An automated shotgun lipidomics platform for high throughput, comprehensive, and quantitative analysis of blood plasma intact lipids. *Eur. J. Lipid Sci. Technol.* 117: 1540–1549. https://doi.org/10.1002/ejlt.201500145.

79 Ni, Z. and Fedorova, M. (2020). Lipidlynxx: a data transfer hub to support integration of large scale lipidomics datasets. *bioRxiv* p. 2020.2004.2009.033894. https://doi.org/10.1101/2020.04.09.033894.

80 Ulmer, C.Z., Ragland, J.M., Koelmel, J.P. et al. (2017). Lipidqc: method validation tool for visual comparison to SRM 1950 using NIST interlaboratory comparison exercise lipid consensus mean estimate values. *Anal. Chem.* 89: 13069–13073. https://doi.org/10.1021/acs.analchem.7b04042.

81 Izumi, Y., Matsuda, F., Hirayama, A. et al. (2019). Inter-laboratory comparison of metabolite measurements for metabolomics data integration. *Metabolites* 9: 257.

82 Burla, B., Arita, M., Arita, M. et al. (2018). MS-based lipidomics of human blood plasma: a community-initiated position paper to develop accepted guidelines. *J. Lipid Res.* 59: 2001–2017. https://doi.org/10.1194/jlr.S087163.

83 CLSI (ed.) (2010). *Defining, Establishing, and Verifying Reference Intervals in the Clinical Laboratory*, 3erd ed. CLSI guideline ep28-a3c, vol. 28. Wayne, Pennsylvania: Clinical and Laboratory Standards Institute.

84 Aarsand, A.K., Díaz-Garzón, J., Fernandez-Calle, P. et al. (2018). The eubivas: within- and between-subject biological variation data for electrolytes, lipids, urea, uric acid, total protein, total bilirubin, direct bilirubin, and glucose. *Clin. Chem.* 64: 1380–1393. https://doi.org/10.1373/clinchem.2018.288415.

85 Carobene, A., Guerra, E., Marqués-García, F. et al. (2020). Biological variation of morning serum cortisol: updated estimates from the European Biological Variation Study (EuBIVAS) and meta-analysis. *Clin. Chim. Acta* 509: 268–272. https://doi.org/10.1016/j.cca.2020.06.038.

86 Bae, Y.J., Zeidler, R., Baber, R. et al. (2019). Reference intervals of nine steroid hormones over the life-span analyzed by LC-MS/MS: effect of age, gender, puberty, and oral contraceptives. *J. Steroid Biochem. Mol. Biol.* 193: 105409. https://doi.org/10.1016/j.jsbmb.2019.105409.

87 Martín-Grau, C., Deulofeu, R., Serrat Orus, N. et al. (2021). Trimester-specific reference ranges for saturated, monounsaturated and polyunsaturated fatty acids in serum of pregnant women: a cohort study from the eclipses group. *Nutrients* 13: 4037.

88 Tsoukalas, D., Alegakis, A.K., Fragkiadaki, P. et al. (2019). Application of metabolomics part II: focus on fatty acids and their metabolites in healthy adults. *Int. J. Mol. Med.* 43: 233–242. https://doi.org/10.3892/ijmm.2018.3989.

89 Tian, G.L., Xu, F., Jiang, K. et al. (2020). Evaluation of a panel of very long-chain lysophosphatidylcholines and acylcarnitines for screening of X-linked adrenoleukodystrophy in China. *Clin. Chim. Acta* 503: 157–162. https://doi.org/10.1016/j.cca.2020.01.016.

90 Klinke, G., Rohrbach, M., Giugliani, R. et al. (2015). LC-MS/MS based assay and reference intervals in children and adolescents for oxysterols elevated in Niemann–Pick diseases. *Clin. Biochem.* 48: 596–602. https://doi.org/10.1016/j.clinbiochem.2015.03.007.

91 Moritz, E., Wegner, D., Groß, S. et al. (2017). Reference intervals for serum sphingosine-1-phosphate in the population-based study of health in pomerania. *Clin. Chim. Acta* 468: 25–31. https://doi.org/10.1016/j.cca.2017.01.029.

92 Dirks, N.F., Vesper, H.W., van Herwaarden, A.E. et al. (2016). Various calibration procedures result in optimal standardization of routinely used 25(OH)D ID-LC-MS/MS methods. *Clin. Chim. Acta* 462: 49–54. https://doi.org/10.1016/j.cca.2016.08.016.

93 CDC (2021). Clinical chemistry branch. Division of Laboratory Sciences. https://www.cdc.gov/nceh/dls/ccb.html (accessed 23 February 2022).

94 Brosco, J.P. and Paul, D.B. (2013). The political history of PKU: reflections on 50 years of newborn screening. *Pediatrics* 132: 987–989. https://doi.org/10.1542/peds.2013-1441.

95 Pitt, J.J. (2010). Newborn screening. *Clin. Biochem. Rev.* 31: 57–68.

96 Lehotay, D.C., Hall, P., Lepage, J. et al. (2011). LC-MS/MS progress in newborn screening. *Clin. Biochem.* 44: 21–31. https://doi.org/10.1016/j.clinbiochem.2010.08.007.

97 Ficicioglu, C. (2017). New tools and approaches to newborn screening: ready to open pandora's box? *Cold Spring Harb. Mol. Case Stud.* 3: a001842–a001842. https://doi.org/10.1101/mcs.a001842.

98 Cicalini, I., Valentinuzzi, S., Pieragostino, D. et al. (2021). Analytical evaluation of the ideal strategy for high-throughput flow injection analysis by tandem mass spectrometry in routine newborn screening. *Metabolites* 11: 473.

99 Janzen, N., Hofmann, A.D., Schmidt, G. et al. (2017). Non-invasive test using palmitate in patients with suspected fatty acid oxidation defects: disease-specific acylcarnitine patterns can help to establish the diagnosis. *Orphanet. J. Rare Dis.* 12: 187. https://doi.org/10.1186/s13023-017-0737-7.

100 Smith, E.H., Thomas, C., McHugh, D. et al. (2010). Allelic diversity in mcad deficiency: the biochemical classification of 54 variants identified during 5 years of acadm sequencing. *Mol. Genet. Metab.* 100: 241–250. https://doi.org/10.1016/j.ymgme.2010.04.001.

101 Laboratories, M.C. (2018). Newborn screening follow-up for elevations of C8, C6 and C10 acylcarnitines. [accessed 02 February 2022].

102 Jones, P., Khushbu, P., and Dinesh, R. (2020). *Disorder: Carnitine Palmitoyltransferase 2 Deficiency*, 141–145. Academic Press.

103 Mak, C.M., Lee, H.C., Chan, A.Y. et al. (2013). Inborn errors of metabolism and expanded newborn screening: review and update. *Crit. Rev. Clin. Lab Sci.* 50: 142–162. https://doi.org/10.3109/10408363.2013.847896.

104 Miller, M.J., Cusmano-Ozog, K., Oglesbee, D. et al. (2021). Laboratory analysis of acylcarnitines, 2020 update: a technical standard of the American College of Medical Genetics and Genomics (ACMG). *Genet. Med.* 23: 249–258. https://doi.org/10.1038/s41436-020-00990-1.

105 Gallagher, R.C., Pollard, L., Scott, A.I. et al. (2018). Laboratory analysis of organic acids, 2018 update: a technical standard of the American College of Medical Genetics and Genomics (ACMG). *Genet. Med.* 20: 683–691. https://doi.org/10.1038/gim.2018.45.

106 Wasserstein, M.P., Caggana, M., Bailey, S.M. et al. (2019). The New York pilot newborn screening program for lysosomal storage diseases: report of the first 65,000 infants. *Genet. Med.* 21: 631–640. https://doi.org/10.1038/s41436-018-0129-y.

107 Tsuji-Hosokawa, A. and Kashimada, K. (2021). Thirty-year lessons from the newborn screening for congenital adrenal hyperplasia (CAH) in Japan. *Int. J. Neonatal. Screen.* 7: 36.

108 Dobrovolny, R., Dvorakova, L., Ledvinova, J. et al. (2005). Relationship between X-inactivation and clinical involvement in fabry heterozygotes. Eleven novel mutations in the alpha-galactosidase a gene in the czech and slovak population. *J. Mol. Med. (Berl.)* 83: 647–654. https://doi.org/10.1007/s00109-005-0656-2.

109 Echevarria, L., Benistan, K., Toussaint, A. et al. (2016). X-chromosome inactivation in female patients with fabry disease. *Clin. Genet.* 89: 44–54. https://doi.org/10.1111/cge.12613.

110 Dinu, I.R. and Firu, Ş.G. (2021). Fabry disease – current data and therapeutic approaches. *Rom. J. Morphol. Embryol.* 62: 5–11. https://doi.org/10.47162/rjme.62.1.01.

111 Arends, M., Wanner, C., Hughes, D. et al. (2017). Characterization of classical and nonclassical Fabry disease: a multicenter study. *J. Am. Soc. Nephrol.* 28: 1631–1641. https://doi.org/10.1681/asn.2016090964.

112 Gaspar, P., Herrera, J., Rodrigues, D. et al. (2010). Frequency of Fabry disease in male and female haemodialysis patients in Spain. *BMC Med. Genet.* 11: 19. https://doi.org/10.1186/1471-2350-11-19.

113 Laboratories, M.C. (2020). Fabry disease diagnostic testing algorithm.

114 Auray-Blais, C., Ntwari, A., Clarke, J.T. et al. (2010). How well does urinary lyso-Gb3 function as a biomarker in Fabry disease? *Clin. Chim. Acta* 411: 1906–1914. https://doi.org/10.1016/j.cca.2010.07.038.

115 Auray-Blais, C., Lavoie, P., Boutin, M. et al. (2017). Biomarkers associated with clinical manifestations in Fabry disease patients with a late-onset cardiac variant mutation. *Clin. Chim. Acta* 466: 185–193. https://doi.org/10.1016/j.cca.2017.01.018.

116 Sakuraba, H., Togawa, T., Tsukimura, T. et al. (2018). Plasma lyso-Gb3: a biomarker for monitoring Fabry patients during enzyme replacement therapy. *Clin. Exp. Nephrol.* 22: 843–849. https://doi.org/10.1007/s10157-017-1525-3.

117 Riboldi, G.M. and Di Fonzo, A.B. (2019). GBA, Gaucher disease, and Parkinson's disease: from genetic to clinic to new therapeutic approaches. *Cells* 8: 364. https://doi.org/10.3390/cells8040364.

118 Roshan Lal, T. and Sidransky, E. (2017). The spectrum of neurological manifestations associated with Gaucher disease. *Diseases* 5: 10. https://doi.org/10.3390/diseases5010010.

119 Ferraz, M.J., Kallemeijn, W.W., Mirzaian, M. et al. (2014). Gaucher disease and fabry disease: new markers and insights in pathophysiology for two distinct glycosphingolipidoses. *Biochim. Biophys. Acta* 1841: 811–825. https://doi.org/10.1016/j.bbalip.2013.11.004.

120 Revel-Vilk, S., Fuller, M., Zimran, A. et al. (2020). Value of glucosylsphingosine (lyso-Gb1) as a biomarker in gaucher disease: a systematic literature review. *Int. J. Mol. Sci.* 21: 7159.

121 Raghavan, S.S., Mumford, R.A., Kanfer, J.N. et al. (1974). Isolation and characterization of glucosylsphingosine from Gaucher's spleen. *J. Lipid Res.* 15: 484–490.

122 Groener, J.E., Poorthuis, B.J., Kuiper, S. et al. (2007). HPLC for simultaneous quantification of total ceramide, glucosylceramide, and ceramide trihexoside concentrations in plasma. *Clin. Chem.* 53: 742–747. https://doi.org/10.1373/clinchem.2006.079012.

123 Chipeaux, C., de Person, M., Burguet, N. et al. (2017). Optimization of ultra-high pressure liquid chromatography – tandem mass spectrometry determination in plasma and red blood cells of four sphingolipids and their evaluation as biomarker candidates of gaucher's disease. *J. Chromatogr. A* 1525: 116–125. https://doi.org/10.1016/j.chroma.2017.10.038.

124 Arkadir, D., Dinur, T., Revel-Vilk, S. et al. (2018). Glucosylsphingosine is a reliable response biomarker in gaucher disease. *Am. J. Hematol.* 93: E140–E142. https://doi.org/10.1002/ajh.25074.

125 Mayo, R., Crespo, J., Martínez-Arranz, I. et al. (2018). Metabolomic-based noninvasive serum test to diagnose nonalcoholic steatohepatitis: results from discovery and validation cohorts. *Hepatol. Commun.* 2: 807–820. https://doi.org/10.1002/hep4.1188.

126 Rosato, V., Masarone, M., Aglitti, A. et al. (2020). The diagnostic conundrum in non-alcoholic fatty liver disease. *Exp. Med.* 1: 259–286. https://doi.org/10.37349/emed.2020.00018.

127 Bril, F., Millán, L., Kalavalapalli, S. et al. (2018). Use of a metabolomic approach to non-invasively diagnose non-alcoholic fatty liver disease in patients with type 2 diabetes mellitus. *Diabetes Obes. Metab.* 20: 1702–1709. https://doi.org/10.1111/dom.13285.

128 Piechota, J. and Jelski, W. (2020). Intrahepatic cholestasis in pregnancy: review of the literature. *J. Clin. Med.* 9: 1361.

129 Bull, L.N., Hu, D., Shah, S. et al. (2015). Intrahepatic cholestasis of pregnancy (ICP) in U.S. Latinas and Chileans: clinical features, ancestry analysis, and admixture mapping. *PLoS One* 10: e0131211. https://doi.org/10.1371/journal.pone.0131211.

130 Pusl, T. and Beuers, U. (2007). Intrahepatic cholestasis of pregnancy. *Orph. J. Rare Dis.* 2: 26. https://doi.org/10.1186/1750-1172-2-26.

131 Li, T. and Apte, U. (2015). Bile acid metabolism and signaling in cholestasis, inflammation, and cancer. *Adv. Pharmacol.* 74: 263–302. https://doi.org/10.1016/bs.apha.2015.04.003.

132 Labcorp (2019). Bile acids, fractionated and total, LC/MS-MS. https://www.labcorp.com/tests/503640/bile-acids-fractionated-and-total-lc-ms-ms (accessed 02 February 2022).

133 Mayo Clinic Laboratories (1995–2022). Bile acids, fractionated and total, serum.https://www.mayocliniclabs.com/test-catalog/Overview/62234#Overview (accessed 02 February 2022).

134 Holst, J.P., Soldin, O.P., Guo, T. et al. (2004). Steroid hormones: relevance and measurement in the clinical laboratory. *Clin. Lab. Med.* 24: 105–118. https://doi.org/10.1016/j.cll.2004.01.004.

135 Shackleton, C. (2010). Clinical steroid mass spectrometry: a 45-year history culminating in HPLC-MS/MS becoming an essential tool for patient diagnosis. *J. Steroid Biochem. Mol. Biol.* 121: 481–490. https://doi.org/10.1016/j.jsbmb.2010.02.017.

136 Yalow, R. and Berson, S. (1959). Assay of plasma insulin in human subjects by immunological methods. *Nature* 184: 1648–1649.

137 Concolino, P. and Costella, A. (2018). Congenital adrenal hyperplasia (CAH) due to 21-hydroxylase deficiency: a comprehensive focus on 233 pathogenic variants of cyp21a2 gene. *Mol. Diagn. Ther.* 22: 261–280. https://doi.org/10.1007/s40291-018-0319-y.

138 Pang, S., Hotchkiss, J., Drash, A.L. et al. (1977). Microfilter paper method for 17α-hydroxyprogesterone radioimmunoassay: its application for rapid screening for congenital adrenal hyperplasia. *J. Clin. Endocrinol. Metabol.* 45: 1003–1008. https://doi.org/10.1210/jcem-45-5-1003.

139 Olgemöller, B., Roscher, A.A., Liebl, B. et al. (2003). Screening for congenital adrenal hyperplasia: adjustment of 17-hydroxyprogesterone cut-off values to both age and birth weight markedly improves the predictive value. *J. Clin. Endocrinol. Metabol.* 88: 5790–5794.

140 Speiser, P.W., Arlt, W., Auchus, R.J. et al. (2018). Congenital adrenal hyperplasia due to steroid 21-hydroxylase deficiency: An Endocrine Society* clinical practice guideline. *J. Clin. Endocrinol. Metabol.* 103: 4043–4088. https://doi.org/10.1210/jc.2018-01865.

141 Lacey, J.M., Minutti, C.Z., Magera, M.J. et al. (2004). Improved specificity of newborn screening for congenital adrenal hyperplasia by second-tier steroid profiling using tandem mass spectrometry. *Clin. Chem.* 50: 621–625. https://doi.org/10.1373/clinchem.2003.027193.

142 Gaudl, A., Kratzsch, J., Ceglarek, U. et al. (2019). Advancement in steroid hormone analysis by lc-ms/ms in clinical routine diagnostics - a three year recap from serum cortisol to dried blood 17α-hydroxyprogesterone. *J. Steroid Biochem. Mol. Biol.* 192: 105389. https://doi.org/10.1016/j.jsbmb.2019.105389.

143 Cutolo, M., Paolino, S., Sulli, A. et al. (2014). Vitamin D, steroid hormones, and autoimmunity. *Ann. N.Y. Acad. Sci.* 1317: 39–46. https://doi.org/10.1111/nyas.12432.

144 Galior, K., Ketha, H., Grebe, S. et al. (2018). 10 years of 25-hydroxyvitamin-d testing by LC-MS/MS-trends in vitamin-d deficiency and sufficiency. *Bone Rep.* 8: 268–273. https://doi.org/10.1016/j.bonr.2018.05.003.

145 Makris, K., Sempos, C., Cavalier, E. et al. (2020). The measurement of vitamin D metabolites: part I-metabolism of vitamin d and the measurement of 25-hydroxyvitamin D. *Hormones (Athens)* 19: 81–96. https://doi.org/10.1007/s42000-019-00169-7.

146 Li, D., Jeffery, L.E., Jenkinson, C. et al. (2019). Serum and synovial fluid vitamin D metabolites and rheumatoid arthritis. *J. Steroid Biochem. Mol. Biol.* 187: 1–8. https://doi.org/10.1016/j.jsbmb.2018.10.008.

147 Singh, R.J., Taylor, R.L., Reddy, G.S. et al. (2006). C-3 epimers can account for a significant proportion of total circulating 25-hydroxyvitamin d in infants, complicating accurate measurement and interpretation of vitamin d status. *J. Clin. Endocrinol. Metabol.* 91: 3055–3061. https://doi.org/10.1210/jc.2006-0710.

148 Laboratories, M.C. (1995–2022). 25-hydroxyvitamin D2 and D3, serum. https://www.mayocliniclabs.com/test-catalog/overview/83670#Overview (accessed 02 February 2022).

149 Diagnostics, Q. Questassured 25-hydroxyvitamin D (D2,D3). [02/2/2022]. https://testdirectory.questdiagnostics.com/test/test-detail/92888/questassured-25-hydroxyvitamin-d-d2-d3?cc=MASTER.

150 Laboratories, A. 25-hydroxyvitamin D2 and D3 by tandem mass spectrometry, serum. [02/2/2022]. https://ltd.aruplab.com/Tests/Pub/2002343.

151 Labcorp (2021). 25-hydroxy vitamin D (D2+D3 fractionated), LC/MS-MS. https://www.labcorp.com/tests/504115/25-hydroxy-vitamin-d-d-sub-2-sub-d-sub-3-sub-fractionated-lc-ms-ms (accessed 02 February 2022).

152 Strobel, N.A., Fassett, R.G., Marsh, S.A. et al. (2011). Oxidative stress biomarkers as predictors of cardiovascular disease. *Int. J. Cardiol.* 147: 191–201. https://doi.org/10.1016/j.ijcard.2010.08.008.

153 Sharifi-Rad, M., Anil Kumar, N.V., Zucca, P. et al. (2020). Lifestyle, oxidative stress, and antioxidants: back and forth in the pathophysiology of chronic diseases. *Front. Physiol.* 11: https://doi.org/10.3389/fphys.2020.00694.

154 Milne, G.L., Sanchez, S.C., Musiek, E.S. et al. (2007). Quantification of F_2-isoprostanes as a biomarker of oxidative stress. *Nat. Protoc.* 2: 221–226. https://doi.org/10.1038/nprot.2006.375.

155 Morrow, J.D., Hill, K.E., Burk, R.F. et al. (1990). A series of prostaglandin F_2-like compounds are produced in vivo in humans by a non-cyclooxygenase, free radical-catalyzed mechanism. *Proc. Natl. Acad. Sci. U.S.A.* 87: 9383–9387. https://doi.org/10.1073/pnas.87.23.9383.

156 Kadiiska, M.B., Gladen, B.C., Baird, D.D. et al. (2005). Biomarkers of oxidative stress study II: are oxidation products of lipids, proteins, and DNA markers of CCL4 poisoning? *Free Radic. Biol. Med.* 38: 698–710. https://doi.org/10.1016/j.freeradbiomed.2004.09.017.

157 Il'yasova, D., Spasojevic, I., Wang, F. et al. (2010). Urinary biomarkers of oxidative status in a clinical model of oxidative assault. *Cancer Epidemiol. Biomarkers Prev.* 19: 1506–1510. https://doi.org/10.1158/1055-9965.Epi-10-0211.

158 Heartlab, C. (2013). F2-isoprostane/creatinine. https://www.clevelandheartlab.com/tests/f2-isoprostanecreatinine.

159 Marrocco, I., Altieri, F., Peluso, I. et al. (2017). Measurement and clinical significance of biomarkers of oxidative stress in humans. *Oxidat. Med. Cell. Long.* 2017: 6501046. https://doi.org/10.1155/2017/6501046.

160 Greaves, R.F., Bernardini, S., Ferrari, M. et al. (2019). Key questions about the future of laboratory medicine in the next decade of the 21st century: a report from the ifcc-emerging technologies division. *Clin. Chim. Acta* 495: 570–589. https://doi.org/10.1016/j.cca.2019.05.021.

161 Vvedenskaya, O., Holčapek, M., Vogeser, M. et al. (2022). Clinical lipidomics – a community-driven roadmap to translate research into clinical applications. *J. Mass Spectrom. Adv. Clin. Lab* 24: 1–4. https://doi.org/10.1016/j.jmsacl.2022.02.002.

21

Lipidomics in Food Industry and Nutrition

Danilo Donnarumma[1], Giuseppe Micalizzi[1], Luigi Mondello[1,2,3], and Paola Dugo[1,2]

[1] *University of Messina, Department of Chemical, Biological, Pharmaceutical and Environmental Sciences, Viale Palatucci snc, 98168, Messina, Italy*
[2] *Chromaleont s.r.l., c/o Department of Chemical, Biological, Pharmaceutical and Environmental Sciences, University of Messina, Viale Palatucci snc, 98168, Messina, Italy*
[3] *University Campus Bio-Medico of Rome, Department of Sciences and Technologies for Human and Environment, Via Alvaro del Portillo 21, 00128, Rome, Italy*

21.1 Introduction

Foodomics and lipidomics are interconnected disciplines that study in synergic manner food constituents and lipid components, respectively, with the common objective to improve consumer's well-being, health, and knowledge. The term foodomics was coined by Cifuentes's group [1] and denotes the application and integration of advanced omics technologies, separately or in combination, not only for the investigation of food constituents, but also in food authentication and evaluation of food safety and quality [2]. On the other hand, lipidomics took the place of the "old" lipidology, with the aim to provide a dynamic vision of lipid classes and their exchange during various metabolic conditions within the complex scenario of the cell compartments and the whole organism [3]. In the last years, foodomics and lipidomics disciplines were particularly correlated, considering that lipidomics approaches have been widely applied in food lipid analysis in order to trace the origin of food, to investigate food adulteration, and to determine harmful lipids in food processing and storage. Dietary lipids are involved in many biological processes, including energy storage, formation of cellular membrane, and signal transduction [4]. Consequently, due to the vital roles that lipids play in biological systems, it is clear that the lipidome monitoring provides new insights into how foods and nutrients influence health and diseases. This is an important requisite that allows not only to establish the health state of cells and entire organism, but eventually to direct nutritionist and clinical practices toward specific nutritional and therapeutic choices, in order to bring significant benefits to patients. The development of foodomics and lipidomics has been widely driven

Mass Spectrometry for Lipidomics: Methods and Applications, First Edition.
Edited by Michal Holčapek and Kim Ekroos.

by rapid advances in mass spectrometry (MS)-based technologies, alone or in tandem with chromatographic methodologies such as gas chromatography (GC) and high-performance liquid chromatography (HPLC). Generally, a complete MS-based lipidomics workflow involves sample preparation, MS-based analysis, and data processing. Of all these steps, MS-based analysis is cardinal as datasets for lipid qualification and quantitation are generated in this segment [5].

In this chapter the basic concepts of chemistry and biology related to the lipid molecules in terms of fatty acids in cell membranes will be elucidated, bringing the reader to a revisiting of their biological and metabolic importance. Then, an overview of current applications of MS-based foodomics will be described emphasizing the necessity to maintain a correct dietary intake.

21.2 Lipids in Nutrition and Human Health

Although the lipid research has a long history, only in the last two decades there was crucial developments to define the scenario of lipids in nutrition and human health. Nowadays, lipids are considered multifunctional molecules, and it is very difficult to find an area of life sciences where lipids are not playing a fundamental role. In this context, lipidomics is defined as the discipline that studies lipids from a physiological or pathological point of view in living systems. The comprehension of lipid behavior under specific conditions could open new frontiers toward better health prevention and disease treatment [3]. In this sense, lipidomics can be considered as a valid molecular diagnostic tool in clinical practices in order to interpret the lifestyle and health state of patients and to intervene with adequate medical care.

As fundamental components of an organism, lipids play important biological roles including their functions in cell membrane, energy storage, and signaling. In some cases, the function of the lipid is not influenced by its composition and molecular structure. For example, triacylglycerols are used as energy source independently by the fatty acid composition. In other cases, the functions of lipids depend on the chemical structures and can therefore be different for lipid species belonging to the same class [6]. This concept highlights the importance of investigating the nature and chemical structure of lipids in order to elucidate the function within a biological context. However, this objective is not simple considering that the number of the lipid species is very high (tens of thousands), and the concentration levels of different lipids may vary thousand- or million-fold [6]. Based on biological function, lipids are grouped in eight categories in accordance with their chemical structures: fatty acids, triacylglycerols, phospholipids, sphingolipids, sterols, prenol lipids, saccharolipids, and polyketides. A total of 43.659 lipid molecules (21.706 curated and 21.953 computationally generated) are listed in the LIPID MAPS database, the most accurate and reliable database in lipidomics field [7]. MS-based lipidomics techniques are currently the most effective instruments for the identification and quantification of all lipid entities as well as for the determination of their biological functions in cell processes. The attention of this paragraph is focused on the importance and roles of fatty acids involved in phospholipidic structures, from their biosynthetic origin to dietary intake.

Cell membranes represent the fundamental structures through which all communications and exchanges useful to cell life occur. To date, cell membranes are considered to be the most direct and innovative site for a correlation with the health condition [3]. The fundamental unit of cell membranes is the phospholipid with its typical amphipathic structure. The amphipathic nature of the phospholipid allows the molecules to form a particular rearrangement called double layer in which two phospholipids are positioned one in front of the other, with the hydrophilic parts disposed outward, establishing interactions with water molecules. Focusing on fatty acids molecules, fatty acid synthase is the multienzymatic system responsible for the synthesis of saturated fatty acids (SFAs). The biosynthetic mechanism is based on the assembly of several two-carbon units, starting from acetyl-coenzyme A, and shuttling the products from one to the other enzymatic subunits The final product of this process is palmitic acid (C16:0) containing 16-carbon atoms along the chain. Another important SFA is stearic acid (C18:0) containing 18-carbon atoms. Its formation requires the intervention of the elongase enzyme, which adds two-carbon units to palmitic acid. This biosynthetic scheme brings to the formation of all SFAs with an even carbon number, extending the chain up to 26 carbon atoms [8]. On the other hand, the affinity of SFA versus desaturase enzyme (that adds a double bond along carbon chain) is very high; consequently, very-long-chain SFA cannot be formed in high quantity, and their excess could indicate a lipid pathway failure. SFAs are also the constituents of many foods. In this respect, palmitic and stearic acids represent the principal components of animal fats and vegetable oils. SFAs intake should give a negative feedback to biosynthesis regulating in this manner their content; however, considering that fatty acid synthase system is regulated by numerous cofactors, it exists the serious risk of an accumulation of SFAs. In light of the metabolism and food intake so far described, the World Health Organization (WHO) and the European Food Safety Agency (EFSA) recommended to maintain at a minimum level the SFA intake (maximum value of 33% of the total fats) [9, 10]. Even if they mostly distributed in triacylglycerols as circulating lipids, SFAs can also be included in membrane phospholipids. Their presence results into the formation of a compartment with a more rigid structure, affecting the normal functions of cell membranes. In order to maintain the well-being and health of cells and whole organism, it is advisable to increase the intake of unsaturated FAs.

The degree of unsaturation of a fatty acid is defined by the presence of double bonds along the carbon chain. In membranes, the most abundant monounsaturated fatty acid (MUFA) is oleic acid (C18:1ω9). As previously described for SFA class, the presence of oleic acid into cell membranes can be ascribed to biosynthetic pathways and dietary intake. The former is due to the interevent of delta-9 desaturase enzyme, which introduce a double bond in *cis* configuration in position 9 of stearic acid carbon chain. With respect to dietary intake, oleic acid is the primary constituent of olive oil in form of triolein, particularly consumed in the countries of Mediterranean area. When oleic acid is supplied by the diet, the cell economy is activated and its biosynthesis is halted. The contribution of oleic acid to cell membranes is fundamental to obtain the proper fluidity. Moreover, it is less sensitive to oxidative stress conditions, compared to polyunsaturated fatty acids (PUFAs) containing a higher number of double bonds. Another important MUFA is palmitoleic acid (C16:1ω7) obtained by enzymatic action of delta-9 desaturase on the palmitic acid. Compared to oleic acid, palmitoleic acid is not particularly abundant in food; therefore, a direct measure of its concentration in human tissues can give important insights on the metabolic processes in cell.

PUFA compounds are defined as essential fatty acids (EFAs) due to the fact that eukaryotic cells cannot synthetize them; therefore, they derive only from nutritional intake. PUFAs are conventionally classified in two different subclasses based on the position of the double bonds along the carbon chain: omega-6 and omega-3. Omega-6 fatty acids have linoleic acid (C18:2ω6) as precursor, while omega-3 derives from α-linolenic acid (C18:3ω3). Concerning omega-6 pathway, the first step involves the intervention of delta-6 desaturase enzyme by adding a third double bond in linoleic acid carbon chain and the formation of γ-linolenic acid (C18:3ω6) with relevant functions at biological level. For example, it contributes to the correct functionality of the epidermis and hydration [3]. Some different cofactors influence heavily the activity of this enzyme (e.g. Fe, Zn, Mg, vitamin B). In this context, a dysregulated intake of alcohol and *trans* fatty acids leads to inhibition of delta-6 desaturase. Afterward, the carbon chain of γ-linolenic acid is elongated of two-carbon units by activity of elongase enzyme with formation of di-homo-γ-linolenic acid (C20:3ω6). This fatty acid performs various control functions as reported by Wang et al. [11], including the proper functioning of immune system. Finally, delta-5 desaturase enzyme introduces the fourth double bond along the carbon chain leading to the synthesis of arachidonic acid (C20:4ω6), the final product of omega-6 pathway. Arachidonic acid is the precursor of some inflammatory mediators such as prostaglandins, leukotrienes, and lipoxins [3], and for this reason omega-6 pathways are conventionally defined as pro-inflammatory. Omega-3 metabolic pathway starts from α-linolenic acid that is initially converted by the delta-6 desaturase enzyme into corresponding derivative with four double bonds, namely stearidonic acid (C18:4ω3). This is the substrate of elongase and delta-5 desaturase for the synthesis of eicosapentaenoic acid or commonly named EPA (C20:5ω3), a precursor of prostanoids and leukotrienes with well-known anti-inflammatory effects. Finally, the intervention of the elongase and delta-6 desaturase enzymes, and β-oxidation process form docosahexaenoic acid or DHA (C22:6ω3), the final products of omega-3 pathway.

The role of EPA and DHA is enormously important in cells. In fact, after release from phospholipid membranes, they are converted by oxidative enzymatic pathways in neuroprotectine (from DHA) and resolvins (from EPA) with strong protective properties in tissues [12, 13]. Examining Figure 21.1, it is clear that initial precursors of two metabolic pathways compete each other for the same enzyme (delta-6 desaturase). Consequently, an omega-6-rich diet will shift equilibrium in arachidonic acid direction reducing the possibility of omega-3 competition. This means an increase of omega-6/omega-3 ratio in cells and a higher predisposition to inflammatory state. In order to contrast these effects, a careful and personalized choice of nutrients must be made for restoring the correct balance of the omega-6/omega-3 ratio of about

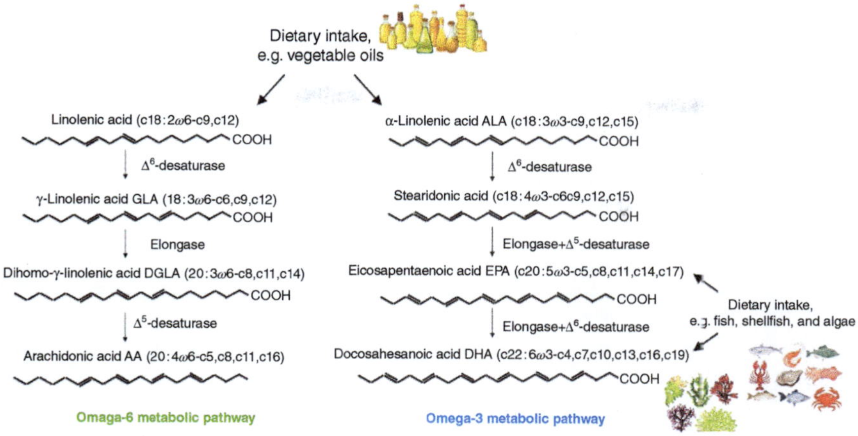

Figure 21.1 The enzymatic pathways of omega-6 and omega-3 fatty acids occurring in cells. Omega-6 and omega-3 lipids can also be provided by dietary intake.

5:1. In conclusion, when nutritional intake does not include fish and algae consumption, there is a risk of EPA and DHA deficiency compromising the proper metabolic pathways of cells and whole organism. To complicate matters, the conversion of α-linolenic acid into EPA and DHA takes place in human organism with low percentage values, about 8% in the case of EPA and about 0.1% for DHA [14, 15].

In order to determine a clear picture of the patient status, almost all lipidomics approaches are MS-based. Due to recent technological advances, today several different types of instruments in lipidomics portfolio are available. However, quite widespread single quadrupole (qMS) systems continue to be particularly useful for lipidomic investigations, especially when coupled with chromatographic separations. GC coupled with qMS (GC-MS) is an effective and suitable platform in analysis of fatty acids. Derivatization step is mandatory to convert the intact lipids into more volatile and less polar components, usually fatty acids methyl esters (FAMEs) [16]. GC-MS methods present a higher resolution and sensitivity with respect to other analytical techniques. However, this technique has some limitations such as the indistinguishable mass spectra of double-bond positional isomers (Figure 21.2), which raises difficulties in the peak identifications. The combination of retention time information in terms of linear retention index (LRI) with MS spectra may support the identification by MS library search [17]. The use of GC-MS in lipidomics analysis has been abundantly reported in literature. Sánchez-Avila et al. in 2009 [18] described an interesting approach for the determination of fatty acid content in serum by using GC-MS with micro-selected ion storage (micro-SIS) detection mode, an approach similar to selected-ion-monitoring (SIM) mode. The proposed method allowed to detect 39 FAMEs in approximately one hour. A rapid and sensitive GC-MS methodology was developed in 2012 by Ecker et al. for the determination of positional and geometrical (*cis*/*trans*) isomers with chain lengths between C_8 and C_{28} [19]. A micro-bore capillary column allowed the separation of analytes in very short time (ca. 17.2 minutes). The authors confirmed that optimized method was suitable for the quantification not

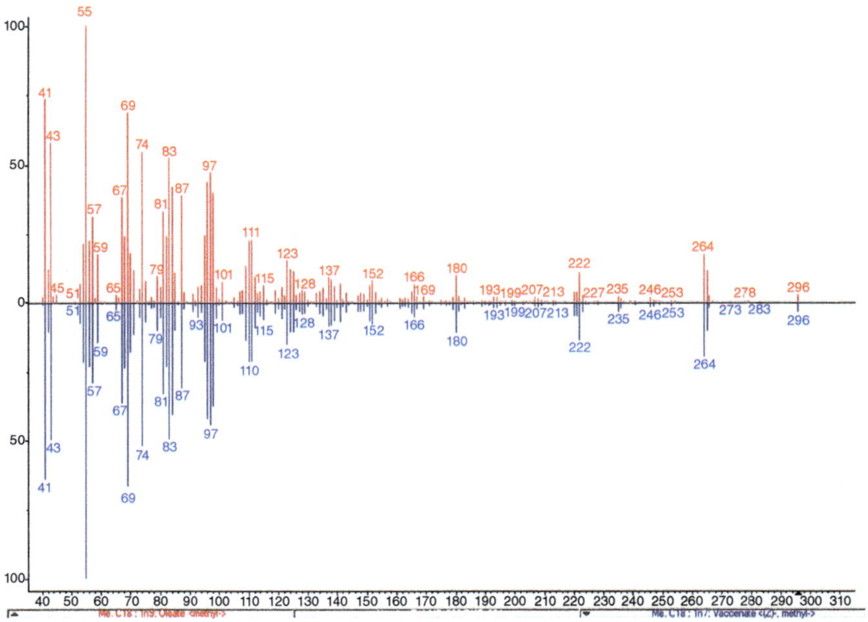

Figure 21.2 Indistinguishable MS spectra of oleic acid (C18:1ω9, red) and *cis*-vaccenic acid (C18:1ω7, blue) methyl esters listed in available commercial LIPIDS MS library.

only of positional and geometrical isomers, but also conjugated fatty acids, such as conjugated linoleic acid (CLA) isomers. Recently, Micalizzi et al. published a rapid, miniaturized, and sensitive strategy for the determination of FAMEs profile in a single blood drop spotted on DBS (dried blood spot) collection paper cards [16]. Thirty-two fatty acids were univocally identified by using a qMS system in scan mode and the search of MS spectra was carried out using a dedicated MS library. Peak assignment was confirmed by the match of LRIs listed in library with the experimental ones. The developed method allowed the authors to determine the total content of SFA, MUFAs, and PUFA, as well as omega-6 and omega-3 levels. Also, the analytical strategy allowed the determination of different indices for evaluating the enzymatic activity of delta-9 desaturase, delta-6 desaturase, elongase, and delta-5 desaturase.

Trans geometry of lipids became a hot topic in lipidomics after the discovery that *trans* fatty isomers were found in deodorized and partially hydrogenated oils [3]. From health point of view, *trans* lipid structures have a profound effect on phospholipids bilayers resulting in a significant reduction in membranes fluidity and permeability. This is due to the disposition of substituents in the *trans* configuration that determines an almost linear hydrocarbon chain, structurally similar to the saturated acids. This means that a dysregulation intake of foods containing *trans* fatty acids affects dangerously health state, coronary heart disease in primis [3]. In New York and many other localities of US, the control of *trans* fatty acids in foods is regulated by a legislation that restricts their amount to 0.5 g per serving [20]. In Europe, the maximum limit of *trans* species has been recently introduced (Commission Regulation (EU) 2019/649) to limit their presence in foods at 2 g for 100 g of fats. However, several research studies have revealed that functional foods (e.g. dietary supplements and fortified foods) may not reflect the real *trans* fatty acids content [21]. This can represent a serious problem since the deodorization process, largely applied to fish oils in order to remove their unpleasant smells, could bring to the formation of *trans* fatty acids [22]. It is worth noting that the presence of *trans* isomers in cells is connected not only with dietary intake and high temperatures processing of oils, but also with endogenous processes due to free radicals during cellular stress [23]. In fact, free radicals are able to perform *cis*/*trans* isomerization of double bonds of unsaturated lipids present in phospholipids membranes.

Cis/*trans* isomers are undistinguishable in MS dimension due to identical fragmentation patterns; therefore, chromatographic separation step is mandatory. For accurate and reliable *cis/trans* fatty acids measurements, polar and long capillary columns are conventionally used. In such respect, Del Monte et al. utilized a 200 m SLB-IL111 (200 m × 0.25 mm ID × 0.20 μm d_f) to resolve most of the unsaturated positional and geometrical isomers contained in milk fat [24]. In contrast, Srigley and Rader evaluated fatty acids content including *trans* isomers in omega-3 dietary supplements from fish source containing high levels of EPA and DHA [21]. The same capillary column previously described was used by the authors for the GC separation. The detection of FAMEs derivatives was carried out by acquiring MS spectra through an ion-trap MS instrument. FAME components were identified by covalent adduct chemical ionization (CACI) tandem MS using acetonitrile as the reagent gas [25].

21.3 Fish, Shellfish, and Algae: Main Food Sources of Omega-3

Fish, shellfish, and fish products are universally recognized as one of the most important "functional foods," being an important source of minerals, antihypertensive proteins, and antioxidants long-chain omega-3 fatty acids; for this reason, their consumption in the last decades is continuously rising. In the past, several studies showed how the consumption of omega-3-based fatty acids present in fish, mainly EPA and DHA, was protective against cardiovascular diseases, such as heart attacks and strokes [26, 27]. Nevertheless, recent studies demonstrated that dietary supplementation with omega-3 did not show any positive effect on decreasing the risk of cancer or heart disease [28]. These findings were also in agreement with the studies conducted by Aung et al. [29] and Grey et al. [30] that showed how the supplementation with fish oil did not prevent heart attacks or any other vascular disease outcomes. However, the uptake of marine omega-3 fatty acids is still highly recommended, given their importance in brain development [31], and their anti-inflammatory activity in contrast with the pro-inflammatory omega-6 fatty acids.

According to "The State of World Fisheries and Aquaculture 2020. Sustainability in action" [32], released in 2020 by the Food and Agriculture Organization of the United Nations (FAO), the global seafood industry production has reached 179 million tons in 2018, worth 401 billion USD, with 82 million tons deriving from aquaculture. Roughly 87% of the total production (156 millions of tons) was used for human consumption, with half of it coming from aquaculture, while the remaining 13% (22 million of tons) were used for the production of fishmeal and fish oil. Along with the rapid increase of the fish global demand, also seafood adulteration and sophistication are in continuous growth, with misrepresented or mislabeled commercial fish products estimated to be about 30% of the total [33]. Usually such fish frauds are detected using genomic profiling, since DNA markers are unique for each species. The main drawback of this approach is the complexity of the sample preparation and the amount of time required to perform the polymerase chain reaction (PCR) assay. In the last years, ambient mass spectrometry (AMS), in particular rapid evaporative ionization mass spectrometry (REIMS), has been used for such scope with very promising results. Black et al. [34] successfully used REIMS to differentiate between five genetically similar white fish species (cod, coley, haddock, pollock, and whiting) in an accurate and near-instantaneous way. Rigano et al. [35] used the same approach to allow the speciation of 18 species typical of the Mediterranean Sea. The authors created a model containing samples taken by all the 18 species, as well as dedicated sub-models to avoid identification problems within the same class, order, or family. Moreover, they build the models considering also the seasonal variability and the maturation stage. The ability of this approach to distinguish between different fish species has been attributed by the authors mainly to the information obtained from the lipid fraction, in particular fatty acids (in the 200–350 m/z region) and phospholipids (in the 600–950 m/z region).

Since the amount of wild caught-fish keeps decreasing, more than 50% of the fish used for human consumption is produced by aquaculture. Even though marine fishes are the main human dietary source of omega-3 fatty acids, they are not able to synthesize them; rather they introduce EPA and DHA with the diet, mainly through algae and plankton. Therefore, farmed marine fish need to be supplemented with an omega-3 fatty acids source, usually fish oil, in order to have an amount of omega-3 fatty acids comparable with the wild ones. For this reason, in 2009, more than 81% of the global fish oil supply was used for aquaculture [36]. Several studies already reported how the use of vegetable oils instead of fish oils in aquaculture altered the total fatty acid composition of the farmed marine fish, with a significant decrease in EPA and DHA amount and an increase of the linoleic acid [37, 38]. In this context, Hvattum et al. [39] used normal-phase liquid chromatography coupled to a triple quadrupole MS (QqQ) to compare the phospholipids composition of Atlantic salmon head kidney from fish fed with fish oil or vegetable oil. They found significant changes in the phospholipid's distribution depending on the diet. Vegetable oil-based diet resulted in higher amounts of linoleic acid, and consequently of all the 18:2 fatty acids containing phospholipid species. They also found a significant increase of the 20:3 containing phospholipid species, and the 20:4 containing species in the phosphatidylethanolamine and phosphatidylserine family, when comparing the vegetable oil diet group with the fish oil diet group. On the other hand, the vegetable lipid diet significantly decreased all the EPA-containing species when compared to the fish-based diet, while the DHA-containing species did not show significant differences.

In recent years, the application of a circular economy model of production and consumption arose from the necessity to maximize the use of resources and minimize waste production [40]. The re-utilization of industrial by-products as starting material for obtaining other products of economic interest became the standard, more than an exception, also in the fishery industries. For example, the amount of wastes generated from the processing of raw tuna fish into edible products can be as high as 70% of the starting material [41]. Several research studies focused on the retrieval of high-value generating molecules, particularly omega-3 fatty acids, from tuna industrial by-products. Saliu et al. [42] used a multi-analytical approach, based on infrared (IR), nuclear magnetic resonance (NMR), and MS to investigate the lipid composition of bluefin tuna perigonadal fat, usually discarded during the evisceration process. They demonstrated by high-performance liquid chromatography coupled with MS (HPLC-MS) analysis that this type of waste consists mainly of PUFA triacylglycerols (76.2% of dry mass), with low amount of cholesterol. Moreover, EPA and DHA were preferentially located at the *sn*-2 position, which is considered the optimum for intestinal absorption. Rigano et al. [43] took advantage of fully automated preparative stations in order to perform the extraction and trans-esterification (for the GC analysis) of the lipid fraction obtained from a mix of wastes derived from the tuna fishery industry, prior to the injection into the HPLC and GC instrumental setup, respectively. The use of robotic platforms increased the reproducibility of the analyses, reduced the amount of solvents, and minimized the errors due to the handling of the samples by the operators. Thanks to the combination of GC and HPLC data, the authors demonstrated the presence of a higher content of PUFA-containing triacylglycerols (mainly EPA and DHA) in the waste products compared to the muscle. Furthermore, the main lysophospholipids and phospholipids families were characterized by performing neutral loss and precursor ion scan experiments by means of a QqQ instrument, showing significant quantitative differences.

As previously stated, fish are not able to synthesize omega-3 fatty acids, but they need to introduce them with the diet, mainly through algae and plankton. Moreover, the supply of omega-3 compounds from these traditional sources is quickly becoming insufficient to satisfy human nutritional requirements [44]. For this reason, in the last decades, the use of marine microalgae as supplements for human dietary has rapidly grown. In 2018, the total algae industry production reached 32.4 million tons, with 97.1% coming from farmed seaweeds, for an estimated value of 10.8 billion USD [32]. The amount of lipids in microalgae greatly varies depending on the species, ranging between 1% and 70% of dry weight, with some species (e.g. *Nannochloropsis* sp. for EPA, *Schizochytrium limacinum* for DHA, and *Phaeodactylum tricornutum*) showing a content of omega-3 between 30% and 40% of their total fatty acids, as reported by Poblete-Castro et al. [45] by GC-MS approach. In this context, several studies have been published regarding the isolation of new microalgae species and to evaluate their lipid production and suitability for human consumption. Haoujar et al. isolated different species of microalgae from Moroccan Mediterranean seawater, and characterized them in terms of lipid content and lipid production by HPLC-MS approach, concluding that algae from the *Nannochloris* sp. had the highest lipid productivity, reaching 15.93 mg/l/day [46]. Recently, it has been shown that the non-enzymatic oxidation of different PUFAs (NEO-PUFAs) can result in other compounds of high-added value known as isoprostanoids, such as phytoprostanes (from α-linolenic acid), series 2 isoprostanes (from arachidonic acid), series 3 isoprostane (from EPA), dihomo-isoprostanes and dihomo-isofurans (from adrenic acid, C22:4ω-6), and neuroprostanes (from DHA). All these molecules are produced in plants and animals under oxidative stress conditions, and have been suggested to act as lipid mediators with potential beneficial effect on human health [47]. Vigor et al. [48] used an HPLC-MS/MS approach in multiple reaction monitoring (MRM) mode in order to quantify the isoprostanoids production of four marine microalgae under oxidative stress conditions. For each microalgal species investigated, the isoprostanoid profile followed the most abundant PUFAs. No significant variations in NEO-PUFAs amount were observed for *Rhodomonas salina* and *Chaetoceros gracilis* using copper as oxidative stress, whereas *Tisochrysis lutea* and *P. tricornutum* showed an increase in the production of C18-, C20-, and C22-derived isoprostanoids. On the other hand, the use of hydrogen peroxidase oxidative stress resulted in variation in the NEO-PUFAs profile of *R. salina* and *P. tricornutum*, while no changes variations were detected for the other two algae.

21.4 Edible Plants and Vegetable Oils: Main Food Sources of Omega-6

It is well known that a diet rich in fruits and vegetables is associated with lower risks of numerous chronic diseases, including cancer and cardiovascular disease [49, 50]. For this reason, consumption of both fresh and frozen vegetables is growing globally, even if the goal to eat them at least five times per day is still far to be reached. Vegetables are a good source of lipids, simple and complex carbohydrates (especially soluble and insoluble fiber), proteins, minerals, and secondary metabolites for human nutrition. The fatty acid composition of vegetables is mainly composed of MUFAs and PUFAs, rather than SFAs, with lauric (C12:0), myristic (C14:0), palmitic (C16:0), stearic (C18:0), oleic (C18:1ω9), linoleic (C18:2ω6), and α-linolenic (C18:3ω3) acids being the main ones. Generally, lipids from vegetable sources are liquid (oil) at room temperature and less saturated than animal fats, and contain a significant amount of omega-6 fatty acids.

Among the different vegetable oils, olive oil has been used for millennia in the Mediterranean diet as a source of healthy fat, and has gained nowadays global importance thanks to its use in gourmet cuisine, with an estimated value of 11.2 billion USD in the year 2020, and expected to reach a value of 13.8 billion USD by 2027 [51]. Olive oil is mainly composed of triacylglycerols (about 98%) mainly formed by oleic, linoleic, and palmitic acids, with a smaller fraction of polar lipids and other components such as phenols and lipid soluble vitamins. Montealegre et al. analyzed the phospholipidic profile of olive fruits and olive oils from different Spanish cultivars and regions by nonaqueous capillary electrophoresis coupled to electrospray-MS, highlighting several differences in the relative abundance of the phospholipid classes as a function of the botanical and geographical origins of the fruits analyzed [52]. Moreover, only five phospholipids families (lysophosphatidic acid, phosphatidylcholine, phosphatidylethanolamine, lysophosphatidylethanolamine, and phosphatidylglycerols) were identified in the oils. The International Olive Council (IOC) classifies the olive oils depending on the method of extraction and the free acidity (percentage of free fatty acids) in categories, where the Extra Virgin Olive Oils (EVOOs) represent the highest grade and the most valuable one [53]. For this reason, EVOOs are highly susceptible to frauds and adulterations, mainly by the addition of other less valuable vegetable oils [54]. In this context, Mangraviti et al. in 2021 [55] applied the REIMS technology to differentiate between several Italian EVOOs. The authors were able to adapt a sampling method, usually applied for the analysis of solid or semi-solid samples, to achieve mass spectral data from a liquid and poorly conductive sample, such as olive oils. This goal was reached by freezing a mixture of oil and saline solution, in order to modify its physical state and obtain a sample of proper consistency for the sampling technique. The approach used by the authors worked very well for the discrimination of mono-cultivar oils, while the recognition of high-quality oils labeled as PDO (Protected Denomination of Origin) was more challenging, probably because these oils are often produced by using different cultivars in different percentages. In another work, Alves et al. [56] found that each olive oil possesses a different phosphocholine profile, and hypothesized that such feature can be used in future for olive oil's identity and traceability. In the case of adulteration, Calvano et al. [57] used matrix-assisted laser desorption and ionization time-of-flight mass spectrometry (MALDI-ToF MS) with an ionic liquid as matrix and extraction solvent to detect adulterated olive oil with levels as low as 1–5% of hazelnut oil, using phospholipids as markers. This is a very important milestone since EVOOs and hazelnut oil have similar triacylglycerol, fatty acid, and sterol profiles; therefore, this kind of adulteration cannot be easily detected by other well-established techniques.

In the last decade, the effects of the circular economy model, already applied all over the world, are starting to impact also the fruits and vegetables processing industries. Many researches have been done in the last years in order to establish procedures to reduce the waste production and use it to extract, in particular from fruits seeds, value-added products, such as phytochemicals or oils for human nutrition [58]. Fruit seed oils are mainly composed of linoleic acid, usually the most abundant one, oleic, palmitic, and α-linolenic acids. A remarkable C18:3ω3 content was found in some fruit seed oils as the kiwi fruit, and in berries, such as bilberry, cranberry, gooseberry, and strawberry [59]. Holčapek et al. [60] used non-aqueous reverse-phase liquid chromatography coupled to atmospheric pressure chemical ionization MS (NARP-HPLC-APCI-MS) to characterize the triacylglycerols content of nine plant oils (walnut, hazelnut, cashew nut, almond, poppy seed, yellow melon, mango, fig, and date), and quantify them using a method based on the use of response factors. A total of 133 triacylglycerols, containing 22 fatty acids with carbon atoms ranging from 8 to 25 and up to 3 double bonds, were successfully identified and quantified with this approach. Loutelier-Bourhis et al. [61] used two different MS-based approaches, namely direct infusion multi-stage MS (MSn) and NARP liquid chromatography/tandem MS (NARP-HPLC-MS/MS), for the complete characterization of the linseed oil triacylglycerols composition. The use of both approaches was essential for the discrimination of regioisomers, through MS3 data, and isobaric species, separated and identified by HPLC-MS/MS, leading to the identification of 26 triacylglycerols. Finally, Arena et al. [62] used silver thiolate × NARP comprehensive liquid chromatography coupled to APCI-MS, to characterize the lipid profile of a borage oil, characterized by high levels of omega-6 fatty acids, including linoleic acid and γ-linolenic acid. The coupling of orthogonal separation techniques, since Ag+-HPLC discriminates triacylglycerols for unsaturation degree and double bonds (DB) position while NARP-HPLC elutes according to increasing partition number (PN=C atoms number-2DBs), and MS detection allowed the reliable identification of 94 triacylglycerols.

21.5 Concluding Remarks

A complete analysis of the role of lipids in cells is the primary objective of lipidomics. The comprehension of the behavior of lipids is crucial since it could open new frontiers in the field of precise medicine for the health prevention and disease treatments. Considering that dietary lipids are involved in biological processes, it becomes necessary to monitor lipidomics status of cells for establishing their state of health, and consequently those of entire organism. Lipidomics has the role to direct nutritionist and clinical practices toward specific nutritional and therapeutic choices. The foodomics state of art of principal functional foods such as fish, algae, and vegetable oils has been elucidated in this chapter, as well as the most recent MS-based analytical instrumentations. In conclusion, lipidomics MS-based technologies described are essential for understanding the complex scenario of lipid molecules and their importance in food industry and nutrition.

References

1 Cifuentes, A. (2009). Food analysis and foodomics. *J. Chromatogr. A* 1216 (43): 7109.

2 Balkir, P., Kemahlioglu, K., and Yucel, U. (2021). Foodomics: a new approach in food quality and safety. *Trends Food Sci. Technol.* 108: 49–57.

3 Ferreri, C. and Chatgilialoglu, C. (2015). *Membrane Lipidomics for Personalized Health*. Southern Gate: Wiley.

4 Han, X. (2016). Lipidomics for studying metabolism. *Nat. Rev. Endocrinol.* 12 (11): 668–679.

5 Hu, T. and Zhang, J.L. (2018). Mass-spectrometry-based lipidomics. *J. Sep. Sci.* 41 (1): 351–372.

6 Ekroos, K. (2012). *Lipidomics: technologies and applications*. Weinheim: Wiley-VCH Verlag & Co. KGaA.

7 Züllig, T. and Köfeler, H.C. (2021). High resolution mass spectrometry in lipidomics. *Mass Spectrom. Rev.* 40 (3): 162–176.

8 Jakobsson, A., Westerberg, R., and Jacobsson, A. (2006). Fatty acid elongases in mammals: their regulation and roles in metabolism. *Prog. Lipid Res.* 45 (3): 237–249.

9 EFSA Panel on Dietetic Products (2011). Nutrition and Allergies (NDA) Scientific Opinion on the substantiation of health claims related to the replacement of mixtures of saturated fatty acids (SFAs) as present in foods or diets with mixtures of monounsaturated fatty acids (MUFAs) and/or mixtures of polyunsaturated fatty acids (PUFAs), and maintenance of normal blood LDL-cholesterol concentrations (ID 621, 1190, 1203, 2906, 2910, 3065) pursuant to Article 13(1) of Regulation (EC) No 1924/2006. *EFSA J.* 9 (4): 2069.

10 WHO/FAO (2003). Joint WHO/FAO Expert Consultation. Diet, Nutrition and the Prevention of Chronic Diseases. *World Health Organization technical report series*, No. 916.

11 Wang, X., Lin, H., and Gu, Y. (2012). Multiple roles of dihomo-γ-linolenic acid against proliferation diseases. *Lipids Health Dis.* 11: 25.

12 Lankinen, M., Schwab, U., Erkkilä, A. et al. (2009). Fatty fish intake decreases lipids related to inflammation and insulin signaling – a lipidomics approach. *PLoS One* 4 (4): e5258.

13 Bannenberg, G. and Serhan, C.N. (2010). Specialized pro-resolving lipid mediators in the inflammatory response: an update. *Biochim. Biophys. Acta* 1801 (12): 1260–1273.

14 Burdge, G. (2004). Alpha-linolenic acid metabolism in men and women: nutritional and biological implications. *Curr. Opin. Clin. Nutr. Metab. Care* 7 (2): 137–144.

15 Brenna, J.T., Salem, N., Sinclair, A.J., and Cunnane, S.C. (2009). α-Linolenic acid supplementation and conversion to *n*-3 long-chain polyunsaturated fatty acids in humans. *Prostaglandins Leukot. Essent. Fatty Acids* 80 (2): 85–91.

16 Micalizzi, G., Ragosta, E., Farnetti, S. et al. (2020). Rapid and miniaturized qualitative and quantitative gas chromatography profiling of human blood total fatty acids. *Anal. Bioanal. Chem.* 412 (10): 2327–2337.

17 Trovato, E., Micalizzi, G., Dugo, P. et al. (2020). Use of linear retention indices in GC-MS libraries for essential oil analysis. In: *Handbook of Essential Oils: Science, Technology, and Applications*, 3e (ed. Baser KHsC and G. Buchbauer), 229–251. Boca Raton: CRC Press.

18 Sánchez-Avila, N., Mata-Granados, J.M., Ruiz-Jiménez, J., and Luque de Castro, M.D. (2009). Fast, sensitive and highly discriminant gas chromatography-mass spectrometry method for profiling analysis of fatty acids in serum. *J. Chromatogr. A* 1216 (40): 6864–6872.

19 Ecker, J., Scherer, M., Schmitz, G., and Liebisch, G. (2012). A rapid GC-MS method for quantification of positional and geometric isomers of fatty acid methyl esters. *J. Chromatogr. B* 897: 98–104.

20 Angell, S.Y., Cobb, L.K., Curtis, C.J. et al. (2012). Change in trans fatty acid content of fast-food purchases associated with New York City's restaurant regulation. *Ann. Int. Med.* 157 (2): 81–86.

21 Srigley, C.T. and Rader, J.I. (2014). Content and composition of fatty acids in marine oil omega-3 supplements. *J. Agric. Food Chem.* 62 (29): 7268–7278.

22 Ferreri, C., Grabovskiy, S.A., Aoun, M. et al. (2012). Trans fatty acids: chemical synthesis of eicosapentaenoic acid isomers and detection in rats fed a deodorized fish oil diet. *Chem. Res. Toxicol.* 25 (3): 687–694.

23 Chatgilialoglu, C., Ferreri, C., Melchiorre, M. et al. (2014). Lipid geometrical isomerism: from chemistry to biology and diagnostics. *Chem. Rev.* 114 (1): 255–284.

24 Delmonte, P., Fardin-Kia, A.R., Kramer, J.K. et al. (2012). Evaluation of highly polar ionic liquid gas chromatographic column for the determination of the fatty acids in milk fat. *J. Chromatogr. A* 1233: 137–146.

25 Van Pelt, C.K. and Brenna, J.T. (1999). Acetonitrile chemical ionization tandem mass spectrometry to locate double bonds in polyunsaturated fatty acid methyl esters. *Anal. Chem.* 71 (10): 1981–1989.

26 Sirot, V., Oseredczuk, M., Bemrah-Aouachria, N. et al. (2008). Lipid and fatty acid composition of fish and seafood consumed in France: CALIPSO study. *J. Food Compos. Anal.* 21 (1): 8–16.

27 Mori, T.A. (2017). Marine OMEGA-3 fatty acids in the prevention of cardiovascular disease. *Fitoterapia* 123: 51–58.

28 MacLean, C.H., Newberry, S.J., Mojica, W.A. et al. (2006). Effects of omega-3 fatty acids on cancer risk: a systematic review. *JAMA* 295 (4): 403–415.

29 Aung, T., Halsey, J., Kromhout, D. et al. (2018). Associations of omega-3 fatty acid supplement use with cardiovascular disease risks: meta-analysis of 10 trials involving 77917 individuals. *JAMA Cardiol.* 3 (3): 225–234.

30 Grey, A. and Bolland, M. (2014). Clinical trial evidence and use of fish oil supplements. *JAMA Int. Med.* 174 (3): 460–462.

31 Bradbury, J. (2011). Docosahexaenoic acid (DHA): an ancient nutrient for the modern human brain. *Nutrients* 3 (5): 529–554.

32 FAO (2020). *The State of World Fisheries and Aquaculture 2020. Sustainability in Action*. FAO.

33 Miguel, Á.P. (2016). Misdescription incidents in seafood sector. *Food Control* 62: 277–283.

34 Black, C., Chevallier, O.P., Haughey, S.A. et al. (2017). A real time metabolomic profiling approach to detecting fish fraud using rapid evaporative ionisation mass spectrometry. *Metabolomics* 13 (12): 153.

35 Rigano, F., Mangraviti, D., Stead, S. et al. (2019). Rapid evaporative ionization mass spectrometry coupled with an electrosurgical knife for the rapid identification of Mediterranean Sea species. *Anal. Bioanal. Chem.* 411 (25): 6603–6614.

36 FAO (2018). Farmed fish: a major provider or a major consumer of omega-3 oils? GLOBEFISH – Information and Analysis on World Fish Trade. https://www.fao.org/in-action/globefish/fishery-information/resource-detail/en/c/338773 (accessed 23 June 2022).

37 Karalazos, V., Treasurer, J., Cutts, C.J. et al. (2007). Effects of fish meal replacement with full-fat soy meal on growth and tissue fatty acid composition in Atlantic cod (*Gadus morhua*). *J. Agric. Food Chem.* 55 (14): 5788–5795.

38 Torstensen, B.E., Bell, J.G., Rosenlund, G. et al. (2005). Tailoring of Atlantic Salmon (*Salmo salar* L.) flesh lipid composition and sensory quality by replacing fish oil with a vegetable oil blend. *J. Agric. Food Chem.* 53 (26): 10166–10178.

39 Hvattum, E., Røsjø, C., Gjøen, T. et al. (2000). Effect of soybean oil and fish oil on individual molecular species of Atlantic salmon head kidney phospholipids determined by normal-phase liquid chromatography coupled to negative ion electrospray tandem mass spectrometry. *J. Chromatogr. B Biomed. Sci. Appl.* 748 (1): 137–149.

40 Morseletto, P. (2020). Targets for a circular economy. *Resour. Conserv. Recycl.* 153: 104553.

41 Herpandi, N.H., Rosma, A., and Wan Nadiah, W.A. (2011). The tuna fishing industry: a new outlook on fish protein hydrolysates. *Compr. Rev. Food Sci. Food Saf.* 10 (4): 195–207.

42 Saliu, F., Magoni, C., Lasagni, M. et al. (2019). Multi-analytical characterization of perigonadal fat in bluefin tuna: from waste to marine lipid source. *J. Sci. Food Agric.* 99 (10): 4571–4579.

43 Rigano, F., Arena, P., Mangraviti, D. et al. (2021). Identification of high-value generating molecules from the wastes of tuna fishery industry by liquid chromatography and gas chromatography hyphenated techniques with automated sample preparation. *J. Sep. Sci.* 44 (8): 1571–1580.

44 Tocher, D.R., Betancor, M.B., Sprague, M. et al. (2019). Omega-3 long-chain polyunsaturated fatty acids, EPA and DHA: bridging the gap between supply and demand. *Nutrients* 11 (1): 89–108.

45 Poblete-Castro, I., Escapa, I.F., Jäger, C. et al. (2012). The metabolic response of P. putida KT2442 producing high levels of polyhydroxyalkanoate under single- and multiple-nutrient-limited growth: highlights from a multi-level omics approach. *Microb. Cell Fact.* 11 (1): 34.

46 Haoujar, I., Cacciola, F., Manchado, M. et al. (2020). Isolation of microalgae from mediterranean seawater and production of lipids in the cultivated species. *Foods* 9 (11): 1601–1613.

47 Galano, J.-M., Lee, Y.Y., Oger, C. et al. (2017). Isoprostanes, neuroprostanes and phytoprostanes: an overview of 25 years of research in chemistry and biology. *Prog. Lipid Res.* 68: 83–108.

48 Vigor, C., Oger, C., Reversat, G. et al. (2020). Isoprostanoid profiling of marine microalgae. *Biomolecules* 10 (7): 1073.

49 Ness, A.R. and Powles, J.W. (1997). Fruit and vegetables, and cardiovascular disease: a review. *Int. J. Epidemiol.* 26 (1): 1–13.

50 Block, G., Patterson, B., and Subar, A. (1992). Fruit, vegetables, and cancer prevention: a review of the epidemiological evidence. *Nutr. Cancer* 18 (1): 1–29.

51 Global Olive Oil Market Report 2021–2027: Market to Reach $13.8 Billion – Growing Adoption of Olive Oil in the Cosmetic Industry is Driving the Market 2021. Research and Markets, Dublin, Ireland. https://www.globenewswire.com/news-release/2021/07/01/2256340/28124/en/Global-Olive-Oil-Market-Report-2021-2027-Market-to-Reach-13-8-Billion-Growing-Adoption-of-Olive-Oil-in-the-Cosmetic-Industry-is-Driving-the-Market.html (accessed 03 June 2022).

52 Montealegre, C., Sánchez-Hernández, L., Crego, A.L., and Marina, M.L. (2013). Determination and characterization of glycerophospholipids in olive fruit and oil by nonaqueous capillary electrophoresis with electrospray-mass spectrometric detection. *J. Agric. Food Chem.* 61 (8): 1823–1832.

53 IOC. Designation and definitions of olive oils. International Olive Council. https://www.internationaloliveoil.org/olive-world/olive-oil (accessed 23 June 2022).

54 Jabeur, H., Zribi, A., and Bouaziz, M. (2016). Extra-virgin olive oil and cheap vegetable oils: distinction and detection of adulteration as determined by GC and chemometrics. *Food Anal. Methods* 9 (3): 712–723.

55 Mangraviti, D., Rigano, F., Arigò, A. et al. (2021). Differentiation of Italian extra virgin olive oils by rapid evaporative ionization mass spectrometry. *LWT* 138: 110715.

56 Alves, E., Melo, T., Rey, F. et al. (2016). Polar lipid profiling of olive oils as a useful tool in helping to decipher their unique fingerprint. *LWT* 74: 371–377.

57 Calvano, C.D., Ceglie, C.D., D'Accolti, L., and Zambonin, C.G. (2012). MALDI-TOF mass spectrometry detection of extra-virgin olive oil adulteration with hazelnut oil by analysis of phospholipids using an ionic liquid as matrix and extraction solvent. *Food Chem.* 134 (2): 1192–1198.

58 Fidelis, M., de Moura, C., Kabbas Junior, T. et al. (2019). Fruit seeds as sources of bioactive compounds: sustainable production of high value-added ingredients from by-products within circular economy. *Molecules* 24 (21): 3854.

59 Alves, E., Simoes, A., and Domingues, M.R. (2021). Fruit seeds and their oils as promising sources of value-added lipids from agro-industrial byproducts: oil content, lipid composition, lipid analysis, biological activity and potential biotechnological applications. *Crit. Rev. Food Sci. Nutr.* 61 (8): 1305–1339.

60 Holčapek, M., Lísa, M., Jandera, P., and Kabátová, N. (2005). Quantitation of triacylglycerols in plant oils using HPLC with APCI-MS, evaporative light-scattering, and UV detection. *J. Sep. Sci.* 28 (12): 1315–1333.

61 Loutelier-Bourhis, C., Zovi, O., Lecamp, L. et al. (2009). Contribution of two approaches using electrospray ionization with multi-stage mass spectrometry for the characterization of linseed oil. *Rapid Commun. Mass Spectrom.* 23: 3743–3752.

62 Arena, P., Sciarrone, D., Dugo, P. et al. (2021). Pattern-type separation of triacylglycerols by silver thiolate×non-aqueous reversed phase comprehensive liquid chromatography. *Separations* 8 (6): 88.

22

Lipidomics in Plant Science

Zolian Zoong Lwe[1,2], Yu Song[1,2], P. A. D. B. Vinusha Wickramasinghe[2,3], and Ruth Welti[2,3]

[1] *Kansas State University, Department of Biochemistry and Molecular Biophysics, 1711 Claflin Rd., Manhattan, KS, 66506, USA*
[2] *Kansas State University, Kansas Lipidomics Research Center, 1717 Claflin Rd., Manhattan, KS, 66506, USA*
[3] *Kansas State University, Division of Biology, 1717 Claflin Rd., Manhattan, KS, 66506, USA*

Zolian Zoong Lwe and Yu Song contributed equally to this work.

22.1 Introduction

In the latter part of the twentieth century, mass spectrometry was employed to elucidate the chemical structures of plant lipids, and gas chromatography-mass spectrometry was used for analysis of simple lipids from plants [1–6]. Quantitative analysis of a wide range of complex plant lipids by mass spectrometry, or plant "lipidomics," began in the early 2000s [7, 8], with the first plant analyses taking advantage of the ground-breaking methods developed for mammalian lipid analysis (e.g. [9, 10]). Lipidomics has increased the ease and speed of lipid analysis for plant scientists, and mass spectrometry has enabled the discovery of numerous new plant lipids. Many hundreds of papers utilizing plant lipid profiling using mass spectrometry have been published. Here, we highlight several examples, from the last 20 years, of how lipidomics has shed light on plant lipid metabolism and the roles of lipids in plants.

22.2 The Role of Phosphatidic Acid in Plant Response to Nutrients and Stress

Phosphatidic acids (PA) are lipid molecules of low abundance in plants. In leaf tissues of *Arabidopsis thaliana*, the concentration of PA is much lower than that of the major phospholipids, phosphatidylcholine (PC), phosphatidylethanolamine (PE), and phosphatidylglycerol (PG). PA was measured in 2002 by a direct-infusion electrospray ionization triple quadrupole mass spectrometry

Mass Spectrometry for Lipidomics: Methods and Applications, First Edition.
Edited by Michal Holčapek and Kim Ekroos.

method [7]. PA is often detected by direct infusion mass spectrometry, because negatively charged PA molecular species tend to smear on chromatography columns rather than eluting in sharp peaks. The Arabidopsis PA profile of 10 molecular species was quantitatively determined in leaves after cold and freezing treatments.

Phospholipase Ds (PLDs) are enzymes that catalyze the hydrolysis of phospholipids, such as PC, PE, and PG, to PA. Comparative profiling of wild-type (WT) and mutant plants defective in two of the 12 PLD gene products in Arabidopsis was employed to investigate the function of particular PLDs in generating PA under various conditions [7, 11, 12]. Sublethal freezing treatment at −8 °C induced an increase in PA levels in both WT and plants deficient in PLDα. However, as shown in Figure 22.1a, the PA levels in the PLDα-deficient plants after freezing were lower than those in WT plants. During the freezing treatment, it was also demonstrated that the level of PC dropped more in WT plants than in PLDα-deficient plants, while drops in the levels of PE and PG were similar in WT and PLDα-deficient plants. These results implied that PC, rather than the other phospholipids, was the major substrate of PLDα during freezing treatment [7]. An analysis of PA species during post-freezing recovery indicated that the levels of PA 34:4 increased more in recovery than in freezing, while other PA species did not show obvious differences between freezing and post-freezing recovery [12]. Quantitative changes in PA 34:4 and PG 34:4 of WT and PLDα-deficient plants, measured by lipidomics, supported the hypothesis that PA is produced from PG during post-freezing recovery and that PLDα is responsible for post-freezing PG hydrolysis, as well as PC hydrolysis during freezing.

Analysis of knockout (KO) and overexpression (OE) mutants of a different PLD, PLDδ, indicated that PLDδ contributed about 20% of the PA production under freezing stress [11]. However, PLDδ displayed an opposite role compared to PLDα in freezing tolerance. PLDδ protects plants from freezing damage, whereas PLDα action increases freezing damage [7, 11].

Lipidomic analysis also demonstrated an increase in PA 34:6, but this species increased to a similar extent in WT and PLDα-deficient plants upon freezing (Figure 22.1a), suggesting that this PA species is not derived from PLDα activity [7]. It was also noted that the "34:6" fatty acid combination, which in Arabidopsis represents 18:3/16:3, typically occurs primarily in plastidically localized galactolipids. Analysis of lipidomic data on plants subjected to mechanical wounding showed that PA levels also increased under wounding stress, and that the levels of PA 34:6, trigalactosyldiacylglycerol 34:6 (TrGDG 34:6), and tetragalactosyldiacylglycerol 34:6 (TeGDG 34:6) were highly correlated, both over time and plant sample, indicating that these compounds are produced by the same metabolic pathway [13, 14]. Based on these data, a model of PA production under freezing is proposed (Figure 22.1b), suggesting that diacylglycerol 34:6 (DG 34:6) formed by processive galactosylation of monogalactosyldiacylglycerol 34:6 (MGDG 34:6) is subsequently converted into PA 34:6 through the action of a DG kinase.

Besides their roles in freezing and wounding stress, PLDs also have essential functions for plant survival under phosphorus-limited conditions. Phospholipid and galactolipid molecular species changes during phosphorus starvation were analyzed

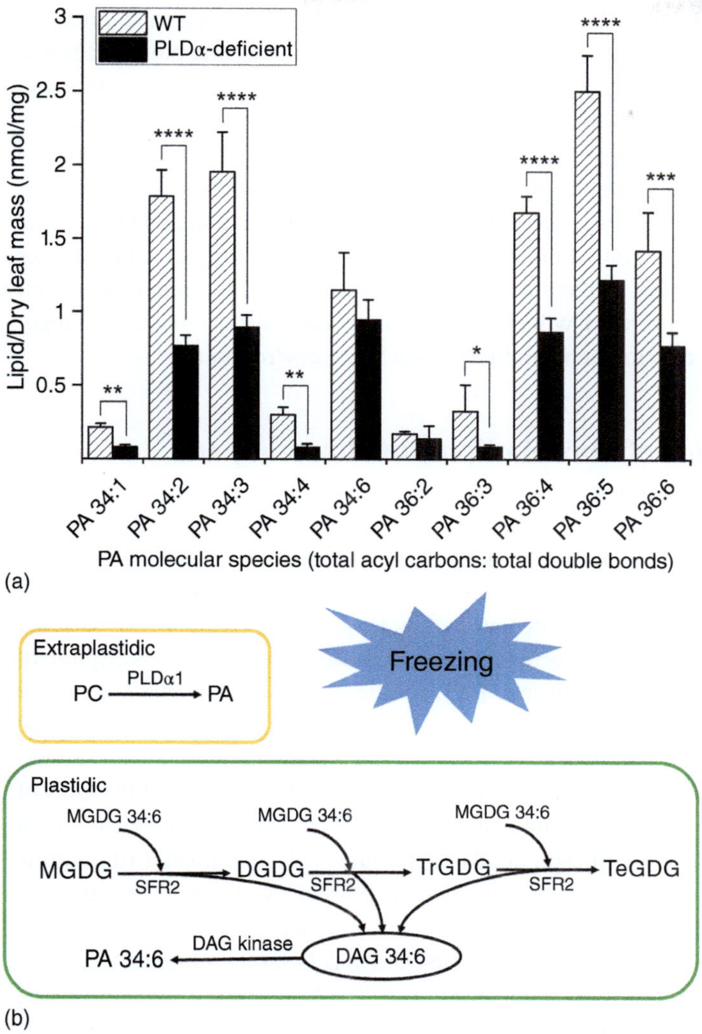

Figure 22.1 PLDα1 and SENSITIVE TO FREEZING 2 (SFR2) regulate PA production during freezing stress. (a) Freezing-induced changes in PA molecular species in Arabidopsis leaves measured by direct infusion electrospray ionization triple quadrupole mass spectrometry. Source: Data from Welti et al. [7]. The hatched bars represent wild-type Col-0, and the black bars represent PLDα1-deficient mutants; both were subjected to −8 °C. The values are the means + S.D. ($n = 5$). * $p < 0.05$, ** $p < 0.01$, *** $p < 0.001$, **** $p < 0.0001$. (b) Proposed model of PA production in response to freezing in Arabidopsis leaves. SFR2 catalyzes the processive galactosylation of MGDG.

by direct infusion electrospray ionization triple quadrupole mass spectrometry [15]. Lipidomics analysis of PA, PC, and digalactosyldiacylglycerol (DGDG) in rosettes and roots of WT Arabidopsis and *pldζ1*, *pldζ2*, and *pldζ1pldζ2* double mutants (defective in these PLD isoforms) suggested that PLDζ1 and PLDζ2 regulate hydrolysis of PC in response to low phosphorus levels and that this hydrolysis supplies DG moieties for the synthesis of DGDG, which can compensate for loss of phospholipids.

Analysis of genetically altered plant lines under varied physiological conditions, including stress conditions, has led to characterization of substrate preferences and physiological roles of many members of the large PLD family [16]. Lipidomic analysis has opened the door for many additional studies on the roles of PLDs (e.g. the role of PLDs in plant immunity [17]). Lipidomics, combined with other biochemical methods, has also led to an appreciation of the varied functions of specific PA molecular species, which are produced in response to varied stimuli and by multiple enzyme pathways [18]. Many examples now exist of specific PA molecular species modulating the function of proteins by affecting their catalytic properties or their localization [18]. Future work will expand our understanding of PA action in regulating metabolism, development, and responses to the environment.

22.3 The Roles of Phospholipids in Flowering and Diurnal Metabolism

The onset of flowering in Arabidopsis and many other plant species is regulated by photoperiod, i.e. day length [19]. Changes in photoperiod are detected in the leaves of Arabidopsis, where increased day length and circadian signals are integrated into a cascade of molecular events that ultimately lead to transcription of the *FLOWERING LOCUS T* (FT) gene [19]. The FT protein, or florigen, is a mobile signal synthesized in phloem companion cells in leaves. FT protein is exported into phloem sieve cells and then travels through the phloem to the shoot apical meristem where it induces flowering [20]. Given that FT protein belongs to the family of "phosphatidylethanolamine-binding proteins" (PEBPs) and that its mammalian homolog binds to different phospholipids, Nakamura et al. [21] sought to determine the FT–lipid interaction(s) in Arabidopsis and the physiological role of these interactions *in vivo*. The authors first established *in vitro* that FT is a PC-binding protein. Considering that PC is a major phospholipid class in cellular membranes, that FT is key to inducing flowering, and that there are FT–PC binding interactions, the authors tested whether altering PC levels affected flowering time. This was done by creating a transgenic plant that harbored an artificial microRNA designed to suppress gene expression of the enzyme CTP:phosphorylethanolamine cytidylyltransferase1 (PECT1), resulting in reduced PE levels with a concomitant increase in PC levels. The altered plants exhibited an early-flowering phenotype and no other growth defects. The correlation between increased PC levels and early flowering was observed when the construct was expressed under a promoter specific to the shoot apical meristem but not when expressed using a promoter specific to companion cells. Furthermore, when the authors overexpressed *PECT1* in the shoot apical meristem, they observed an opposite, late-flowering phenotype associated with decreased PC levels. The authors hypothesized that the positive correlation between increased PC levels and the early-flowering phenotype was related to FT activity. They showed that plants with increased PC levels and early-flowering phenotype also had increased expression levels of effectors of FT. However, the question remained as to whether specific PC species were associated with the early-flowering phenotype and with FT protein. Lipidomics profiling using a quadrupole/time-of-flight

(Q-TOF) mass spectrometer outfitted with a nanospray ion source allowed the authors to detect in leaves a diurnally dependent change in lipid molecular species composition of PC, but not of PE. They found that saturated and less-than-fully unsaturated PC molecular species were predominant in the daytime, and polyunsaturated molecular species were predominant at night. The authors tested FT binding to PC molecular species and found stronger binding to saturated and less-than-fully unsaturated PC lipid species. The results suggest that the increased presence of PC lipid species unique to daytime facilitates the activity of FT protein in inducing flowering [21]. This conclusion was corroborated by the finding that a transgenic plant line with increased levels of polyunsaturated lipids showed a late-flowering phenotype.

It was later discovered by Susila et al. [22] that FT protein can also bind to PG and that FT is sequestered under suboptimal temperature in cellular membranes of companion cells in a PG-dependent manner, resulting in delay in flowering. Lipidomics profiling using electrospray ionization triple quadrupole mass spectrometry, with or without chromatography, of lipids coimmunoprecipitated with FT-green fluorescent protein (GFP)-His, in plants overexpressing FT with GFP and His tags, demonstrated that FT binds with PG *in vivo*. Subsequent analysis of PG biosynthesis mutants in Arabidopsis showed an early-flowering phenotype under suboptimal temperature, a condition which normally delays flowering. Addition of exogenous PG, but not PC or PE, caused later flowering of both WT plants and of a PG-biosynthesis mutant. Lipidomics profiling of the mutants revealed a significant decrease in endogenous PG levels under both optimal and suboptimal temperatures. Using confocal laser scanning microscopy and transmission electron microscopy, the mutants were found to have a higher ratio of soluble FT protein to membrane-bound FT protein, and the intracellular localization of FT protein to membranes was increased by addition of exogenous PG. However, levels of PG in whole seedlings and membrane fractions of WT were unaffected by change in temperature, indicating that the delay in flowering under suboptimal temperature was not due to a change in total PG levels. Instead, lipidomics revealed that temperature altered the unsaturation of PG molecular species. Thus, it seems likely that FT protein interacts with specific PG molecular species present at higher levels under suboptimal temperature, similar to what Nakamura et al. [21] reported for PC. The authors acknowledge that spatial and temporal lipidomics are necessary to determine if changes in PG unsaturation levels in response to temperature cause the PG- and temperature-dependent changes in FT protein sequestration in companion cell membranes, which are correlated with changes in flowering time.

Whether or how cellular metabolism affects the circadian clock in plants was largely unknown until Kim et al. [23] discovered that PA interacts with LATE ELONGATED HYPOCOTYL (LHY) and CIRCADIAN CLOCK ASSOCIATED (CCA1) proteins. Both proteins are core clock regulators, and PA inhibits both from binding to their target gene, *TIMING OF CAB EXPRESSION1* (*TOC1*). Lipidomics profiling using electrospray ionization triple quadrupole mass spectrometry of PA species coprecipitated with LHY showed that PA molecular species with C16 acyl chains, such as PA 32:0 and PA 34:2, but not all PA species, bind with LHY *in vivo*. This finding corroborated *in vitro* results showing that both LHY and CCA1 proteins bind preferentially with C16-containing PA species. Furthermore, electrophoretic mobility shift assays demonstrated that C16-containing PA species, specifically, cause the inhibition of LHY binding to the *TOC1* promoter. Using a PA-hydrolyzing double mutant whose lipidomic profile shows an increase in total PA and C16-containing PA species, the authors isolated *TOC1* promoters bound to LHY proteins using ChIP-PCR and quantified the amount with qPCR. The mutant had significantly fewer LHY–*TOC1* interactions compared to WT, which was attributed to the increase in PA levels and specifically C16-containing PA species in the mutant. Conversely, the addition of 1-butanol and a diacylglycerol kinase inhibitor, which both decrease total PA and C16-containing PA species *in planta*, caused an effect opposite to that of the PA-hydrolyzing double mutant. Another double mutant *lhy cca1* not only disrupted the diurnal fluctuation of *TOC1* expression, but also altered the levels of many phospholipid species and caused a significant reduction in total seed oil content. Taken together, the results of this study suggest that the direction of interaction between the circadian clock and cellular metabolism is two-way and that the LHY–PA interaction may integrate the two [23].

22.4 Sphingolipid Analysis Has Facilitated the Discovery of Pathways Regulating Important Plant Cell Functions

Sphingolipids are a structurally diverse group of lipids that function in biological processes ranging from development and stress responses to cellular processes [24]. The structural diversity of sphingolipids comes from variety in each of the component parts (long chain base, fatty acid chain, and head group moiety) that make up a sphingolipid and all the possible combinations of sphingolipid components [25–28]. The development of a high-throughput method by Markham and Jaworski [26] to measure sphingolipids in Arabidopsis paved the way for multiple sphingolipidomic studies in plants that have elucidated important biological functions. These studies describe sphingolipid involvement in growth and organogenesis [29, 30], cell death [31, 32], temperature response [33], mycotoxin sensitivity [34], very-long-chain fatty acid (VLCFA)-ceramide protection against hypoxia [35], intracellular trafficking [36], and root sphingolipids regulating the leaf ionome [37].

Using reversed-phase liquid chromatography electrospray ionization triple quadrupole mass spectrometry, Markham et al. [38] discovered that VLCFA-containing sphingolipids and not long-chain fatty acid-containing sphingolipids are integral in lateral root growth and the endosomal pathway. They found that a reduction in ceramide levels in mutants resulted in a defective secretory pathway for polar auxin carriers, PIN1 and AUX1. Ito et al. [39] discovered, using liquid chromatography electrospray ionization triple quadrupole mass spectrometry, that it is VLCFA-containing glycosylated sphingolipids, and not ceramides, that are necessary in regulating the homeostasis of phosphoinositides (PI) at the *trans*-Golgi network.

Reduction in sphingolipids resulted in an increase in phosphatidylinositol-4-phosphate (PI4P) at the *trans*-Golgi network. The authors discovered that sphingolipids mediate the activity of PI-specific phospholipase Cs (PI-PLCs) that consume PI4P at the *trans*-Golgi network, maintaining PI homeostasis, which is essential in properly sorting the auxin efflux carrier PIN2 at the *trans*-Golgi network.

Glycosylinositol phosphorylceramides (GIPCs) are sphingolipids found in the outer leaflet of the plasma membrane, where they interact with the extracellular environment. Lenarčič et al. [40] reported that GIPCs act as receptors binding the microbial toxin necrosis and ethylene-inducing peptide 1-like (NLP) proteins. To understand why NLP toxins affect only eudicots and not monocots, the authors examined the GIPC composition, using liquid chromatography electrospray ionization triple quadrupole mass spectrometry. They determined that, generally, monocots produce series B GIPC molecules that carry three hexoses, whereas eudicots produce series A GIPC that carry two hexoses. Arabidopsis mutants defective in GIPC synthesis are not susceptible to NLP toxicity, while an unusual monocot species that produces substantial amounts of both GIPC series is susceptible to NLP toxicity, demonstrating that the presence of series A GIPCs explains the difference in NLP toxicity between the two plant clades.

GIPC sphingolipids also bind to salt ions as a mechanism for salt perception. GIPC-Na^+ interactions trigger Ca^{2+} spikes and waves [41]. Jiang et al. [41] identified a mutant that was sensitive to growth in salt, but otherwise had no growth defects. Profiling with matrix-assisted laser desorption ionization (MALDI) mass spectrometry indicated that the mutant had significantly lower amounts of GIPCs and had a defect in sphingolipid biosynthesis. Based on their results, the authors hypothesize that sodium binding by GIPCs regulates proteins involved in calcium influx through the plasma membrane, leading to intracellular events that result in salt tolerance.

High levels of sphingolipids can be just as detrimental as reductions in sphingolipids, because sphingolipids play roles in both cell growth and initiation of programmed cell death. OROSOMUCOID-LIKE PROTEINS (ORMs) are proteins that interact with the first enzyme in sphingolipid biosynthesis, serine palmitoyltransferase, and regulate plant sphingolipid synthesis in response to variations in cellular sphingolipid concentrations. Gonzalez-Solis et al. [42] recently discovered, in mutants altered in ORMs, a disruption in sphingolipid homeostasis. The *orm* mutants had nonviable seeds, abnormal embryo development, and a substantial reduction in seed oil content. Measurement of sphingolipid levels in the *orm* mutants, using liquid chromatography electrospray ionization triple quadrupole mass spectrometry, showed a significant increase in long chain bases and ceramides, early products in sphingolipid biosynthesis, reflecting the absence of the negative regulatory effect of ORMs on the initial step in sphingolipid biosynthesis. This phenotype was mimicked in a mutant of *ORM1* with a deleted amino acid that resulted in reduced protein–protein interaction. Taken together, the data provide the basis for a model of ORM regulation of sphingolipid biosynthesis [42]. In future work, the authors plan to improve their understanding of how ORMs respond to cellular sphingolipid concentration and to characterize the details of the interaction between ORMs and serine palmitoyltransferase.

22.5 Identification of a New Lipid Class in Plants Under Phosphate Stress

Phosphorus is essential in healthy plant growth and development, as it is a key component of biomolecules, including nucleic acids and phospholipids. However, P limitation in the soil can lead to P deficiency in plants. Under such circumstances, plants mobilize P by lipid remodeling, in which membrane phospholipids are replaced by non-P lipids such as sulfoquinovosyldiacylglycerol (SQDG) in the chloroplast and DGDG in the chloroplast and extraplastidic membranes [43].

Okazaki et al. [44] discovered a novel lipid in *Arabidopsis thaliana* by monitoring the total ion counts as leaf lipid extracts, subjected to hydrophilic interaction chromatography (HILIC), were eluted and detected by mass spectrometry in the negative ion mode. The unknown lipid was indicated by a peak that appeared only in plants subjected to phosphorus depletion. Further mass spectrometric analysis of the unknown in both positive and negative ion modes revealed that it belongs to a lipid class consisting of molecular species that differ by 2 Da, hinting at the presence of fatty acids. Further investigation of its fragmentation pattern using ion trap time-of-flight mass spectrometry indicated that the unknown contained two fatty acids and a glucuronosylglycerol moiety. The lipid, which is found in some microbes but had not been identified in plants, was later identified as 1,2-diacyl-3-*O*-alpha-glucuronosylglycerol (GlcADG).

Analysis of the molecular species compositions indicated that GlcADG and SQDG have very similar DG molecular species compositions. Additionally, fragmentation analysis of GlcADG and SQDG showed high compositional similarities, suggesting palmitic acid (16:0) in the *sn*-2 position, which is characteristic of plastid compartmentalization. This finding led Okazaki and coworkers [44] to hypothesize that GlcADG and SQDG share a biosynthetic pathway. To test this hypothesis, the authors examined levels of GlcADG in plants with mutations in the synthetic pathway for SQDG, which involves the enzymes UDP-Glc pyrophosphorylase (UGP3), UDP-sulfoquinovose synthase (SQD1), and SQDG synthase (SQD2) (Figure 22.2). Arabidopsis *sqd2* mutants were severely damaged under phosphorus stress, with impaired growth, and lacked both SQDG and GlcADG, indicating that SQD2 was involved in GlcADG synthesis. In contrast, the *ugp3* and *sqd1* mutants showed a moderate increase of GlcADG compared to WT. The data imply that SQD2 is common to both SQDG and GlcADG synthesis, while UGP3 and SQD1 are not involved in synthesis of GlcADG. It seems likely that SQD2 can use a UDP-sugar other than UDP-sulfoquinovose (SQ) to produce GlcADG by glycosylation of DG; the UDP-sugar is likely to be UDP-GlcA.

Since GlcADG was discovered in Arabidopsis, its presence has been demonstrated in other plant species. Turtlegrass (*Thalassia testudinum*), a seagrass, also upregulates GlcADG levels in leaves under low phosphorus concentrations [46]. Okazaki et al. [44] observed that GlcADG was elevated in rice leaves (*Oryza sativa* L.) during phosphorus deficiency. Further investigation of two rice cultivars varying in

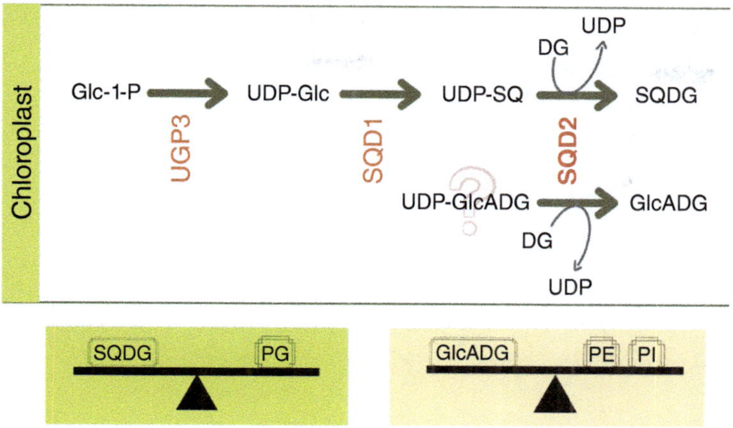

Figure 22.2 GlcADG biosynthetic pathway. The SQDG biosynthesis pathway, including the three enzymes involved and SQD2 synthase engagement in GlcADG synthesis described in the text and by Okazaki et al. [44]. The unidentified sugar donor is hypothesized to be UDP-GlcA. The lower portion of the figure indicates that Okazaki et al. [44] determined during phosphate deficiency that PG levels are reciprocally regulated with SQDG levels, and PI and PE levels together are reciprocally regulated with GlcADG levels. Abbreviations: Uridine diphosphate (UDP), Glucose-1-phosphate (Glc-1-P), UDP-glucose (UDP-Glc), UDP-sulfoquinovose (UDP-SQ), diacylglycerol (DG), UDP-Glc pyrophosphorylase (UGP3), UDP-SQ synthase (SQD1), and SQDG synthase (SQD2).

sensitivity to P deficiency found that GlcADG increased under phosphorus deficiency in both cultivars, but to the highest extent in the cultivar that was tolerant of low phosphorus conditions [47]. Okazaki et al. [48] found that when soybean plants were grown under P-limitation, the older leaves showed decreases in phospholipids and increases in GlcADG, whereas the younger leaves did not, suggesting that the older leaves break down phospholipids to supply the new leaves with phosphorus.

22.6 Oxidation and Head-Group Acylation of Membrane Lipids in Plant Stress

Mass spectrometry has played a prominent role in structural elucidation of oxidized free fatty acids (oxylipins). Before the advent of full-fledged lipidomics, quantitative high-performance liquid chromatography and gas chromatography–mass spectrometry methods were developed for many oxylipin classes in plants (e.g. [49–59]). These analyses provide accurate measurement of oxylipins, and, when coupled with stress treatments and/or genetic manipulation, provide information about the role of specific oxylipins in plant stress responses.

The application of lipidomics in plant science has identified many oxylipins esterified in membrane lipids as well in free fatty acid form. These oxylipin-containing membrane lipids, or "ox-lipids," whose functions remain to be fully characterized, have been structurally identified in various plant species and are typically

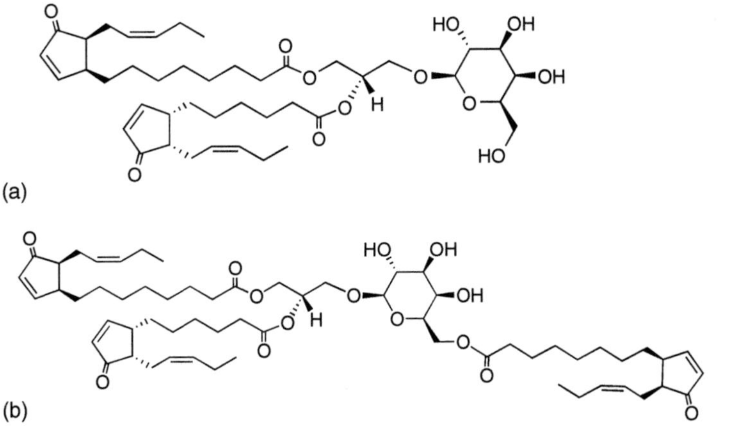

Figure 22.3 The molecular structures of Arabidopsides A and E. (a) Arabidopside A:1-((1*S*,2*S*)-3-oxo-2-(2′*Z*-pentenyl)cyclopent-4-eneoctanoyl)-2-(1*S*,2*S*)-3-oxo-2-(2′*Z*-pentenyl)cyclopent-4-enehexanoyl)-3-*O*-β-D-galactosyl-*sn*-glycerol, (b) Arabidopside E:1-((1*S*,2*S*)-3-oxo-2-(2′*Z*-pentenyl)cyclopent-4-eneoctanoyl)-2-(1*S*,2*S*)-3-oxo-2-(2′*Z*-pentenyl)cyclopent-4-enehexanoyl)-3-*O*-(6′-*O*-((1*S*,2*S*)-3-oxo-2-(2′*Z*-pentenyl)cyclopent-4-eneoctanoyl)-β-D-galactosyl-*sn*-glycerol. Source: Adapted from Sud et al. [45].

associated with plant stress. Arabidopsides, a group of ox-lipids first discovered in *Arabidopsis thaliana*, are galactolipids with the 18C oxylipin,12-oxophytodienoic acid (OPDA), or its 16C analog, dinor-oxophytodienoic acid (dnOPDA) [60–62]. Arabidopsides A, B, C, and D were characterized in Arabidopsis using electrospray ionization Q-TOF mass spectrometry. Arabidopsides A (Figure 22.3a) and B are MGDGs containing OPDA and/or dnOPDA, while Arabidopsides C and D are DGDGs consisting of OPDA and/or dnOPDA. An ox-lipid called Arabidopside E (Figure 22.3b) contains one OPDA and one dnOPDA esterified to the glycerol of MGDG and an additional OPDA esterified to the 6-position of the MGDG galactose. An all-OPDA acylated MGDG analog, Arabidopside G, was also identified [63, 64]. Additionally, OPDA-containing PG, acylated OPDA-containing PG, and OPDA-containing SQDG species were characterized by electrospray ionization Q-TOF mass spectrometry [65–67]. Scanning for precursors of oxidized fatty acid anions and fatty acid anions minus water, many more ox-lipid species containing oxylipins other than OPDA and dnOPDA have been identified. The chemical formulas of detectable oxidized lipids have been discerned by accurate mass measurements or liquid chromatography coupled with product ion analysis using an ion trap [65, 67–69].

Many varied esterified oxidized lipid molecular species, particularly from leaf tissue, have been quantitatively measured using electrospray ionization triple quadrupole mass spectrometry with or without chromatography [13, 14, 63–70]. The rapid accumulation of high levels of OPDA and dnOPDA containing galactolipids in Arabidopsis was observed under wounding stress, supporting the hypothesis that the accumulation of esterified oxylipins is a major signature of wounding [13, 14, 65, 67–69]. The rapid accumulation in Arabidopsis of OPDA or dnOPDA-containing

MGDG, DGDG, acylated MGDG, SQDG, and acylated PG species was also observed during the hypersensitive response triggered by bacterial infection [63, 64, 66, 68, 69]. Oxylipin profiling using mass spectrometry confirms and extends our knowledge that mechanical wounding and bacterial infection induce the oxylipin biosynthetic pathway, generating esterified, as well as free, oxylipin compounds.

Head-group-acylated lipids, which can include all oxidized or all normal-chain fatty acids, or a mixture of the two, have also been extensively analyzed. To date, more than 100 molecular species of acylated MGDG, acylated DGDG, acylated SQDG, acylated PG, and acylated PE (i.e. *N*-acylphosphatidylethanolamine or NAPE) have been characterized in plants using mass spectrometry [13, 14, 63, 64, 66–72]. Lipidomics data show that mechanical wounding, pathogen infection, freezing, and heat stress all induce significant increases in head-group-acylated, plastidic lipids [13, 14, 63, 64, 66–69, 71]. Analysis of lipids from WT Arabidopsis and plants with mutations in the gene, *Acylated Galactolipid Associated Phospholipase 1* (*AGAP1*), extracted after a freeze–thaw treatment, revealed that AGAP1 was required for accumulation of acylated MGDG, acylated DGDG, and acylated PG [70].

Lipidomics was also applied to explore the regulation of NAPE and *N*-acyl ethanolamine (NAE) metabolism in WT Arabidopsis and in plants with defects in NAE metabolism. Kilaru et al. [72, 73] used electrospray ionization triple quadrupole methods to profile NAPE molecular species in seeds and seedlings of Arabidopsis. Knockout of fatty acid amide hydrolase (FAAH), which hydrolyzes NAE, and exogenous NAE 12:0 treatment caused elevated NAE levels and a very dramatic increase of NAPE, suggesting NAE contributes to negative feedback regulation of the NAPE hydrolysis that produces NAE. Elevated NAE levels inhibited seedling establishment and growth in FAAH knockout mutants, while lower-than-WT levels of NAE in FAAH overexpression lines resulted in accelerated growth and early flowering [74, 75]. Alterations in NAPE and NAE metabolism also resulted in changes in the balance of membrane lipids assembled via the plastidic and endoplasmic reticulum pathways, as well as alterations in seedling growth and development [72].

22.7 Triacylglycerols in Seeds and Leaves

Plant oils, consisting mainly of triacylglycerols (TGs), are valued as components of the human diet. Plant oils that contain fatty acids modified with certain functional groups, such as hydroxyls, can serve as feedstocks for industry. For plants, seed oils are energy reserves for seedling growth. Lipidomics has aided our ability to rapidly assess seed oil compositional details. In plant tissues other than seeds, the roles of lipid droplets, composed mainly of triacylglycerols, are increasingly appreciated, largely due to our ability to measure their component triacylglycerol via lipidomics. Quantitative analysis of seed or other tissue extracts containing triacylglycerols by mass spectrometry, with or without chromatography, most often is performed using electrospray ionization in the positive mode [76–79]. Many methods take advantage of the neutral loss of a fatty acid to form a positively charged diacylglycerol fragment [76–79]. Combining information from multiple neutral loss scans or monitoring of multiple transitions (from intact ion to diacylglycerol cation) can be used to define the triacylglycerols in terms of all three fatty acids [76–79]. However, sometimes only the total carbon number: total double bonds or the identity of one fatty acid with the other two combined have been reported (e.g. [71, 80]). In addition to analysis of lipid extracts, triacylglycerol localization has also been imaged using tandem mass spectrometry with MALDI ionization (e.g. [81–83]).

Lipidomic analysis of seed oils has resulted in improvements in our fundamental understanding of oil composition and fatty acid arrangements. In the twentieth century, analysis by hydrolysis with position-specific enzymes demonstrated that the different positions on the glycerol molecule of TGs varied in fatty acid composition. Yet it remained for lipidomics to clarify how fatty acids were combined into a population of triacylglycerol molecules. Figure 22.4 shows an analysis of data from lipidomic profiling that defined and quantified the fatty acids on TG molecules [78]. The analysis shows that the fatty acids in Arabidopsis TGs are not randomly combined in TG molecules. In particular, TGs with elongated chains (C in Figure 22.4) combine with two chains that are not elongated (X,X,C) more readily than predicted based on random chain arrangement, and combinations with two or three elongated chains (X,C,C or C,C,C) are less common than would be found in a random arrangement. In another demonstration of non-randomness in the formation of TGs, a large body of lipid imaging data on plant seeds demonstrates that specific TG molecular species, as well as specific phospholipid molecular species, are not uniformly localized within the seeds. For example, Horn et al. [81] demonstrated that TGs with linoleic acid (18:2) were preferentially localized to the embryonic axis, while those containing gadoleic acid (20:1) were found in the cotyledons. Horn et al. [81] demonstrated that when they altered the overall fatty acid composition by suppressing desaturases or elongases, or by transgenic expression of an acyl–acyl carrier protein, specific molecular species were also spatially redistributed. These data indicate that spatial variation in lipid composition, and potentially of enzymes, needs to be considered in our understanding of seed lipid metabolism.

Lipidomic TG profiling of leaf extracts has led to the notion that increasing leaf TG is part of an adaptive stress response. When plants are subjected to heat, membrane lipids in leaves become less unsaturated, and polyunsaturated fatty acids removed from membrane lipids accumulate in TGs [71, 84–88]. Mueller et al. [71] demonstrated that the accumulation of TG in leaves under heat stress was dependent on the enzyme phospholipid: diacylglycerol acyltransferase (PDAT) and that PDAT deficiency results in decreased basal thermotolerance. It has also been suggested that fatty acids, removed from membrane lipids by lipolytic activity activated in response to heat, may be incorporated into TG to avoid potential membrane damage from high free fatty acid levels and their potential oxidation products. Indeed, recently Demski et al. [89] demonstrated that increasing PDAT levels can protect plants from heat-induced damage.

When plants are subjected to cold temperature, membrane lipids are also remodeled. This remodeling process has been documented by lipidomic studies and includes increases in TG levels [90, 91]. When a close relative of Arabidopsis,

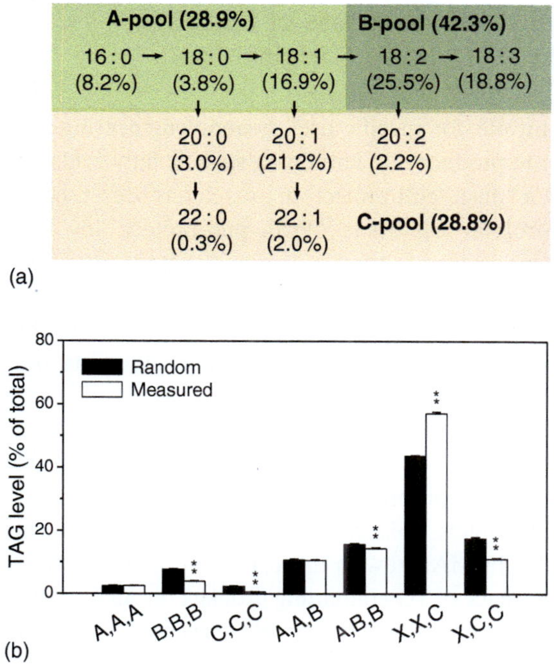

Figure 22.4 Distribution of fatty acyl chains among TG groups. (a) Definition of three hypothetical pools of fatty acyl chains destined to TG assembly. The fatty acyl chains 16:0, 18:0, and 18:1 are collectively designated as the A-pool; the fatty acyl chains 18:2 and 18:3 are collectively designated as the B-pool; and the fatty acyl chains 20:0, 20:1, 20:2, 22:0, and 22:1 are collectively designated as the C-pool. (b) TG pattern comparison between random fatty acyl selection from the fatty acyl pools and the measured pattern. A,A,A refers to TGs with three fatty acyl chains from A-pool; B,B,B refers to TGs which have three fatty acyl chains from B-pool, and so on. X refers to either A or B. A chi-square test has been performed to compare the difference between observed group pattern and expected group pattern, and the p-value is 0.0098, which is smaller than 0.05 and indicates that the difference is significant. "**" refers to $P < 0.01$ as compared with the randomized acyl combination and measured combination, based on Student's t test. Values were means ± SE ($n = 5$). Source: Figure and legend from Li et al. [78] (Figure 5) under agreement for reuse by co-contributor (RW).

Boechera stricta, was subjected to freezing after cold acclimation, TG molecular species increased markedly [91]. In particular, TG molecular species containing the fatty acid 16:3, which is primarily a component of plastidic galactolipids, were the most elevated. Analysis of the freezing tolerance of recombinant inbred lines from a cross between Boechera accessions that differed in freezing tolerance identified one major quantitative trait locus (QTL), which included the gene for the enzyme diacylglycerol acyltransferase 1 (DGAT1), which transfers a fatty acid from acyl-CoA to DG to make TG. Similarly, loss of DGAT1 function in Arabidopsis has been demonstrated to reduce freezing tolerance [92]. Indeed, overexpression of DGAT1 increased freezing tolerance both in Boechera and in Arabidopsis [91]. Taken together, the lipidomic, genetic, and biochemical data indicate that fatty acids or DG derived from plastidic galactolipids accumulate in TG when plants are subjected to freezing stress.

22.8 Lipidomics to Monitor the Progress of Genetic Engineering to Alter Plant TG Level or Composition

In the quest for biofuels and plant oils for specific uses, genetic engineering has been used to increase oil levels or to produce oils containing specific fatty acids. In general, to increase TG synthesis, a "push–pull-protect" approach has been used to obtain high levels of triacylglycerols in leaf tissue. "Push–pull-protect" involves engineering combinations of genes designed to express enzymes that increase synthesis of specific fatty acids, increase their incorporation into TG, and protect TG-containing lipid droplets from degradation. Lipidomics has been added to the toolbox to increase the ease of monitoring the outputs of genetic engineering and identifying metabolic bottlenecks (e.g. [82, 93–99]).

Among the most successful approaches for production of TG in leaves has been the expression in tobacco of the Wrinkled1 (WRI1) transcription factor to increase fatty acid synthesis, DGAT1 to increase incorporation of fatty acids in TG, and genes encoding oleosins to stabilize lipid droplets. This strategy led to oil accumulation at ~15% of leaf dry mass [82]. Imaging by MALDI with linear ion trap-Orbitrap mass spectrometry demonstrated the presence of TG-containing droplets throughout the leaf tissue, and particularly in mesophyll cells, whereas WT leaves had very few lipid droplets.

In the development of an industrial feedstock oil, Yu et al. [99] used lipidomics to monitor the production of triacylglycerols containing cyclopropane fatty acids, which can be used in the production of coatings, lubricants, plastics, and cosmetics. They wanted insight into factors that limit the production of cyclopropane fatty-acid-containing TGs in plant seeds. These workers generated lines of the crop plant *Camelina sativa* expressing the *Escherichia coli* fatty acid cyclopropane synthase along with a lysophosphatidic acid acyltransferase to facilitate incorporation of the cyclopropane into complex lipids. They found that cyclopropane fatty acids accumulated in camelina seeds, but the cyclopropane-containing chains were found to the largest extent in PC and PE, to a lesser extent in DG, and to the least extent in TG. These data suggested a bottleneck in the transfer of cyclopropane fatty acids from PC to TG. Use of MALDI mass spectral imaging showed that while PC containing a single cyclopropane fatty acyl chain was concentrated in the cotyledons, PC containing two cyclopropane acyl chains was concentrated in the embryonic axis. Comparison of enzyme expression in the cotyledons and embryonic axis, which also has a higher overall level of cyclopropane fatty acids than the cotyledons, may yield additional insight into factors that restrict accumulation of cyclopropane fatty-acid-containing TG. Yu et al. [99] used expression data on lipid metabolic enzymes in different parts of the seeds, overall lipid compositional data, and the lipid localization data to develop plans for future work to overcome the metabolic bottleneck by suppressing the camelina acyltransferases catalyzing the formation of TG (DGAT and/or PDAT), while adding acyltransferases with more specificity for cyclopropane fatty acids.

Similarly, the endeavor of developing a plant-based source of TG containing long-chain ω-3 polyunsaturated fatty acids, known as "fish oil," has relied on lipidomics to inform iterative approaches to improving the genetic engineering of this pathway [93–97]. The conventional engineered pathway for TG containing eicosapentaenoic acid is a collection of elongation and desaturation reactions, catalyzed by enzymes from lower plants and microbes, called the "Δ6-pathway" in which a Δ6-desaturase acts on 18:2 or 18:3 followed by elongation and then Δ5-desaturation to form arachidonic acid (20:4) or eicosapentaenoic acid (20:5). An alternative pathway is initiated with C2 elongation of 18:2 or 18:3, then desaturation (Δ8-desaturase and Δ5-desaturase). In these pathways, desaturation reactions happen in the PC pool, while elongation reactions happen on acyl-CoAs. When Ruiz-Lopez et al. [94] used the alternative pathway to produce eicosapentaenoic (20:5) acid-containing TG in the seeds of *Camelina sativa*, liquid chromatography triple quadrupole mass spectrometry of acyl-CoAs showed that 20:2 and 20:3 acyl-CoAs were building up. This and other evidence suggested a bottleneck in incorporation of the 20C elongation products into PC for further desaturation. Findings such as these form the basis for iterative improvements in identifying the optimal genes and promoters to add and the native gene expression that needs modification to produce the desired oil products.

22.9 The Future of Lipidomics in Plant Science

The recent increases in speed and sensitivity of mass spectrometers have increased the options for their uses in lipid analysis. Lipidomics is being employed in a variety of large-scale experiments on plant species, including those of agronomic importance. Using lipidomics in combination with genetic data via QTL or genome-wide association studies has begun (e.g. [100–102]), and this area will expand as researchers seek to identify genes that affect plant production of lipid products or plant responses associated with lipid traits. It is likely that there will be greater use of lipidomics for analysis of subcellular and extracellular components, as well as in measuring lipid–protein interactions (e.g. [23, 103–107]). Undoubtedly, the area of lipid imaging will also continue to grow as its resolution increases. Finally, the increased archiving of lipidomic data in public databases will increase access (e.g. [79]). In the toolbox of plant lipid scientists, lipidomics is now a mainstay, which will support and drive advances in plant biology.

Acknowledgments

The authors are grateful for a summer stipend award to YS from the Johnson Cancer Research Center, for support to YS from the China Scholarship Council, and for support from the USDA National Institute of Food and Agriculture (Hatch/Multi-State project 1013013). Contribution no. 22-107-B from the Kansas Agricultural Experiment Station.

References

1 Hsieh TC, Kaul K, Laine RA, Lester RL. Structure of a major glycophosphoceramide from tobacco leaves, PSL-I: 2-deoxy-2-acetamido-d-glucopyranosyl(alpha1 leads to 4)-d-glucuronopyranosyl(alpha1 leads to 2) myoinositol-1-*O*-phosphoceramide. *Biochemistry* 1978;17(17): 3575–3581. doi:https://doi.org/10.1021/bi00610a024.

2 Hsieh, T.C., Lester, R.L., and Laine, R.A. (1981). Glycophosphoceramides from plants. Purification and characterization of a novel tetrasaccharide derived from tobacco leaf glycolipids. *J. Biol. Chem.* 256 (15): 7747–7755.

3 Ravi, K., Rip, J.W., and Carroll, K.K. (1983). Characterization of dolichol and dolichyl phosphate phosphatase from soya beans (*Glycine max*). *Biochem. J* 213 (2): 513–518. https://doi.org/10.1042/bj2130513.

4 Wissing, J.B. and Behrbohm, H. (1993). Diacylglycerol pyrophosphate, a novel phospholipid compound. *FEBS Lett.* 315 (1): 95–99. https://doi.org/10.1016/0014-5793(93)81141-l.

5 Sandoval, J.A., Huang, Z.H., Garrett, D.C. et al. (1995). *N*-acylphosphatidylethanolamine in dry and imbibing cottonseeds. Amounts, molecular species, and enzymatic synthesis. *Plant Physiol.* 109 (1): 269–275. https://doi.org/10.1104/pp.109.1.269.

6 Chapman, K.D., Tripathy, S., Venables, B., and Desouza, A.D. (1998). *N*-acylethanolamines: formation and molecular composition of a new class of plant lipids. *Plant Physiol.* 116 (3): 1163–1168. https://doi.org/10.1104/pp.116.3.1163.

7 Welti, R., Li, W., Li, M. et al. (2002). Profiling membrane lipids in plant stress responses. Role of phospholipase D alpha in freezing-induced lipid changes in Arabidopsis. *J. Biol. Chem.* 277 (35): 31994–32002. https://doi.org/10.1074/jbc.M205375200.

8 Welti, R. and Wang, X. (2004). Lipid species profiling: a high-throughput approach to identify lipid compositional changes and determine the function of genes involved in lipid metabolism and signaling. *Curr. Opin. Plant Biol.* 7 (3): 337–344. https://doi.org/10.1016/j.pbi.2004.03.011.

9 Han, X. and Gross, R.W. (1994). Electrospray ionization mass spectroscopic analysis of human erythrocyte plasma membrane phospholipids. *Proc. Natl. Acad. Sci. U.S.A.* 91 (22): 10635–10639. https://doi.org/10.1073/pnas.91.22.10635.

10 Brügger, B., Erben, G., Sandhoff, R. et al. (1997). Quantitative analysis of biological membrane lipids at the low picomole level by nano-electrospray ionization tandem mass spectrometry. *Proc. Natl. Acad. Sci. U.S.A.* 94 (6): 2339–2344. https://doi.org/10.1073/pnas.94.6.2339.

11 Li, W., Li, M., Zhang, W. et al. (2004). The plasma membrane-bound phospholipase D delta enhances freezing tolerance in *Arabidopsis thaliana*. *Nat. Biotechnol.* 22 (4): 427–433. https://doi.org/10.1038/nbt949.

12 Li, W., Wang, R., Li, M. et al. (2008). Differential degradation of extraplastidic and plastidic lipids during freezing and post-freezing recovery in *Arabidopsis thaliana*. *J. Biol. Chem.* 283 (1): 461–468. https://doi.org/10.1074/jbc.M706692200.
13 Vu, H.S., Shiva, S., Roth, M.R. et al. (2014). Lipid changes after leaf wounding in *Arabidopsis thaliana*: expanded lipidomic data form the basis for lipid co-occurrence analysis. *Plant J.* 80 (4): 728–743. https://doi.org/10.1111/tpj.12659.
14 Vu, H.S., Roston, R., Shiva, S. et al. (2015). Modifications of membrane lipids in response to wounding of *Arabidopsis thaliana* leaves. *Plant Signaling Behav.* 10 (9): e1056422. https://doi.org/10.1080/15592324.2015.1056422.
15 Li, M., Welti, R., and Wang, X. (2006). Quantitative profiling of Arabidopsis polar glycerolipids in response to phosphorus starvation. Roles of phospholipases D zeta1 and D zeta2 in phosphatidylcholine hydrolysis and digalactosyldiacylglycerol accumulation in phosphorus-starved plants. *Plant Physiol.* 142 (2): 750–761. https://doi.org/10.1104/pp.106.085647.
16 Li, M., Hong, Y., and Wang, X. (2009). Phospholipase D- and phosphatidic acid-mediated signaling in plants. *Biochim. Biophys. Acta* 1791 (9): 927–935. https://doi.org/10.1016/j.bbalip.2009.02.017.
17 Li, J. and Wang, X. (2019). Phospholipase D and phosphatidic acid in plant immunity. *Plant Sci.* 279: 45–50. https://doi.org/10.1016/j.plantsci.2018.05.021. Epub 2018 May 25.
18 Kim, S.C. and Wang, X. (2020). Phosphatidic acid: an emerging versatile class of cellular mediators. *Essays Biochem.* 64 (3): 533–546. https://doi.org/10.1042/EBC20190089.
19 Andrés, F. and Coupland, G. (2012). The genetic basis of flowering responses to seasonal cues. *Nat. Rev. Genet.* 13 (9): 627–639. https://doi.org/10.1038/nrg3291.
20 Corbesier, L., Vincent, C., Jang, S. et al. (2007). FT protein movement contributes to long-distance signaling in floral induction of Arabidopsis. *Science* 316 (5827): 1030–1033. https://doi.org/10.1126/science.1141752.
21 Nakamura, Y., Andrés, F., Kanehara, K. et al. (2014). Arabidopsis florigen FT binds to diurnally oscillating phospholipids that accelerate flowering. *Nat. Commun.* 5: 3553. https://doi.org/10.1038/ncomms4553.
22 Susila, H., Jurić, S., Liu, L. et al. (2021). Florigen sequestration in cellular membranes modulates temperature-responsive flowering. *Science* 373 (6559): 1137–1142. https://doi.org/10.1126/science.abh4054.
23 Kim, S.C., Nusinow, D.A., Sorkin, M.L. et al. (2019). Interaction and regulation between lipid mediator phosphatidic acid and circadian clock regulators. *Plant Cell* 31 (2): 399–416. https://doi.org/10.1105/tpc.18.00675.
24 Mamode Cassim, A., Grison, M., Ito, Y. et al. (2020). Sphingolipids in plants: a guidebook on their function in membrane architecture, cellular processes, and environmental or developmental responses. *FEBS Lett.* 594 (22): 3719–3738. https://doi.org/10.1002/1873-3468.13987.
25 Markham, J.E., Li, J., Cahoon, E.B., and Jaworski, J.G. (2006). Separation and identification of major plant sphingolipid classes from leaves. *J. Biol. Chem.* 281 (32): 22684–22694. https://doi.org/10.1074/jbc.M604050200.

26 Markham, J.E. and Jaworski, J.G. (2007). Rapid measurement of sphingolipids from *Arabidopsis thaliana* by reversed-phase high-performance liquid chromatography coupled to electrospray ionization tandem mass spectrometry. *Rapid Commun. Mass Spectrom.* 21 (7): 1304–1314. https://doi.org/10.1002/rcm.2962.

27 Buré, C., Cacas, J.L., Wang, F. et al. (2011). Fast screening of highly glycosylated plant sphingolipids by tandem mass spectrometry. *Rapid Commun. Mass Spectrom.* 25 (20): 3131–3145. https://doi.org/10.1002/rcm.5206.

28 Cacas, J.L., Buré, C., Furt, F. et al. (2013). Biochemical survey of the polar head of plant glycosylinositolphosphoceramides unravels broad diversity. *Phytochemistry* 96: 191–200. https://doi.org/10.1016/j.phytochem.2013.08.002.

29 Chen, M., Markham, J.E., Dietrich, C.R. et al. (2008). Sphingolipid long-chain base hydroxylation is important for growth and regulation of sphingolipid content and composition in Arabidopsis. *Plant Cell* 20 (7): 1862–1878. https://doi.org/10.1105/tpc.107.057851.

30 Msanne, J., Chen, M., Luttgeharm, K.D. et al. (2015). Glucosylceramides are critical for cell-type differentiation and organogenesis, but not for cell viability in Arabidopsis. *Plant J.* 84 (1): 188–201. https://doi.org/10.1111/tpj.13000.

31 Saucedo-García, M., Guevara-García, A., González-Solís, A. et al. (2011). MPK6, sphinganine and the LCB2a gene from serine palmitoyltransferase are required in the signaling pathway that mediates cell death induced by long chain bases in Arabidopsis. *New Phytol.* 191 (4): 943–957. https://doi.org/10.1111/j.1469-8137.2011.03727.x.

32 Zienkiewicz, A., Gömann, J., König, S. et al. (2020). Disruption of Arabidopsis neutral ceramidases 1 and 2 results in specific sphingolipid imbalances triggering different phytohormone-dependent plant cell death programmes. *New Phytol.* 226 (1): 170–188. https://doi.org/10.1111/nph.16336.

33 Chen M, Markham JE, Cahoon EB. Sphingolipid **Δ**8 unsaturation is important for glucosylceramide biosynthesis and low-temperature performance in Arabidopsis. *Plant J.* 2012;69(5): 769–781. https://doi.org/10.1111/j.1365-313X.2011.04829.x.

34 Kimberlin, A.N., Majumder, S., Han, G. et al. (2013). Arabidopsis 56-amino acid serine palmitoyltransferase-interacting proteins stimulate sphingolipid synthesis, are essential, and affect mycotoxin sensitivity. *Plant Cell* 25 (11): 4627–4639. https://doi.org/10.1105/tpc.113.116145.

35 Xie, L.J., Chen, Q.F., Chen, M.X. et al. (2015). Unsaturation of very-long-chain ceramides protects plant from hypoxia-induced damages by modulating ethylene signaling in Arabidopsis. *PLoS Genet.* 11 (3): e1005143. https://doi.org/10.1371/journal.pgen.1005143.

36 Iswanto, A.B.B., Shon, J.C., Liu, K.H. et al. (2020). Sphingolipids modulate secretion of glycosylphosphatidylinositol-anchored plasmodesmata proteins and callose deposition. *Plant Physiol.* 184 (1): 407–420. https://doi.org/10.1104/pp.20.00401.

37 Chao, D.Y., Gable, K., Chen, M. et al. (2011). Sphingolipids in the root play an important role in regulating the leaf ionome in *Arabidopsis thaliana*. *Plant Cell* 23 (3): 1061–1081. https://doi.org/10.1105/tpc.110.079095.

38 Markham, J.E., Molino, D., Gissot, L. et al. (2011). Sphingolipids containing very-long-chain fatty acids define a secretory pathway for specific polar plasma membrane protein targeting in Arabidopsis. *Plant Cell* 23 (6): 2362–2378. https://doi.org/10.1105/tpc.110.080473.

39 Ito, Y., Esnay, N., Platre, M.P. et al. (2021). Sphingolipids mediate polar sorting of PIN2 through phosphoinositide consumption at the *trans*-Golgi network. *Nat. Commun.* 12 (1): 4267. https://doi.org/10.1038/s41467-021-24548-0.

40 Lenarčič, T., Albert, I., Böhm, H. et al. (2017). Eudicot plant-specific sphingolipids determine host selectivity of microbial NLP cytolysins. *Science* 358 (6369): 1431–1434. https://doi.org/10.1126/science.aan6874.

41 Jiang, Z., Zhou, X., Tao, M. et al. (2019). Plant cell-surface GIPC sphingolipids sense salt to trigger Ca^{2+} influx. *Nature* 572 (7769): 341–346. https://doi.org/10.1038/s41586-019-1449-z.

42 Gonzalez-Solis, A., Han, G., Gan, L. et al. (2020). Unregulated sphingolipid biosynthesis in gene-edited Arabidopsis *ORM* mutants results in nonviable seeds with strongly reduced oil content. *Plant Cell* 32 (8): 2474–2490. https://doi.org/10.1105/tpc.20.00015.

43 Dörmann, P. and Benning, C. (2002). Galactolipids rule in seed plants. *Trends Plant Sci.* 7 (3): 112–118. https://doi.org/10.1016/s1360-1385(01)02216-6.

44 Okazaki, Y., Otsuki, H., Narisawa, T. et al. (2013). A new class of plant lipid is essential for protection against phosphorus depletion. *Nat. Commun.* 4: 1510. https://doi.org/10.1038/ncomms2512.

45 Sud, M., Fahy, E., Cotter, D. et al. (2007). LMSD: LIPID MAPS structure database. *Nucleic Acids Res.* 35 (Database issue): D527–D532. https://doi.org/10.1093/nar/gkl838.

46 Koelmel, J.P., Campbell, J.E., Guingab-Cagmat, J. et al. (2019). Re-modeling of foliar membrane lipids in a seagrass allows for growth in phosphorus-deplete conditions. *PLoS One* 14 (11): e0218690. https://doi.org/10.1371/journal.pone.0218690.

47 Tawaraya, K., Honda, S., Cheng, W. et al. (2018). Ancient rice cultivar extensively replaces phospholipids with non-phosphorus glycolipid under phosphorus deficiency. *Physiol. Plant.* 163 (3): 297–305. https://doi.org/10.1111/ppl.12699. Epub ahead of print.

48 Okazaki, Y., Takano, K., and Saito, K. (2017). Lipidomic analysis of soybean leaves revealed tissue-dependent difference in lipid remodeling under phosphorus-limited growth conditions. *Plant Biotechnol.* 34 (1): 57–63. https://doi.org/10.5511/plantbiotechnology.17.0113a.

49 Kühn, H., Wiesner, R., Rathmann, J., and Schewe, T. (1991). Formation of ketodienoic fatty acids by the pure pea lipoxygenase-1. *Eicosanoids* 4 (1): 9–14.

50 Hamberg, M. and Gardner, H.W. (1992). Oxylipin pathway to jasmonates: biochemistry and biological significance. *Biochim. Biophys. Acta* 1165 (1): 1–18. https://doi.org/10.1016/0005-2760(92)90069-8.

51 Grechkin, A.N. and Hamberg, M. (1997). Biosynthesis of novel divinyl ether oxylipins by enzyme from garlic (*Allium sativum* L.) bulbs. In: *Recent Advances in Prostaglandin, Thromboxane, and Leukotriene Research*, Advances in

Experimental Medicine and Biology, vol. 433 (ed. B. Samuelsson, J.R. Vane, R. Paoletti, et al.), 61–64. Boston, MA: Springer https://doi.org/10.1007/978-1-4899-1810-9_11.

52 Grechkin, A.N., Ilyasov, A.V., and Hamberg, M. (1997). On the mechanism of biosynthesis of divinyl ether oxylipins by enzyme from garlic bulbs. *Eur. J. Biochem.* 245 (1): 137–142. https://doi.org/10.1111/j.1432-1033.1997.00137.x.

53 Hamberg, M. (1997). Avenoleic acid: a new oxylipin from oat seeds. In: *Recent Advances in Prostaglandin, Thromboxane, and Leukotriene Research*, Advances in Experimental Medicine and Biology, vol. 433 (ed. B. Samuelsson, J.R. Vane, R. Paoletti, et al.), 69–72. Boston, MA: Springer https://doi.org/10.1007/978-1-4899-1810-9_13.

54 Hamberg, M. (1999). An epoxy alcohol synthase pathway in higher plants: biosynthesis of antifungal trihydroxy oxylipins in leaves of potato. *Lipids* 34 (11): 1131–1142. https://doi.org/10.1007/s11745-999-0464-7.

55 Weber, H., Vick, B.A., and Farmer, E.E. (1997). Dinor-oxo-phytodienoic acid: a new hexadecanoid signal in the jasmonate family. *Proc. Natl. Acad. Sci. U.S.A.* 94 (19): 10473–10478. https://doi.org/10.1073/pnas.94.19.10473.

56 Imbusch, R. and Mueller, M.J. (2000). Analysis of oxidative stress and wound-inducible dinor isoprostanes F_1 (phytoprostanes F_1) in plants. *Plant Physiol.* 124 (3): 1293–1304. https://doi.org/10.1104/pp.124.3.1293.

57 Feussner, I. and Wasternack, C. (2002). The lipoxygenase pathway. *Annu. Rev. Plant Biol.* 53: 275–297. https://doi.org/10.1146/annurev.arplant.53.100301.135248.

58 Mueller, M.J., Mène-Saffrané, L., Grun, C. et al. (2006). Oxylipin analysis methods. *Plant J.* 45 (4): 472–489. https://doi.org/10.1111/j.1365-313X.2005.02614.x.

59 Göbel, C. and Feussner, I. (2009). Methods for the analysis of oxylipins in plants. *Phytochemistry* 70 (13, 14): 1485–1503. https://doi.org/10.1016/j.phytochem.2009.07.040.

60 Stelmach, B.A., Müller, A., Hennig, P. et al. (2001). A novel class of oxylipins, *sn*1-*O*-(12-oxophytodienoyl)-*sn*2-*O*-(hexadecatrienoyl)-monogalactosyl diglyceride, from *Arabidopsis thaliana*. *J. Biol. Chem.* 276 (16): 12832–12838. https://doi.org/10.1074/jbc.M010743200.

61 Hisamatsu, Y., Goto, N., Hasegawa, K., and Shigemori, H. (2003). Arabidopsides A and B, two new oxylipins from *Arabidopsis thaliana*. *Tetrahedron Lett.* 44 (29): 5553–5556.

62 Hisamatsu, Y., Goto, N., Sekiguchi, M. et al. (2005). Oxylipins arabidopsides C and D from *Arabidopsis thaliana*. *J. Nat. Prod.* 68 (4): 600–603. https://doi.org/10.1021/np0495938.

63 Andersson, M.X., Hamberg, M., Kourtchenko, O. et al. (2006). Oxylipin profiling of the hypersensitive response in *Arabidopsis thaliana*. Formation of a novel oxo-phytodienoic acid-containing galactolipid, arabidopside E. *J. Biol. Chem.* 281 (42): 31528–31537. https://doi.org/10.1074/jbc.M604820200.

64 Kourtchenko, O., Andersson, M.X., Hamberg, M. et al. (2007). Oxo-phytodienoic acid-containing galactolipids in Arabidopsis: jasmonate signaling dependence. *Plant Physiol.* 145 (4): 1658–1669. https://doi.org/10.1104/pp.107.104752.

65 Buseman, C.M., Tamura, P., Sparks, A.A. et al. (2006). Wounding stimulates the accumulation of glycerolipids containing oxophytodienoic acid and dinor-oxophytodienoic acid in Arabidopsis leaves. *Plant Physiol.* 142 (1): 28–39. https://doi.org/10.1104/pp.106.082115.

66 Nilsson, A.K., Johansson, O.N., Fahlberg, P. et al. (2014). Formation of oxidized phosphatidylinositol and 12-oxo-phytodienoic acid containing acylated phosphatidylglycerol during the hypersensitive response in Arabidopsis. *Phytochemistry* 101: 65–75. https://doi.org/10.1016/j.phytochem.2014.01.020.

67 Ibrahim, A., Schütz, A.L., Galano, J.M. et al. (2011). The alphabet of galactolipids in *Arabidopsis thaliana*. *Front. Plant Sci.* 2: 95. https://doi.org/10.3389/fpls.2011.00095.

68 Vu, H.S., Tamura, P., Galeva, N.A. et al. (2012). Direct infusion mass spectrometry of oxylipin-containing Arabidopsis membrane lipids reveals varied patterns in different stress responses. *Plant Physiol.* 158 (1): 324–339. https://doi.org/10.1104/pp.111.190280.

69 Vu, H.S., Roth, M.R., Tamura, P. et al. (2014). Head-group acylation of monogalactosyldiacylglycerol is a common stress response, and the acyl-galactose acyl composition varies with the plant species and applied stress. *Physiol. Plant.* 150 (4): 517–528. https://doi.org/10.1111/ppl.12132.

70 Nilsson, A.K., Johansson, O.N., Fahlberg, P. et al. (2015). Acylated monogalactosyl diacylglycerol: prevalence in the plant kingdom and identification of an enzyme catalyzing galactolipid head group acylation in *Arabidopsis thaliana*. *Plant J.* 84 (6): 1152–1166. https://doi.org/10.1111/tpj.13072.

71 Mueller, S.P., Unger, M., Guender, L. et al. (2017). Phospholipid: diacylglycerol acyltransferase-mediated triacylglyerol synthesis augments basal thermotolerance. *Plant Physiol.* 175 (1): 486–497. https://doi.org/10.1104/pp.17.00861.

72 Kilaru, A., Tamura, P., Isaac, G. et al. (2012). Lipidomic analysis of *N*-acylphosphatidylethanolamine molecular species in Arabidopsis suggests feedback regulation by *N*-acylethanolamines. *Planta* 236 (3): 809–824. https://doi.org/10.1007/s00425-012-1669-z.

73 Kilaru, A. and Chapman, K.D. (2012). *N*-acylated phospholipid metabolism and seedling growth: insights from lipidomics studies in Arabidopsis. *Plant Signaling Behav.* 7 (9): 1200–1202. https://doi.org/10.4161/psb.21314.

74 Cotter, M.Q., Teaster, N.D., Blancaflor, E.B., and Chapman, K.D. (2011). *N*-acylethanolamine (NAE) inhibits growth in *Arabidopsis thaliana* seedlings via ABI3-dependent and -independent pathways. *Plant Signaling Behav.* 6 (5): 671–679. https://doi.org/10.4161/psb.6.5.14704.

75 Teaster, N.D., Keereetaweep, J., Kilaru, A. et al. (2012). Overexpression of fatty acid amide hydrolase induces early flowering in *Arabidopsis thaliana*. *Front. Plant Sci.* 3: 32. https://doi.org/10.3389/fpls.2012.00032.

76 Han, X. and Gross, R.W. (2001). Quantitative analysis and molecular species fingerprinting of triacylglyceride molecular species directly from lipid extracts of biological samples by electrospray ionization tandem mass spectrometry. *Anal. Biochem.* 295 (1): 88–100. https://doi.org/10.1006/abio.2001.5178.

77 McAnoy, A.M., Wu, C.C., and Murphy, R.C. (2005). Direct qualitative analysis of triacylglycerols by electrospray mass spectrometry using a linear ion trap. *J. Am. Soc. Mass Spectrom.* 16 (9): 1498–1509. https://doi.org/10.1016/j.jasms.2005.04.017.

78 Li, M., Baughman, E., Roth, M.R. et al. (2014). Quantitative profiling and pattern analysis of triacylglycerol species in Arabidopsis seeds by electrospray ionization mass spectrometry. *Plant J.* 77 (1): 160–172. https://doi.org/10.1111/tpj.12365.

79 Kehelpannala, C., Rupasinghe, T., Pasha, A. et al. (2021). An Arabidopsis lipid map reveals differences between tissues and dynamic changes throughout development. *Plant J.* 107 (1): 287–302. https://doi.org/10.1111/tpj.15278.

80 Lee, J., Welti, R., Schapaugh, W.T., and Trick, H.N. (2011). Phospholipid and triacylglycerol profiles modified by PLD suppression in soybean seed. *Plant Biotechnol. J.* 9 (3): 359–372. https://doi.org/10.1111/j.1467-7652.2010.00562.x.

81 Horn, P.J., Silva, J.E., Anderson, D. et al. (2013). Imaging heterogeneity of membrane and storage lipids in transgenic *Camelina sativa* seeds with altered fatty acid profiles. *Plant J.* 76 (1): 138–150. https://doi.org/10.1111/tpj.12278. Epub.

82 Vanhercke, T., El Tahchy, A., Liu, Q. et al. (2014). Metabolic engineering of biomass for high energy density: oilseed-like triacylglycerol yields from plant leaves. *Plant Biotechnol. J.* 12 (2): 231–239. https://doi.org/10.1111/pbi.12131.

83 Feenstra, A.D., Alexander, L.E., Song, Z. et al. (2017). Spatial mapping and profiling of metabolite distributions during germination. *Plant Physiol.* 174 (4): 2532–2548. https://doi.org/10.1104/pp.17.00652.

84 Mueller, S.P., Krause, D.M., Mueller, M.J., and Fekete, A. (2015). Accumulation of extra-chloroplastic triacylglycerols in Arabidopsis seedlings during heat acclimation. *J. Exp. Bot.* 66 (15): 4517–4526. https://doi.org/10.1093/jxb/erv226.

85 Narayanan, S., Tamura, P.J., Roth, M.R. et al. (2016). Wheat leaf lipids during heat stress: I. High day and night temperatures result in major lipid alterations. *Plant Cell Environ.* 39 (4): 787–803. https://doi.org/10.1111/pce.12649.

86 Higashi, Y., Okazaki, Y., Myouga, F. et al. (2015). Landscape of the lipidome and transcriptome under heat stress in *Arabidopsis thaliana*. *Sci. Rep.* 5: 10533. https://doi.org/10.1038/srep10533.

87 Higashi, Y. and Saito, K. (2019). Lipidomic studies of membrane glycerolipids in plant leaves under heat stress. *Prog. Lipid Res.* 75: 100990. https://doi.org/10.1016/j.plipres.2019.100990.

88 Shiva, S., Samarakoon, T., Lowe, K.A. et al. (2020). Leaf lipid alterations in response to heat stress of *Arabidopsis thaliana*. *Plants (Basel)* 9 (7): 845. https://doi.org/10.3390/plants9070845.

89 Demski, K., Łosiewska, A., Jasieniecka-Gazarkiewicz, K. et al. (2020). Phospholipid:diacylglycerol acyltransferase1 overexpression delays senescence and enhances post-heat and cold exposure fitness. *Front. Plant Sci.* 11: 611897. https://doi.org/10.3389/fpls.2020.611897.

90 Degenkolbe, T., Giavalisco, P., Zuther, E. et al. (2012). Differential remodeling of the lipidome during cold acclimation in natural accessions of *Arabidopsis thaliana*. *Plant J.* 72 (6): 972–982. https://doi.org/10.1111/tpj.12007.

91 Arisz, S.A., Heo, J.Y., Koevoets, I.T. et al. (2018). DIACYLGLYCEROL ACYLTRANSFERASE1 contributes to freezing tolerance. *Plant Physiol.* 177 (4): 1410–1424. https://doi.org/10.1104/pp.18.00503.

92 Tan, W.J., Yang, Y.C., Zhou, Y. et al. (2018). DIACYLGLYCEROL ACYLTRANSFERASE and DIACYLGLYCEROL KINASE modulate triacylglycerol and phosphatidic acid production in the plant response to freezing stress. *Plant Physiol.* 177 (3): 1303–1318. https://doi.org/10.1104/pp.18.00402.

93 Abbadi, A., Domergue, F., Bauer, J. et al. (2004). Biosynthesis of very-long-chain polyunsaturated fatty acids in transgenic oilseeds: constraints on their accumulation. *Plant Cell* 16 (10): 2734–2748. https://doi.org/10.1105/tpc.104.026070.

94 Ruiz-Lopez, N., Haslam, R.P., Usher, S. et al. (2015). An alternative pathway for the effective production of the omega-3 long-chain polyunsaturates EPA and ETA in transgenic oilseeds. *Plant Biotechnol. J.* 13 (9): 1264–1275. https://doi.org/10.1111/pbi.12328.

95 Haslam, R.P., Ruiz-Lopez, N., Eastmond, P. et al. (2013). The modification of plant oil composition via metabolic engineering – better nutrition by design. *Plant Biotechnol. J.* 11 (2): 157–168. https://doi.org/10.1111/pbi.12012.

96 Napier, J.A., Haslam, R.P., Beaudoin, F., and Cahoon, E.B. (2014). Understanding and manipulating plant lipid composition: metabolic engineering leads the way. *Curr. Opin. Plant Biol.* 19 (100): 68–75. https://doi.org/10.1016/j.pbi.2014.04.001.

97 Napier, J.A., Olsen, R.E., and Tocher, D.R. (2019). Update on GM canola crops as novel sources of omega-3 fish oils. *Plant Biotechnol. J.* 17 (4): 703–705. https://doi.org/10.1111/pbi.13045.

98 Bansal, S., Kim, H.J., Na, G. et al. (2018). Towards the synthetic design of camelina oil enriched in tailored acetyl-triacylglycerols with medium-chain fatty acids. *J. Exp. Bot.* 69 (18): 4395–4402. https://doi.org/10.1093/jxb/ery225.

99 Yu, X.H., Cahoon, R.E., Horn, P.J. et al. (2018 Apr). Identification of bottlenecks in the accumulation of cyclic fatty acids in camelina seed oil. *Plant Biotechnol. J.* 16 (4): 926–938. https://doi.org/10.1111/pbi.12839.

100 Riedelsheimer, C., Brotman, Y., Méret, M. et al. (2013). The maize leaf lipidome shows multilevel genetic control and high predictive value for agronomic traits. *Sci. Rep.* 3: 2479. https://doi.org/10.1038/srep02479.

101 de Abreu, E., Lima, F., Li, K. et al. (2018). Unraveling lipid metabolism in maize with time-resolved multi-omics data. *Plant J.* 93 (6): 1102–1115. https://doi.org/10.1111/tpj.13833.

102 Chernova, A.I., Gubaev, R.F., Singh, A. et al. (2021). Genotyping and lipid profiling of 601 cultivated sunflower lines reveals novel genetic determinants of oil fatty acid content. *BMC Genomics* 22 (1): 505. https://doi.org/10.1186/s12864-021-07768-y.

103 Grison, M.S., Brocard, L., Fouillen, L. et al. (2015). Specific membrane lipid composition is important for plasmodesmata function in Arabidopsis. *Plant Cell* 27 (4): 1228–1250. https://doi.org/10.1105/tpc.114.135731.

104 Coulon, D. and Bréhélin, C. (2021). Isolation of plastoglobules for lipid analyses. In: *Plant Lipids*, Methods in Molecular Biology, vol. 2295 (ed. D. Bartels and

P. Dörmann), 321–335. New York, NY: Humana https://doi.org/10.1007/978-1-0716-1362-7_17.

105 Leterme, S., Michaud, M., and Jouhet, J. (2021). Isolation of mitochondria for lipid analysis. In: *Plant Lipids*, Methods in Molecular Biology, vol. 2295 (ed. D. Bartels and P. Dörmann), 337–349. New York, NY: Humana https://doi.org/10.1007/978-1-0716-1362-7_18.

106 Hoffmann-Benning, S. (2021). Collection and analysis of phloem lipids. In: *Plant Lipids*, Methods in Molecular Biology, vol. 2295 (ed. D. Bartels and P. Dörmann), 351–361. New York, NY: Humana https://doi.org/10.1007/978-1-0716-1362-7_19.

107 Schenk, H.J., Michaud, J.M., Mocko, K. et al. (2021). Lipids in xylem sap of woody plants across the angiosperm phylogeny. *Plant J.* 105 (6): 1477–1494. https://doi.org/10.1111/tpj.15125.

23

Lipidomics in Multi-Omics Studies

Bjoern Titz, Oksana Lavrynenko, and Nikolai V. Ivanov

PMI R&D, Philip Morris Products S.A., Quai Jeanrenaud 5, CH-2000, Neuchâtel, Switzerland

23.1 Introduction

Omics analyses have provided us with a wealth of information about the functioning of biological systems in their healthy state – and their dysfunction when entering disease states. Complex adaptive biological systems are based on sophisticated molecular circuitry, and systems biology, in particular, has argued that a detailed understanding of this molecular wiring is required in order to generate actionable biological insights. While genomics and proteomics approaches point the way, lipidomics approaches are well developed and enable quantitative profiling of diverse lipid classes [1].

By design, every omics approach can only offer a single and incomplete perspective of a biological system. Biological processes usually bridge several molecular layers, and the most interesting biology often takes place at the interfaces, e.g. enzymes that produce compounds that affect other protein activities or lipid and protein mediators that synergistically control biological processes such as inflammation. For this reason – as well as increased technical feasibility – multi-omics approaches have quickly emerged, and they combine two or more omics measurements with suitable integrative data analysis strategies. The first publications that mentioned "multi-omics" appeared only in 2006; however, after rapid growth, >1000 multi-omics articles were published in 2020, reaching similar numbers as publications on lipidomics (Figure 23.1a).

The benefits of multi-omics approaches have been demonstrated in a variety of applications, including mechanistic disease research, toxicity assessment, and biomarker discovery. In the context of cancer research, it has been pointed out, for example, that a single (omics) platform does not adequately capture the complexity of the

Mass Spectrometry for Lipidomics: Methods and Applications, First Edition.
Edited by Michal Holčapek and Kim Ekroos.

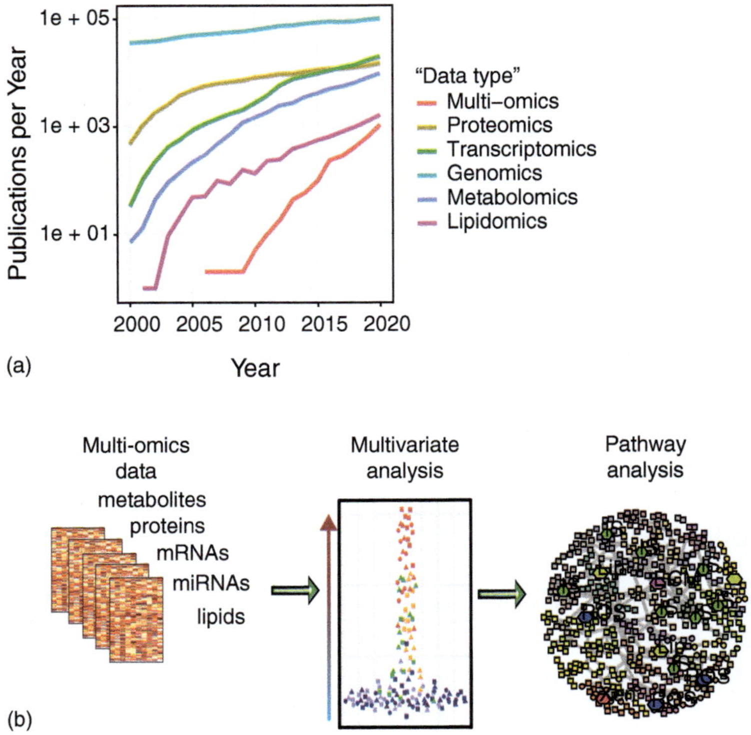

Figure 23.1 Multi-omics. (a) Number of publications per year that mention different omics data modalities. (b) Example of a multi-omics study that integrated five omics data modalities (metabolomics, proteomics, mRNA transcriptomics, miRNA transcriptomics, and lipidomics). Multivariate analysis using the MOFA approach was combined with pathway/network analysis using the prize-collecting Steiner forest (PCSF) algorithm. Source: Adapted from Titz et al. [2].

underlying cancer genomes. At the same time, integrative multi-omics approaches can be effectively used to identify new cancer subtypes, elucidate the pathophysiology of cancer, identify drug targets, and enable clinical decision-making [3]. Similarly, toxicology deals with complex biological responses to exposure that span multiple molecular levels and encompass multiple levels of biological organization (from cells to tissues to organs to negative effects at the organism level) [4].

In this chapter, we specifically highlight the role that lipidomics data can play in multi-omics studies. First, we will exemplify how lipidomics data have contributed to insights in recent multi-omics studies. We describe specific considerations on the experimental methods and give an overview of relevant integrative computational analysis approaches that enable the integration of lipidomics data. The experimental design must be adapted to the multi-omics context, and the challenges must be taken into account. Finally, we summarize the great potential of lipidomics data in multi-omics and suggest ways in which the knowledge from such multimodal analysis sets can be further increased.

23.2 Lipidomics in Multi-Omics Studies

Given the general relevance of lipids in biology and the emergence of lipidomics methods, we are already seeing an increasing number of studies that are integrating lipidomics data into multi-omics contexts. These studies cover a wide range of application areas, including disease research, drug discovery, and biomarker applications.

Infections cause complex reactions in the immune system and the entire physiological system. Coronavirus disease 2019 (COVID-19), which was declared a global pandemic by the World Health Organization in March 2020, triggered several studies for elucidating such complex biological responses to a viral infection. Sun et al. performed longitudinal multi-omics profiling in critically ill COVID-19 patients [5]. In total, the authors collected a dataset from 33 patients, which contained 113 transcriptomes, 85 proteomes, 85 metabolomes, and 85 lipidomes from blood samples. Lipidomics analysis was performed by liquid chromatography–mass spectrometry (LC-MS) on a Q Exactive mass spectrometer (Thermo Scientific, San Jose, USA), and the lipidome was analyzed by Lipidsearch (Thermo Fisher Scientific, USA). The integrated analysis of their multi-omics dataset included establishment of differential co-expression networks between the study groups and identification of patient clusters [6]. One of the identified molecular clusters was centered on L-kynurenine and contained several changes in the correlation structure of phospholipids as well as triacylglycerols, and the authors linked these changes to the observed enrichment of the erythrocyte differentiation pathway in the transcriptomics data. However, it is important to note that, in this analysis of COVID-19-related changes, the clinical, transcriptomic, and metabolomic data collected guided the data interpretation more than the lipidomics data.

In a second recent COVID-19 study, He et al. followed up on the gastrointestinal symptoms that are commonly observed in COVID-19 patients [7]. The authors performed a multi-omics analysis of stool samples from 13 COVID-19 patients and 21 healthy controls. Metaproteome, glycoproteome, metabolome, and lipidome analyses were performed by using an LC-MS-based lipidomics approach with an Orbitrap Fusion Lumos Tribrid mass spectrometer (Thermo Fisher Scientific, USA) and Lipidsearch (Thermo Fisher Scientific, USA) for the computational analysis. Ceramides were the most frequently identified class of lipids, followed by triacylglycerol and monohexosylceramide, with several lipid types showing significant differences in abundance between COVID-19 patients and healthy controls. Interestingly, several of the identified ceramides (odd-chain-length and hydroxylated) have been assigned a likely microbial origin.

The integration of lipidomic data into a multi-omics context has also been applied to the pathophysiology of Alzheimer's disease (AD) [8]. In addition to the well-known amyloid pathology and tau-related neurodegeneration, several other molecular changes have been reported in AD, including changes in lipid metabolism [9]. To gain further insight into how molecular patterns relate to the core features of AD, Clark et al. profiled the molecular changes in cerebrospinal fluid (CSF) from a well-characterized cohort of 120 elderly subjects with normal cognition, mild cognitive impairment, or mild AD. This molecular profiling comprised antibody- and mass spectrometry-based proteomics, metabolomics, and lipidomics analyses. The lipidomics analysis was performed using a direct injection (shotgun) approach in a QExactive mass spectrometer (Thermo Fisher Scientific) with a TriVersa NanoMate ion source (Advion Biosciences), and data analysis was performed using the LipidXplorer software [10, 11]. Integrative computational analysis included Multi-Omics Factor Analysis (MOFA), which can be seen as a principal component analysis (PCA)-like approach adapted for multi-omics datasets [12]. Across the identified latent factors – corresponding to principal components in a PCA – the CSF AD biomarkers accounted for 38.5%, proteins for 39.8%, lipids for 10.3%, neuroinflammation markers for 10.3%, one-carbon metabolites for 9%, and other metabolites for 3.7% of the explained variance. Among the analytes with the largest overall contributions to the model were five lipid species (PC 32:0, SE 27:1 18:2, SE 27:1 20:4, SE 27:1 16:0, LPG 20:1), of which only PC 32:0 had been associated with AD previously. Of note, when evaluating these identified analytes in the context of predictive models, an extended reference model for cognitive decline that included protein 14-3-3 zeta/delta, clusterin, SE (cholesteryl ester) 27:1 16:0, and monocyte chemoattractant protein-1 demonstrated significantly improved predictions (sensitivity of 0.80; specificity of 0.88). Overall, this study uncovered both known and novel molecular patterns associated with AD, and the authors highlighted its value in developing new combinations of biomarkers, personalized diagnoses, and tailored therapeutic interventions for AD.

Odenkirk et al. applied a multi-omics approach to investigate two predominant maternal conditions that can occur during pregnancy, preeclampsia (PRE) and gestation diabetes mellitus (GDM) [13]. Clinically, PRE presents with new-onset hypertension and proteinuria or systemic organ dysfunction after 20 weeks of gestation and can only be addressed by premature delivery. GDM presents with new onset of glucose intolerance during pregnancy. To gain a deeper understanding of these conditions, the authors used proteomics and lipidomics analyses to profile plasma samples from 191 subjects (48 PRE, 45 GDM, and 98 healthy controls), collected at admission for labor and delivery. Lipidomics analysis was performed by LC-IMS-MS (IMS, ion-mobility spectrometry) with an Agilent 6560 IM-QTOF instrument (Santa Clara, CA, USA). The authors reported 81 and 28 significant lipid changes in PRE and GDM, respectively – of note, not fully controlling for multiple hypotheses testing across all quantified lipid species. The authors emphasized the challenge of integrating protein and lipid data owing to the limited details of their association reported in the literature, e.g. on the distinguishing functional features of isomeric lipids. Nevertheless, they used an integration approach guided by the available literature and, for example, suggested a functional relationship between significant downregulation of phosphatidylethanolamine-binding protein 1 and peroxiredoxin-6 and upregulation of fatty acid 22:6-containing alkenyl-glycerophosphoethanolamines species in GDM. Future studies could further investigate whether the reported observations could ultimately support the development of PRE- and GDM-specific biomarkers.

Multi-omics studies are also becoming increasingly popular in cancer research. Vantaku et al. combined metabolomics, lipidomics, and (public) transcriptomics data to predict patient survival and potentially identify new drug targets for bladder cancer [14]. The authors performed LC-MS-based lipidomics analysis with a TripleTOF 5600 instrument (AB Sciex). A total of 519 metabolites and 19 lipids were expressed differently between low- and high-grade bladder cancer tissues. Using associations from the Human Metabolome Database (HMDB) database [15], the authors identified 27 genes that were expressed differently and were also associated with an identified lipid or metabolite. First, the authors determined the relevance of the 27-gene signature for survival and then focused on the gene carnitine palmitoyltransferase I (*CPT1B*), which is involved in fatty acid transport and the rate-limiting step of fatty acid oxidation in the mitochondria. Through additional experiments, the authors associated high-grade bladder cancer with lower levels of CPT1B and fatty acid oxidation, providing further evidence that overexpression of CPT1B in a high-grade bladder cancer cell line reduces its epithelial–mesenchymal transition phenotype and metastatic potential.

Multi-omics analyses are also becoming increasingly popular for deciphering complex toxicological reactions. Systematic mode-of-action studies for xenobiotics are essential for guiding the development of effective, mechanism-based risk-assessment approaches. Across nine published studies, multi-omics assessment of exposure responses in macrophages has been highlighted as a viable approach for such mechanism-of-action studies [16]. A study by Bannuscher et al. combined proteomics and metabolomics to study the effects of nanomaterials on alveolar macrophages [17]. Metabolomics analysis was performed using the AbsoluteIDQ 180 kit (Biocrates, Innsbruck, Austria), which also targets 15 lysophosphatidylcholines, 76 phosphatidylcholines, and 15 sphingolipids. The integrated enrichment analyses in this study revealed the effects of nanomaterials on a wide range of biological processes, including oxidative stress, cytokine signaling, and mitochondrial function.

As a final example, we had applied a multi-omics approach to assess the effects of aerosols from two heat-not-burn tobacco products and cigarette smoke in mouse lungs (Figure 23.1b) [2]. For this, we analyzed each lung sample by MS-based proteomics, lipidomics, and metabolomics analyses as well as microarray-based mRNA and miRNA transcriptomics analyses. Lipidomics analysis was performed by a direct injection (shotgun) approach using a Q Exactive™ Plus Orbitrap (Thermo Scientific, Germany) equipped with a Triversa NanoMate robotic interface (Advion, USA) and with LipidXplorer as the data analysis software. Integrative data analysis included MOFA and multi-modality functional network interpretation. When the five data modalities were combined, the data demonstrated an increased inflammatory and oxidative stress response as well as lipid/surfactant alterations upon cigarette smoke exposure, with the effects upon exposure to aerosol from the tested heat-not-burn tobacco products being much more limited to absent. In addition to the effects of cigarette smoke on the above-mentioned molecular surfactant cluster, our analyses also identified an increased abundance of phospholipase D3 (Pld3) as a component of the lipid metabolism cluster upon cigarette smoke exposure. Given Pld3's immunoregulatory role, this finding further demonstrated the potential of multi-omics approaches to discover relevant biologies that are active at the interfaces of the different molecular layers.

Overall, the few examples described above show that multi-omics studies that include lipidomics have potential for a wide range of biological applications. In Section 23.3, we highlight a few points that are important to bring multi-omics studies to life.

23.3 Planning and Conducting Multi-Omics Studies

A few special features must be considered when planning and carrying out a multi-omics study. The most important perhaps is access to high-quality methods for the various analytical modalities. It is rare that a single laboratory has all of the multi-omics methods considered ready to use by itself. Cooperation with other laboratories, core omics facilities, and commercial external partners can also bring in the required portfolio of methods. Building up multi-omics expertise in-house or in close cooperation is essential for the success of a multi-omics study – especially to support the best possible interpretation of the complex datasets generated. In the case of one of our most recent multi-omics studies, for example, we performed proteomics, lipidomics, and transcriptomics analyses in our research facility, while the metabolomics data were generated by an external commercial partner [2].

Choosing relevant omics data types and the best possible study design is critical. As with any experimental study, it is essential to bring the biology of interest together with the technical possibilities of the available omics methods. For example, a liver phospholipidosis and cholestasis study would benefit from the inclusion of phospholipid (e.g. shotgun lipidomics) and bile acid profiling together with, for example, proteomics analysis to capture changes in enzyme levels. For a study focused on immunoregulatory mechanisms in macrophages, it might be beneficial to combine lipid mediator profiling by lipidomics analysis, cytokine profiling by proteomics analysis, and profiling of signal transduction mechanisms by phosphoproteomics analysis. Ultimately, the final selection of the omics modalities will always represent a compromise between biological relevance and technical feasibility. But in any case, from the beginning, it is important to plan an effective data analysis approach and design the experiments accordingly.

"Garbage In, Garbage Out," an idiom commonly used in computer science, naturally also applies to bioanalytical measurements and is even more relevant in the multi-omics context. While the stringent requirements of bioanalytical method validation cannot be directly applied to high-dimensional omics measurements, especially when non-targeted methods are used [18], a strict quality framework is essential to adequately characterize and the analytical performance of omics methods and keep them under control. The International Lipidomics Society (ILS), the Lipidomics Standards Initiative (LSI), and LIPID MAPS are actively promoting the improvement of lipidomics methods and data interpretation and reporting and provide useful guidance documents that should be considered [19–21].

Having high-quality multi-omics methods available then raises the question of how to design the study and sample collection for integrative analysis of the data. Power calculation – i.e. rationally deciding on the number of required biological replicates – is essential for any bioanalytical measurement, but the multi-omics context brings its own challenges [22, 23]. Traditional power calculations deal with a single type of measurement and require solid estimates of the expected effect size and variability for a single predefined statistical test. All this information is rarely available in the (multi-)omics context, and the integrative calculation methods that are used have very complex dependencies on the real data structure. Thus, while a real power calculation would be preferred, realistically, only "power guesstimates" can be achieved for the design of real studies. For example, results from previous (internal or external) similar studies, a dedicated pilot study, and/or relevant data simulations can inform the appropriate design of a study. However, dedicated approaches for power calculations in multi-omics studies are also emerging. Tarazona et al. have published the MultiPower method, which supports power calculations for multi-omics data [24]. In any case, an independent study for validating the results should be considered from the start. For this, targeted lipid and protein quantitation assays could be established to confirm multi-omics results from non-targeted methods.

Another important consideration that comes up in multi-omics studies is about how to generate and distribute the samples for the multitude of planned measurements. Generally, the most powerful integrative analyses require complete datasets, with all planned measurements conducted for each sample. This is easily achieved when dealing with biofluids of sufficient volume, like human blood plasma. Also, when analyzing bulk animal or plant tissue, samples can be pulverized or sliced and then distributed for the different analyses. However, when sample volumes are limited or difficult to split physically before extraction, measurements must be prioritized and bioanalytical methods miniaturized or specifically adapted for multi-omics analyses. Several protocols for multi-omics analysis of proteins, lipids, and metabolites have been published. The "Metabolite, protein, and lipid extraction" (MPLEx) by Nakayasu et al. is based on a simple chloroform–methanol–water extraction method [25]. Salem et al. proposed a protocol based on a methyl *tert*-butyl ether – methanol–water extraction [26]. Recently, Kang et al. published a slight variation of the chloroform–methanol–water extraction procedure by Nakayasu et al. [27].

Taking these considerations into account, a large number of molecular layers can be assessed and brought together by integrative computational analyses. For example, as discussed above, we combined five omics datasets in our analyses of mouse lung tissue [2]. However, it is clear that, owing to experimental constraints, it is often not possible to generate the best possible integrative dataset, and the computational analyses need to account for this. For example, it may not be possible to analyze all the molecular layers within a single experimental set of samples, and it might be necessary to bridge between datasets. Moreover, in a systems toxicology study with A/J mice, we were unable to obtain histopathology and multi-omics data from the same mouse cohort [28]. Thus, to establish links between omics data and histopathological endpoints, we had to resort to a special procedure in which we matched histopathology scores to molecular samples at the group level by employing random sampling from the empirical score distribution of the respective group. However, in such cases, the information about the variation within the group is lost, which limits the informative value of such multi-omics analyses that span several experimental sets.

All in all, multi-omics analyses present their own implementation challenges. Nevertheless, it is worth going through them step by step while taking into account general study restrictions in order to ensure the best possible multilayered datasets for subsequent computational analysis.

23.4 Analyzing Multi-Omics Data

With the rise of multi-omics methods, the number of available computational analysis methods is also increasing. An analyst can choose from a broad variety of approaches but needs to identify the most appropriate methods for the data at hand. Integration of lipidomics data comes with its own challenges, and – owing to gaps in knowledge on how specific lipid species contribute to specific biological contexts, for example – approaches that rely on extensive prior functional annotations might be less appropriate. Also, the structure of a dataset will determine which integrative approaches are appropriate – e.g. whether frequently complete datasets across all data modalities are required. Finally, and most importantly, the biological question should drive the analysis (and overall design) of a multi-omics study, as different approaches answer different questions, e.g. identification of sample clusters versus functional molecular modules.

In the following, we present a small selection of multi-omics analysis approaches (Table 23.1). Different authors have categorized integrative data analysis approaches in different ways.

Ebels et al. identified three levels of data integration [37, 38]:

1) Conceptual integration: Multiple omics datasets are analyzed and interpreted independently; the derived conclusions are then combined and their relationships discussed.
2) Statistical integration: Multi-omics data sets are analyzed jointly, and statistical associations across the omics data modalities are identified.
3) Model-based integration: Data from multiple omics measurements are directly integrated into a mathematical model, such as a fully parameterized enzyme–metabolite metabolic network.

Cavill et al. further defined the different methods of statistical integration [38]:

1) Correlation-based integration: Correlative associations between analytes of the different omics data modalities are identified.
2) Concatenation-based integration: Multiple omics datasets are combined into a single matrix for further analysis, which can then be analyzed with standard (single omics) methods, such as k-means clustering.

Table 23.1 Examples of multi-omics statistical analysis approaches.

Category	Method	Principle	Primary output	References
Concatenation	iCluster	Gaussian latent variable model	Shared factors/ clusters	[29]
Pathway	Paintomics 3.0	Integrated visualization of multiple omic data types into KEGG pathway diagrams	Pathway visualization	[30]
Pathway	OmicsNet	Pathway analysis in multilayer networks	Co-regulated genes and pathways	[31]
Multivariate	sGCCA	Canonical correlation analysis	Omics-specific factors	[32]
Multivariate	MOFA	Bayesian factor analysis	Shared factors	[12]
Multivariate	Similarity network fusion (SNF)	Aggregation of similarity-based networks by message-passing approach	Fused similarity network/ clusters	[33]
Multivariate	intNMF	Non-negative matrix factorization	Shared factors/ clusters	[34]
Multivariate	MCIA	Co-inertia analysis	Omics-specific factors	[35]
Multivariate	MoCluster	Multi-block multivariate analysis and clustering	Clusters	[36]

3) Multivariate-based integration: Multivariate statistical approaches (such as PCA-like approaches) are adapted for the multi-omics context.
4) Pathway-based integration: Multiple omics data modalities are integrated, leveraging prior biological knowledge captured in pathway or functional interaction databases.

Picard et al. (and others) classified multi-omics analysis approaches according to the data processing stage, at which the integration is performed [39]:

1) Early integration: All data are concatenated into single matrix, and multivariate analysis are applied to this combined matrix.
2) Mixed integration: Each data modality is transformed independently and then fused for downstream analysis.
3) Intermediate integration: Individual omics data are transformed concomitantly to derive a common representation.
4) Late integration: Separate analysis of each omics data modality, then integration of final results.
5) Hierarchical integration: Prior knowledge on functional relationships across omics layers is leveraged for integrative analysis.

Table 23.1 provides a few examples of commonly used multi-omics analysis approaches. As discussed above, with regard to lipidomics data, we have leveraged the MOFA and sGCCA methods to unravel exposure effects in a systems toxicology study in mouse lungs [2]. Clark et al. used the MOFA approach to analyze multi-omics changes, including in the lipidome, in the context of AD [8].

Given the large list of integrative analytical methods available, as well as published examples including lipidomics methods, it is clear that nothing should stand in the way of including lipidomics data in multi-omics studies. However, lipidomics data present their own challenges, and few resources have been dedicated to integrative lipid analyses. For example, Kopczynski et al. have highlighted that, in comparison to transcriptomics and proteomics data, lipidomics data lack comprehensive quality-controlled databases and visualization tools [40].

Nevertheless, these gaps are slowly being filled. Kohler et al. have developed the Lipid Network Explorer (LINEX), a web-based tool that supports the visualization and analysis of functional lipid metabolism networks [41]. Pathway-based integrative analysis of lipid mediator data is facilitated by LimeMap, a comprehensive map of lipid mediator metabolic pathways [42]. LimeMap includes 282 elements (222 lipid mediators, together with 60 enzymes, receptors, and ion channels) and 279 enzymatic reactions that were aggregated from 102 studies. This metabolic map is easily available for integrated analyses via the CellDesigner and VANTED software on the GARUDA platform (http://www.garuda-alliance.org) [43]. A caveat to keep in mind is the focus on lipid mediator metabolism in mice, which can differ from the situation in humans, for example, with respect to the role of lipoxygenases [44].

Given the large number of available multi-omics analysis methods, a data analyst is not limited by the available data analysis options. Often, however, more guidance would be necessary for selecting the most effective analysis methods. Mangul et al. have emphasized the need for a systematic evaluation of calculation methods by using benchmarking approaches [45]. However, the common challenges of these benchmarking studies – such as the definition of a reliable gold standard dataset and the translatability of benchmarking results to real-world data – are commonly amplified in the multi-omics context.

This means that there are only a limited number of benchmarking studies available for multi-omics data. Most of the time, these studies assessed the methods using data from the Cancer Genome Atlas (TCGA), which includes genomics, epigenomics, transcriptomics, proteomics, and clinical data for various cancers [46]. This makes it difficult to extrapolate these results to other use cases. Rappoport et al. benchmarked multi-omics clustering algorithms by evaluating the differential survival or clinical parameters associated with different clusters [47]. Among the nine evaluated algorithms, rMKL-LPP performed best in terms of clinical enrichment and outperformed all methods except MCCA and MultiNMF with respect to survival. Cantini et al. evaluated nine joint dimensionality reduction approaches in the context of cancer [48]. Among these evaluated methods, intNMF performed the best for clustering, and MCIA produced the most consistent results across the different evaluated contexts. Pierre-Jean et al. evaluated 13 unsupervised methods for multi-omics data integration and highlighted MoCluster as the best-performing method, although several other methods showed comparable performance; the authors, overall, concluded that results depend on the specific design and features of the multi-omics dataset [49].

23.5 Current Challenges

As we have seen, lipidomics methods and the available integration methods have already reached sufficient maturity to be considered for large-scale multi-omics studies. However, several challenges remain – some specific to lipidomics data, others more generally applicable across omics data types [23, 50]:

- A general challenge for multi-omics data integration is identifier mapping, e.g. from the study dataset to a pathway database or across different publicly available datasets. However, for lipidomics data, efforts to harmonize the lipid nomenclature [20] and converter tools [51] can be expected to ameliorate this issue over time.
- Different omics technologies, including lipidomics, demonstrate heterogeneity in the signal-to-noise ratio [50]. With this, different omics data modalities can come with different statistical power to detect changes, which is often not considered in the study design or integrative analyses. To address this issue, novel data analyses methods that directly take these differences into account would be beneficial.
- Missing values are also a common issue, especially when it comes to large multi-omics datasets, for which different samples might have failed the quality control criteria across the different omics types [50]. However, some integrative analysis approaches, such as MOFA, are already designed to deal with missing information [12].
- Insufficient curation and information on the role of specific lipid species is another challenge. There is a general lack of comprehensive pathway-based lipid databases that describe the regulatory and functional interactions of specific lipid species [40].
- As discussed, only a limited number of benchmarking studies have been performed for the large number of multi-omics analyses, and the test datasets only cover a narrow application space [46]. This is especially true for lipidomics data, which are not considered in these benchmark studies.

23.6 Conclusions and Outlook

Both lipidomics methods and multi-omics analysis approaches have matured considerably in recent years. Although published multi-omics studies that contain lipidomics data are limited in number, they already provide insights into how multi-omics approaches can provide a deeper understanding of the role of lipidomics in health and disease. Going forward, it will be critical to address the challenges that limit the inclusion of lipidomics data in such integrative analyses, including a lack of standardization and insufficient information on specific molecular lipid species. With this and a growing understanding in the scientific community of what lipidomics data can add, it is likely that we will see many more lipidomics-centered multi-omics studies in the future.

Competing Interests and Funding

Philip Morris International is the sole source of funding and sponsor of this work. All authors are employees of Philip Morris International.

References

1 O'Donnell, V.B., Ekroos, K., Liebisch, G., and Wakelam, M. (2020). Lipidomics: current state of the art in a fast moving field. *Wiley Interdiscip. Rev. Syst. Biol. Med.* 12 (1): e1466.

2 Titz, B., Szostak, J., Sewer, A. et al. (2020). Multi-omics systems toxicology study of mouse lung assessing the effects of aerosols from two heat-not-burn tobacco products and cigarette smoke. *Comput. Struct. Biotechnol. J.* 18: 1056–1073.

3 Heo, Y.J., Hwa, C., Lee, G.-H. et al. (2021). Integrative multi-omics approaches in cancer research: from biological networks to clinical subtypes. *Mol. Cells* 44 (7): 433.

4 Canzler, S., Schor, J., Busch, W. et al. (2020). Prospects and challenges of multi-omics data integration in toxicology. *Arch. Toxicol.* 94 (2): 371–388.

5 Sun, C., Sun, Y., Wu, P. et al. (2021). Longitudinal multi-omics transition associated with fatality in critically ill COVID-19 patients. *Int. Care Med. Exp.* 9 (1): 13.

6 McKenzie, A.T., Katsyv, I., Song, W.-M. et al. (2016). DGCA: a comprehensive R package for differential gene correlation analysis. *BMC Syst. Biol.* 10 (1): 1–25.

7 He, F., Zhang, T., Xue, K. et al. (2021). Fecal multi-omics analysis reveals diverse molecular alterations of gut ecosystem in COVID-19 patients. *Anal. Chim. Acta* 1180: 338881.

8 Clark, C., Dayon, L., Masoodi, M. et al. (2021). An integrative multi-omics approach reveals new central nervous system pathway alterations in Alzheimer's disease. *Alzheimers Res. Ther.* 13 (1): 71.

9 Wong, M.W., Braidy, N., Poljak, A. et al. (2017). Dysregulation of lipids in Alzheimer's disease and their role as potential biomarkers. *Alzheimers Dementia* 13 (7): 810–827.

10 Surma, M.A., Herzog, R., Vasilj, A. et al. (2015). An automated shotgun lipidomics platform for high throughput, comprehensive, and quantitative analysis of blood plasma intact lipids. *Eur. J. Lipid Sci. Technol.* 117 (10): 1540–1549.

11 Herzog, R., Schuhmann, K., Schwudke, D. et al. (2012). LipidXplorer: a software for consensual cross-platform lipidomics. *PLoS One* 7 (1): e29851.

12 Argelaguet, R., Velten, B., Arnol, D. et al. (2018). Multi-omics factor analysis – a framework for unsupervised integration of multi-omics data sets. *Mol. Syst. Biol.* 14 (6): e8124.

13 Odenkirk, M.T., Stratton, K.G., Gritsenko, M.A. et al. (2020). Unveiling molecular signatures of preeclampsia and gestational diabetes mellitus with multi-omics and innovative cheminformatics visualization tools. *Mol. Omics* 16 (6): 521–532.

14 Vantaku, V., Dong, J., Ambati, C.R. et al. (2019). Multi-omics integration analysis robustly predicts high-grade patient survival and identifies CPT1B effect on fatty acid metabolism in bladder cancer. *Clin. Cancer Res.* 25 (12): 3689–3701.

15 Wishart, D.S., Tzur, D., Knox, C. et al. (2007). HMDB: the human metabolome database. *Nucleic Acids Res.* 35 (suppl_1): D521–D526.

16 Karkossa, I., Raps, S., von Bergen, M., and Schubert, K. (2020). Systematic review of multi-omics approaches to investigate toxicological effects in macrophages. *Int. J. Mol. Sci.* 21 (24): 9371.

17 Bannuscher, A., Karkossa, I., Buhs, S. et al. (2020). A multi-omics approach reveals mechanisms of nanomaterial toxicity and structure-activity relationships in alveolar macrophages. *Nanotoxicology* 14 (2): 181–195.

18 Kaza, M., Karaźniewicz-Łada, M., Kosicka, K. et al. (2019). Bioanalytical method validation: new FDA guidance vs. EMA guideline. Better or worse? *J. Pharm. Biomed. Anal.* 165: 381–385.

19 Liebisch, G., Ahrends, R., Makoto, A. et al. (2019). Lipidomics needs more standardization. *Nat. Metab.* 1 (November): 745–747.

20 Liebisch, G., Fahy, E., Aoki, J. et al. (2020). Update on LIPID MAPS classification, nomenclature, and shorthand notation for MS-derived lipid structures. *J. Lipid Res.* 61 (12): 1539–1555.

21 Köfeler, H.C., Eichmann, T.O., Ahrends, R. et al. (2021). Quality control requirements for the correct annotation of lipidomics data. *Nat. Commun.* 12 (1): 1–4.

22 Graw, S., Chappell, K., Washam, C.L. et al. (2021). Multi-omics data integration considerations and study design for biological systems and disease. *Mol. Omics* 170–185.

23 Krassowski, M., Das, V., Sahu, S.K., and Misra, B.B. (2020). State of the field in multi-omics research: from computational needs to data mining and sharing. *Front. Genet.* 11.

24 Tarazona, S., Balzano-Nogueira, L., Gómez-Cabrero, D. et al. (2020). Harmonization of quality metrics and power calculation in multi-omic studies. *Nat. Commun.* 11 (1): 1–13.

25 Nakayasu, E.S., Nicora, C.D., Sims, A.C. et al. (2016). MPLEx: a robust and universal protocol for single-sample integrative proteomic, metabolomic, and lipidomic analyses. *mSystems* 1 (3): e00043–e00016.

26 Salem, M., Bernach, M., Bajdzienko, K., and Giavalisco, P. (2017). A simple fractionated extraction method for the comprehensive analysis of metabolites, lipids, and proteins from a single sample. *J. Vis. Exp.* 124: 55802.

27 Kang, J., David, L., Li, Y. et al. (2021). Three-in-one simultaneous extraction of proteins, metabolites and lipids for multi-omics. *Front. Genet.* 12: 313.

28 Titz, B., Sewer, A., Luettich, K. et al. (2020). Respiratory effects of exposure to aerosol from the candidate modified-risk tobacco product THS 2.2 in an 18-month systems toxicology study with A/J mice. *Toxicol. Sci.* 178 (1): 138–158.

29 Shen, R., Mo, Q., Schultz, N. et al. (2012). Integrative subtype discovery in glioblastoma using iCluster. *PLoS One* 7 (4): e35236.

30 Hernández-de-Diego, R., Tarazona, S., Martínez-Mira, C. et al. (2018). PaintOmics 3: a web resource for the pathway analysis and visualization of multi-omics data. *Nucleic Acids Res.* 46 (W1): W503–W509.

31 Akhmedov, M., Arribas, A., Montemanni, R. et al. (2017). OmicsNet: integration of multi-omics data using path analysis in multilayer networks. *bioRxiv* 238766.

32 Rohart, F., Gautier, B., Singh, A., and Lê Cao, K.-A. (2017). mixOmics: an R package for 'omics feature selection and multiple data integration. *PLoS Comput. Biol.* 13 (11): e1005752.

33 Wang, B., Mezlini, A.M., Demir, F. et al. (2014). Similarity network fusion for aggregating data types on a genomic scale. *Nat. Methods* 11 (3): 333–337.

34 Chalise, P. and Fridley, B.L. (2017). Integrative clustering of multi-level 'omic data based on non-negative matrix factorization algorithm. *PLoS One* 12 (5): e0176278.

35 Meng, C., Kuster, B., Culhane, A.C., and Gholami, A.M. (2014). A multivariate approach to the integration of multi-omics datasets. *BMC Bioinf.* 15 (1): 1–13.

36 Meng, C., Helm, D., Frejno, M., and Kuster, B. (2016). moCluster: identifying joint patterns across multiple omics data sets. *J. Proteome Res.* 15 (3): 755–765.

37 Ebbels, T.M. and Cavill, R. (2009). Bioinformatic methods in NMR-based metabolic profiling. *Prog. Nucl. Magn. Reson. Spectrosc.* 4 (55): 361–374.

38 Cavill, R., Jennen, D., Kleinjans, J., and Briedé, J.J. (2016). Transcriptomic and metabolomic data integration. *Briefings Bioinf.* 17 (5): 891–901.

39 Picard, M., Scott-Boyer, M.-P., Bodein, A. et al. (2021). Integration strategies of multi-omics data for machine learning analysis. *Comput. Struct. Biotechnol. J.* 19: 3735–3746.

40 Kopczynski, D., Coman, C., Zahedi, R.P. et al. (2017). Multi-OMICS: a critical technical perspective on integrative lipidomics approaches. *Biochim. Biophys. Acta* 1862 (8): 808–811.

41 Köhler, N., Rose, T.D., Falk, L., and Pauling, J.K. (2021). Investigating global lipidome alterations with the lipid network explorer. *Metabolites* 11 (8): 488.

42 Nishi, A., Ohbuchi, K., Kaifuchi, N. et al. (2021). LimeMap: a comprehensive map of lipid mediator metabolic pathways. *npj Syst. Biol. Appl.* 7 (1): 6.

43 Ghosh, S., Matsuoka, Y., Asai, Y. et al. (2011). Software for systems biology: from tools to integrated platforms. *Nat. Rev. Genet.* 12 (12): 821–832.

44 Kuhn, H., Banthiya, S., and Van Leyen, K. (2015). Mammalian lipoxygenases and their biological relevance. *Biochim. Biophys. Acta* 1851 (4): 308–330.

45 Mangul, S., Martin, L.S., Hill, B.L. et al. (2019). Systematic benchmarking of omics computational tools. *Nat. Commun.* 10 (1): 1393.

46 Blum, A., Wang, P., and Zenklusen, J.C. (2018). SnapShot: TCGA-analyzed tumors. *Cell* 173 (2): 530.

47 Rappoport, N. and Shamir, R. (2018). Multi-omic and multi-view clustering algorithms: review and cancer benchmark. *Nucleic Acids Res.* 46 (20): 10546–10562.

48 Cantini, L., Zakeri, P., Hernandez, C. et al. (2021). Benchmarking joint multi-omics dimensionality reduction approaches for the study of cancer. *Nat. Comm.* 12 (1): 1–2.

49 Pierre-Jean, M., Deleuze, J.-F., Le Floch, E., and Mauger, F. (2019). Clustering and variable selection evaluation of 13 unsupervised methods for multi-omics data integration. *Briefings Bioinf.* 21 (6): 2011–2030.

50 Tarazona, S., Arzalluz-Luque, A., and Conesa, A. (2021). Undisclosed, unmet and neglected challenges in multi-omics studies. *Nat. Comput. Sci.* 1 (6): 1–8.

51 Ni, Z. and Fedorova, M. (2020). LipidLynxX: lipid annotations converter for large scale lipidomics and epilipidomics datasets. *bioRxiv*:2020.04.09.033894.

24

Tracer Lipidomics

Jonas Dehairs[1], Ine Koeken[1,2], Lake-Ee Quek[2], Andrew Hoy[2], Bart Ghesquière[3], and Johannes V. Swinnen[1]

[1] KU Leuven, Lipidomics Core Facility Lipometrix, Herestraat 49, 3000 Leuven, Belgium
[2] The University of Sydney, Charles Perkins Centre, John Hopkins Drive, Camperdown, NSW 2006, Australia
[3] Metabolomics Expertise Center, VIB Center for Cancer Biology, VIB, Herestraat 49, 3000 Leuven, Belgium

In standard lipidomics studies, the levels of different lipid species are measured across experimental and control conditions, and these levels are compared to study the effect of disease, treatment, or other metabolic events on lipid metabolism. Although lipid metabolism is a very dynamic process, these experiments inherently provide only very momentary, static snapshots of lipid composition. Of equal importance is the activity of metabolic pathways, which can be expressed as the amount of material flow per unit of time, i.e. flux. Static levels do not provide insight into the fluxes of the lipid synthesis, degradation, and remodeling pathways, and a high lipid concentration is not necessarily indicative of a high metabolic flux. Indeed, classical lipidomics fails to reveal the cause of the relative changes in the abundance of molecular targets, e.g. is the increase in the abundance of a target due to the fact that there is more synthesis or is it simply due to the fact that there is less degradation of that specific target. A popular analogy to demonstrate this concept is the one of highway traffic, where flux tends to increase with the amounts of cars on the road until a traffic jam occurs resulting in a high car concentration but a low flux. Similarly, adipocytes may have huge triacylglycerol reservoirs that can be relatively inert, depending on the nutritional status, while phosphatidic acid (PA) is low abundant but has a high flux as it is constantly synthesized and quickly converted as it serves as the precursor to the synthesis of other phospholipids [1]. Changes in the lipids levels also come short in revealing the specific pathways that are involved. For example, an increase in the intracellular free fatty acid levels may be the result of increased synthesis, increased uptake, decreased degradation, or the release from membrane or lipid droplet stores.

Mass Spectrometry for Lipidomics: Methods and Applications, First Edition.
Edited by Michal Holčapek and Kim Ekroos.

Therefore, to get a better insight into the activity of the various lipid metabolic pathways, tracer lipidomics studies are gaining popularity. Remarkably, the earliest studies that used stable isotope labeled precursors to study the activity of lipid synthesis and breakdown, were *in vivo* studies that date back to the 1930s. In one pioneering study, Schoenheimer and Rittenberg used mice on a whole-wheat bread diet and replaced their drinking water with a dilute solution of deuterated water. They found that, after six to eight days, the deuterium incorporation into the fatty acids of the adipose tissue reached a plateau and upon removal of the deuterium source, the labeling disappeared in a similar amount of time. This demonstrated for the first time that a continuous and rapid turnover of fatty acids takes place in a living organism [2]. Due to the experimental complexity of deuterium detection, which involved oxidizing the fats to water and measuring the specific gravity or refractive index of this deuterated water, this methodology failed to gain traction and was in subsequent years replaced by radioactive isotope labeling [3]. Since then, experiments using radioactive tritiated water became the standard approach to study lipogenic synthesis rates and formed the basis for our current understanding of the regulation of *de novo* lipogenesis. And while radioactive tracers allowed for very sensitive detection, this approach was not well suited to study individual lipid species. Some degree of specificity was possible with high-performance liquid chromatography (HPLC) methods, but this remained limited and was experimentally complex [4]. For these reasons, and owing to advancements in mass spectrometry (MS) instruments, the past two decades have seen a resurgence of targeted stable isotope-based tracer experiments, which have provided new insights into the synthesis and turnover of individual lipid species (*vide infra*). More recently, the advent of fast, sensitive, high-resolution MS instruments has brought on the dawn of lipidome-wide tracer lipidomics.

24.1 Flux Analysis and Stable Isotope Labeling Patterns

Stable heavy isotopes are atoms that contain a higher number of neutrons in their nucleus than the most common isotope of that element, resulting in a higher mass. In contrast to radioactive heavy isotopes, stable isotopes do not spontaneously decay into other elements. Molecules that have the same chemical formula and structure but only differ from each other in isotopic composition are called isotopologues (often mistakenly referred to as isotopomers, which are molecules with the same isotopic composition, but which differ in the location of the isotopes within the molecule).

Flux analysis of lipid metabolism is made possible by the uptake of heavy isotope-labeled metabolic precursors into the cell (most commonly, uniformly labeled ^{13}C substrates), resulting in the labeling of lipids as the labeled precursor is metabolized. It is worth pointing out that formal flux analysis entails the calculation of turnover rate constants and half-life times for each lipid species of interest, an approach that requires a priori knowledge of the participating metabolic pathways, as well as mathematical modeling and knowledge on the uptake and secretion rate of the precursor [5]. Most studies, however, opt for a direct interpretation of isotopic labeling patterns, as this is sufficient to derive information on the relative activity of different pathways and the contribution of precursors to the synthesis of different lipids. This approach is what we refer to in this chapter as ^{13}C tracer lipidomics.

The isotopic labeling patterns of lipid species can be described by so called Mass Distribution Vectors (MDVs). The MDV effectively collects for each lipid species the abundance of all possible isotopologues, whereby each abundance is normalized to the total sum of all isotopologues of that species, such that the fractions add up to 100%. Theoretically, a lipid with n carbons can have anywhere from zero to n carbons that are ^{13}C and as such the MDV describing that lipid will contain n fractional abundances for each possible isotopologue. For example, the measurement of ^{13}C labeled palmitate (where $n = 16$ since palmitate has 16 carbons) would be described by an MDV of the form:

$$\vec{X} = \{X_0, X_1, X_2, X_3, X_4, X_5, X_6, \ldots, X_{16}\}$$

where for instance $X_2 = 0.25$ would mean that 25% of the total palmitate pool carries precisely two ^{13}C carbons. Note that the MDV does not consider the location of the labeled carbons within the molecules, but most experimental setups do not provide this information anyway, as resolving the isotopomers of complex lipid species is challenging and requires complementary techniques such as high-resolution 2D nuclear magnetic resonance (NMR) [6]. The MDV is typically sufficient however to draw conclusions on relative pathway activity and nutrient contributions [5].

Because stable isotopes are naturally occurring (for example, 1.07% of all carbon on earth is ^{13}C, 0.0115% of all hydrogen is ^{2}H) [7] lipid molecules have a high probability of containing one or more ^{13}C isotopes, even in the absence of ^{13}C labeled precursor. Therefore, the measured isotopic labeling patterns need to be corrected for the presence of the natural abundances of the isotopomers (so-called type I isotopic effects). While the mass spectra of unlabeled controls can serve as a useful reference to determine if observed isotopologues truly reflect labeling or simply reflect natural isotope abundance, it has been pointed out in the literature that subtracting the mass distribution vector (MVD) of an unlabeled control from the MVD of the labeled sample is not a valid way to correct for natural isotope abundance [5]. Instead, the correction should be based on considering the natural distribution of all possible theoretical MVDs when zero to all carbons are ^{13}C. Several programs are available that can convert raw MVDs into corrected ones [8, 9]. Figure 24.1 shows an example of an MVD of a PC(32:1) species in cells that were cultured with [U-^{13}C]-glucose, before and after correction for natural ^{13}C abundances, as well as an unlabeled control.

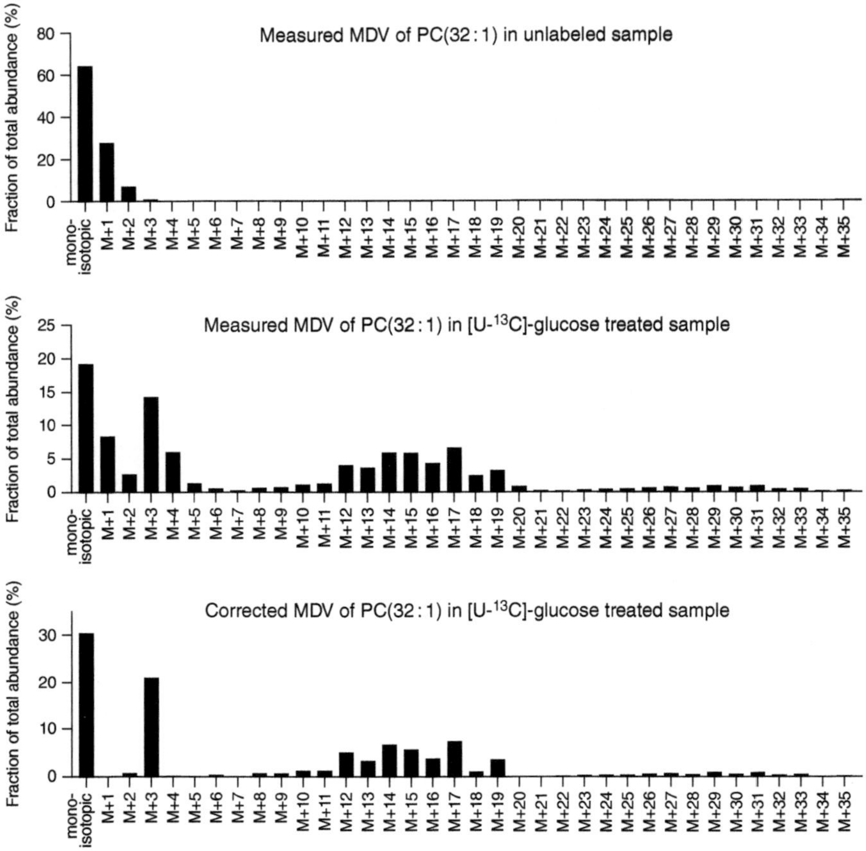

Figure 24.1 Mass distribution vectors (MDVs) of PC(32:1) of cells cultured with [U-^{13}C]-glucose for 24 hours. The unlabeled sample serves as an essential control, as the MDV should closely match the theoretical expectation based on the presence of naturally occurring ^{13}C. A labeled sample can only be correctly interpreted after correction for the natural isotope presence (here corrected by IsoCor [9]). The labeled sample shows a distinct peak at M+3, a distribution of peaks around M+16, and a weaker distribution of peaks around M+30. This suggests a high turnover of the glycerol backbone (M+3), a moderate turnover for one of the acyl chains, and a low turnover for the other acyl chain. These observations can be corroborated with MS/MS experiments (*vide infra*).

24.2 Experimental Conditions and Selecting the Right Tracer

When both the intracellular levels and fluxes of metabolites remain constant, the cells are at metabolic steady state. In culture conditions, true metabolic steady state is rarely achieved, as this requires specialized culture systems that keep the nutrient conditions and the number of cells in the system constant [10]. In practice, in conventional cell culture, the cells are assumed to be in a pseudo-steady state when they are in the exponential growth phase, on the condition that nutrient levels have not become limiting. To allow an appropriate interpretation of isotopic labeling patterns, it is important that cells are at metabolic pseudo-steady state. If this is not the case, for example in response to acute events such as signaling or quickly changing nutrient conditions, specialized approaches with high-frequency sampling at multiple time points and complex computational analysis are required, which are not further discussed here [11]. In general, to reduce the likelihood of introducing metabolic perturbations, it is preferable that from the start of the experiment medium is used where the labeled metabolite has replaced the regular unlabeled metabolite at the same concentration, and that the medium is otherwise identical to the regular medium. Of equal importance is that the cells reach isotopic steady state. As ^{13}C labeled nutrients get metabolized, the amount of ^{13}C label in the lipids of interest increases over time until the amount of label incorporation reaches a plateau. How long it takes to reach isotopic steady state depends both on the tracer that is used and on the lipids of interest. For example, the incorporation of [U-^{13}C]-palmitate into phosphatidylcholine (PC) species can often reach isotopic steady state within 24 hours, while the incorporation of ^{13}C originating from [U-^{13}C]-glucose into various PC species will typically take several days to reach isotopic steady state. It is therefore recommended practice to run an initial time course experiment to determine when isotopic steady is reached.

In tracer metabolomics studies (non-lipid polar metabolites), much emphasis is put on fast harvesting procedures, as many metabolites have turnover rates in the minutes to seconds range. For example, for adherent cultured cells the medium is removed after which cold organic solvent is added directly to the culture dish to immediately quench any metabolic activity. In contrast, tracer lipidomics studies have not put such emphasis on fast quenching and typically harvest and extract lipids according to the same protocols as standard non-tracer lipidomics. Indeed, ultra-fast harvesting is less critical for tracer lipidomics as complex lipids have turnover rates in the range of hours to days [12]. Nevertheless, as lipids remain sensitive to oxidation, chemical degradation or enzymatic conversion, adhering to the sample handling guidelines of standard lipidomics experiments, remains important [13].

Another important consideration is the choice of tracer molecule, which depends on the research question and on which pathway(s) are of interest. The use of labeled glucose, glutamine, or heavy water for example can lead to very widespread labeling of the lipidome which can be highly informative for a wide range of pathways, but this comes at the cost of highly complex mass spectra. In contrast, more pathway-specific tracers can be used, such as glycerol to study glycerolipid metabolism or polar headgroups such as choline to study the synthesis of PC or sphingomyelin. Labeled fatty acids can be used to study fatty acid uptake and to study specific desaturation, elongation or beta-oxidation pathways. Note that fatty acids are best added to the medium by albumin complexation to avoid the toxicity of adding solvents to the medium [14]. The most common choice of isotope is ^{13}C, owing to its ubiquitous presence in lipids and other biomolecules. ^{15}N can also be used but is of course limited to nitrogen-containing lipids such as sphingolipids, PC, phosphatidylethanolamine (PE), and phosphatidylserine (PS). ^{2}H, in the form of heavy water, is used for very generalized incorporation into lipids but has the disadvantage that desaturation of fatty acids can lead to the loss of labeling, complicating the analysis and interpretation.

24.3 Targeted Tracer Analysis

24.3.1 Fatty Acids

Gas chromatography–mass spectrometry (GC-MS) is the most commonly used technique for fatty acid analysis, owing to its excellent sensitivity and ability to provide rich structural information. Numerous studies have used GC-MS to study ^{13}C incorporation from polar precursors into palmitic acid as a measure for the activity of *de novo* fatty acid synthesis [15–17]. [U-^{13}C]-glucose and [U-^{13}C]-acetate are commonly used tracers in this case. Glucose enters the glycolytic pathway in the cytosol to produce pyruvate and adenosine triphosphate (ATP). Pyruvate can then be taken up by the mitochondria, where it enters the tricarboxylic acid (TCA) cycle. Citrate can leave the TCA cycle and can be transported back to the cytoplasm, where ATP citrate lyase (ACLY) can convert it to acetyl-CoA, the basic building block for fatty acid synthesis by the enzyme fatty acid synthase (FASN). Acetate that enters the cell can also be converted to acetyl-CoA by the enzyme acetyl-CoA synthetase (ACS). As illustrated in Figure 24.2, when [U-^{13}C]-glucose or [U-^{13}C]-acetate are used as tracers, both carbons of the acetyl group of acetyl-CoA derived from these tracers will be ^{13}C carbons, resulting in a ^{13}C labeling pattern of palmitic acid that is a multiple of 2.

If the fatty acid desaturation and elongation pathways are also of interest, the analysis should include a range of different fatty acids that vary in chain length and unsaturation degree. It is worth pointing out that to allow the measurement of medium and long chain fatty acids with GC-MS, derivatization is required to increase the volatility and to improve the chromatography. Derivatization of the fatty acid (FA) carboxyl group with methyl ester to fatty acid methyl esters (FAMEs) is by far the most common choice in GC-MS fatty acid analysis and has also been used successfully in ^{13}C tracing GC-MS studies [15–18]. The electron ionization (EI)

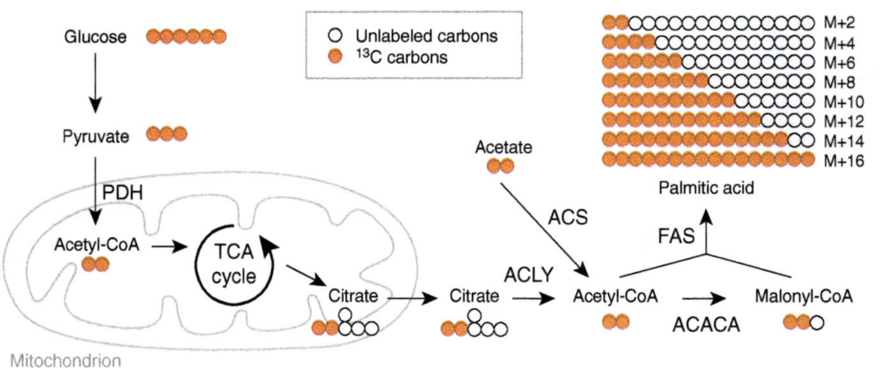

Figure 24.2 Schematic of ^{13}C carbon incorporation into palmitic acid from carbon originating from uniformly labeled glucose and acetate tracers. PDH, pyruvate dehydrogenase; ACS, acetyl-CoA synthetase; ACLY, ATP citrate lyase; FAS, fatty acid synthase; ACACA, acetyl-CoA carboxylase. Note that many variations are possible in the position of the 2-unit labeled ^{13}C carbons in the M+2, M+4, etc. isotopologues.

mode that is typically used for FAME analysis can result in extensive fragmentation and a low signal for the intact molecular ion however [19], which can be detrimental to identifying the ^{13}C labeled fatty acid isotopologues. To mitigate this issue, alternative derivatization protocols such as with pentafluorobenzyl (PFT) combined with soft chemical ionization (CI) have been proposed [20].

24.3.2 Phospholipids

Phospholipid synthesis (Figure 24.3) [21, 22] starts with the transfer of a fatty acid to glycerol-3-phosphate to generate lysophosphatidic acid (LPA), to which a second acyl chain can be added to generate PA. PA can then be converted to cytidine diphosphate-diacylglycerol (CDP-DG), which is the precursor for phosphatidylinositol (PI), PS, and phosphatidylglycerol (PG) synthesis. Alternatively, PA can be dephosphorylated to form diacylglycerol (DG) which is the precursor to the synthesis of PC and PE in the Kennedy pathway, in which choline and ethanolamine are subsequently

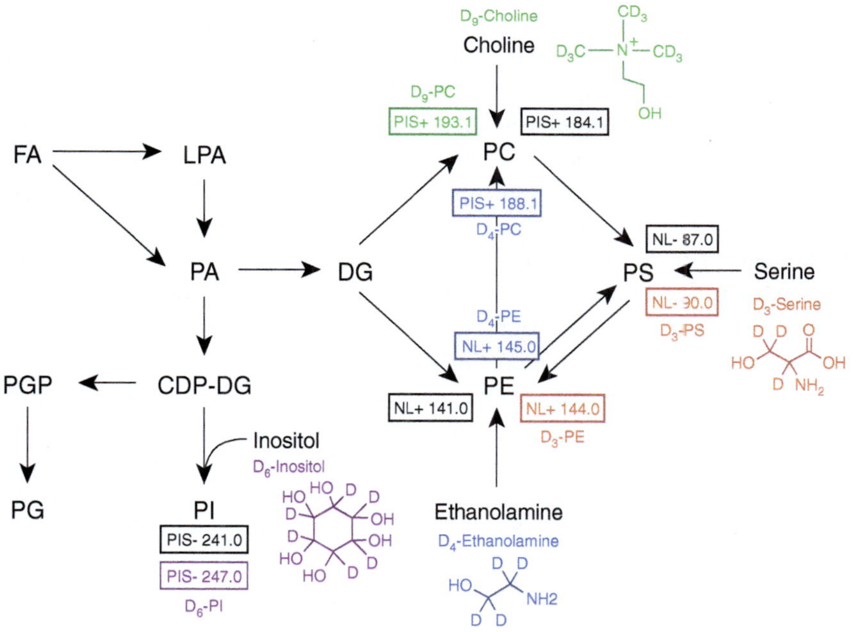

Figure 24.3 Overview of polar headgroup precursor tracers that can be used to study phospholipid synthesis. The black boxes show the precursor ion scan (PIS) and neutral loss scans (NL) that can be used to analyze the various phospholipid classes. A D_3-serine precursor can be traced to PS and PE with NL scan for *m/z* 90.0 in negative-ion mode and for *m/z* 144.0 in positive mode, respectively. The Kennedy pathways for PC and PE synthesis can be studied with a PIS in positive-ion mode for *m/z* 193.1 using a D_9-choline as a tracer and a NL scan in positive-ion mode for *m/z* 145.0 using a D_4-ethanolamine tracer, respectively. A D_4-ethanolamine tracer can also be used to study the PEMT pathway of PC synthesis with a PIS in positive-ion mode for *m/z* 188.1. Finally, PI synthesis can be traced with a D_6-inositol tracer and a PIS in negative-ion mode for *m/z* 247.0. PGP: phosphatidylglycerol phosphate.

phosphorylated and converted to cytidine 5-difosfo (CDP)- choline/ethanolamine before being added as headgroups on DG to yield PC and PE, respectively. Both PC and PE can be converted to PS with a serine headgroup swap, and PS can be converted back to PE by mitochondrial PS decarboxylases. PE can also be converted to PC by methylation through the PE-*N*-methyltransferase (PEMT) pathway.

The most common approach to study the phospholipid synthesis pathways is through the use of polar headgroup precursor tracers and analysis with liquid chromatography (LC) or direct infusion MS. This approach has the advantage that all the species of the targeted phospholipid class can easily be traced by precursor or neutral loss (NL) scans for the labeled headgroups. For example, in positive-ion mode collision-induced dissociation (CID) of unlabeled PC species yields a phosphocholine fragment of *m/z* 184.1. PC species that have incorporated labeling through a D_9-choline tracer, however, will generate a fragment of *m/z* 193.1. Similarly, unlabeled PE species are characterized by a NL *m/z* of 141.0, while species labeled with D_4-ethanolamine will have a NL *m/z* of 145.0. For the study of PC synthesis, D_9-choline and D_4-ethanolamine tracers can even be multiplexed to differentiate between the Kennedy (precursor *m/z* 193.1) and PEMT (precursor *m/z* 188.1) pathways of PC synthesis (Figure 24.3) [23].

Headgroup labeled phospholipid species themselves can also be useful tracers, in particular to study phospholipid remodeling [24]. In Land's cycle, phospholipids successively undergo a deacylation and reacylation reaction which can rapidly modulate their acyl chain composition as the acyl chains are exchanged with cytoplasmic acyl-CoAs [25]. By scanning for the labeled headgroups, headgroup labeled phospholipid species make it straightforward to trace remodeling across all species of the targeted class. Moreover, in the cases where phospholipid class conversion happens without headgroup transfer (for example PS to PE conversion by decarboxylation, and PE to PC conversion by N-methylation), these tracers allow phospholipid class conversion and remodeling to be monitored simultaneously.

Labeled fatty acids can be used to study the incorporation of exogenously supplied fatty acids into complex lipids [21, 26]. Typically, uniformly labeled ^{13}C fatty acids are used as this results in substantial mass shifts which minimizes the likelihood that the labeled species have *m/z*'s that overlap with other significantly abundant species of the same class. These species can therefore be analyzed with scans for the class specific phospholipid headgroup fragments just like unlabeled species. On low-resolution instruments it is advisable to confirm the measurement with a NL scan for the labeled fatty acid, since isobaric overlap with unlabeled species remains possible. For example, when [U-^{13}C]-palmitate is used as a tracer, PC(16:0/18:1) with a labeled palmitate differs in mass from PC(14:1/22:6) by *m/z* 0.1162. PC(14:1/22:6) is typically a low abundant species however, and with mass shifts of M+16 and higher, high abundant labeled species are typically only isobaric with low abundant endogenous species. While fatty acids tracers can undergo metabolic conversion through desaturation, elongation and beta oxidization, these labeled derivatives are typically not considered in targeted analyses as they significantly complicate the analysis.

Polar tracer precursors such as [U-^{13}C]-glucose and [U-^{13}C]-acetate are typically not used in targeted phospholipid analysis since the resulting labeling patterns are so complex and isobaric they cannot be resolved with nominal mass instruments. A feasible strategy is to separate the phospholipid classes with HPLC, hydrolyze the isolated fractions, and subsequently preform GC-MS analysis on the liberated fatty acids. A more direct analysis of intact labeled phospholipid species (as demonstrated in Figure 24.1) calls for an untargeted approach on high-resolution MS instruments.

24.4 Toward Untargeted Lipidome-Wide Tracer Analysis

While the tracing of polar headgroup precursors into complex lipids is well established, tracing ^{13}C carbon originating from polar precursors such as glucose, glutamine, and acetate into complex lipids remains a very challenging task. The labeling patterns become so complex that isobaric overlap becomes a major limitation in the ability to correctly identify lipid species and their isotopologues. Likewise, a comparable level of complexity can arise if a fatty acid tracer is used and the objective is to trace the metabolic destiny of this tracer beyond simple direct incorporation into complex lipids. Fatty acid tracers could for example undergo elongation and desaturation processes prior to incorporation into complex lipids, or if the tracer undergoes beta oxidation the labeling could even appear in entirely different acyl chains as the carbon of the tracer gets recycled.

24.4.1 Isotopic Effects and the Complexity of Tracer Analysis Mass Spectra

So-called type I and II isotopic effects need to be considered in standard lipidomics approaches, and in tracer lipidomics studies it becomes even more imperative that these effects are appropriately dealt with. Type I isotopic effects in lipidomics are caused by the natural presence of ^{13}C and result in significant levels of M+1, M+2, etc. isotopologues even in the absence of a ^{13}C tracer. These effects have already been discussed in the section on MDVs and need to be corrected for with appropriate algorithms [8, 9]. Type II isotopic effects deal with isobaric overlap between isotopologues of different lipid species and typically arise from having one double bond less versus having two extra ^{13}C carbons in the acyl chain, which results in a mass difference of only 0.0089 Da (($^{12}C_2{}^1H_2$ versus $^{13}C_2$) for example PC(34:0) versus PC(34:1)[$^{13}C_2$]). Standard lipidomics approaches only have to consider isobaric overlap resulting from the natural abundance of ^{13}C, and if the resolution of the MS instrument isn't enough to resolve a m/z difference of 0.0089 (a full width half maximum [FWHM] resolution of about 180 000 is needed) [27], algorithms are available that sequentially correct the signal intensities for type II overlap [28, 29]. These algorithms calculate probability distributions of the ^{13}C isotopologues for each lipid species and operate on the assumption that 1.07% of the carbons in the sample are expected to be ^{13}C. This approach is thus unsuitable if exogenously supplied ^{13}C is incorporated into the acyl chains. Therefore, if ^{13}C from a supplied tracer is incorporated into the acyl chains of complex lipids, a theoretical resolution of at least 180 000 is needed to be able to correctly measure the isotopologue abundances.

At a resolution of 180 000 most of the isobaric overlaps that have been well documented in standard lipidomics studies can be fully resolved. For example, in positive-ion mode a PS species with one double bond more than the corresponding PC species has a mass difference of 0.0727 Da, which can already be resolved with a resolution of 25 000 (for example PC 34:1 versus PS 34:2). Sodium adducts could potentially be problematic, as for example the sodium adduct of PC(34:1) differs in mass by only 0.0024 Da from the proton adduct of PC(36:4) ($^{1}H_{1}{}^{23}Na_{1}$ versus $^{12}C_{2}$), which would require a resolution of about 700 000 to resolve. In practice this problem can be avoided by adding an ammonium salt to the run buffer which suppresses the presence of the sodium adducts [30].

It would seem then that a resolution of 180 000 is sufficient for untargeted tracer lipidomics analysis. However, specifically for ^{13}C tracer experiments, beyond the isobaric overlap between a lipid species and the M+2 peak of that same species with one extra double bond, not much has been published on other possible isobaric overlaps that may arise if a high number of carbons in the acyl chains are ^{13}C carbons. To address this, we created an *in silico* database that contains over 800 lipid species across the major classes of phospholipids and neutral lipids (at sum notation level aka species level aka MS1 level). We determined the exact masses of all possible ^{13}C isotopologues of these species (over 35 000 isotopologues in total), and we then calculated how much percentage of these isotopologues could be resolved at different mass resolutions (Figure 24.4). This simulation demonstrates that even with a resolution of one million, which is about the maximum achievable resolution on current Orbitrap instruments, approximately 80% of the isotopologues in the database could be resolved. Only with a resolution of seven million could more than 95% of the species in the database be fully resolved. It should be noted that these numbers will be slightly too pessimistic for tracer experiments that use common tracers such as glucose, acetate, glutamine, and palmitate. The simulation naively

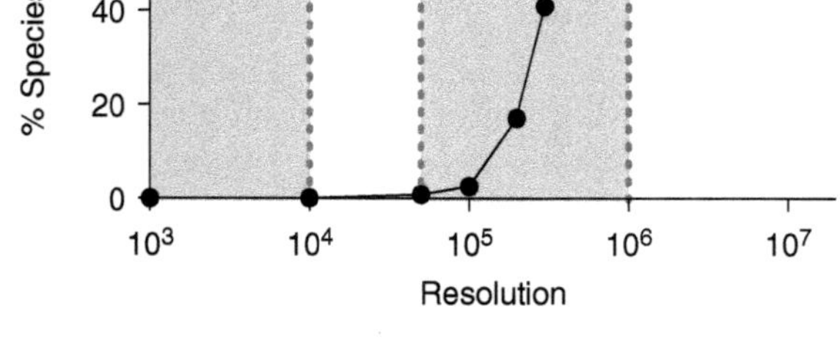

Figure 24.4 Simulation of the ability of MS instruments at various resolutions to resolve a database of 800 lipid species and 35 000 isotopologues by their *m/z* value. Included lipid classes were: PC $[M+H]^+$, PE $[M+H]^+$, PS $[M-H]^-$, PI $[M-H]^-$, PG $[M-H]^-$, PA $[M-H]^-$, lysophosphatidylcholine (LPC) $[M+H]^+$, lysophosphatidylethanolamine (LPE) $[M+H]^+$, LPA $[M-H]^-$, sphingomyelin (SM) $[M+H]^+$, ceramide (CER) $[M+H]^+$, triacylglycerol (TAG) $[M+NH_4]^+$, DG $[M+NH_4]^+$, monoacylglycerol (MAG) $[M+H]^+$, cholesteryl ester (CE) $[M+NH_4]^+$. TOF: time of flight.

considers isotopologues such as for example PE(44:12)[$^{13}C_{34}$] which can in principle not occur in mammalian cells under these tracer conditions because such highly unsaturated species can only be synthesized originating from the 18 carbon essential fatty acids linoleic and linolenic acid, making it impossible for such a high number of carbons to be ^{13}C in this species. Only in specialized applications where [U-^{13}C]-linoleate or [U-^{13}C]-linolenate tracers are used can such highly unsaturated species contain such high numbers of ^{13}C carbons.

Nevertheless, the simulation revealed a range of different patterns that create isobaric overlap between isotopologues, for example the pattern $^{12}C_9HN$ versus $O_2{}^{13}C_7$ which results in a mass difference of 0.0024 and requires a resolution of 700 000 to resolve. This pattern was observed 290 times, for example in SM(d18:1/18:2) versus PE(34:0)[$^{13}C_7$]. Another example is $^{12}C_{14}$ versus $HPO_2{}^{13}C_8$ which results in a mass difference of 0.0017 and requires a resolution of one million to resolve. This pattern occurred in 275 species, for example TAG(46:6) versus PE(38:0)[$^{13}C_8$].

24.4.2 Technical Considerations

It should be clear that if untargeted lipidome-wide tracer analysis is attempted with a direct infusion MS setup, there is very little room for compromise on the mass resolution of the MS instrument. This stringent requirement can be alleviated to some extent by bringing liquid chromatography into the equation. Hydrophilic Interaction Liquid Chromatography (HILIC), which can separate lipids into classes based on their polarity, can be a viable option, but it does not resolve the isobaric overlap originating from having one double bond less versus having two extra ^{13}C carbons and thus a resolution of at least 180 000 would still be required. Reverse-phase chromatography has the ability to separate lipids based on chain length and unsaturation degree and could therefore separate the double bond versus two ^{13}C isobars, but it is less obvious how much other potential isobars may be co-eluting. Regardless of whether or not LC separation is used, untargeted tracer lipidomics needs specialized software that can not only identify lipid species at the appropriate level of detail (depending on the availability of MS/MS data) but that can also automatically identify isotopologues and inform the user about potential isobaric overlap. While there are many software packages that cover the needs of standard lipidomics approaches, the field of tracer lipidomics is lagging behind in this regard.

So far, few studies have attempted untargeted lipidome-wide tracer analysis. One of the most elaborate studies used reverse-phase ultra high performance liquid chromatography (UHPLC) coupled to an Orbitrap (resolution 600 000) to trace [U-^{13}C]-palmitate in cultured primary muscle cells [31]. This study was able to detect more than 200 lipid species and nearly 700 isotopologues across 12 different lipid classes. Publicly available databases were used to identify unlabeled species, which was followed up with an analysis with a custom software program that looked for M+12, M+14, M+16, M+18, M+20, M+28, M+30, M+34, M+36, M+44, M+46, M+50, and M+52 mass shifts originating from the identified species. While this approach does consider elongation and to some extent beta-oxidation of the palmitate tracer, the custom software did not capture the full MDVs and was fixed to consider only 13 specific mass shifts, which would not be appropriate if a different tracer is used.

24.5 MS/MS Analysis as a Unique Approach to Study Fluxes at the Molecular Species Level

High-resolution MS scans of complex lipids only give access to the so-called "species level" or "sum notation" level of identification, which does not include the identity of the acyl chains. For example PC(36:1) could be PC(18:1_18:0), PC(16:0_20:1) or various other species because these are all isomers with identical chemical formulas. The identity of the acyl chains or "molecular species level identification," becomes accessible with tandem MS fragmentation experiments, which produce fragments that allow identification of the various lipid building blocks. Figure 24.5

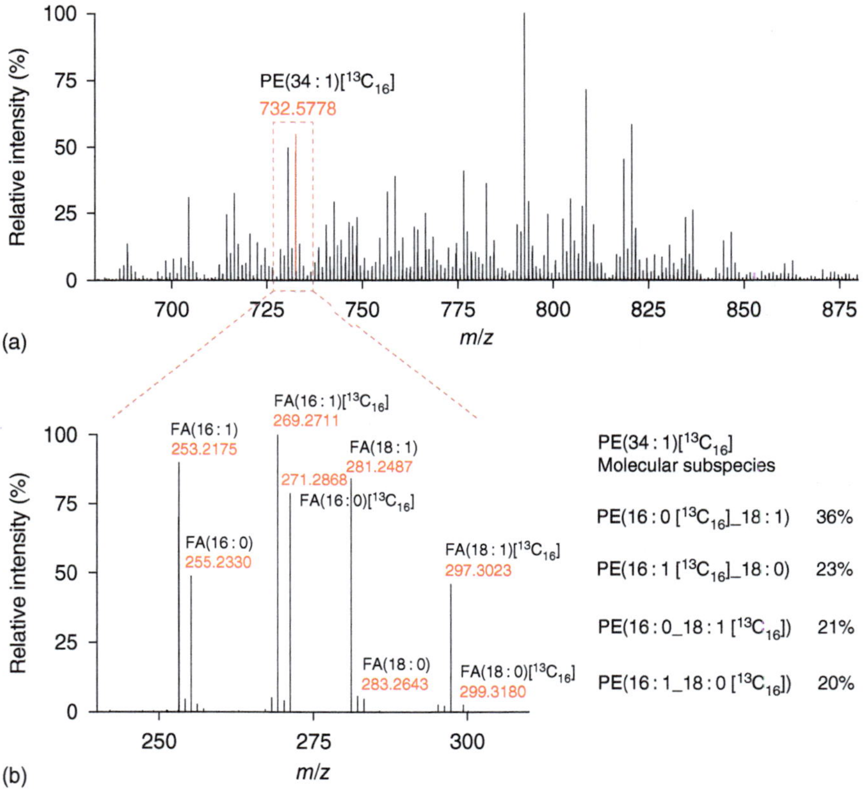

Figure 24.5 Tandem MS experiment of cultured cells treated with [U-^{13}C]-Palmitate. (a) High-resolution (Q-Exactive HF-X, R = 240 000) MS scan in negative-ion mode, zoomed in on the phospholipid region. The peak at 732.5778 could be identified as PE(34:1)[$^{13}C_{16}$] with an error of 0.74 ppm. (b) An MS/MS scan with higher-energy C-trap dissociation (HCD) reveals a complex labeling pattern of the acyl chains of PE(34:1)[$^{13}C_{16}$]. The labeled acyl chains could be traced to four different molecular subspecies.

shows MS/MS data of a [U-^{13}C]-palmitic acid tracing experiment. One of the major PE species that showed label incorporation was PE(34:1)[$^{13}C_{16}$]. Based on MS1 data only, one may have easily made the assumption that PE(34:1)[$^{13}C_{16}$] combines a labeled palmitic acid (the tracer) with an unlabeled oleic acid (18:1). The MS/MS data revealed however that only about a third of the PE(34:1)[$^{13}C_{16}$] molecules contained a labeled palmitic acid, while the remaining parts were accounted for by 16:1 [$^{13}C_{16}$], 18:0 [$^{13}C_{16}$], and 18:1 [$^{13}C_{16}$], which shows that the majority of the tracer underwent elongation and desaturation reactions prior to incorporation into this PE species (16:0 to 18:0 is catalyzed by ELOVL6 and 16:0 to 16:1 and 18:0 to 18:1 are catalyzed by SCD2). This example illustrates the wealth of extra information MS/MS data can provide in tracer experiments and highlights the need for lipidomics software that is capable of integrating MS1 and MS2 data of labeled spectra to enable identification of lipid isotopologues at the molecular species level.

24.6 Concluding Remarks

Stable isotope tracer lipidomics is a powerful tool to study the activity of the various lipid metabolic pathways. While targeted lipid headgroup tracing experiments are easily accessible, even on nominal mass instruments, untargeted tracing of polar precursors into a wide range of different lipid classes and species requires high-resolution accurate mass instruments. A great challenge that lies ahead is the bioinformatic processing of the raw data files thereof. While to some extent bioinformatic tools from the tracer metabolomics field could be leveraged, tracer lipidomics tools should incorporate rule-based lipid identifications that consider MS/MS data when available to identify lipid species at the appropriate level of detail. Moreover, the technical advances and increasing accessibility of matrix assisted laser desorption ionisation (MALDI)-Fourier-transform ion cyclotron resonance (FTICR) instruments is now opening the path to additionally study the spatial distribution of stable isotope tracers and their metabolic derivatives.

All these advancements will allow investigators to study lipid metabolism in unprecedented detail, undoubtedly greatly advancing our understanding of lipids in health and disease.

References

1 Athenstaedt, K. and Daum, G. (1999). Phosphatidic acid, a key intermediate in lipid metabolism. *Eur. J. Biochem.* 266 (1): 1–16.

2 Schoenheimer, R. and Rittenberg, D. (1936). Deuterium as an indicator in the study of intermediary metabolism: VI. Synthesis and destruction of fatty acids in the organism. *J. Biol. Chem.* 114: 381–396.

3 Postle, A.D. and Hunt, A.N. (2009). Dynamic lipidomics with stable isotope labelling. *J. Chromatogr. B Analyt. Technol. Biomed. Life Sci.* 877 (26): 2716–2721.

4 Robins, S.J. and Patton, G.M. (1986). Separation of phospholipid molecular species by high performance liquid chromatography: potentials for use in metabolic studies. *J. Lipid Res.* 27 (2): 131–139.

5 Buescher, J.M., Antoniewicz, M.R., Boros, L.G. et al. (2015). A roadmap for interpreting ^{13}C metabolite labeling patterns from cells. *Curr. Opin. Biotechnol.* 34: 189–201.

6 Lane, A.N., Fan, T.W.-M., Xie, Z. et al. (2009). Isotopomer analysis of lipid biosynthesis by high resolution mass spectrometry and NMR. *Anal. Chim. Acta* 651 (2): 201–208.

7 Fernandez, C.A., Des Rosiers, C., Previs, S.F. et al. (1996). Correction of ^{13}C mass isotopomer distributions for natural stable isotope abundance. *J. Mass Spectrom.* 31 (3): 255–262.

8 Moseley, H.N. (2010). Correcting for the effects of natural abundance in stable isotope resolved metabolomics experiments involving ultra-high resolution mass spectrometry. *BMC Bioinf.* 11 (1): 139.

9 Millard, P., Delépine, B., Guionnet, M. et al. (2019). IsoCor: isotope correction for high-resolution MS labeling experiments. *Bioinformatics* 35 (21): 4484–4487.

10 Stephanopoulos, G., Aristidou, A.A., and Nielsen, J. (1998). *Metabolic Engineering: Principles and Methodologies*, 749. Elsevier.

11 Wiechert, W. and Nöh, K. (2013). Isotopically non-stationary metabolic flux analysis: complex yet highly informative. *Curr. Opin. Biotechnol.* 24 (6): 979–986.

12 Goh, B., Kim, J., Seo, S., and Kim, T.-Y. (2018). High-throughput measurement of lipid turnover rates using partial metabolic heavy water labeling. *Anal. Chem.* 90 (11): 6509–6518.

13 Liebisch, G., Ahrends, R., Arita, M. et al. (2019). Lipidomics needs more standardization. *Nat. Metab.* 1 (8): 745–747.

14 Oliveira, A.F., Cunha, D.A., Ladriere, L. et al. (2015). in vitro use of free fatty acids bound to albumin: a comparison of protocols. *BioTechniques* 58 (5): 228–233.

15 Chu, F.-L.E., Lund, E., Soudant, P., and Harvey, E. (2002). De novo arachidonic acid synthesis in *Perkinsus marinus*, a protozoan parasite of the eastern oyster *Crassostrea virginica*. *Mol. Biochem. Parasitol.* 119 (2): 179–190.

16 Collins, J.M., Neville, M.J., Pinnick, K.E. et al. (2011). De novo lipogenesis in the differentiating human adipocyte can provide all fatty acids necessary for maturation. *J. Lipid Res.* 52 (9): 1683–1692.

17 Perez, C.L. and Van Gilst, M.R. (2008). A ^{13}C isotope labeling strategy reveals the influence of insulin signaling on lipogenesis in *C. elegans*. *Cell Metab.* 8 (3): 266–274.

18 Lorendeau, D., Rinaldi, G., Boon, R. et al. (2017). Dual loss of succinate dehydrogenase (SDH) and complex I activity is necessary to recapitulate the metabolic phenotype of SDH mutant tumors. *Metab. Eng.* 43 (Pt B): 187–197.

19 Xu, Y.-J. and Zhang, J. (2013). Determination of fatty acid methyl esters by GC–triple quadrupole MS using electron and chemical ionization. *Bioanalysis* 5 (12): 1527–1543.

20 Stuani, L., Riols, F., Millard, P. et al. (2018). Stable isotope labeling highlights enhanced fatty acid and lipid metabolism in human acute myeloid leukemia. *Int. J. Mol. Sci.* 19 (11): E3325.

21 Ecker, J. and Liebisch, G. (2014). Application of stable isotopes to investigate the metabolism of fatty acids, glycerophospholipid and sphingolipid species. *Prog. Lipid Res.* 54: 14–31.

22 Vance, J.E. (2015). Phospholipid synthesis and transport in mammalian cells. *Traffic* 16 (1): 1–18.

23 DeLong, C.J., Shen, Y.-J., Thomas, M.J., and Cui, Z. (1999). Molecular distinction of phosphatidylcholine synthesis between the CDP-choline pathway and phosphatidylethanolamine methylation pathway. *J. Biol. Chem.* 274 (42): 29683–29688.

24 Kainu, V., Hermansson, M., and Somerharju, P. (2008). Electrospray ionization mass spectrometry and exogenous heavy isotope-labeled lipid species provide detailed information on aminophospholipid acyl chain remodeling. *J. Biol. Chem.* 283 (6): 3676–3687.

25 Wang, B. and Tontonoz, P. (2019). Phospholipid remodeling in physiology and disease. *Annu. Rev. Physiol.* 81 (1): 165–188.

26 McLaren, D.G., He, T., Wang, S.-P. et al. (2011). The use of stable-isotopically labeled oleic acid to interrogate lipid assembly in vivo: assessing pharmacological effects in preclinical species. *J. Lipid Res.* 52 (6): 1150–1161.

27 Züllig, T. and Köfeler, H.C. (2021). High resolution mass spectrometry in lipidomics. *Mass Spectrom. Rev.* 40 (3): 162–176.

28 Höring, M., Ejsing, C.S., Krautbauer, S. et al. (2021). Accurate quantification of lipid species affected by isobaric overlap in Fourier-transform mass spectrometry. *J. Lipid Res.* 62: 100050.

29 Liebisch, G., Lieser, B., Rathenberg, J. et al. (2004). High-throughput quantification of phosphatidylcholine and sphingomyelin by electrospray ionization tandem mass spectrometry coupled with isotope correction algorithm. *Biochim. Biophys. Acta* 1686 (1, 2): 108–117.

30 Brügger, B., Erben, G., Sandhoff, R. et al. (1997). Quantitative analysis of biological membrane lipids at the low picomole level by nano-electrospray ionization tandem mass spectrometry. *Proc. Natl. Acad. Sci. U.S.A.* 94 (6): 2339–2344.

31 Li, J., Hoene, M., Zhao, X. et al. (2013). Stable isotope-assisted lipidomics combined with nontargeted isotopomer filtering, a tool to unravel the complex dynamics of lipid metabolism. *Anal. Chem.* 85 (9): 4651–4657.

25

Mass Spectrometry for Lipidomics: Methods and Applications – Aging and Alzheimer's Disease

Kevin Huynh, Habtamu B. Beyene, Tingting Wang, Corey Giles, and Peter J. Meikle

Baker Heart and Diabetes Institute, Metabolomics, Commercial Road, Melbourne 3004, Australia

25.1 Introduction

Over the past several decades, the average human life span has steadily increased. With an aging population, understanding the biological processes of natural and pathological aging will become a critical challenge for our health systems. Throughout life, an individual may experience and accumulate numerous metabolic perturbations, culminating in altered disease risk, particularly for cardiometabolic [1–3] and neuropathological [4, 5] diseases. While aging ultimately leads to a series of metabolic changes eventuating in death, a large variation in the trajectory of aging exists between individuals. A growing body of evidence suggests that metabolic health may reside at the heart of an individual's risk of age-related diseases. This chapter will summarize the interplay between metabolic health, aging, and neurodegenerative disorders, specifically Alzheimer's disease (AD); how age-related changes in metabolism can be examined across populations through lipidomic profiling; and the potential of targeting metabolic health to alter the trajectory of age-related diseases.

25.2 Diversity in the Aging Process

Aging is a complex process that involves the gradual and progressive loss in metabolic homeostasis. The exact processes through which aging increases the risk of age-related diseases are not fully understood, but various metabolic processes such cell senescence [6], mitochondrial failures [7], waning immune response [8], and/or systemic inflammation [9] are thought to be major causes among others.

Mass Spectrometry for Lipidomics: Methods and Applications, First Edition.
Edited by Michal Holčapek and Kim Ekroos.

Aging adversely affects all levels of structural organization: cells suffer from telomere attrition, epigenetic alterations, and mitochondrial dysfunction; tissues undergo substantiative remodeling and accumulate lipids, proteins, and other waste metabolites; and organs experience atrophy, decreases in flexibility, and infiltration of immune cells.

Superficial changes associated with aging, such as the appearance of wrinkles and graying of hair, show considerable differences between individuals and populations in the age of appearance and progression, illustrating the diversity in the aging process. While these superficial changes are easily observed, underlying metabolic changes are less visible and can have dire consequences for risk of disease. Before the onset of symptoms, aberrant metabolism and its sequelae can progress for many years without notice. Capturing metabolic markers of the aging process will enable the evaluation of pathological aging and, putatively, therapeutic targeting to reduce the burden of age-related diseases.

To explore the metabolic underpinnings of aging, studies examining large human population cohorts are necessary. One of the most common biospecimens collected in biobanks is peripheral blood because of its relative ease of collection, simplicity of storage, and long history of usefulness for biomarker discovery. Human blood provides a unique, low-risk biofluid to examine the metabolic changes in large human populations, enabling us to define changes seen in healthy or pathological aging, where the risk profile for age-related diseases can differ dramatically.

25.3 Using Lipidomics as a Tool to Examine the Diversity in Aging

Lipids are fundamental to every living system. These diverse and biologically important molecules comprise thousands of individual species, spanning multiple distinct – but structurally related – categories. A key attribute of lipids is their relative insolubility in water because of their hydrophobic structure and composition. In plasma, the majority of lipids are transported in association with lipoprotein particles, such as chylomicrons and high-density and low-density lipoproteins. Amphiphilic lipids envelope the surface of lipoproteins, providing a hydrophilic shell for interaction with the aqueous environment. Strongly non-polar lipids – principally cholesteryl esters and triacylglycerols – form the core of lipoproteins. In tissues, lipids contribute to every aspect of the cellular system, serving as structural components, energy stores, and essential signaling molecules. Cellular membranes function with highly structured micro- and macrocomposition. Lateral heterogeneity and leaflet asymmetry are essential for membrane curvature, cell signaling, and protein trafficking. Quantification and characterization of these distinct lipid molecules makes up the field known as "lipidomics."

With recent advances to high-performance liquid chromatography and mass spectrometry, it is now practicable to examine a large portion of the plasma lipidome. The potential high-throughput nature of these techniques enables large population and clinical studies to address complex biological questions.

Lipidomics covers a broad range of techniques that allow profiling of lipid and lipid-like molecules, typically performed via mass spectrometry, and is covered in previous chapters. A broad number of experimental setups have been detailed for mass spectrometry-based lipidomics, either in tandem with liquid chromatography, alternative separation techniques, or directly infused into the instrument [10]. To fully elucidate the metabolic changes associated with aging, the ability to cover a broad biochemical range of lipid classes, maintain a high-throughput, and achieve highly precise measurements is crucial.

The natural variation in human populations necessitate studies be conducted with large sample numbers to account for the numerous possible confounders when examining an outcome, particularly with age. Small studies have limited ability to control for confounding factors and provide only a limited snapshot of the links between phenotype and disease, limiting interpretation. While many lipidomic methodologies have been described, techniques that are amenable for high-throughput analysis and are suitable for large cohorts would be most useful to study human disease and anthropometry-related outcomes.

25.4 Age-Related Changes to the Plasma Lipidome

Leveraging off extensive plasma lipidomic profiles, changes to the lipidome associated with the aging process can be defined. In the Genetics of Lipid-Lowering Drugs and Diet Network study, Slade et al. identified significant changes to nearly all glycerophospholipid and sphingolipid classes in human serum (n = 980), but not cholesterol [11]. Similarly, in a study comparing young vs aged adults (n = 150), Carrard et al. identified elevations to plasma glycerophospholipids and sphingolipids in the aged group [12]. Using plasma lipidome profiles from the Australian Diabetes, Obesity and Lifestyle Study (AusDiab, n = 640), chronological age was found to strongly and independently associate with numerous lipid classes and species, with the largest effect sizes observed in acylcarnitines [13]. Following this study, the entire AusDiab baseline cohort [14] – comprising over 10 300 individuals – was profiled. This study remains one of the largest population studies profiled to date and has identified clear and distinct age-dependent lipid signatures that spanned nearly all lipid classes (472 of 706 species), independent of common clinical variables such as body mass index (BMI) and sex. Specifically, accumulation of long-chain acylcarnitines, deoxyceramides, and declines in majority of ether lipids were found to track with increased age. Sex-specific associations with aging were also observed, where strong non-linear inflections were observed in women at menopause.

The role of lipids in various mammalian animals and their relationship to expected life spans have been previously extensively reviewed [15] and highlight how differing lipid compositions correlate with longevity. In humans, peripheral lipid signatures for longevity can be linked with several subclasses including ether lipids [16, 17]. Conversely, risk of neurodegenerative disease was significantly linked with pathological aging [18]. Together, these studies highlight the utility of the plasma lipidome to delineate pathological and healthy aging.

25.5 Age Is the Biggest Risk Factor for Alzheimer's Disease

AD is one of the most common neurodegenerative diseases, affecting around 50 million people worldwide [19]. AD is characterized by the progressive death of neurons and loss of brain structure, initially presenting as memory loss and behavioral disturbances [20, 21]. The neuropathological hallmarks of AD include extracellular plaques composed of misfolded amyloid-beta (Aβ) protein, intracellular neurofibrillary tangles composed of hyperphosphorylated Tau, and the accumulation of lipid deposits in glial cells.

Relative to other common age-related diseases, late-onset AD shows the strongest association with increasing age. The prevalence of AD increases exponentially from 3% to 32% between 65 and 85 years of age [22]. In contrast to late-onset AD, which has no single etiology, familial (early-onset) AD is generally a consequence of genetic mutations in specific proteins. The near-exclusive prevalence of late-onset AD in only the elderly (typically >65 years old) highlights the important role of age in disease pathophysiology. However, while the mechanism through which ageing affects AD risk has not been fully elucidated, there have been several proposed hypothesizes (Figure 25.1).

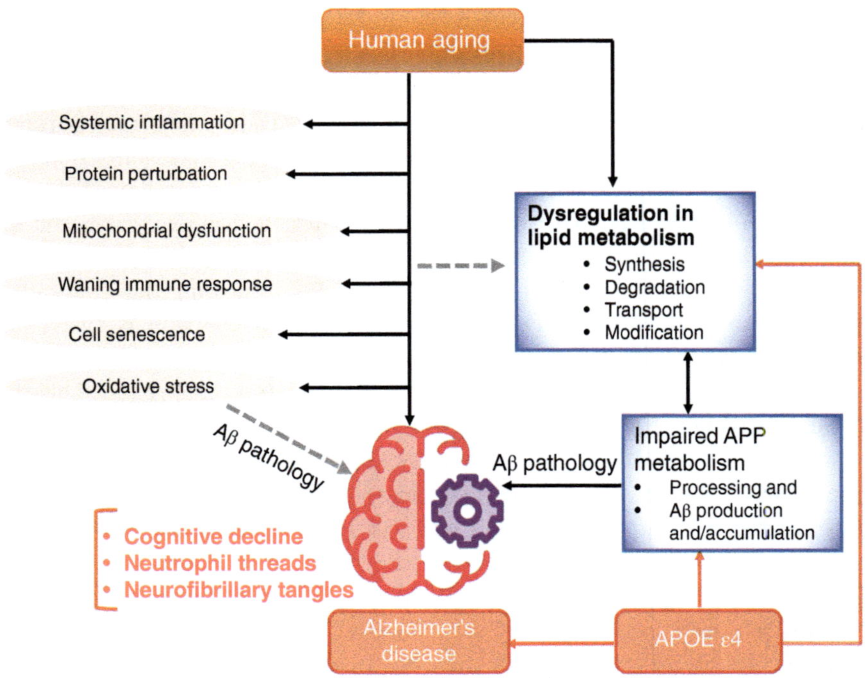

Figure 25.1 The link between ageing, lipid metabolism and Alzheimer's disease (AD). Aβ, β-amyloid peptide; APP, β-amyloid precursor protein.

The major genetic risk factor for late-onset AD is the apolipoprotein E gene (*APOE*). Polymorphisms in this gene were the first identified and remain the largest genetic risk factor in the development of AD [23, 24]. The encoded protein (Apolipoprotein E, APOE) has a well-established role in lipoprotein transport and metabolism, supporting the idea that lipid metabolism plays an important role in AD etiology. The mechanism through which *APOE* influences late-onset AD risk and progression remains to be fully determined. There are two major alleles of interest, the ε4 allele increases risk for AD, whereas the risk of AD is lower with the ε2 allele. In addition to heightened risk of AD development, individuals with the ε4 allele are much younger when symptoms of AD become evident [25]. Polymorphisms in *APOE* have been linked to altered risk of cardiovascular disease and overall life span [26–28], with the ε2 allele particularly associated with extreme longevity (>100 years old), independent of AD [29, 30]. More recently, it has been identified that the APOE ε2 allele may mediate resilience via changes to the peripheral lipidome, particularly through the modulation of ether lipids [31].

25.6 Interplay Between Lipids and Alzheimer's Disease

Lipids play important roles in many tissues, but this is nowhere more epitomized than in the human brain. Over 50% of the dry weight of the human brain is composed of lipids [32], with a large fraction forming the myelin sheath around neuronal axons. Increasingly, many studies have demonstrated an intimate link between lipid profiles and AD pathology. Post mortem and cross-sectional studies have shown that numerous lipid classes and subclasses are altered in AD, notably phospholipids [33–35], plasmalogens [36], ceramides [37], gangliosides [38], and sulfatides.

Several studies have explored whether blood-based biomarker signatures are able to differentiate individuals who have or ultimately develop AD. Mapstone et al. [35] performed a 5-year follow-up study on a defined small group of subjects. A panel of 10 lipids were observed to differentiate pre-converter and control groups. Similarly, Orešič et al. [39] developed a four-lipid metabolite panel with age and APOE that differentiated between AD patients and controls. A three-lipid panel was subsequently used to classify the likelihood of individuals with mild-cognitive impairment (MCI) progressing to AD. Liu et al. [40] demonstrated the ability of lipid species from various classes, particularly triacylglycerols, cholesteryl esters, and sphingomyelins, to discriminate AD from controls. Using two large prospective studies of AD [41], Huynh et al. identified lipid species showing concordant associations with both prevalent and incident AD, independent of traditional clinical risk factors such as age and APOE. Subsequent modeling identified improved classification for both prevalent and incident AD over conventional clinical markers. Nho et al. highlighted serum lipid signatures, specifically bile acids that associate with multiple aspects of AD pathology, including cerebral spinal fluid measures of amyloid, phosphorylated tau, and markers of neurodegeneration [42]. While many of these are smaller studies, together they highlight that an altered peripheral lipid metabolism is observable in AD and is potentially perturbed before disease onset.

25.7 Concept of Chronological and Metabolic Age

"Chronological age" is a term used to define an individual's age based solely on the time since birth. "Metabolic age," on the other hand, attempts to quantify an individual's metabolic state, providing a metric on how those individuals are compared to metabolic changes expected with the normal aging process. Typically, metabolic age is normalized to the population such that the mean is aligned with the chronological age and a higher or lower metabolic age defined an individual's position relative to the population Figure 25.2. Lifelong and transient metabolic perturbations can be driven by a combination of genetics, diet, lifestyle, and other environmental factors/exposures. An individual with an elevated metabolic age – compared to their chronological age – will have increased risk for age-related disease. Importantly, an individual's metabolic age is not a fixed trajectory, such as chronological age. Rather, through appropriate interventions, it should be possible to alter an individual's metabolic age and therefore potentially reducing one's risk of age-related disease.

Several attempts have been made to derive an individual's metabolic age. Typically produced from high-dimensional data, metabolic age has been modeled using DNA

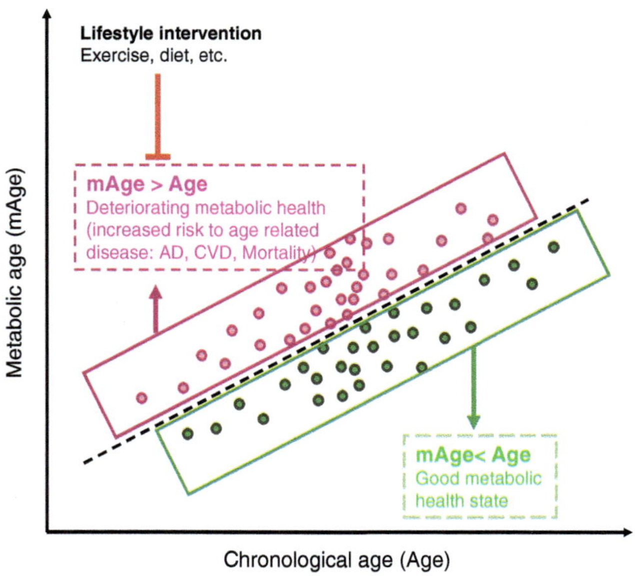

Figure 25.2 The concept of metabolic age (mAge)

methylation profiles [43], telomerase length [44, 45], transcriptomic profiles [46], proteome [47], metabolome [48–52], transcriptomic profiles [53], electrocardiogram traces [54], multi-omics signatures [55], composite measures of telomerase length and epigenetic clock [56], and other biomarkers indexing cardiometabolic, renal, inflammatory/immune, urinary, pulmonary, and hepatic systems [57–59]. Many of these studies have utilized the information generated from blood. However, tissue-specific derivatives of metabolic age have also been proposed, such as a brain-specific metabolic age estimated using magnetic resonance imaging, for risk assessment of AD and dementia [60–62].

25.8 Development and Application of a Lipidomic Metabolic Age Score: The Next Steps

Despite an enormous investment into drug development and disease-modifying treatments, AD clinical trials have repeatedly not met their endpoints in reducing cognitive decline. Several hypotheses have been proposed for these failures, with the major hypothesis involving the late timing of the intervention. It has long been known that neurodegeneration, the formation of amyloid plaques, and accumulation of neurofibrillary tau tangles, start many years before symptom onset [63]. The symptomless nature of the prodromal stage of AD, coupled with high invasiveness of assays for measuring brain amyloid plaque burden, prevents the wide-spread screening of pre-symptomatic AD.

With the high-throughput nature of lipidomics in conjunction with the strong relationship between aging, AD, and lipid metabolism, a metabolic age derived from lipidomic profiling is an attractive approach that would yield an informative tool for the identification of high-risk individuals. In a hypothetical situation, a snapshot of the lipidome can be captured following routine clinical screening for other risk factors such as high density lipoprotein (HDL) and low density lipoprotein (LDL). Mass spectrometry-based assays are currently used for drug screening; thus, the technology exists for lipidomic screening in non-research environments.

The development of a clinically viable metabolic age score is still in its infancy. The main conceptual underpinning of a metabolic age score is to identify individuals who deviate from the norm. Derivation of metabolic age can follow numerous methodologies, but typically fitting linear multivariable models and predicting chronological age is the most common approach, but more complex approaches are becoming available with improvements to machine learning approaches. Deviations from the line of best fit (residuals) are interpreted as an indication of a difference between chronological and metabolic age, i.e. an individual with a positive residual would have a metabolic age greater than their chronological age. Subsequent transformations of this score can result in a more easily interpretable value that can dictate the direction and magnitude of risk reduction interventions.

As previously described, the large diversity in the human population necessitates analysis in multiple populations to ensure the robustness of the approach in capturing deviations from normal aging. Validation of the statistical models is an important, yet underappreciated component of developing metabolic age scores to ensure that the approach is easily translatable to newly measured samples. This requires large samples that are independent of the study used to generate the model.

To demonstrate whether the metabolic age truly captures the underlying differences in metabolic state related to age, several approaches can be used for validation. Comparisons with known interventions that are associated with increased longevity and/or decreased risk of age-related diseases [64] is one such approach. Hypothetically, interventions that are protective of age-related diseases should show a concordant decrease in an individual's metabolic age; similarly, poor lifestyle choices associated with age-related diseases would result in an elevated score.

While tremendous progress has been made, we are still some years out from widespread adoption of a mass spectrometry-based clinical lipidomics platform. The foundations are being set, with several large consortia's and research initiatives exploring this concept [65–70].

25.9 Conclusion

The scalability of mass spectrometry assays, particularly lipidomics, coupled with the large amount of biological information captured in such a high-throughput manner, provides the opportunity to truly capture the diversity in the human aging process. With the prevalence of age-related diseases expected to rise dramatically in the coming decades, approaches such as the development of a metabolic age score are needed to provide a means for identifying high-risk individuals for early intervention in reducing age-related disease and supporting healthy aging.

References

1 Lakatta, E.G. (2002). Age-associated cardiovascular changes in health: impact on cardiovascular disease in older persons. *Heart Fail. Rev.* 7 (1): 29–49.

2 Rodgers, J.L., Jones, J., Bolleddu, S.I. et al. (2019). Cardiovascular risks associated with gender and aging. *J. Cardiovasc. Dev. Dis.* 6 (2): 19.

3 Tang, Z., Wang, C., Song, X. et al. (2013). Co-occurrence of cardiometabolic diseases and frailty in older Chinese adults in the Beijing longitudinal study of ageing. *Age Ageing* 42 (3): 346–351.

4 Hebert, L.E., Scherr, P.A., Beckett, L.A. et al. (1995). Age-specific incidence of Alzheimer's disease in a community population. *JAMA* 273 (17): 1354–1359.

5 Jakovcevski, M. and Akbarian, S. (2012). Epigenetic mechanisms in neurological disease. *Nat. Med.* 18 (8): 1194–1204.

6 Campisi, J. and Robert, L. (2014). Cell senescence: role in aging and age-related diseases. *Interdiscip. Top. Gerontol.* 39: 45–61.

7 Sastre, J., Pallardó, F.V., de la Asunción, G., J. et al. (2000). Mitochondria, oxidative stress and aging. *Free Radic. Res.* 32 (3): 189–198.

8 Woodland, D.L. and Blackman, M.A. (2006). Immunity and age: living in the past? *Trends Immunol.* 27 (7): 303–307.

9 Pawelec, G., Goldeck, D., and Derhovanessian, E. (2014). Inflammation, ageing and chronic disease. *Curr. Opin. Immunol.* 29: 23–28.

10 Yang, K. and Han, X. (2016). Lipidomics: techniques, applications, and outcomes related to biomedical sciences. *Trends Biochem. Sci* 41 (11): 954–969.

11 Slade, E., Irvin, M.R., Xie, K. et al. (2021). Age and sex are associated with the plasma lipidome: findings from the GOLDN study. *Lipids Health Dis.* 20 (1): 1–12.

12 Carrard, J., Gallart-Ayala, H., Infanger, D. et al. (2021). Metabolic view on human healthspan: a lipidome-wide association study. *Metabolites* 11 (5): 287.

13 Huynh, K., Barlow, C.K., Jayawardana, K.S. et al. (2019). High-throughput plasma lipidomics: detailed mapping of the associations with cardiometabolic risk factors. *Cell Chem. Biol.* 26 (1): 71–84.e4.

14 Beyene, H.B., Olshansky, G., AAT, S. et al. (2020). High-coverage plasma lipidomics reveals novel sex-specific lipidomic fingerprints of age and BMI: evidence from two large population cohort studies. *PLoS Biol.* 18 (9): e3000870.

15 Jové, M., Mota-Martorell, N., Pradas, I. et al. (2020). The lipidome fingerprint of longevity. *Molecules* 25 (18): 4343.

16 Gonzalez-Covarrubias, V., Beekman, M., Uh, H.W. et al. (2013). Lipidomics of familial longevity. *Aging Cell* 12 (3): 426–434.

17 Pradas, I., Jové, M., Huynh, K. et al. (2019). Exceptional human longevity is associated with a specific plasma phenotype of ether lipids. *Redox Biol.* 21: 101127.

18 Lefèvre-Arbogast, S., Hejblum, B.P., Helmer, C. et al. (2021). Early signature in the blood lipidome associated with subsequent cognitive decline in the elderly: a case-control analysis nested within the Three-City cohort study. *EBioMedicine* 64: 103216.

19 Breijyeh, Z. and Karaman, R. (2020). Comprehensive review on Alzheimer's disease: causes and treatment. *Molecules* 25 (24): 5789.

20 Niccoli, T. and Partridge, L. (2012). Ageing as a risk factor for disease. *Curr. Biol.* 22 (17): R741–R752.

21 Scheltens, P. et al. (2021). Alzheimer's disease. *Lancet* 397 (10284): 1577–1590.

22 Alzheimer's Association (2019). 2019 Alzheimer's disease facts and figures. *Alzheimers Dement.* 15 (3): 321–387.

23 Farrer, L.A., Cupples, L.A., Haines, J.L. et al. (1997). Effects of age, sex, and ethnicity on the association between apolipoprotein E genotype and Alzheimer disease: a meta-analysis. *JAMA* 278 (16): 1349–1356.

24 Corder, E.H., Saunders, A.M., Strittmatter, W.J. et al. (1993). Gene dose of apolipoprotein E type 4 allele and the risk of Alzheimer's disease in late onset families. *Science* 261 (5123): 921–923.

25 Blacker, D., Haines, J., Rodes, L. et al. (1997). ApoE-4 and age at onset of Alzheimer's disease: the NIMH genetics initiative. *Neurology* 48 (1): 139–147.

26 Deelen, J., Beekman, M., Uh, H.W. et al. (2011). Genome-wide association study identifies a single major locus contributing to survival into old age; the APOE locus revisited. *Aging Cell* 10 (4): 686–698.

27 Ewbank, D.C. (2007). Differences in the association between apolipoprotein E genotype and mortality across populations. *J. Gerontol. A Biol. Sci. Med. Sci.* 62 (8): 899–907.

28 Gerdes, L.U., Jeune, B., Ranberg, K.A. et al. (2000). Estimation of apolipoprotein E genotype-specific relative mortality risks from the distribution of genotypes in centenarians and middle-aged men: apolipoprotein E gene is a "frailty gene," not a "longevity gene". *Genet. Epidemiol.* 19 (3): 202–210.

29 Schächter, F., Faure-Delanef, L., Guénot, F. et al. (1994). Genetic associations with human longevity at the APOE and ACE loci. *Nat. Genet.* 6 (1): 29–32.

30 Shinohara, M., Kanekiyo, T., Tachibana, M. et al. (2020). APOE2 is associated with longevity independent of Alzheimer's disease. *eLife* 9: e62199.

31 Wang, T., Huynh, K., Giles, C. et al. (2022). APOE ε2 resilience for Alzheimer's disease is mediated by plasma lipid species: analysis of three independent cohort studies. *Alzheimers Dement.* PMID: 35077012 https://doi.org/10.1002/alz.12538.

32 Bazenet, C. and Lovestone, S. (2012). Plasma biomarkers for Alzheimer's disease: much needed but tough to find. *Biomark. Med.* 6 (4): 441–454.

33 Goodenowe, D.B., Cook, L.L., Liu, J. et al. (2007). Peripheral ethanolamine plasmalogen deficiency: a logical causative factor in Alzheimer's disease and dementia. *J. Lipid Res.* 48 (11): 2485–2498.

34 Igarashi, M., Ma, K., Gao, F. et al. (2011). Disturbed choline plasmalogen and phospholipid fatty acid concentrations in Alzheimer's disease prefrontal cortex. *J. Alzheimers Dis.* 24 (3): 507–517.

35 Mapstone, M., Cheema, A.K., Fiandaca, M.S. et al. (2014). Plasma phospholipids identify antecedent memory impairment in older adults. *Nat. Med.* 20 (4): 415–418.

36 Han, X., Holtzman, D.M., and McKeel, D.W. Jr. (2001). Plasmalogen deficiency in early Alzheimer's disease subjects and in animal models: molecular characterization using electrospray ionization mass spectrometry. *J. Neurochem.* 77 (4): 1168–1180.

37 Cheng, H., Zhou, Y., Holtzman, D.M. et al. (2010). Apolipoprotein E mediates sulfatide depletion in animal models of Alzheimer's disease. *Neurobiol. Aging* 31 (7): 1188–1196.

38 Yamamoto, N., Igbabvoa, U., Shimada, Y. et al. (2004). Accelerated Abeta aggregation in the presence of GM1-ganglioside-accumulated synaptosomes of aged apoE4-knock-in mouse brain. *FEBS Lett.* 569 (1–3): 135–139.

39 Orešič, M., Hyötyläinen, T., Herukka, S. et al. (2011). Metabolome in progression to Alzheimer's disease. *Transl. Psychiatry* 1 (12): e57.

40 Liu, Y., Thalamuthu, A., Mather, K.A. et al. (2021). Plasma lipidome is dysregulated in Alzheimer's disease and is associated with disease risk genes. *Transl. Psychiatry* 11 (1): 344.

41 Huynh, K., WLF, L., Giles, C. et al. (2020). Concordant peripheral lipidome signatures in two large clinical studies of Alzheimer's disease. *Nat. Commun.* 11 (1): 5698.

42 Nho, K., Kueider-Paisley, A., Mahmoudian Dehkordi, S. et al. (2019). Altered bile acid profile in mild cognitive impairment and Alzheimer's disease: relationship to neuroimaging and CSF biomarkers. *Alzheimers Dement.* 15 (2): 232–244.

43 Hannum, G., Guinney, J., Zhao, L. et al. (2013). Genome-wide methylation profiles reveal quantitative views of human aging rates. *Mol. Cell* 49 (2): 359–367.

44 Khalangot, M., Krasnienkov, D., and Vaiserman, A. (2020). Telomere length in different metabolic categories: clinical associations and modification potential. *Exp. Biol. Med. (Maywood)* 245 (13): 1115–1121.

45 Peters, M.J., Joehanes, R., Pilling, L.C. et al. (2015). The transcriptional landscape of age in human peripheral blood. *Nat. Commun.* 6: 8570–8570.

46 Sanders, J.L. and Newman, A.B. (2013). Telomere length in epidemiology: a biomarker of aging, age-related disease, both, or neither? *Epidemiol. Rev.* 35 (1): 112–131.

47 Menni, C., Kiddle, S.J., Mangino, M. et al. (2014). Circulating proteomic signatures of chronological age. *J. Gerontol. A Biol. Sci. Med. Sci.* 70 (7): 809–816.

48 Tomás-Loba, A., Bernardes de Jesus, B., Mato, J.M. et al. (2013). A metabolic signature predicts biological age in mice. *Aging Cell* 12 (1): 93–101.

49 Hertel, J., Friedrich, N., Wittfeld, K. et al. (2016). Measuring biological age via metabonomics: the metabolic age score. *J. Proteome Res.* 15 (2): 400–410.

50 van den Akker, E.B., Trompet, S., JJH, B.W. et al. (2020). Metabolic age based on the BBMRI-NL ^{1}H-NMR metabolomics repository as biomarker of age-related disease. *Circ. Genom. Precis. Med.* 13 (5): 541–547.

51 Bucaciuc Mracica, T., Anghel, A., Ion, C.F. et al. (2020). MetaboAge DB: a repository of known ageing-related changes in the human metabolome. *Biogerontology* 21 (6): 763–771.

52 Davies, S.K., Bundy, J.G., and Leroi, A.M. (2015). Metabolic youth in middle age: predicting aging in *Caenorhabditis elegans* using metabolomics. *J. Proteome Res.* 14 (11): 4603–4609.

53 Wu, J.W., Yaqub, A., Ma, Y. et al. (2021). Biological age in healthy elderly predicts aging-related diseases including dementia. *Sci. Rep.* 11 (1): 15929–15929.

54 Lima, E.M., Ribeiro, A.H., GMM, P. et al. (2021). Deep neural network-estimated electrocardiographic age as a mortality predictor. *Nat. Commun.* 12 (1): 5117.

55 Zierer, J., Pallister, T., Tsai, P.C. et al. (2016). Exploring the molecular basis of age-related disease comorbidities using a multi-omics graphical model. *Sci. Rep.* 6: 37646.

56 Belsky, D.W., Moffitt, T.E., Cohen, A.A. et al. (2017). Eleven telomere, epigenetic clock, and biomarker-composite quantifications of biological aging: do they measure the same thing? *Am. J. Epidemiol.* 187 (6): 1220–1230.

57 Belsky, D.W., Caspi, A., Houts, R. et al. (2015). Quantification of biological aging in young adults. *Proc. Natl. Acad. Sci. U.S.A.* 112 (30): E4104–E4110.

58 Bae, C.-Y., Piao, M., Kim, M. et al. (2021). Biological age and lifestyle in the diagnosis of metabolic syndrome: the NHIS health screening data, 2014–2015. *Sci. Rep.* 11 (1): 444.

59 Elliott, M.L., Caspi, A., Houts, R.M. et al. (2021). Disparities in the pace of biological aging among midlife adults of the same chronological age have implications for future frailty risk and policy. *Nat. Aging* 1 (3): 295–308.

60 Wrigglesworth, J., Yaacob, N., Ward, P. et al. (2022). Brain-predicted age difference is associated with cognitive processing in later-life. *Neurobiol. Aging* 109: 195–203.

61 Boyle, R., Jollans, L., Rueda-Delgado, L.M. et al. (2021). Brain-predicted age difference score is related to specific cognitive functions: a multi-site replication analysis. *Brain Imaging Behav.* 15 (1): 327–345.

62 Wang, J., Knol, M.J., Tiulpin, A. et al. (2019). Gray matter age prediction as a biomarker for risk of dementia. *Proc. Natl. Acad. Sci. U.S.A.* 116 (42): 21213–21218.

63 Sperling, R., Mormino, E., and Johnson, K. (2014). The evolution of preclinical Alzheimer's disease: implications for prevention trials. *Neuron* 84 (3): 608–622.

64 Lee, M.B., Hill, C.M., Bitto, A. et al. (2021). Antiaging diets: separating fact from fiction. *Science* 374 (6570): eabe7365.

65 Meikle, T.G., Huynh, K., Giles, C. et al. (2021). Clinical lipidomics: realizing the potential of lipid profiling. *J. Lipid Res.* 62: 100127.

66 Hyötyläinen, T. and Orešič, M. (2015). Optimizing the lipidomics workflow for clinical studies – practical considerations. *Anal. Bioanal. Chem.* 407 (17): 4973–4993.

67 Wood, P.L. (2014). Mass spectrometry strategies for clinical metabolomics and lipidomics in psychiatry, neurology, and neuro-oncology. *Neuropsychopharmacology* 39 (1): 24–33.

68 Triebl, A., Burla, B., Selvalatchmanan, J. et al. (2020). Shared reference materials harmonize lipidomics across MS-based detection platforms and laboratories. *J. Lipid Res.* 61 (1): 105–115.

69 Burla, B., Arita, M., Arita, M. et al. (2018). MS-based lipidomics of human blood plasma: a community-initiated position paper to develop accepted guidelines. *J. Lipid Res.* 59 (10): 2001–2017.

70 Bowden, J.A., Heckert, A., Ulmer, C.Z. et al. (2017). Harmonizing lipidomics: NIST interlaboratory comparison exercise for lipidomics using SRM 1950-metabolites in frozen human plasma. *J. Lipid Res.* 58 (12): 2275–2288.

26

Lipidomics in Cell Biology

Noemi Jiménez-Rojo, Fabrizio Vacca, and Howard Riezman

University of Geneva, NCCR Chemical Biology, Department of Biochemistry, 30 quai E.Ansermet, 1211, Geneva 4, Switzerland

Lipids are essential cellular components making up the membrane barriers that define the exterior and interior of the cell and of its internal compartments. Although it would be theoretically possible to make a lipid bilayer fulfilling barrier functions with only one glycerophospholipid (GPL), nature has evolved diverse and numerous different lipid species. This most likely reflects the many other specific membrane functions in solute transport, signaling, adherence, fission, and fusion. Not only are specific lipids directly required for membrane protein function and protein recruitment to membranes, but they are also indirectly involved through their effects on membrane biophysical properties such as tension, bending rigidity, and viscosity. Therefore, the cell needs to control its lipid composition and its intracellular distribution in order to fulfill all of these functions. The difficulty of this task comes also from the fact that lipid composition can change depending on the external environment (medium or diet) and has also been shown to change according to circadian rhythm and age. Therefore, cells need to be able to detect changes in membrane properties and adjust the lipid composition of their organelles accordingly but still, very little is known about regulatory mechanisms that allow lipid composition sensing and fine tuning. Nevertheless, we have learnt a lot in the last decades about particular lipid signatures relevant for subcellular structures. In this chapter we will revisit those studies in which lipidomic analysis has helped to understand basic lipid cell biology and intracellular trafficking events, including a future perspective on where the field is heading.

26.1 Lipid Composition of Organelles

Thirteen years have passed since the seminal review "Membrane lipids: where they are and how they behave" by van Meer et al. [1]. Afterwards and, to a great extent, thanks to advances in the lipidomic field, other studies and reviews have been

Mass Spectrometry for Lipidomics: Methods and Applications, First Edition.
Edited by Michal Holčapek and Kim Ekroos.

published trying to explain the heterogeneous distribution of lipids within the eukaryotic cell but, despite the efforts, fundamental questions of lipid cell biology remain still unknown [2–4]. The main reason for this has been the technical challenge that isolating pure organelles entails. All of the techniques for determination of the total lipid composition of subcellular organelles depend indeed upon their isolation, extraction and analysis by mass spectrometry. For this, various techniques have been used, including differential centrifugation, gradient ultracentrifugation, immunoisolation, or combinations of these. We know quite a bit about the lipid composition of the plasma membrane (PM) of red blood cells because it is essentially the only membrane of the cells. In more complex cells there are lipidomics data for various subcellular compartments including nuclei, mitochondria, endoplasmic reticulum (ER), Golgi compartment, endosomes, and the plasma membrane (PM). As part of the LIPID MAPS project, lipidomes of nuclei, mitochondria, ER and PM were determined in macrophages in basal and activated states [5]. Differential and gradient centrifugation methods were used. As seen in many other studies, the amounts of sterols and sphingolipid were highly enriched in the plasma membrane when compared to the ER. Even lower amounts were found in mitochondria and the nucleus. Their work confirmed the idea of organelle-specific lipid composition. They found that the two most abundant GPL in mitochondria were (phosphatidylcholine) PC (36:2) and PC (36:1), whereas in the ER they were phosphatidylethanolamine (PE) (36:2) and ether-linked PE (38:4), and in the PM phosphatidylserine (PS) (36:1) and PC (36:1). In general, one notes a higher degree of GPL saturation in the PM than in the ER (Figure 26.1). While it is often mentioned that there is a gradient of saturation as well as cholesterol concentration along the secretory pathway there are few data to support this. Experimental data with clean isolation of the various compartments together with careful quantitative analysis are lacking. The importance of the methods used for organelle isolation is nicely illustrated in a study using different techniques to isolate mitochondria [6], which is one of the easier organelles to purify. Comparing differential centrifugation with gradient ultracentrifugation and immunoisolation they concluded that the gradient ultracentrifugation yields the most reliable results. Systematic studies like this are sorely needed to investigate the lipidomes of other subcellular organelles. In some cases, immunoisolation may be the only method of choice due to very similar properties of several organelles in terms of density and sedimentation speed.

When comparing organelle compositions between cell types, as a function of environmental conditions or genotype, one must also take into account that the lipidome is affected by the diet of the animal or the composition of the medium, time of culture and confluency of the tissue culture cells [7]. Lipidomes are inherently flexible as shown by studies in mice varying diet and genotype [8]. Furthermore, lipid composition varies in animals and in tissue culture according to circadian rhythm and age in complex ways [9–11]. Circadian changes even occur at the subcellular level, in mitochondria in particular [12]. Therefore, understanding the lipid composition and its function in subcellular organelles will require carefully executed studies with controlled environments and rigorous organelle purification and determination of purity/contaminants, which has not been achieved yet.

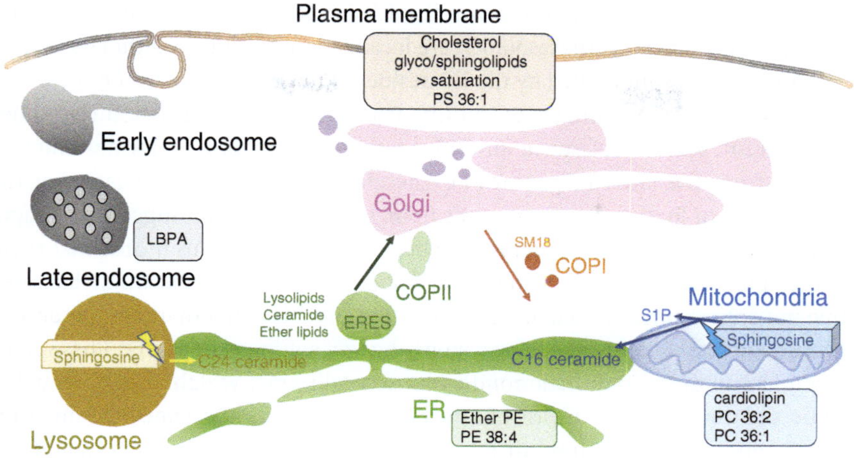

Figure 26.1 Lipidomics in cell biology. Lipid composition of organelles as learned through approaches that include lipidomics analysis. The concept of metabolic bias is represented by a different metabolic fate of "caged" sphingosine released either at the lysosomes or at the mitochondria. ER, endoplasmic reticulum; ERES, endoplasmic reticulum exit sites.

These studies would be important additions to our knowledge of lipid metabolism and function. Many studies use lipid binding protein domains, toxins, or other molecules that bind lipids to localize these molecules in cells. While useful to address some questions, these tools have not been shown to be quantitative and in most cases the precise conditions and context for binding are not understood [13–16].

26.1.1 Metabolic Bias Depending Upon Subcellular Location

One of the important reasons why we need to know the subcellular distribution of lipids to understand how they are transported within cells is because recent experiments show that the initial fates of lipid precursors depend upon their subcellular location. For the moment, the clearest examples of this come from chemical biology studies on the fate of photocaged sphingosine, which has been either delivered to the whole cell or specifically targeted to mitochondria and lysosomes [17, 18]. Upon photo-release, the metabolism of isotope-labeled sphingosine could be followed by lipidomics using mass spectrometry. The metabolism was strikingly different depending upon where the sphingosine was released. If it was released in the mitochondria or globally it could be quickly phosphorylated to S1P (sphingosine-1-phosphate) by sphingosine kinases, which are localized in the ER and some other compartments, including the mitochondria [19]. Interestingly, no S1P was detected after release from lysosomes. These results teach us a fundamental point about the link between lipid metabolism and trafficking. Indeed, the metabolic fate of lipids very often depends on the accessibility to enzymes and other metabolic intermediates, which is strictly dependent on subcellular localization. In this example, the sphingosine pool from lysosomes does not gain access to sphingosine kinases like the globally or mitochondria-released pool. Moreover, we know that the lysosome-released sphingosine reaches the ER because it is converted into ceramide, by the ceramide synthases which are localized there, but nevertheless, it does not seem to be phosphorylated by the ER-resident kinases. Strikingly, the species of ceramides made from sphingosine coming from the mitochondria and lysosomes were different, with a preference for C16 ceramide made from the former and C24 ceramide from the latter (Figure 26.1). Therefore, these results indicate that the metabolic fate of lipids strongly depends upon their location and suggests that lipid transport is a process that is more hard-wired than diffusive. This means that there is a metabolic bias in lipid metabolism that depends upon subcellular location. While not entirely unexpected, the extent of this effect is still remarkable, as shown by the undetectable conversion of lysosome-derived sphingosine to S1P. These types of chemical tools in conjunction with photocrosslinking, which allow organelle-specific and time-resolved release of lipids in cells, show great promise for uncovering the mechanisms of lipid transport and localization-dependent metabolic pathways [20].

26.1.2 Correlation Between Lipid Composition and Membrane Biophysical Properties

As mentioned above, organellar identity is defined, among other aspects, by a particular lipid composition. This distinct lipid stoichiometry, in turn, establishes specific membrane organization patterns that potentially evolved to support specialized functions. A recent example is the effect of membrane lipid composition and viscosity in regulating energy production by the electron transport chain in mitochondria [21]. Organelles are relatively independent and large entities that have distinct lipid requirements in time and space to perform their various cellular and molecular functions. A major example for this paradigm is the ER that supports crucial processes such as lipid synthesis, protein translocation, and export through vesicular transport and offers at the same time sites for lipid droplet nucleation and formation. Many of these functions have been shown to need membranes with different properties, including bilayer thickness, fluidity, and curvature among others [22–24]. However, technical advances are needed in the field to be able to address these questions *in vivo*. Along these lines, a recent study shows how the role of a particular ceramide species at ER exit sites allows the selective export of a particular set of proteins, with a glycosylphosphatidylinositol-anchored protein (GPI-AP) in yeast [25]. A complementary study indicates a similar mechanism of action for ether lipids in mammalian cells and lysolipids have been found to facilitate COPII vesicle formation (Figure 26.1) [23, 26]. All these three types of lipids have a very particular membrane behavior due to their special physico-chemical characteristics. Ceramides have been found to phase separate from other lipids and to permeabilize membranes [27, 28]. They are also able to induce membrane curvature and the flip-flop of surrounding lipids, and thus these molecules can be potent disruptors of membrane structure when present at high concentrations [29–31]. This could explain, at least in part, the toxicity induced by the accumulation of certain ceramide species in cells such as 1-deoxyceramides that have been associated with neurodegeneration and metabolic disease [32, 33]. More detailed information on the biophysical properties of sphingolipids is available [27, 34–38]. Lysolipids are also able to induce membrane curvature due to their inverted conical shape, and they have been also shown to decrease the bending rigidity which could explain their enrichment in ER/endocytic sites where membrane bending is needed [23, 39]. Ether lipids, a lipid subclass comprised by both alkyl and alkenyl ether lipids, show a membrane behavior highly dependent on their headgroup. PE plasmalogens, for example, are very well known to stabilize hexagonal nonlamellar phases decreasing the phase transition temperatures to around 38 °C (lamellar to hexagonal). More recently ether PC has been shown to decrease membrane packing in model membranes. A more detailed description about the physico-chemical properties of ether lipids is available [40–43]. However, the mechanisms by which these lipids are enriched in certain membrane subdomains in order to exert their effects still remain elusive due to technical limitations. Probably, specific lipid–protein interactions within the membrane drive this enrichment of lipids as it has been proposed for sphingolipids and ether lipids to support protein secretion and recycling of cargo receptors [26, 44].

Another degree of complexity arises from trying to understand the contribution of individual lipid species to membrane fitness *in vivo*. The difficulty to track lipid molecules using microscopy and the limitation of biophysical techniques mostly for *in vitro* studies have set up a gap between the fields of membrane biophysics and the cell biology and physiology communities. For decades we have been able to analyze the properties of lipids in reconstituted membrane systems, which has greatly advanced our knowledge, for instance, on lipid–lipid interactions, melting temperatures, and, their effect on membrane packing. Nevertheless, as an example of the limitations of these studies, it is still not completely established how melting temperatures and packing can be correlated and how we can extrapolate those results to the behavior of cellular membranes *in vivo*. Studies trying to correlate changes in lipid composition with changes in membrane packing and physiological responses *in cellulo* are scarce [26, 45–47]. But, why is this important? As mentioned below, structure and biophysical properties of organellar membranes are dictated, to a great extent, by the chemical diversity and the stoichiometry of their building blocks, namely their lipid composition. However, the mechanisms by which cells regulate the abundance of different lipids to maintain adaptive and functional membranes remain largely unknown and represent a major challenge in membrane research. Under pathologically or environmentally challenging conditions, cells need to maintain cellular membrane fitness in order to support life. An example of this is the lipid remodeling that happens during homeoviscous adaptation that poikilotherms undergo in order to maintain their membrane homeostasis in the cold [48, 49]. Lipidomic analysis in combination with experiments in reconstituted membrane systems has been also used to show that in bacteria, headgroup-specific acyl chain remodeling is associated with growth at different temperatures, emphasizing the importance of membrane adaptation mechanisms through evolution [50]. Another example of this is the mechanism that yeast use to detect changes in the plasma membrane and connect this to new lipid synthesis. When sphingolipids are depleted, changes in plasma membrane tension are detected, most likely via the Slm proteins which change their location and activate TORC2 [51]. This activation leads to phosphorylation of the ORM proteins and ceramide synthases and subsequent upregulation of sphingolipid synthesis [52, 53].

The fact that cell membranes adapt their lipid composition upon environmental cues or cellular damage implies the presence of sensing mechanisms, most of which remain still unidentified. Mechanisms of membrane sensing prior to adaptation involve chemical sensing (charge) as well as biophysical sensing (packing, interdigitation) [54]. Successful membrane adaptation mechanisms imply the acquisition of membrane repair mechanisms, especially important in animal cells which, lacking a cell wall, are more vulnerable to mechanical and chemical stressors. These mechanisms include changes in intracellular trafficking routes (exo/endocytosis) and the activity of proteins such as the evolutionary conserved multiprotein complex called endosomal sorting complexes required for transport (ESCRT) complex [55]. Yet to be fully understood is whether there are universal membrane repair mechanisms, how they are modulated to ensure membrane fitness under different conditions and in different cell types, and how important lipid–protein interactions are for the activity or activation of such machineries [56]. What seems clear is that membrane lipid composition cannot be disentangled from biophysical properties such as fluidity and thus, transversal studies that include biophysical approaches and quantitative lipid analysis are needed to fully understand how cells are able to maintain the homeostasis of such complex and important structures.

26.2 Lipid Composition Dictates Mechanisms of Intracellular Trafficking

Most of the organelles within eukaryotic cells, which constitute the endomembrane system, are in continuous intercommunication exchanging material through carriers of different shapes and sizes. Lipid composition dictates to a great extent the characteristics of those carriers as well as their specificity and probably their protein composition through specific lipid–protein interactions. In this section we will revisit those studies in which lipidomic analysis has helped to understand intracellular trafficking events with a particular focus on the endocytic and the early secretory pathway.

26.2.1 The Endocytic Pathway

In eukaryotic cells the endocytic pathway represents the main entry route for certain essential nutrients from the extracellular space and plays a central role in fundamental processes such as signaling regulation and turnover of lipid and protein components of the plasma membrane. In mammalian cells, the pathway can be functionally separated into two different parts: The early endosome (EE) and the late endosome (LE)/lysosome. The EE represent the first destination of material internalized by endocytosis and can be viewed as a sorting station differentiating the components that will be recycled to the PM or other destinations, like *trans* Golgi network (TGN) from those that are directed to lysosomes for degradation [57]. The LE and lysosome represent final degradative stations of the cell, but as it emerged in recent years, they also represent a signaling hub for metabolic regulation, being the platform for mammalian target of rapamycin (mTOR) complex activation. Again, in mammalian cells, this organelle is on the main route for cholesterol uptake from endocytosed lipoprotein particles.

Most of the above-mentioned processes require complex rearrangements of lipid membranes including fusion and fission events, tubulation and invagination processes such as the production of intralumenal vesicles that are typical in the morphology of the late stages of the pathway and essential for sorting functions. Despite the essential and well-known role of protein mediators in these processes, like the ESCRT complex [58], the lipid composition, by determining the biophysical properties of the endosomal membranes, also has a great importance in these processes. Indeed, it has been recently shown that membrane tension, a parameter strictly correlated with lipid biophysical interactions, plays a major role in the invagination process in early endosomes [59]. Moreover, in later stages of the pathway, the endosome-specific phospholipid lysobisphosphatidic acid/bis(monoacylglycerol) phosphate (LBPA/BMP) has been shown to promote the formation of intralumenal vesicles even *in vitro* in the absence of proteins, indicating a possible role in the same process *in vivo* [60].

When analyzing the role of lipids in the pathway one should clearly distinguish between signaling lipids, which are generally low-abundant and short-lived, from the main constituents that determine the biophysical properties of membranes. The prototype of the first class is represented in the endocytic pathway by (phospho) inositides. These lipids, which are produced by the addition of phosphate groups in the 3-, 4-, and 5-OH position on the inositol ring of phosphatidylinositol and are finely regulated by the action of lipid kinases and phosphatases, play a central role in protein docking and organellar identity [61]. For this reason, these lipids have been the object of extensive investigations over the past decades. Nevertheless, given also the difficulties in the analytical quantification of this class of lipids, most of the information comes from detecting the lipids with specific probes or antibodies [62].

When discussing about lipid composition, it is generally assumed that endosomes, especially in the early stages of the pathway, have a lipid composition similar to the PM from which they originate (i.e. rich in sphingomyelin [SM], cholesterol, and PS). Moving to later stages the composition gradually changes with cholesterol and sphingomyelin (SM), mostly confined to the intralumenal space of the organelles [63]. It is worth mentioning the case of the atypical phospholipid LBPA which is specifically localized in the lumen of late endosomes and not found elsewhere in the cell (Figure 26.1). This suggests that it is synthesized locally [64]. For both types of lipid, a large amount of information that has accumulated over the years comes from imaging techniques making use of specific probes or antibodies [61]. Different from organelles that can be more readily isolated, like mitochondria and nuclei, endosomes are generally difficult to isolate from other organelles with similar sizes and densities, and moreover they represent a heterogeneous population whose characteristics can strongly differ in different cell types and tissues. The classical techniques that have been developed over several decades rely on separation of the organelles on density gradients exploiting the different density of organelles [65]. Alternative techniques like affinity purification and immuno-isolation have been also used for the lipidomic analysis of endosomal organelles [66]. Recent advances in the field of magnetic nanoparticles have extended their applications to organelle purification. The last generation of superparamagnetic iron oxide nanoparticles (SPIONs), given their small diameter (<50 nm), can be taken up by cells by fluid-phase endocytosis and have been successfully used for the isolation of endosomes and lysosomes [67–69] and for the analysis of protein and lipid composition of plasma membrane in mammalian cells [70]. More recently the same strategy has been used to perform parallel lipidomic and proteomic analysis of plasma membrane and endo-lysosomal compartments of wild-type and Niemann–Pick C1 (NPC1)-deficient Hela cells. This study confirms the lipid alterations in the endo-lysosomal compartment while PM remains unaffected in this lysosomal disorder [71].

As mentioned above, the main limitation of subcellular fractionation approaches coupled to analytical lipidomics comes from the presence of contaminants from other fractions. In fact, even relatively small contaminations from fractions that are highly abundant in some specific lipid can strongly undermine the quantification in fractions where the same lipid is less abundant. The same problem was already faced by other -omics techniques, mostly proteomics, aiming at the determination of the protein composition of subcellular compartments. Several groups have made the effort of creating organellar maps of protein markers through protein correlation profiling techniques [72, 73]. There are several methods for creating organellar maps, but in the simplest form the starting point is creating a list of "known" marker proteins, and then train machine learning algorithms to assign the rest of the protein to one or more organellar clusters. In the case of lipidomics, with some exceptions (like LBPA for late endosomes/lysosomes or cardiolipin for mitochondria) it would be more difficult to find specific organellar markers to create similar maps based only on lipid data. On the other hand, in the near future we will probably see more and more studies combining proteomics, lipidomics, and metabolomics quantifications on the same subcellular fraction that will allow the use of organellar maps to define more precisely the molecular composition of the organelles in normal and perturbed states.

26.2.2 The Early Secretory Pathway

Lipidomic analysis has been historically used in combination with immuno-isolation in the field of membrane trafficking to detect the enrichment of specific lipids in particular cellular subdomains such as the TGN [74, 75]. Actually, most of the lipid-based sorting mechanisms that have been proposed take place post-Golgi apparatus and were described using detergent-based techniques now known to be highly artifactual [76]. Nonetheless, those studies helped emphasize the role of lipids as active contributors to protein sorting and function.

A role for lipids in early stages of the secretory pathway has recently become apparent. It was long thought, at least in mammalian cells, that all proteins that leave the ER in COPII vesicles travel together to the Golgi apparatus, where they are sorted for retrieval back to the ER or further transport to the TGN, where they are sorted for further destinations in the secretory pathway [77–79]. It now seems that this was an oversimplification. The presence of a protein sorting mechanism upon exit from the ER was proposed in the early 2000s by Muñiz et al. who immuno-isolated vesicles that were formed by *in vitro* budding assays from the ER and showed that they were enriched with different types of cargoes. In particular, they found that GPI-AP were separated from other plasma membrane proteins upon exit from the ER demonstrating that cargo sorting can actually occur before the proteins arrive to the Golgi. Lipid-based sorting of GPI-AP would be a smart mechanism to separate those lipid-linked membrane proteins from other proteins, and it has been recently demonstrated. Mass spectrometry of lipids has helped establish that in yeast, the GPI moiety of some of these GPI-APs contain a ceramide molecule while in mammalian cells it contains an ether lipid [80]. Lipidomics analysis also suggested that GPI regulates the transport of very long chain ceramide from the ER to Golgi, and thus they could be co-transported [81]. Moreover, GPI-AP sorting at the ER depends on the acyl chain length of the surrounding ceramide molecules [25]. A related mechanism is still to be found in mammalian cells where ether lipids are also potentially involved [26]. The cargo receptor for the exit of GPI-AP from the ER is the p24 complex. Interestingly the transmembrane domain of p24 has been found to specifically bind ceramide in yeast and ether lipids in mammalian cells. It seems clear that the enrichment of a particular lipid in a specific location (ER exit site in this particular case) might be driven by direct lipid–protein interactions which have evolved to meet the needs of different organisms. After its transport with GPI-AP to the Golgi compartment, the p24 complex needs to be retrieved, and this is done using COPI vesicles. A remarkable work by Brugger et al. showed how those COPI vesicles were depleted from all measured sphingomyelins but one single species, SM 18 (18 carbons in the acyl chain) (Figure 26.1) [82]. This work was the precursor of the recent discovery of the first direct and highly specific interaction of a sphingolipid species (SM 18) with a protein transmembrane domain (p24) where they also show that this interaction is crucial for an efficient anterograde transport of COPI vesicles [44].

In summary, quantitative lipid analysis using various mass spectrometry approaches has helped to uncover important lipid functions such as protein sorting in the early secretory pathway opening new avenues to think of lipids as tools to modulate the secretion of specific proteins and thus their activity in cells. A summary of this aspect and this chapter regarding the role of lipids in the early secretory pathway is illustrated in Table 26.1.

Table 26.1 Role of lipids in the early secretory pathway as discussed in this chapter.

LIPID name	Lipid–protein interaction	Effect on membranes		Citation
Lysolipids	Not described	Decrease bending rigidity	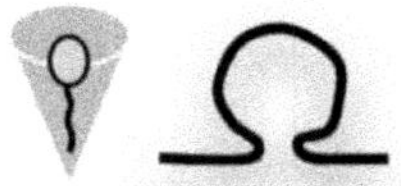	Melero et al. [23]
Ceramides	Yes Yeast p24 cargo receptor for GPI-AP	Ceramide-dependent sorting	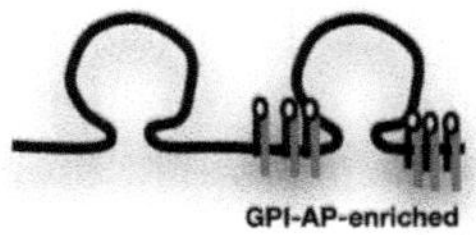	Rodriguez-Gallardo et al. [25]
Ether PC	Yes Mammalian p24 cargo receptor for GPI-AP	Decrease packing/tension	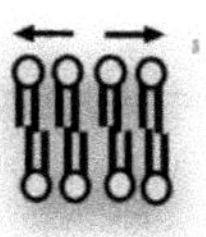	Jiménez-Rojo, Leonetti et al. [26]
SM 18	Yes Mammalian p24 cargo receptor for GPI-AP	Dimer stabilization		Contreras, Ernst et al. [44]

26.3 Multiomic Approaches to Investigate Cell Biology

Since the establishment of lipidomics, hundreds of datasets have been produced, describing the lipidomes of organisms, tissues, or cell types which has progressively advanced our understanding on the complexity and diversity of membrane lipid composition and the role of lipids in certain processes and diseases. However, in most cases, these observations only provide a static picture and do not clarify the underlying molecular mechanisms. New approaches are required to obtain insights into the mechanisms of lipid function. Thanks to the development of novel techniques to study lipid metabolism and function, including high-throughput genetic and biochemical approaches as well as computational studies using molecular dynamics simulations [83, 84], novel directions are being opened. In the era of big data in science, the power of *omics* analysis has called the attention of many scientists looking for big-picture answers to fundamental questions. Genomic, proteomic, and metabolic studies have flooded the scientific literature in the quest to understand not only basic cell biology but also the molecular basis of pathologies such as cancer metabolic and neurodegenerative diseases [85, 86]. These approaches have delineated the discovery of new biological processes such as ferroptosis, a unique type of cell death driven by iron-dependent phospholipid peroxidation. Functional interactions between different lipid pathways have been also found combining transcriptomics with lipidomics as is the case for sterols and sphingolipids [87]. Along these lines, a groundbreaking study established a circular network of co-regulated lipids that was transcriptionally related to innate immune responses [88]. The same strategy has been widely used in the search for biomarkers in cancer [85, 89] and, for example, to establish differences between extracellular vesicles (EVs) of various origins (liquid versus matrix-bound) searching for extracellular matrix (ECM)-based biomaterials with high therapeutic potential [90]. A similar study, which also included proteomic analysis, was carried out to understand mTORC2-driven tumorigenesis. The authors concluded that its oncogenicity relies on fatty acid and lipid synthesis and that glucosylceramides are needed to induce tumorigenesis [91]. In conclusion, advances in computational sciences in parallel with approaches that combine various *omics* have pushed forward data-driven science and have been proven useful for yielding unexpected results and, thus, accelerating discovery research.

Perhaps more relevant to understand basic lipid cell biology are those studies using systematic and unbiased lipidomic approaches together with high-throughput genetic screens aiming to uncover basic aspects of lipid metabolism. Along these lines, da Silveira Dos Santos et al. performed a lipidomics-based screen of yeast phosphate and kinase mutants revealing novel insights into lipid coregulation and homeostasis [92]. Clustered regularly interspaced short palindromic repeats (CRISPR)-based genetic screens relevant for the lipid field can be divided into two groups: those that aimed to study processes in principle non-related to lipid biology but that uncovered novel lipid functions and, those directed toward discovering new aspects of lipid metabolism. As an example of the first group Jain et al. used a CRISPR screen of cells in high and low oxygen conditions to find particular lipid pathways involved in oxygen sensing, signaling, and metabolism. The findings were validated using lipidomics, and accumulation of peroxisome-derived ether lipids was found as a characteristic feature of hypoxia [93]. More recently and using a very similar approach, Zou et al. found that peroxisomes and ether lipids drive the susceptibility to ferroptosis using an untargeted approach [94]. Piccolis et al. evaluated the toxicity induced by palmitate treatment in mammalian cells. The authors found pathways and genes mediating this lipotoxicity, which is thought to be the underlying cause of many metabolic diseases [95]. In the same issue, Zhu et al. using as well as CRISPR screen strategy combined with lipidomic analysis identified a regulator of glycerolipid metabolism (CHP1) and gained insights into adaptation mechanisms as they found the peroxisomal enzyme GNPAT compensated for the loss of CHP1 [96]. Finally, in our own study, we systematically evaluated the cellular responses to sphingolipid depletion using a CRISPRi screen in combination with a quantitative targeted lipidomic approach and found a co-regulation between that class of lipids and ether lipids that revealed common roles for these lipids in membrane trafficking [26]. Altogether, these studies show the utility of using untargeted and multidisciplinary strategies that include mass spectrometry of lipids to uncover novel aspects of lipid biology, a research field that still lags behind many other disciplines but whose importance for physiology cannot be ignored. High-throughput lipidomic analysis should allow scientists and companies to draw maps containing vast amounts of information about the lipid distribution in different tissues [8]. For instance, in the last year, a very comprehensive reference lipidome of human white adipose tissue had been published by Lange et al. [97]. This so-called *AdipoAtlas* has helped to highlight differences between subcutaneous and visceral tissue depots and represents a very good example of a lipidomics study that can be used for systems medicine data integration.

26.4 Perspectives

As mentioned above, several studies, with the systematic analysis of the lipid composition of tissues, cells, and subcellular fractions, often in combination with other omics technologies, helped to uncover the role of lipid complexity in cell biology. One great merit of these studies has been to underscore the importance of analyzing lipids at the systems level. However, as we extensively discussed over this chapter, it is now clear that to fully understand the role of lipids in cell physiology, it is crucial to analyze with better definition the spatial distribution of lipids in the cell and their dynamic regulation over time. New technologies will hopefully help us to gain more information about lipid composition and their functions at the cellular and subcellular levels *in vivo*. New improvements to imaging techniques will be essential tools for this, and hopefully they can increase their resolution and sensitivity in the near future. Techniques that show promise are matrix assisted laser desorption/ionization (MALDI) imaging and secondary ion mass spectrometry-time-of-flight (SIMS-ToF) [98]. Finally, chemical biology techniques, with the possibility to deliver lipids in the cells in a spatial and time-resolved manner, have already shown the potential to give a great contribution to uncover lipid function and regulation. Likely, an increased contribution of these techniques in the near future combined with other functionalities will help relieve some of the deficiencies of genetics in unveiling the mysteries of lipid diversity.

Abbreviations

GPL	glycerophospholipid
ER	endoplasmic reticulum
PM	plasma membrane
PC	phosphatidylcholine
PE	phosphatidylethanolamine
PS	phosphatodylserine
S1P	sphingosine-1-phosphate
GPI-AP	glycosylphosphatidylinositol-anchored protein
TORC	target of rapamycin complex
ESCRT	endosomal sorting complexes required for transport
EE	early endosome

LE	late endosome
TGN	*trans* Golgi network
mTOR	mammalian target of rapamycin
LBPA/BMP	lysobisphosphatidic acid/ bis(monoacylglycerol)phosphate
SM	sphingomyelin
SPION	superparamagnetic iron oxide nanoparticles
NPC1	Niemann–Pick C1
EV	extracellular vesicles
ECM	extracellular matrix
CRISPR	clustered regularly interspaced short palindromic repeats
CHP1	calcineurin B homologous protein 1
GNPAT	dihydroxyacetonephosphate acyltransferase
MALDI	matrix assisted laser desorption/ionization
SIMS	secondary ion mass spectrometry
ToF	time-of-flight

References

1 Van Meer, G., Voelker, D.R., and Feigenson, G.W. (2008). Membrane lipids: where they are and how they behave. *Nat. Rev. Mol. Cell Biol.* 9 (2): 112–124.

2 Harayama, T. and Riezman, H. (2018). Understanding the diversity of membrane lipid composition. *Nat. Rev. Mol. Cell Biol.* 19 (5): 281–296.

3 Klose, C., Surma, M.A., and Simons, K. (2013). Organellar lipidomics – background and perspectives. *Curr. Opin. Cell Biol.* 25 (4): 406–413.

4 Brügger, B. (2014). Lipidomics: analysis of the lipid composition of cells and subcellular organelles by electrospray ionization mass spectrometry. *Annu. Rev. Biochem.* 83: 79–98.

5 Andreyev, A.Y., Fahy, E., Guan, Z. et al. (2010). Subcellular organelle lipidomics in TLR-4-activated macrophages 1 [S]. *J. Lipid Res.* 51 (9): 2785–2797.

6 Kappler, L., Li, J., Häring, H.-U. et al. (2016). Purity matters: a workflow for the valid high-resolution lipid profiling of mitochondria from cell culture samples. *Sci. Rep.* 6 (1): 1–10.

7 Frechin, M., Stoeger, T., Daetwyler, S. et al. (2015). Cell-intrinsic adaptation of lipid composition to local crowding drives social behaviour. *Nature* 523 (7558): 88–91.

8 Surma, M.A., Gerl, M.J., Herzog, R. et al. (2021). Mouse lipidomics reveals inherent flexibility of a mammalian lipidome. *Sci. Rep.* 11 (1): 1–14.

9 Adamovich, Y., Rousso-Noori, L., Zwighaft, Z. et al. (2014). Circadian clocks and feeding time regulate the oscillations and levels of hepatic triglycerides. *Cell Metab.* 19 (2): 319–330.

10 Loizides-Mangold, U., Perrin, L., Vandereycken, B. et al. (2017). Lipidomics reveals diurnal lipid oscillations in human skeletal muscle persisting in cellular myotubes cultured in vitro. *Proc. Natl. Acad. Sci. U.S.A.* 114 (41): E8565–E8574.

11 Williams, E.G., Pfister, N., Roy, S. et al. (2021). Multi-omic profiling of the liver across diets and age in a diverse mouse population. *bioRxiv* https://doi.org/10.1016/j.cels.2021.09.005.

12 Aviram, R., Manella, G., Kopelman, N. et al. (2016). Lipidomics analyses reveal temporal and spatial lipid organization and uncover daily oscillations in intracellular organelles. *Mol. Cell* 62 (4): 636–648.

13 Balla, T. (2007). *Imaging and Manipulating Phosphoinositides in Living Cells*. Wiley Online Library.

14 Kay, J.G. and Grinstein, S. (2011). Sensing phosphatidylserine in cellular membranes. *Sensors* 11 (2): 1744–1755.

15 Endapally, S., Frias, D., Grzemska, M. et al. (2019). Molecular discrimination between two conformations of sphingomyelin in plasma membranes. *Cell* 176 (5): 1040–53.e17.

16 Hannich, J.T., Umebayashi, K., and Riezman, H. (2011). Distribution and functions of sterols and sphingolipids. *Cold Spring Harbor Perspect. Biol.* 3 (5): a004762.

17 Feng, S., Harayama, T., Montessuit, S. et al. (2018). Mitochondria-specific photoactivation to monitor local sphingosine metabolism and function. *Elife* 7: e34555.

18 Feng, S., Harayama, T., Chang, D. et al. (2019). Lysosome-targeted photoactivation reveals local sphingosine metabolism signatures. *Chem. Sci.* 10 (8): 2253–2258.

19 Gomez, L., Paillard, M., Price, M. et al. (2011). A novel role for mitochondrial sphingosine-1-phosphate produced by sphingosine kinase-2 in PTP-mediated cell survival during cardioprotection. *Basic Res. Cardiol.* 106 (6): 1341–1353.

20 Höglinger, D. (2019). Bi- and trifunctional lipids for visualization of sphingolipid dynamics within the cell. In: *Intracellular Lipid Transport* (ed. G. Drin), 95–103. Springer.

21 Budin, I., de Rond, T., Chen, Y. et al. (2018). Viscous control of cellular respiration by membrane lipid composition. *Science* 362 (6419): 1186–1189.

22 Thiam, A.R., Farese, R.V. Jr., and Walther, T.C. (2013). The biophysics and cell biology of lipid droplets. *Nat. Rev. Mol. Cell Biol.* 14 (12): 775–786.

23 Melero, A., Chiaruttini, N., Karashima, T. et al. (2018). Lysophospholipids facilitate COPII vesicle formation. *Curr. Biol.* 28 (12): 1950–8.e6.

24 Nilsson, I., Ohvo-Rekilä, H., Slotte, J.P. et al. (2001). Inhibition of protein translocation across the endoplasmic reticulum membrane by sterols. *J. Biol. Chem.* 276 (45): 41748–41754.

25 Rodriguez-Gallardo, S., Kurokawa, K., Sabido-Bozo, S. et al. (2020). Ceramide chain length–dependent protein sorting into selective endoplasmic reticulum exit sites. *Sci. Adv.* 6 (50): eaba8237.

26 Jiménez-Rojo, N., Leonetti, M.D., Zoni, V. et al. (2020). Conserved functions of ether lipids and sphingolipids in the early secretory pathway. *Curr. Biol.* 30 (19): 3775–87.e7.

27 Goñi, F.M. and Alonso, A. (2009). Effects of ceramide and other simple sphingolipids on membrane lateral structure. *Biochim. Biophys. Acta* 1788 (1): 169–177.

28 Silva, L., De Almeida, R.F., Fedorov, A. et al. (2006). Ceramide-platform formation and-induced biophysical changes in a fluid phospholipid membrane. *Mol. Membr. Biol.* 23 (2): 137–148.

29 Contreras, F.-X., Basañez, G., Alonso, A. et al. (2005). Asymmetric addition of ceramides but not dihydroceramides promotes transbilayer (flip-flop) lipid motion in membranes. *Biophys. J.* 88 (1): 348–359.

30 Contreras, F.-X., Sánchez-Magraner, L., Alonso, A., and Goñi, F.M. (2010). Transbilayer (flip-flop) lipid motion and lipid scrambling in membranes. *FEBS Lett.* 584 (9): 1779–1786.

31 Montes, L.R., Ruiz-Argüello, M.B., Goñi, F.M., and Alonso, A. (2002). Membrane restructuring via ceramide results in enhanced solute efflux. *J. Biol. Chem.* 277 (14): 11788–11794.

32 Jiménez-Rojo, N., Sot, J., Busto, J.V. et al. (2014). Biophysical properties of novel 1-deoxy-(dihydro) ceramides occurring in mammalian cells. *Biophys. J.* 107 (12): 2850–2859.

33 Santos, T.C.B., Vaz, A., Ventura, A.E. et al. (2020). Canonical and 1-deoxy(methyl) sphingoid bases: tackling the effect of the lipid structure on membrane biophysical properties. *Langmuir* 36 (21): 6007–6016.

34 Castro, B.M., Prieto, M., and Silva, L.C. (2014). Ceramide: a simple sphingolipid with unique biophysical properties. *Prog. Lipid Res.* 54: 53–67.

35 Goñi, F.M. and Alonso, A. (2006). Biophysics of sphingolipids I. Membrane properties of sphingosine, ceramides and other simple sphingolipids. *Biochim. Biophys. Acta* 1758 (12): 1902–1921.

36 Goñi, F.M., Sot, J., and Alonso, A. (2014). Biophysical properties of sphingosine, ceramides and other simple sphingolipids. *Biochem. Soc. Trans.* 42 (5): 1401–1408.

37 Ventura, A.E., Mestre, B., and Silva, L.C. (2019). Ceramide domains in health and disease: a biophysical perspective. *Adv. Exp. Med. Biol.* 1159: 79–108. https://doi.org/10.1007/978-3-030-21162-2_6.

38 Zelnik, I.D., Ventura, A.E., Kim, J.L. et al. (2020). The role of ceramide in regulating endoplasmic reticulum function. *Biochim. Biophys. Acta* 1865 (1): 158489.

39 Ailte, I., Lingelem, A.B.D., Kvalvaag, A.S. et al. (2017). Exogenous lysophospholipids with large head groups perturb clathrin-mediated endocytosis. *Traffic* 18 (3): 176–191.

40 Jiménez-Rojo, N. and Riezman, H. (2019). On the road to unraveling the molecular functions of ether lipids. *FEBS Lett.* 593 (17): 2378–2389.

41 Koivuniemi, A. (2017). The biophysical properties of plasmalogens originating from their unique molecular architecture. *FEBS Lett.* 591 (18): 2700–2713.

42 Lohner, K., Hermetter, A., and Paltauf, F. (1984). Phase behavior of ethanolamine plasmalogen. *Chem. Phys. Lipids* 34 (2): 163–170.

43 Paltauf, F. (1994). Ether lipids in biomembranes. *Chem. Phys. Lipids* 74 (2): 101–139.

44 Contreras, F.-X., Ernst, A.M., Haberkant, P. et al. (2012). Molecular recognition of a single sphingolipid species by a protein's transmembrane domain. *Nature* 481 (7382): 525–529.

45 Levental, K.R., Malmberg, E., Symons, J.L. et al. (2020). Lipidomic and biophysical homeostasis of mammalian membranes counteracts dietary lipid perturbations to maintain cellular fitness. *Nat. Commun.* 11 (1): 1–13.

46 Levental, K.R., Surma, M.A., Skinkle, A.D. et al. (2017). ω-3 polyunsaturated fatty acids direct differentiation of the membrane phenotype in mesenchymal stem cells to potentiate osteogenesis. *Sci. Adv.* 3 (11): eaao1193.

47 Monasterio, B.G., Jiménez-Rojo, N., García-Arribas, A.B. et al. (2021). CHO/LY-B cell growth under limiting sphingolipid supply: correlation between lipid composition and biophysical properties of sphingolipid-restricted cell membranes. *FASEB J.* 35 (6): e21657.

48 Ernst, R., Ejsing, C.S., and Antonny, B. (2016). Homeoviscous adaptation and the regulation of membrane lipids. *J. Mol. Biol.* 428 (24): 4776–4791.

49 Dymond, M.K. (2015). Mammalian phospholipid homeostasis: homeoviscous adaptation deconstructed by lipidomic data driven modelling. *Chem. Phys. Lipids* 191: 136–146.

50 Chwastek, G., Surma, M.A., Rizk, S. et al. (2020). Principles of membrane adaptation revealed through environmentally induced bacterial lipidome remodeling. *Cell Rep.* 32 (12): 108165.

51 Berchtold, D., Piccolis, M., Chiaruttini, N. et al. (2012). Plasma membrane stress induces relocalization of Slm proteins and activation of TORC2 to promote sphingolipid synthesis. *Nat. Cell Biol.* 14 (5): 542–547.

52 Muir, A., Ramachandran, S., Roelants, F.M. et al. (2014). TORC2-dependent protein kinase Ypk1 phosphorylates ceramide synthase to stimulate synthesis of complex sphingolipids. *Elife* 3: e03779.

53 Roelants, F.M., Breslow, D.K., Muir, A. et al. (2011). Protein kinase Ypk1 phosphorylates regulatory proteins Orm1 and Orm2 to control sphingolipid homeostasis in *Saccharomyces cerevisiae*. *Proc. Natl. Acad. Sci. U.S.A.* 108 (48): 19222–19227.

54 Ernst, R., Ballweg, S., and Levental, I. (2018). Cellular mechanisms of physicochemical membrane homeostasis. *Curr. Opin. Cell Biol.* 53: 44–51.

55 Jimenez, A.J., Maiuri, P., Lafaurie-Janvore, J. et al. (2014). ESCRT machinery is required for plasma membrane repair. *Science* 343 (6174): https://doi.org/10.1126/science.1247136.

56. Cooper, S.T. and McNeil, P.L. (2015). Membrane repair: mechanisms and pathophysiology. *Physiol. Rev.* 95 (4): 1205–1240.

57 Gruenberg, J. (2001). The endocytic pathway: a mosaic of domains. *Nat. Rev. Mol. Cell Biol.* 2 (10): 721–730.

58 Vietri, M., Radulovic, M., and Stenmark, H. (2020). The many functions of ESCRTs. *Nat. Rev. Mol. Cell Biol.* 21 (1): 25–42.

59 Mercier, V., Larios, J., Molinard, G. et al. (2020). Endosomal membrane tension regulates ESCRT-III-dependent intra-lumenal vesicle formation. *Nat. Cell Biol.* 22 (8): 947–959.

60 Matsuo, H., Chevallier, J., Mayran, N. et al. (2004). Role of LBPA and Alix in multivesicular liposome formation and endosome organization. *Science* 303 (5657): 531–534.

61 Corvera, S., D'Arrigo, A., and Stenmark, H. (1999). Phosphoinositides in membrane traffic. *Curr. Opin. Cell Biol.* 11: 460–465.

62 Balla, T. and Várnai, P. (2009). Visualization of cellular phosphoinositide pools with GFP-fused protein-domains. *Curr. Protoc. Cell Biol.* 42 (1): 24.4.1–24.4.7.

63 Möbius, W., Van Donselaar, E., Ohno-Iwashita, Y. et al. (2003). Recycling compartments and the internal vesicles of multivesicular bodies harbor most of the cholesterol found in the endocytic pathway. *Traffic* 4 (4): 222–231.

64 Kobayashi, T., Stang, E., Fang, K.S. et al. (1998). A lipid associated with the antiphospholipid syndrome regulates endosome structure and function. *Nature* 392: 193–197.

65 de Araújo, M.E., Lamberti, G., and Huber, L.A. (2015). Isolation of early and late endosomes by density gradient centrifugation. *Cold Spring Harbor Protoc.* 2015 (11):pdb. prot083444.

66 Ito, Y., Grison, M., Esnay, N. et al. (2020). Immunopurification of intact endosomal compartments for lipid analyses in *Arabidopsis*. In: *Plant Endosomes*, Methods in Molecular Biology (Clifton, NJ), vol. 2177, 119–141. New York, NY: Humana Press.

67 Rofe, A.P. and Pryor, P.R. (2016). Purification of lysosomes using supraparamagnetic iron oxide nanoparticles (SPIONs). *Cold Spring Harbor Protoc.* 2016 (4):pdb.prot084822.

68 Walker, M.W. and Lloyd-Evans, E. (2015). A rapid method for the preparation of ultrapure, functional lysosomes using functionalized superparamagnetic iron oxide nanoparticles. *Methods Cell Biol.* 126: 21–43.

69 Bilgin, M., Nylandsted, J., Jäättelä, M., and Maeda, K. (2017). Quantitative profiling of lysosomal lipidome by shotgun lipidomics. In: *Lysosomes*, Methods in Molecular Biology (Clifton, NJ), vol. 1594, 19–34. New York, NY: Humana Press.

70 Thimiri Govinda Raj, D.B., Ghesquiere, B., Tharkeshwar, A.K. et al. (2011). A novel strategy for the comprehensive analysis of the biomolecular composition of isolated plasma membranes. *Mol. Syst. Biol.* 7: 541.

71 Tharkeshwar, A.K., Trekker, J., Vermeire, W. et al. (2017). A novel approach to analyze lysosomal dysfunctions through subcellular proteomics and lipidomics: the case of NPC1 deficiency. *Sci. Rep.* 7: 41408.

72 Foster, L.J., de Hoog, C.L., Zhang, Y. et al. (2006). A mammalian organelle map by protein correlation profiling. *Cell* 125 (1): 187–199.

73 Itzhak, D.N., Tyanova, S., Cox, J., and Borner, G.H. (2016). Global, quantitative and dynamic mapping of protein subcellular localization. *Elife* 5: e16950. https://doi .org/10.7554/eLife.16950.

74 Klemm, R.W., Ejsing, C.S., Surma, M.A. et al. (2009). Segregation of sphingolipids and sterols during formation of secretory vesicles at the trans-Golgi network. *J. Cell Biol.* 185 (4): 601–612.

75 Ito, Y., Esnay, N., Platre, M.P. et al. (2021). Sphingolipids mediate polar sorting of PIN2 through phosphoinositide consumption at the *trans*-Golgi Network. *Nat. Commun.* 12 (1): 1–18.

76 Lichtenberg, D., Goñi, F.M., and Heerklotz, H. (2005). Detergent-resistant membranes should not be identified with membrane rafts. *Trends Biochem. Sci.* 30 (8): 430–436.

77 Fuller, S., Bravo, R., and Simons, K. (1985). An enzymatic assay reveals that proteins destined for the apical or basolateral domains of an epithelial cell line share the same late Golgi compartments. *EMBO J.* 4 (2): 297–307.

78 Orci, L., Ravazzola, M., Amherdt, M. et al. (1987). The trans-most cisternae of the Golgi complex: a compartment for sorting of secretory and plasma membrane proteins. *Cell* 51 (6): 1039–1051.

79 Rindler, M.J., Ivanov, I.E., Plesken, H. et al. (1984). Viral glycoproteins destined for apical or basolateral plasma membrane domains traverse the same Golgi apparatus during their intracellular transport in doubly infected Madin–Darby canine kidney cells. *J. Cell Biol.* 98 (4): 1304–1319.

80 Brewis, I.A., Ferguson, M.A., Mehlert, A. et al. (1995). Structures of the glycosyl-phosphatidylinositol anchors of porcine and human renal membrane dipeptidase: comprehensive structural studies on the porcine anchor and interspecies comparison of the glycan core structures. *J. Biol. Chem.* 270 (39): 22946–22956.

81 Loizides-Mangold, U., David, F.P., Nesatyy, V.J. et al. (2012). Glycosylphosphatidylinositol anchors regulate glycosphingolipid levels. *J. Lipid Res.* 53 (8): 1522–1534.

82 Brügger, B., Sandhoff, R., Wegehingel, S. et al. (2000). Evidence for segregation of sphingomyelin and cholesterol during formation of COPI-coated vesicles. *J. Cell Biol.* 151 (3): 507–518.

83 Saliba, A.-E., Vonkova, I., and Gavin, A.-C. (2015). The systematic analysis of protein–lipid interactions comes of age. *Nat. Rev. Mol. Cell Biol.* 16 (12): 753–761.

84 Vanni, S., Hirose, H., Barelli, H. et al. (2014). A sub-nanometre view of how membrane curvature and composition modulate lipid packing and protein recruitment. *Nat. Commun.* 5 (1): 1–10.

85 Vantaku, V., Dong, J., Ambati, C.R. et al. (2019). Multi-omics integration analysis robustly predicts high-grade patient survival and identifies CPT1B effect on fatty acid metabolism in bladder cancer. *Clin. Cancer Res.* 25 (12): 3689–3701.

86 Xu, J., Bankov, G., Kim, M. et al. (2020). Integrated lipidomics and proteomics network analysis highlights lipid and immunity pathways associated with Alzheimer's disease. *Transl. Neurodegener.* 9 (1): 1–15.

87 Guan, X.L., Souza, C.M., Pichler, H. et al. (2009). Functional interactions between sphingolipids and sterols in biological membranes regulating cell physiology. *Mol. Biol. Cell* 20 (7): 2083–2095.

88 Köberlin, M.S., Snijder, B., Heinz, L.X. et al. (2015). A conserved circular network of coregulated lipids modulates innate immune responses. *Cell* 162 (1): 170–183.

89 Liesenfeld, D.B., Grapov, D., Fahrmann, J.F. et al. (2015). Metabolomics and transcriptomics identify pathway differences between visceral and subcutaneous adipose tissue in colorectal cancer patients: the ColoCare study. *Am. J. Clin. Nutr.* 102 (2): 433–443.

90 Hussey, G.S., Molina, C.P., Cramer, M.C. et al. (2020). Lipidomics and RNA sequencing reveal a novel subpopulation of nanovesicle within extracellular matrix biomaterials. *Sci. Adv.* 6 (12): eaay4361.

91 Guri, Y., Colombi, M., Dazert, E. et al. (2017). mTORC2 promotes tumorigenesis via lipid synthesis. *Cancer Cell* 32 (6): 807–23.e12.

92 da Silveira Dos Santos, A.X., Riezman, I., Aguilera-Romero, M.-A. et al. (2014). Systematic lipidomic analysis of yeast protein kinase and phosphatase mutants reveals novel insights into regulation of lipid homeostasis. *Mol. Biol. Cell* 25 (20): 3234–3246.

93 Jain, I.H., Calvo, S.E., Markhard, A.L. et al. (2020). Genetic screen for cell fitness in high or low oxygen highlights mitochondrial and lipid metabolism. *Cell* 181 (3): 716–27.e11.

94 Zou, Y., Henry, W.S., Ricq, E.L. et al. (2020). Plasticity of ether lipids promotes ferroptosis susceptibility and evasion. *Nature* 585 (7826): 603–608.

95 Piccolis, M., Bond, L.M., Kampmann, M. et al. (2019). Probing the global cellular responses to lipotoxicity caused by saturated fatty acids. *Mol. Cell* 74 (1): 32–44.e8.

96 Zhu, X.G., Puthenveedu, S.N., Shen, Y. et al. (2019). CHP1 regulates compartmentalized glycerolipid synthesis by activating GPAT4. *Mol. Cell* 74 (1): 45–58.e7.

97 Lange, M., Angelidou, G., Ni, Z. et al. (2021). AdipoAtlas: A reference lipidome for human white adipose tissue. *Cell Rep. Med.* 2 (10): 100407. https://doi.org/10.1016/j.xcrm.2021.100407.

98 Castro, D.C., Xie, Y.R., Rubakhin, S.S. et al. (2021). Image-guided MALDI mass spectrometry for high-throughput single-organelle characterization. *Nat. Methods* 18: 1233–1238. https://doi.org/10.1038/s41592-021-01277-2.

27

Microbial Lipidomics

Masahiro Ueda[1,2], Nobuyuki Okahashi[1,3], and Makoto Arita[1,4,5]

[1] Laboratory for Metabolomics, RIKEN Center for Integrative Medical Sciences, 1-7-22 Suehiro-cho, Tsurumi-ku, Yokohama, Kanagawa, 230-0045, Japan
[2] JSR Bioscience and informatics R&D center, JSR Corporation, 3-103-9 Tonomachi, Kawasaki-ku, Kawasaki, Kanagawa, 210-0821, Japan
[3] Osaka University, Graduate School of Information Science and Technology, Department of Bioinformatics Engineering, 1-5 Yamadaoka, Suita, Osaka, 565-0871, Japan
[4] Keio University, Graduate School of Pharmaceutical Sciences, Division of Physiological Chemistry and Metabolism, 1-5-30 Shibakoen, Minato-ku, Tokyo, 105-8512, Japan
[5] Yokohama-City University, Graduate School of Medical Life Science, Cellular and molecular epigenetics laboratory, 1-7-29 Suehiro-cho, Tsurumi-ku, Yokohama, Kanagawa, 230-0045, Japan

27.1 Introduction

The structures of bacterial lipids have distinct characteristics from those of mammals in the lengths and branches of fatty acyl chains, and modification of polar heads, resulting in a unique and massive lipidome. For a long time, bacterial lipids such as lipoteichoic acids and lipopolysaccharides have been used as taxonomic indicators of bacteria [1, 2]. Recent advances in mass spectrometry (MS) have enabled high-throughput detection of pathogenic bacteria based on the fingerprinting of proteins and lipids analyzed by using matrix-assisted laser desorption/ionization time-of-flight mass spectrometry (MALDI-TOF/MS) for clinical examinations and food quality control [3–7]. Measurements of gut bacterial lipids are also emerging to elucidate host–bacteria interactions. Using next-generation sequencing, the human intestinal microbiota have been estimated to comprise at least 1800 genera [8] belonging to the four phyla Firmicutes, Bacteroidetes, Actinobacteria, and Proteobacteria, which have distinct characteristic lipids [9]. As the structural difference in lipids has a great impact on recognition by host receptors [10–12], the separation and identification of the fine structural features is required when using mass spectrometry in gut microbial lipidomics. This chapter describes the current status and limitations of gut microbial lipidomics using representative mass spectrometric methods. In addition, the latest technologies that improve the resolution of microbial lipidomics are introduced.

Mass Spectrometry for Lipidomics: Methods and Applications, First Edition.
Edited by Michal Holčapek and Kim Ekroos.

27.2 Diversity of Lipid Structures in Intestinal Bacteria and Analytical Methods Using Mass Spectrometry

27.2.1 Fatty Acids

Fatty acids biosynthesized by mammals are mainly linear and have an even number of carbons, whereas microbial fatty acids often contain odd-numbered carbon chains and branched structures such as *iso*- and *anteiso*-type [13]. The use of internal standards containing odd-chain fatty acids is not recommended for the analysis of bacterial lipids, although they are frequently used for the analysis of mammalian cells. It should also be noted that the unsaturated bonds are not always carbon–carbon double bonds, as bacteria have lipids with cyclopropane ring structures in their fatty acyl chains [14, 15].

Intestinal bacteria metabolize unsaturated fatty acids into bioactive metabolites [11, 16]. Linoleic acid is converted to a series of hydroxylated and oxo-containing fatty acids via the hydration enzyme in *Lactobacillus plantarum* [16]. 10-Hydroxy-*cis*-12-octadecenoic acid enhances the intestinal epithelial barrier function via GPR40, and 10-oxo-*cis*-12-octadecenoic acid activates the TRPV1 channel to regulate host energy homeostasis (Figure 27.1a) [11, 22]. These bacteria-derived fatty acid metabolites were measured by targeted lipidomics with a liquid chromatography–mass spectrometry (LC-MS) system under a relatively slow LC gradient for the separation of structural isomers, and a multiple reaction monitoring mode for the detection of structure-specific fragments [16, 23].

N-acyl amides, which are molecules comprising fatty acids and amines linked by an amide bond, are produced by the host and microbiota. In host tissues, *N*-arachidonoylethanolamide is endogenously produced and exerts neurological effects by acting on cannabinoid receptors, a member of the G protein-coupled receptor (GPCR) family [24]. In contrast, prokaryotic *N*-acyl amides comprise fatty acids and amino acids, including glycine, alanine, ornithine, serinol, and serine-glycine dipeptide. The fatty acid structures are linear, *iso*-, or *anteiso*-types [17–19, 25–27], and often have a hydroxy group at the β-position, which is sometimes further acylated by a fatty acid (Figure 27.1b). These structural differences affect the affinity and selectivity to GPCRs [19, 27]. The structures of *N*-acylglycine produced by Bacteroidetes species, namely *Bacteroides vulgatus*, *Bacteroides thetaiotaomicron*, and *Porphyromonas gingivalis*, were analyzed using LC-MS/MS [17–19]. Collision-induced dissociation (CID) produces fragment ions corresponding to glycine and acyl chain moieties, whereas nuclear magnetic resonance determines the branching structures. The acyl chain properties differed depending on the bacterial species.

N-acyl homoserine lactones (AHLs), fatty acids conjugated to the amino group of homoserine lactone, function as signaling molecules for bacterial communication (Figure 27.1c). Gram-negative bacteria communicate with each other via AHLs for biofilm formation and virulence factor secretion [28], which is called quorum sensing. Recently, Xue et al. [20] developed targeted lipidomics for 27 AHLs with various

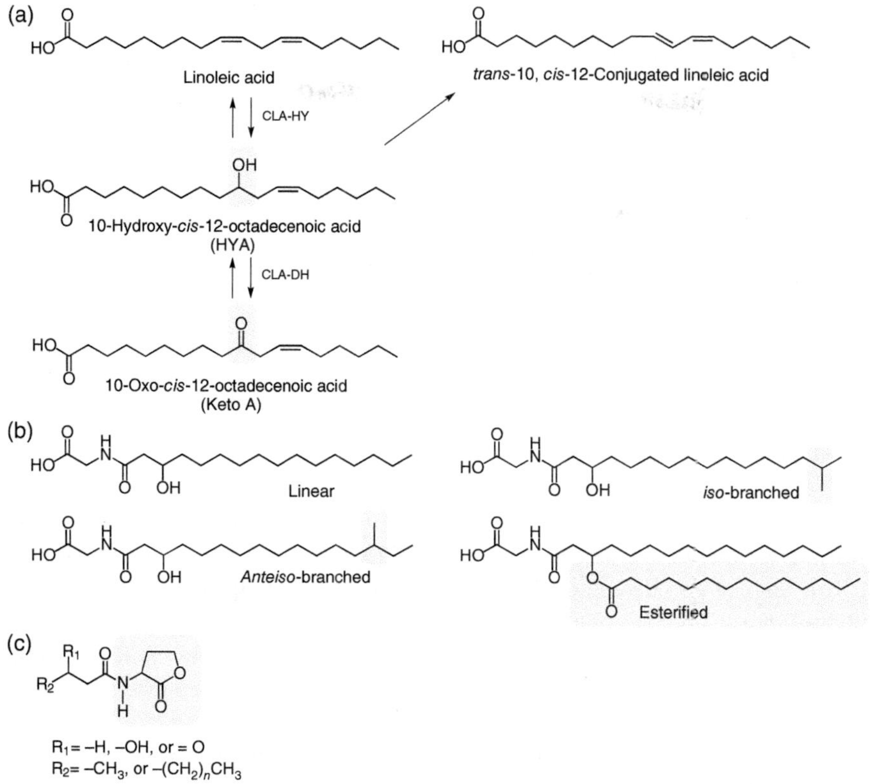

Figure 27.1 Bacterial fatty acid metabolites. (a) Metabolic pathway of linoleicacid-derived metabolites by *Lactobacillus plantarum* [16]. (b) Structures of *N*-acyl glycine produced by different Bacteroidetes [17–19]. (c) Structural diversity of *N*-acyl homoserine lactones in Gram-negative bacteria [20, 21].

acyl chain lengths and hydroxy/oxo modifications using a UPLC-triple quadrupole mass spectrometer equipped with a C18 column. Multiple reaction monitoring of the *m/z* 102 and $[M+H-101]^+$ fragment ions corresponding to the lactone ring and acyl chain, respectively, in a positive-ion mode revealed that intestinal microbiota present in specific pathogen-free mice were involved in AHL production.

27.2.2 Glycerophospholipids

Glycerophospholipids constitute the general classification of phospholipids with a glycerol backbone and are the major components of the lipid bilayer. Bacterial cell membranes consist of phosphatidylethanolamine (PE), phosphatidylglycerol (PG), cardiolipin (CL), phosphatidylinositol (PI), phosphatidic acid (PA), and phosphatidylserine (PS). Although phosphatidylcholine (PC) is the major membrane lipid in eukaryotic cells, only a few environmental bacteria can synthesize PCs [9, 29]. Some Gram-positive bacteria have PGs and CLs modified with aminoacyl groups, such as

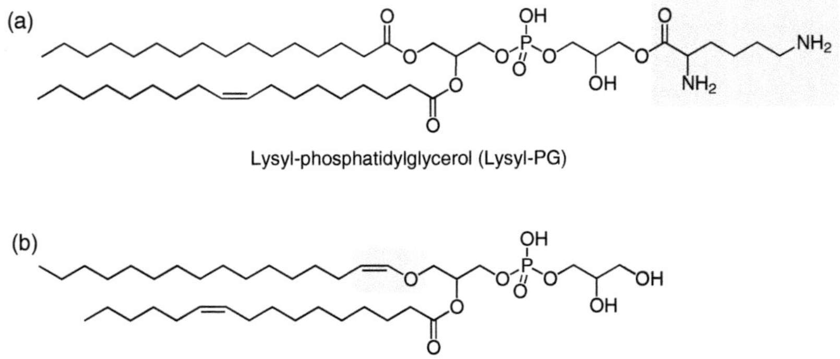

Figure 27.2 Bacterial glycerophospholipids. (a) Lysyl-phosphatidylglycerol in *Enterococcus faecalis* [31]. (b) Plasmenylglycerol in *Clostridioides difficile* [32].

lysine and alanine [30] (Figure 27.2a). Different amounts of lysyl-PG have been found in two *Enterococcus faecalis* strains with different resistance levels to daptomycin. This change also decreases the susceptibility of cells to cationic antimicrobial peptides [31, 33]. Previously, lysyl-PG was purified using thin-layer chromatography before mass spectrometry analysis [34], but in recent years, the total lipid extract of bacteria has been directly analyzed using LC-MS/MS in the negative-ion mode [31, 33].

Plasmalogens, a subtype of glycerophospholipids, are characterized by a long-chain alcohol linked to the *sn*-1 position of glycerol with a vinyl ether bond. The structures can be distinguished from non-vinyl ether-type glycerophospholipids, focusing on the specific mass fragment ions generated in the positive-ion mode [35]. In mammalian plasmalogen-type PE (PlsE) and PC (PlsC), *sn*-1 fatty alcohols and *sn*-2 fatty acids are mainly linear chains, and some bacterial fatty alcohols and fatty acids include odd number of carbons [36–38]. Previously, it was reported that anaerobic enteric bacteria such as *Clostridium sporogenes* [39], *Clostridioides difficile* [32], and *Bacteroides succinogenes* [40] produce plasmalogen (Figure 27.2b). Lipidomic analysis using high-resolution mass spectrometry of glycerophospholipids in *C. difficile*, a causative bacterium of diarrhea and of serious complications owing to antibiotic-induced dysbiosis, revealed the presence of plasmalogen-type CLs and PGs [32]. It has been suggested that anaerobic bacteria directly reduce the ester bond of diacylglycerophospholipids to the vinyl–ether bond, whereas the mammalian plasmalogen is oxidatively biosynthesized via desaturation of the delta 1 position in ether-type PE [41]. The difference in the biosynthetic pathways between mammals and anaerobic bacteria can be a potential therapeutic target for *C. difficile*-associated enteritis [42]. Generally, plasmalogens are susceptible to oxidative degradation of the vinyl ether at the *sn*-1 position [43]; thus, the extraction procedure should be performed in the presence of an antioxidant such as butylated hydroxytoluene for strict quantitative analysis [37].

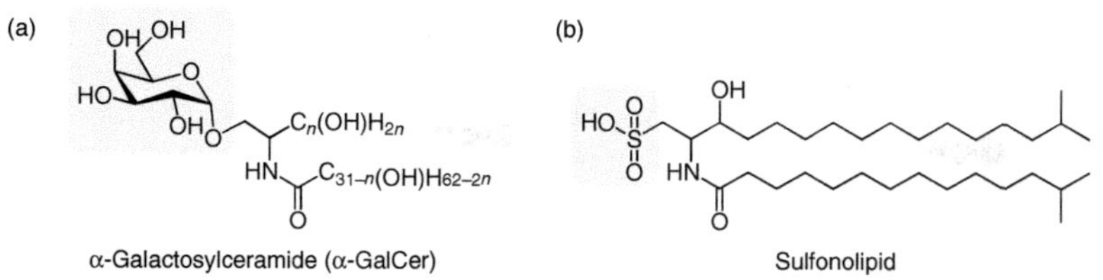

Figure 27.3 Bacterial sphingolipid and sulfonolipid. (a) Alpha-galactosylceramide in *Bacteroides fragilis* [49]. (b) Sulfonolipid in *Alistipes finegoldii* [51].

27.2.3 Sphingolipids

Sphingolipids contain a sphingoid base formed by the condensation of serine and fatty acyl-CoA via serine palmitoyltransferase in prokaryotic and eukaryotic cells. Sphingolipids have diverse structures of sphingoid bases and fatty acids, as well as head group modifications including sugar, phosphate, phosphoethanolamine, phosphocholine, and phosphoinositol [44].

Mammalian cells contain sphingolipids, which are essential components of the cell membrane and intracellular signaling. However, only a limited number of prokaryotic bacteria, Bacteroidetes, and some Proteobacteria produce sphingolipids [45–47]. The hexosyl modifications of the head group in mammals are generally β-linkages [48], whereas *Bacteroides fragilis* produces α-galactosyl ceramide, which modulates the activation of CD1d-restricted invariant natural killer T cells [49, 50] (Figure 27.3a). Recently, Brennan et al. [52] distinguished the glycosidic bond configuration of monohexosylceramide from fragmentation patterns obtained by ion-trap mass spectrometry in a positive mode.

Sulfonolipids are a unique lipid class with a backbone structure similar to that of sphingolipids, but the sphingoid base is substituted by a capnine backbone (Figure 27.3b). Diatoms and soil bacteria are known to produce sulfonolipids [53, 54], although their biosynthetic enzymes are unclear. Intestinal microbiota also produce sulfonolipids [51, 55]. Using a Fourier transform ion cyclotron resonance mass spectrometer, Walker et al. [51] revealed that sulfonolipids are produced by *Alistipes* and *Odoribacter* in Bacteroidetes. The structures ranged from 18 molecular species depending on the variation in chain length and hydroxy group modifications in the capnine backbone and fatty acids.

27.2.4 Bile Acids

Bile acids comprise a sterol backbone and a carboxyl group in the side chain [56]. In humans, primary bile acids such as cholic acid (CA) and chenodeoxycholic acid (CDCA) are biosynthesized from cholesterol via CYP450 enzymes expressed in the liver. They are stored in the gallbladder as conjugated bile acids with taurine or glycine, and then secreted into the duodenum via the bile ducts to facilitate the efficient absorption of dietary fat-soluble ingredients via emulsification. The detergent

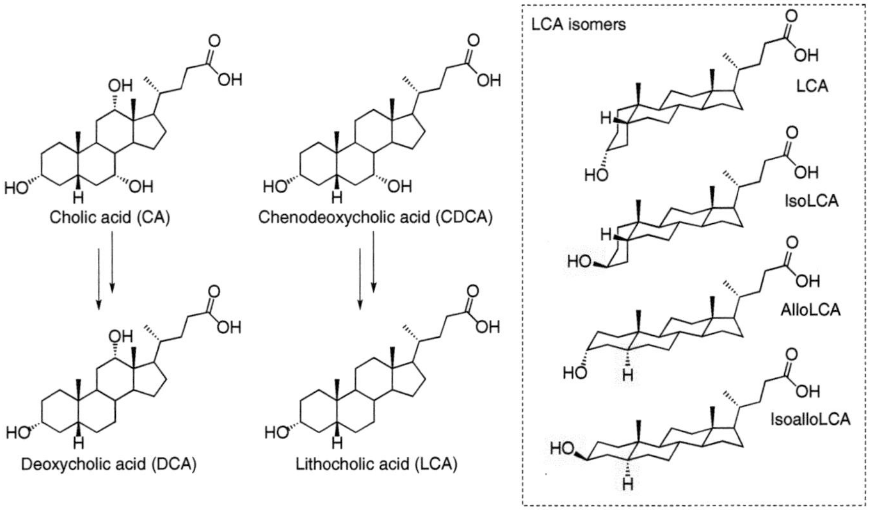

Figure 27.4 Representative bile acids. LCA stereoisomers drawn in boat-chair conformation are shown in the dashed square.

effects of bile acids also cause damage to the bacterial cell membrane and exert strong bactericidal activity [57]. TGR5 and farnesoid X receptor (FXR) are known to be receptors for bile acids and are involved in the regulation of energy homeostasis and cholesterol metabolism [57]. Conjugated bile acids are deconjugated via bacterial bile salt hydrolase and then further converted to secondary bile acids, namely deoxycholic acid and lithocholic acid (LCA), in which the 7-hydroxy group of the sterol backbone is dehydroxylated via multi-step reactions by intestinal bacteria [57–59]. Moreover, intestinal bacteria produce unique structural isomers of LCA, namely isoLCA and alloLCA, which have a 3β-hydroxy group and 5α-hydrogen, respectively (Figure 27.4). IsoalloLCA, a combination of the characteristics of both isomers, is involved in the differentiation of regulatory T cells via H3K27 acetylation at the *Foxp3* promoter region [12] and inhibits the growth of Gram-positive bacteria [60]. Simultaneous analysis of bile acids with diverse structures is challenging because CID rarely yields the structure-specific MS/MS spectra of a rigid cholanic backbone. Hence, chromatographic separation applying a gentle gradient ranging from high to low polarity is important in LC-MS-based analysis [61–63].

27.2.5 Saccharolipids

Saccharolipids are a lipid class containing a glycosidic bond between the sugar moieties and fatty acids. The representative bacterial saccharolipid is lipid A, a hydrophobic anchor of lipopolysaccharides in Gram-negative bacteria. Lipid A comprises a common diglucosamine backbone and various modified phosphate groups and β-hydroxy fatty acids depending on the bacterial species and environmental factors [2]. Structural diversity is closely related to the bioactivity mediated by host pattern recognition receptors [2]. Lipid A of *Escherichia coli* has two phosphate and

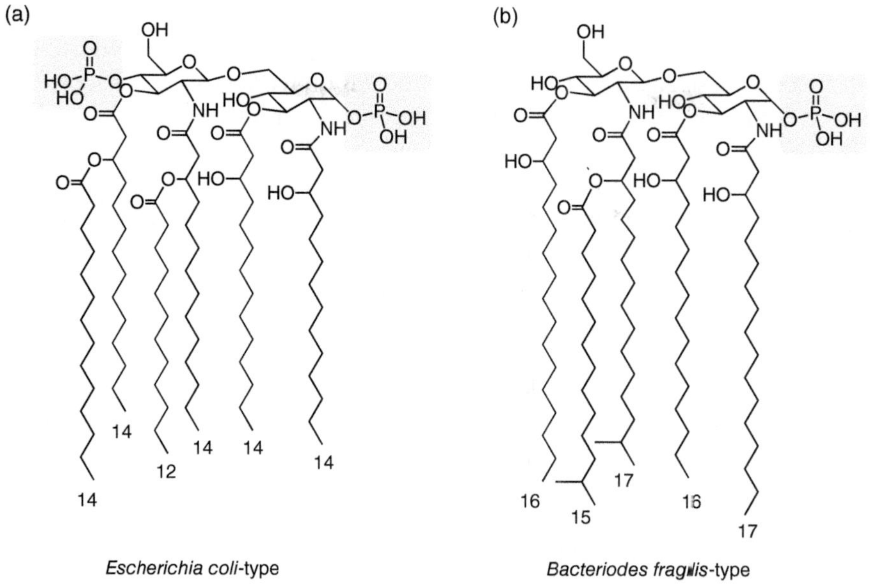

Figure 27.5 Bacterial saccharolipids. (a) Lipid A in *Escherichia coli*. (b) Lipid A in *Bacteroides fragilis*. The numbers near the acyl chains indicate the number of carbons.

six acyl groups (Figure 27.5a) and is a potent agonist of Toll-like receptor 4 (TLR4). Bacteroidetes produces structurally distinct lipid A of a phosphate and five acyl groups, and it antagonizes the activation of TLR4 by *E. coli* lipid A [64, 65]. Although these findings emphasize the importance of the structural analysis of lipid A for each bacterial species, it has been challenging owing to the occurrence of broad lipid A peaks in LC caused by the ionic interaction between phosphate and stainless steel piping. To avoid this limitation, direct detection using MALDI-TOF/MS or MALDI-Fourier transform mass spectrometry is a major method for lipid A analysis [66–68]. Recent studies demonstrated that the use of triethylamine as an ion-pairing reagent [69], and alkaline solvents under a metal-free column are effective for obtaining sharp lipid A peaks [70]. These studies revealed the heterogeneity of lipid A, even in a single bacterial strain [71], and species-specific distribution of the chain length of lipid A in Bacteroidetes [70] (Figure 27.5b).

27.3 New MS Technology

Lipidomics technology using mass spectrometry is useful for high resolution and comprehensive measurement of unique lipids in intestinal bacteria. In particular, the hypothesis-building approach using untargeted analysis is emerging in bacterial lipid research, where novel lipids with unconventional structures and metabolic pathways are being discovered. However, there are many technical obstacles that have not been resolved by current lipidomics. Analysis of the fine structural differences, such as branched chains, double-bond properties, and stereoisomers of fatty acids using chromatography or MS/MS, is difficult in general LC-MS setups. The comprehensiveness of compound detection is a challenge that needs to be addressed. Untargeted analysis based on data-dependent acquisition (DDA) is a major approach in current lipidomics; MS/MS spectra of minor precursor ions are not obtained in DDA. In addition, the MS/MS data acquisition rate in DDA is limited (~60%) [72], implying the presence of unidentified compounds. In this section, remarkable technologies that have been recently developed to address these limitations in bacterial lipidomics are introduced.

27.3.1 Chromatography Technology

The selection of the inner diameter of the column is critical for the separation. The microbore (0.15–0.8 mm), narrow-bore (1–2 mm), and analytical-bore (3–4.6 mm) columns are often used for LC-MS [73]. It is expected that the use of a microbore column improves separation and sensitivity [74]. Danne-Rasche et al. [75] applied nano-LC-based untargeted lipidomics to *Saccharomyces cerevisiae* as a model of eukaryotic lipid metabolism. They demonstrated an increase in the number of identified molecular species and the discovery of putative mono-methylated phosphatidylserine, which has not been reported previously in yeast. Although the throughput of the measurement is reduced in the nanobore column owing to the low flow rates, it is expected to improve the accuracy and depth of the analysis.

27.3.2 Fragmentation

In recent years, technological advances have been made in ion cleavage techniques for MS/MS. In the current CID-based fragmentation, it is difficult to resolve the position of unsaturated double bonds, cyclopropane rings, and branches in acyl chains. To address these issues, two novel fragmentation techniques have been developed. The first is derivatization-based approach using carbon-carbon double-bond-specific reactions, such as the Paternò–Büchi reaction [76], mCPBA epoxidation for lipid double-bond identification (MELDI) [77], and ozone-induced dissociation [78]. Another approach is based on different principles from CID of using oxygen attachment dissociation (OAD) [79, 80], electron impact excitation of ions from organics (EIEIOs) [81], and ultraviolet photodissociation (UVPD) [82–85]. OAD can resolve the position of double bonds in the acyl chain of glycerophospholipids by deciphering the double-bond-specific fragmentation caused by chemically active species such as hydroxyl radicals and atomic oxygens [79, 80]. EIEIO can distinguish the position of the unsaturated double bond as well as the *cis-trans* geometric isomers based on the difference in cleavage energy [81]. UVPD has been applied to the analysis of glycerophospholipids to identify bacteria-specific structures, such as the positions of hydroxy groups, branches, and cyclopropane rings [83–85]. However, a comprehensive analysis method using these techniques has not been developed; thus, further progress in both analytical equipment and software is awaited.

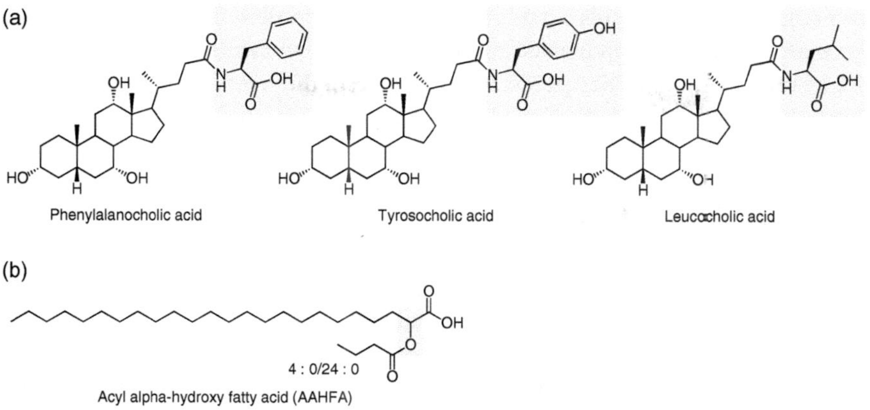

Figure 27.6 Newly identified metabolites using feature-based molecular networking. (a) Cholic acids conjugated by phenylalanine, tyrosine, and leucine produced by *Clostridium bolteae* [88]. (b) Acyl alpha-hydroxy fatty acids found in mice feces [72].

27.3.3 Identification Method for Unknown Structural Molecules

Untargeted lipidomics has the potential to capture unknown molecules, but structure estimation for large amounts of MS/MS spectra is a time-consuming task. Feature-based molecular networking is a technique for classifying MS/MS spectra based on their cosine similarity to promptly discover a series of molecules with similar structures. Jeramie et al. [86] developed a methodology to analyze MS/MS spectra of living microbial colonies grown on agar plates using nanospray desorption electrospray ionization mass spectrometry, and visualized the metabolic landscapes of lipids and peptides. Feature-based molecular networking has now become one of the central methods for MS/MS data and is widely used for natural product discovery [87]. Recently, Quinn et al. [88] applied this technique to the untargeted data of whole-body organs in specific pathogen-free mice and identified CA conjugated with amino acids such as leucine, tyrosine, and phenylalanine produced by the small intestinal bacteria (Figure 27.6a). Yasuda et al. [23, 72] unveiled the complex metabolic network of lipids in mice intestine, and identified a series of novel lipid metabolites including acyl alpha-hydroxy fatty acid (AAHFA) (Figure 27.6b).

27.4 Conclusion and Future Perspective

In this chapter, we summarized the structural features and diversity of gut microbiota-derived lipids and their MS-based analytical methods. Untargeted lipidomics is a powerful method for unbiased profiling, hypothesis building, and identification of unique bacterial lipids. Extraction methods are undoubtedly important for comprehensive detection in untargeted analysis, but because of the diversity of a wide range of physical properties of lipids, no single protocol can provide sufficiently high recovery of all lipid classes simultaneously; thus, appropriate protocols should be selected based on the experimental design [89, 90]. The effect of the volatility of the extraction solvents on the reproducibility of the measurement has also been highlighted in the case of a large-scale analysis [91]. It is necessary to consider extraction protocols that focus on both extraction efficiency and analysis reproducibility.

Standardization of bacterial lipidomes is also an issue that needs to be addressed [92, 93]. The complete chemical assignment of the proposed structures requires authentic standard compounds. In addition, stable isotope-labeled compounds are required as internal standards for quantitative analysis. There are limitations for a single laboratory to prepare such a lineup. To continuously develop highly reliable and quantitative lipidomics, we believe that the establishment of a consortium for sharing standard compounds and their MS/MS fragment information among research institutes is essential. We hope that advanced lipidomics technology to reveal the fine structural differences in microbial lipids will lead to an improved understanding of the host–microbiota interaction.

References

1 Percy, M.G. and Gründling, A. (2014). Lipoteichoic acid synthesis and function in gram-positive bacteria. *Annu. Rev. Microbiol.* 68: 81–100.

2 Steimle, A., Autenrieth, I.B., and Frick, J.-S. (2016). Structure and function: lipid A modifications in commensals and pathogens. *Int. J. Med. Microbiol.* 306 (5): 290–301.

3 Elssner, T., Kostrzewa, M., Maier, T., and Kruppa, G. (2011). Microorganism identification based on MALDI-TOF-MS fingerprints. In: *Detection of Biological Agents for the Prevention of Bioterrorism* (ed. J. Banoub), 99–113. Netherlands: Springer.

4 Mather, C.A., Rivera, S.F., and Butler-Wu, S.M. (2014). Comparison of the Bruker Biotyper and Vitek MS matrix-assisted laser desorption ionization-time of flight mass spectrometry systems for identification of mycobacteria using simplified protein extraction protocols. *J. Clin. Microbiol.* 52 (1): 130–138.

5 Barba, M.J., Fernández, A., Oviaño, M. et al. (2014). Evaluation of MALDI-TOF mass spectrometry for identification of anaerobic bacteria. *Anaerobe* 30: 126–128.

6 Solntceva, V., Kostrzewa, M., and Larrouy-Maumus, G. (2020). Detection of species-specific lipids by routine MALDI TOF mass spectrometry to unlock the challenges of microbial identification and antimicrobial susceptibility testing. *Front. Cell. Infect. Microbiol.* 10: 621452.

7 Gonzalo, X., Broda, A., Drobniewski, F., and Larrouy-Maumus, G. (2021). Performance of lipid fingerprint-based MALDI-ToF for the diagnosis of mycobacterial infections. *Clin. Microbiol. Infect.* 27 (6): 912.e1–912.e5.

8 Hattori, M. and Taylor, T.D. (2009). The human intestinal microbiome: a new frontier of human biology. *DNA Res.* 16 (1): 1–12.

9 Sohlenkamp, C. and Geiger, O. (2015). Bacterial membrane lipids: diversity in structures and pathways. *FEMS Microbiol. Rev.* 40 (1): 133–159.

10 Cohen, L.J., Esterhazy, D., Kim, S.-H. et al. (2017). Commensal bacteria make GPCR ligands that mimic human signalling molecules. *Nature* 549 (7670): 48–53.

11 Miyamoto, J., Igarashi, M., Watanabe, K. et al. (2019). Gut microbiota confers host resistance to obesity by metabolizing dietary polyunsaturated fatty acids. *Nat. Commun.* 10 (1): 1–15.

12 Hang, S., Paik, D., Yao, L. et al. (2019). Bile acid metabolites control T_H17 and T_{reg} cell differentiation. *Nature* 576 (7785): 143–148.

13 Choi, K.H., Heath, R.J., and Rock, C.O. (2000). β-ketoacyl-acyl carrier protein synthase III (FabH) is a determining factor in branched-chain fatty acid biosynthesis. *J. Bacteriol.* 182 (2): 365–370.

14 Grogan, D.W. and Cronan, J.E. Jr. (1997). Cyclopropane ring formation in membrane lipids of bacteria. *Microbiol. Mol. Biol. Rev.* 61 (4): 429–441.

15 Poger, D. and Mark, A.E. (2015). A ring to rule them all: the effect of cyclopropane fatty acids on the fluidity of lipid bilayers. *J. Phys. Chem. B* 119 (17): 5487–5495.

16 Kishino, S., Takeuchi, M., Park, S.B. et al. (2013). Polyunsaturated fatty acid saturation by gut lactic acid bacteria affecting host lipid composition. *Proc. Natl. Acad. Sci. U.S.A.* 110 (44): 17808–17813.

17 Cohen, L.J., Kang, H.-S., Chu, J. et al. (2015). Functional metagenomic discovery of bacterial effectors in the human microbiome and isolation of commendamide, a GPCR G2A/132 agonist. *Proc. Natl. Acad. Sci. U.S.A.* 112 (35): E4825–E4834.

18 Lynch, A., Tammireddy, S.R., Doherty, M.K. et al. (2019). The glycine lipids of Bacteroides thetaiotaomicron are important for fitness during growth in vivo and in vitro. *Appl. Environ. Microbiol.* 85 (10): e02157-18.

19 Nichols, F.C., Clark, R.B., Liu, Y. et al. (2020). Glycine lipids of *Porphyromonas gingivalis* are agonists for toll-like receptor 2. *Infect. Immun.* 88 (4): e00877-19.

20 Xue, J., Chi, L., Tu, P. et al. (2021). Detection of gut microbiota and pathogen produced *N*-acyl homoserine in host circulation and tissues. *npj Biofilms Microbiomes* 7 (1): 1–10.

21 Churchill, M.E.A., Sibhatu, H.M., and Uhlson, C.L. (2011). Defining the structure and function of acyl-homoserine lactone autoinducers. *Methods Mol. Biol.* 692: 159–171.

22 Kim, M., Furuzono, T., Yamakuni, K. et al. (2017). 10-oxo-12(*Z*)-octadecenoic acid, a linoleic acid metabolite produced by gut lactic acid bacteria, enhances energy metabolism by activation of TRPV1. *FASEB J.* 31 (11): 5036–5048.

23 Okahashi, N., Ueda, M., Yasuda, S. et al. (2021). Global profiling of gut microbiota-associated lipid metabolites in antibiotic-treated mice by LC-MS/MS-based analyses. *STAR Protocols* 2 (2): 100492.

24 Hurst, D.P., Schmeisser, M., and Reggio, P.H. (2013). Endogenous lipid activated G protein-coupled receptors: emerging structural features from crystallography and molecular dynamics simulations. *Chem. Phys. Lipids* 169: 46–56.

25 Dees, C. and Shively, J.M. (1982). Localization of quantitation of the ornithine lipid of *Thiobacillus thiooxidans*. *J. Bacteriol.* 149 (2): 798–799.

26 Asselineau, J. (1991). Bacterial lipids containing amino acids or peptides linked by amide bonds. *Fortschr. Chem. Org. Naturst.* 56: 1–85.

27 Clark, R.B., Cervantes, J.L., Maciejewski, M.W. et al. (2013). Serine lipids of *Porphyromonas gingivalis* are human and mouse toll-like receptor 2 ligands. *Infect. Immun.* 81 (9): 3479–3489.

28 Miller, M.B. and Bassler, B.L. (2001). Quorum sensing in bacteria. *Annu. Rev. Microbiol.* 55: 165–199.

29 López-Lara, I.M. and Geiger, O. (2017). Bacterial lipid diversity. *Biochim. Biophys. Acta, Mol. Cell Biol. Lipids* 1862 (11): 1287–1299.

30 Geiger, O., González-Silva, N., López-Lara, I.M., and Sohlenkamp, C. (2010). Amino acid-containing membrane lipids in bacteria. *Prog. Lipid Res.* 49 (1): 46–60.

31 Rashid, R., Cazenave-Gassiot, A., Gao, I.H. et al. (2017). Comprehensive analysis of phospholipids and glycolipids in the opportunistic pathogen *Enterococcus faecalis*. *PLoS One* 12 (4): e0175886.

32 Guan, Z., Katzianer, D., Zhu, J., and Goldfine, H. (2014). Clostridium difficile contains plasmalogen species of phospholipids and glycolipids. *Biochim. Biophys. Acta* 1841 (10): 1353–1359.

33 Rashid, R., Veleba, M., and Kline, K.A. (2016). Focal targeting of the bacterial envelope by antimicrobial peptides. *Front. Cell Dev. Biol.* 4: 55.

34 Tran, T.T., Panesso, D., Mishra, N.N. et al. (2013). Daptomycin-resistant *Enterococcus faecalis* diverts the antibiotic molecule from the division septum and remodels cell membrane phospholipids. *MBio* 4 (4): e00281-13.

35 Berry, K.A.Z. and Murphy, R.C. (2004). Electrospray ionization tandem mass spectrometry of glycerophosphoethanolamine plasmalogen phospholipids. *J. Am. Soc. Mass Spectrom.* 15 (10): 1499–1508.

36 Nagan, N. and Zoeller, R.A. (2001). Plasmalogens: biosynthesis and functions. *Prog. Lipid Res.* 40 (3): 199–229.

37 Řezanka, T., Křesinová, Z., Kolouchová, I., and Sigler, K. (2012). Lipidomic analysis of bacterial plasmalogens. *Folia Microbiol.* 57 (5): 463–472.

38 Řezanka, T., Siristova, L., Matoulková, D., and Sigler, K. (2011). Hydrophilic interaction liquid chromatography: ESI–MS/MS of plasmalogen phospholipids from *Pectinatus bacterium*. *Lipids* 46 (8): 765–780.

39 Guan, Z., Johnston, N.C., Raetz, C.R.H. et al. (2012). Lipid diversity among botulinum neurotoxin-producing clostridia. *Microbiology* 158 (Pt 10): 2577–2584.

40 Wegner, G.H. and Foster, E.M. (1963). Incorporation of isobutyrate and valerate into cellular plasmalogen by *Bacteroides succinogenes*. *J. Bacteriol.* 85: 53–61.

41 Jackson, D.R., Cassilly, C.D., Plichta, D.R. et al. (2021). Plasmalogen biosynthesis by anaerobic bacteria: identification of a two-gene operon responsible for plasmalogen production in *Clostridium perfringens*. *ACS Chem. Biol.* 16 (1): 6–13.

42 Goldfine, H. (2017). The anaerobic biosynthesis of plasmalogens. *FEBS Lett.* 591 (18): 2714–2719.

43 Catalá, A. (2009). Lipid peroxidation of membrane phospholipids generates hydroxy-alkenals and oxidized phospholipids active in physiological and/or pathological conditions. *Chem. Phys. Lipids* 157 (1): 1–11.

44 Futerman, A.H. and Hannun, Y.A. (2004). The complex life of simple sphingolipids. *EMBO Rep.* 5 (8): 777–782.

45 Olsen, I. and Jantzen, E. (2001). Sphingolipids in bacteria and fungi. *Anaerobe* 7 (2): 103–112.

46 Gevers, D., Knight, R., Petrosino, J.F. et al. (2012). The Human Microbiome Project: a community resource for the healthy human microbiome. *PLoS Biol.* 10 (8): e1001377.

47 An, D., Na, C., Bielawski, J. et al. (2011). Membrane sphingolipids as essential molecular signals for Bacteroides survival in the intestine. *Proc. Natl. Acad. Sci. U.S.A.* 108 (Suppl 1): 4666–4671.

48 Merrill, A.H. Jr. (2011). Sphingolipid and glycosphingolipid metabolic pathways in the era of sphingolipidomics. *Chem. Rev.* 111 (10): 6387–6422.

49 An, D., Oh, S.F., Olszak, T. et al. (2014). Sphingolipids from a symbiotic microbe regulate homeostasis of host intestinal natural killer T cells. *Cell* 156 (1, 2): 123–133.

50 Wieland Brown, L.C., Penaranda, C., Kashyap, P.C. et al. (2013). Production of α-galactosylceramide by a prominent member of the human gut microbiota. *PLoS Biol.* 11 (7): e1001610.

51 Walker, A., Pfitzner, B., Harir, M. et al. (2017). Sulfonolipids as novel metabolite markers of Alistipes and Odoribacter affected by high-fat diets. *Sci. Rep.* 7 (1): 11047.

52 Brennan, P.J., Cheng, T.-Y., Pellicci, D.G. et al. (2017). Structural determination of lipid antigens captured at the CD1d-T-cell receptor interface. *Proc. Natl. Acad. Sci. U.S.A.* 114 (31): 8348–8353.

53 Takikawa, H., Nozawa, D., Kayo, A. et al. (1999). Synthesis of sphingosine relatives. Part 22. Synthesis of sulfobacin A, B and flavocristamide A, new sulfonolipids isolated from *Chryseobacterium* sp. *J. Chem. Soc. Perkin 1* (17): 2467–2477.

54 Corcelli, A., Lattanzio, V.M.T., Mascolo, G. et al. (2004). Novel sulfonolipid in the extremely halophilic bacterium *Salinibacter ruber*. *Appl. Environ. Microbiol.* 70 (11): 6678–6685.

55 Radka, C.D., Frank, M.W., Rock, C.O., and Yao, J. (2020). Fatty acid activation and utilization by *Alistipes finegoldii*, a representative Bacteroidetes resident of the human gut microbiome. *Mol. Microbiol.* 113 (4): 807–825.

56 Hagey, L.R., Vidal, N., Hofmann, A.F., and Krasowski, M.D. (2010). Evolutionary diversity of bile salts in reptiles and mammals, including analysis of ancient human and extinct giant ground sloth coprolites. *BMC Evol. Biol.* 10: 133.

57 Ridlon, J.M., Kang, D.-J., and Hylemon, P.B. (2006). Bile salt biotransformations by human intestinal bacteria. *J. Lipid Res.* 47 (2): 241–259.

58 Ridlon, J.M., Harris, S.C., Bhowmik, S. et al. (2016). Consequences of bile salt biotransformations by intestinal bacteria. *Gut Microbes* 7 (1): 22–39.

59 Funabashi, M., Grove, T.L., Wang, M. et al. (2020). A metabolic pathway for bile acid dehydroxylation by the gut microbiome. *Nature* 582 (7813): 566–570.

60 Sato, Y., Atarashi, K., Plichta, D.R. et al. (2021). Novel bile acid biosynthetic pathways are enriched in the microbiome of centenarians. *Nature* 599: 1–10.

61 Kakiyama, G., Muto, A., Takei, H. et al. (2014). A simple and accurate HPLC method for fecal bile acid profile in healthy and cirrhotic subjects: validation by GC-MS and LC-MS. *J. Lipid Res.* 55 (5): 978–990.

62 Choucair, I., Nemet, I., Li, L. et al. (2020). Quantification of bile acids: a mass spectrometry platform for studying gut microbe connection to metabolic diseases. *J. Lipid Res.* 61 (2): 159–177.

63 Taguchi, K., Fukusaki, E., and Bamba, T. (2013). Simultaneous and rapid analysis of bile acids including conjugates by supercritical fluid chromatography coupled to tandem mass spectrometry. *J. Chromatogr. A* 1299: 103–109.

64 Vatanen, T., Kostic, A.D., d'Hennezel, E. et al. (2016). Variation in microbiome LPS immunogenicity contributes to autoimmunity in humans. *Cell* 165 (6): 1551.

65 d'Hennezel, E., Abubucker, S., Murphy, L.O., and Cullen, T.W. (2017). Total lipopolysaccharide from the human gut microbiome silences toll-like receptor signaling. *mSystems* 2 (6): e00046-17.

66 Schumann, P. and Maier, T. (2014). MALDI-TOF mass spectrometry applied to classification and identification of bacteria, Chapter 13. In: *Methods in Microbiology*, vol. 41 (ed. M. Goodfellow, I. Sutcliffe and J. Chun), 275–306. Academic Press.

67 Biswas, S. and Rolain, J.-M. (2013). Use of MALDI-TOF mass spectrometry for identification of bacteria that are difficult to culture. *J. Microbiol. Methods* 92 (1): 14–24.

68 Jones, J.J., Stump, M.J., Fleming, R.C. et al. (2004). Strategies and data analysis techniques for lipid and phospholipid chemistry elucidation by intact cell MALDI-FTMS. *J. Am. Soc. Mass Spectrom.* 15 (11): 1665–1674.

69 Kojima, H., Inagaki, M., Tomita, T. et al. (2010). Improved separation and characterization of lipopolysaccharide related compounds by reverse phase ion pairing-HPLC/electrospray ionization-quadrupole-mass spectrometry (RPIP-HPLC/ESI-Q-MS). *J. Chromatogr. B* 878 (3, 4): 442–448.

70 Okahashi, N., Ueda, M., Matsuda, F., and Arita, M. (2021). Analyses of lipid a diversity in gram-negative intestinal bacteria using liquid chromatography-quadrupole time-of-flight mass spectrometry. *Metabolites* 11 (4): 197.

71 Sándor, V., Dörnyei, Á., Makszin, L. et al. (2016). Characterization of complex, heterogeneous lipid A samples using HPLC-MS/MS technique I. Overall analysis with respect to acylation, phosphorylation and isobaric distribution. *J. Mass Spectrom.* 51 (11): 1043–1063.

72 Yasuda, S., Okahashi, N., Tsugawa, H. et al. (2020). Elucidation of gut microbiota-associated lipids using LC-MS/MS and 16S rRNA sequence analyses. *iScience* 23 (12): 101841.

73 Abian, J., Oosterkamp, A.J., and Gelpí, E. (1999). Comparison of conventional, narrow-bore and capillary liquid chromatography/mass spectrometry for electrospray ionization mass spectrometry: practical considerations. *J. Mass Spectrom.* 34 (4): 244–254.

74 Cajka, T. and Fiehn, O. (2014). Comprehensive analysis of lipids in biological systems by liquid chromatography-mass spectrometry. *TrAC, Trends Anal. Chem.* 61: 192–206.

75 Danne-Rasche, N., Rubenzucker, S., and Ahrends, R. (2020). Uncovering the complexity of the yeast lipidome by means of nLC/NSI-MS/MS. *Anal. Chim. Acta* 1140: 199–209.

76 Zhang, W., Zhang, D., Chen, Q. et al. (2019). Online photochemical derivatization enables comprehensive mass spectrometric analysis of unsaturated phospholipid isomers. *Nat. Commun.* 10 (1): 1–9.

77 Kuo, T.-H., Chung, H.-H., Chang, H.-Y. et al. (2019). Deep lipidomics and molecular imaging of unsaturated lipid isomers: a universal strategy initiated by mCPBA epoxidation. *Anal. Chem.* 91 (18): 11905–11915.

78 Thomas, M.C., Mitchell, T.W., Harman, D.G. et al. (2008). Ozone-induced dissociation: elucidation of double bond position within mass-selected lipid ions. *Anal. Chem.* 80 (1): 303–311.

79 Takahashi, H., Shimabukuro, Y., Asakawa, D. et al. (2018). Structural analysis of phospholipid using hydrogen abstraction dissociation and oxygen attachment dissociation in tandem mass spectrometry. *Anal. Chem.* 90 (12): 7230–7238.

80 Uchino, H., Tsugawa, H., Takahashi, H., and Arita, M. (2021). Computational Mass Spectrometry Accelerates C=C Position-Resolved Untargeted Lipidomics Using Oxygen Attachment Dissociation. *Research Square*. https://doi.org/10.21203/rs.3.rs-727852/v1.

81 Baba, T., Campbell, J.L., Le Blanc, J.C.Y., and Baker, P.R.S. (2017). Distinguishing cis and trans isomers in intact complex lipids using electron impact excitation of ions from organics mass spectrometry. *Anal. Chem.* 89 (14): 7307–7315.

82 Williams, P.E., Klein, D.R., Greer, S.M., and Brodbelt, J.S. (2017). Pinpointing double bond and *sn*-positions in glycerophospholipids via hybrid 193 nm ultraviolet photodissociation (UVPD) mass spectrometry. *J. Am. Chem. Soc.* 139 (44): 15681–15690.

83 Klein, D.R., Blevins, M.S., Macias, L.A. et al. (2020). Localization of double bonds in bacterial glycerophospholipids using 193 nm ultraviolet photodissociation in the negative mode. *Anal. Chem.* 92 (8): 5986–5993.

84 Blevins, M.S., James, V.K., Herrera, C.M. et al. (2020). Unsaturation elements and other modifications of phospholipids in bacteria: new insight from ultraviolet photodissociation mass spectrometry. *Anal. Chem.* 92 (13): 9146–9155.

85 Brodbelt, J.S., Morrison, L.J., and Santos, I. (2020). Ultraviolet photodissociation mass spectrometry for analysis of biological molecules. *Chem. Rev.* 120 (7): 3328–3380.

86 Watrous, J., Roach, P., Alexandrov, T. et al. (2012). Mass spectral molecular networking of living microbial colonies. *Proc. Natl. Acad. Sci. U.S.A.* 109 (26): E1743–E1752.

87 Fox Ramos, A.E., Evanno, L., Poupon, E. et al. (2019). Natural products targeting strategies involving molecular networking: different manners, one goal. *Nat. Prod. Rep.* 36 (7): 960–980.

88 Quinn, R.A., Melnik, A.V., Vrbanac, A. et al. (2020). Global chemical effects of the microbiome include new bile-acid conjugations. *Nature* 579 (7797): 123–129.

89 Cajka, T. and Fiehn, O. (2016). Toward merging untargeted and targeted methods in mass spectrometry-based metabolomics and lipidomics. *Anal. Chem.* 88 (1): 524–545.

90 Sarafian, M.H., Gaudin, M., Lewis, M.R. et al. (2014). Objective set of criteria for optimization of sample preparation procedures for ultra-high throughput untargeted blood plasma lipid profiling by ultra performance liquid chromatography-mass spectrometry. *Anal. Chem.* 86 (12): 5766–5774.

91 Wong, M.W.K., Braidy, N., Pickford, R. et al. (2019). Comparison of single phase and biphasic extraction protocols for lipidomic studies using human plasma. *Front. Neurol.* 10: 879.

92 Lipidomics Standards Initiative Consortium (2019). Lipidomics needs more standardization. *Nat. Metab.* 1 (8): 745–747.

93 O'Donnell, V.B., Ekroos, K., Liebisch, G., and Wakelam, M. (2020). Lipidomics: current state of the art in a fast moving field. *Wiley Interdiscip. Rev. Syst. Biol. Med.* 12 (1): e1466.

Index

Mass Spectrometry for Lipidomics: Methods and Applications, First Edition.
Edited by Michal Holčapek and Kim Ekroos.

C

h